Applications of Synchrotron Radiation Techniques to Materials Science IV

MATERIALS RESEARCH SOCIETY
SYMPOSIUM PROCEEDINGS VOLUME 524

Applications of Synchrotron Radiation Techniques to Materials Science IV

Symposium held April 13–17, 1998, San Francisco, California, U.S.A.

EDITORS:

Susan M. Mini
Northern Illinois University
DeKalb, Illinois, U.S.A.
and
Argonne National Laboratory
Argonne, Illinois, U.S.A.

Stuart R. Stock
Georgia Institute of Technology
Atlanta, Georgia, U.S.A.

Dale L. Perry
Lawrence Berkeley National Laboratory
Berkeley, California, U.S.A.

Louis J. Terminello
Lawrence Livermore National Laboratory
Livermore, California, U.S.A.

Materials Research Society
Warrendale, Pennsylvania

CAMBRIDGE UNIVERSITY PRESS
Cambridge, New York, Melbourne, Madrid, Cape Town,
Singapore, São Paulo, Delhi, Mexico City

Cambridge University Press
32 Avenue of the Americas, New York NY 10013-2473, USA

Published in the United States of America by Cambridge University Press, New York

www.cambridge.org
Information on this title: www.cambridge.org/9781107413665

Materials Research Society
506 Keystone Drive, Warrendale, PA 15086
http://www.mrs.org

© Materials Research Society 1998

First published 1998
First paperback edition 2013

Single article reprints from this publication are available through
University Microfilms Inc., 300 North Zeeb Road, Ann Arbor, MI 48106

CODEN: MRSPDH

ISBN 978-1-107-41366-5 Paperback

CONTENTS

Preface xi

Acknowledgments xiii

Materials Research Society Symposium Proceedings xiv

PART I: MAPPING, MICROPROBES, AND IMAGING

*Expansive Reactions in Concrete Observed by Soft X-ray
Transmission Microscopy 3
   K.E. Kurtis, P.J.M. Monteiro, J.T. Brown, and W. Meyer-Ilse

The First Synchrotron Infrared Beamlines at the Advanced
Light Source: Microspectroscopy and Fast Timing 11
   Michael C. Martin and Wayne R. McKinney

Applications of Synchrotron Infrared Microspectroscopy
to the Study of Inorganic-Organic Interactions at the
Bacterial-Mineral Interface 17
   Hoi-Ying N. Holman, Dale L. Perry, Michael C. Martin,
   and Wayne R. McKinney

High-Resolution X-ray Photoemission Electron Microscopy
at the Advanced Light Source 25
   Thomas Stammler, Simone Anders, Howard A. Padmore,
   Joachim Stöhr, Michael Scheinfein, and Harald Ade

Synchrotron Radiation Hard X-ray Microprobe by
Multilayer Fresnel Zone Plate 31
   S. Tamura, K. Ohtani, M. Yasumoto, K. Murai, N. Kamijo, H. Kihara,
   K. Yoshida, and Y. Suzuki

Micro-, Meso-, and Macrotexture and Fatigue Crack
Roughness in Al-Li 2090 T8E41 37
   J.D. Haase, A. Guvenilir, J.R. Witt, and S.R. Stock

X-ray Microbeam Quantification of Grain Subdivision
Accompanying Large Deformations of Copper 43
   G.C. Butler, A. Guvenilir, D.L. McDowell, and S.R. Stock

Automated Indexing of Laue Images From
Polycrystalline Materials 49
   Jin-Seok Chung and Gene E. Ice

Grain Orientation Mapping of Passivated Aluminum
Interconnect Lines With X-ray Microdiffraction 55
   A.A. MacDowell, C.H. Chang, H.A. Padmore, J.R. Patel,
   and A.C. Thompson

*Invited Paper

Synchrotron Radiation Diffraction Imaging Study
of the Magnetoacoustically-Induced X-ray Focusing
Effect in FeBO₃ 59
    I. Matsouli, V.V. Kvardakov, J.I. Espeso, L. Chabert, and J. Baruchel

The Mechanism of Twinning in Zinc-Blende Structure
Crystals: New Insights on Polarity Effects From a Study
of Magnetic-Liquid-Encapsulated Czochralski Grown InP
Single Crystals 65
    M. Dudley, B. Raghothamachar, Y. Guo, X.R. Huang, H. Chung,
    D.J. Larson, Jr., D.T.J. Hurle, D.F. Bliss, V. Prasad, and Z. Huang

Contrast Mechanism in Superscrew Dislocation Images on
Synchrotron Back-Reflection Topographs 71
    X.R. Huang, M. Dudley, W.M. Vetter, W. Huang, S. Wang,
    and C.H. Carter, Jr.

Twin and Grain Boundary in InP:
A Synchrotron Radiation Study 77
    Yujie Han, Jianhua Jiang, Zhouguang Wang, Xunlang Liu,
    Jinghua Jiao, Yulian Tian, and Lanying Lin

Real-Time In Situ X-ray Topographic Observation of
Deformation of Single Crystals and Thin Films 81
    Z.B. Zhao, J. Hershberger, A. Chiaramonti,
    Z.U. Rek, and J.C. Bilello

PART II: DIFFRACTION AND SCATTERING

Strain and Shape in Self-Assembled Quantum Dots
Studied by X-ray Grazing Incidence Diffraction 89
    I. Kegel, T.H. Metzger, J. Peisl, P. Fratzl, A. Lorke, J.P. Kotthaus,
    J.M. Garcia, and P.M. Petroff

X-ray Diffuse Scattering Investigation of Defects in Ion-
Implanted and Annealed Silicon 95
    C.H. Chang, U. Beck, T.H. Metzger, and J.R. Patel

Growth and Structure of Nanometric Iron Oxide Films 101
    E. Guiot, S. Gota, M. Henriot, M. Gautier-Soyer, and S. Lefebvre

Degree of Crystallinity and Strain in B₄C and SiC Thin Films
as a Function of Processing Conditions 109
    J. Hershberger, Z.U. Rek, F. Kustas, S.M. Yalisove, and J.C. Bilello

Probing Stress State and Phase Content
in Ultrathin Ta Films 115
    J.F. Whitacre, Z.U. Rek, S.M. Yalisove, and J.C. Bilello

Crystallographic Analysis of CVD Films by Using X-ray
Polychromatic Radiation 121
    B. Lavelle, L. Brissonneau, E. Baggot, and C. Vahlas

In Situ Monitoring of the Electrochemical Absorption of
Deuterium Into Palladium by X-ray Diffraction Using
Synchrotron-Wiggler Radiation 127
  *D.D. Dominguez, P.L. Hagans, E.F. Skelton, S.B. Qadri,
  and D.J. Nagel*

High-Pressure Synchrotron Diffraction of $KCa_2Nb_3O_{10}$, a
Layered Perovskite Compound 133
  K.A. Steiner and W.T. Petuskey

High-Temperature X-ray Diffraction in Transmission Under
Controlled Environment 139
  *L. Margulies, M.J. Kramer, J.J. Williams, E.M. Deters, R.W. McCallum,
  D.R. Haeffner, J.C. Lang, S. Kycia, and A.I. Goldman*

Interfacial Effects in Multilayers 145
  Troy W. Barbee, Jr.

Comparison of a Mosaic-Crystal Spectrometer to a
High-Performance Solid-State Detector for X-ray
Microfluorescence Analysis 153
  *J-S. Chung, S. Isa, C.J. Sparks, G.E. Ice, S. McHugo,
  and A. Thompson*

Anisotropy of NH_4AP Crystal X-ray Susceptibility
for Bragg Reflection Near CK Absorption Edge 161
  A.V. Okotrub, G.S. Belikova, T.N. Turskaya, and L.N. Mazalov

PART III: PHOTOEMISSION AND MICROPHOTOEMISSION;
FLUORESCENCE AND MICROFLUORESCENCE

*Spectroscopic Studies of Low-Dielectric-Constant
Fluorinated Amorphous Carbon Films for ULSI
Integrated Circuits 169
  *Yanjun Ma, Hongning Yang, J. Guo, C. Sathe, A. Agui,
  and J. Nordgren*

Angle-Resolved Photoemission Study of the Electronic
Structures of $AuAl_2$ and $PtGa_2$ 179
  L-S. Hsu, J.D. Denlinger, and J.W. Allen

Preliminary Results From a New Spin Spectrometer 185
  *J.G. Tobin, P.J. Bedrossian, T.R. Cummins, G.D. Waddill,
  S. Mishra, P. Larson, R. Negri, M. Miller, E. Peterson,
  P. Boyd, and R. Gunion*

The Spin-Polarized Band Structure of Strained
Thin Films of Gadolinium 191
  C. Waldfried, E. Vescovo, and P.A. Dowben

*Invited Paper

Evidence for the Photoemission Nature of Gd 4f
Resonant Photoemission 197
S.R. Mishra, T.R. Cummins, W.J. Gammon, G.D. Waddill,
G. van der Laan, K.W. Goodman, and J.G. Tobin

*ESCA Microscopy on ELETTRA: Chemical
Characterization of Surfaces and Interfaces With
Submicron Spatial Resolution 203
M. Kiskinova, L. Casalis, L. Gregoratti, S. Gunther, and M. Marsi

MAXIMUM AT ALS: A Powerful Tool to Investigate Open
Problems in Micro- and Optoelectronics 215
G.F. Lorusso, H. Solak, S. Singh, P.J. Batson, J.H. Underwood,
and F. Cerrina

State-of-the-Art X-ray Photoelectron Spectroscopy (XPS):
Conventional and Synchrotron X-ray Sources for Micro-XPS 221
E.L. Principe, R.W. Odom, A.L. Johnson, G.D. Ackermann, Z. Hussain,
and H. Padmore

Chemical Analysis of Particles and Semiconductor
Microstructures by Synchrotron Radiation Soft X-ray
Photoemission Spectromicroscopy 227
F. Gozzo, B. Triplett, H. Fujimoto, R. Ynzunza, P. Coon, C. Ayre,
P.D. Kinney, Y.S. Uritsky, G. Ackermann, A. Johnson, H. Padmore,
T. Renner, B. Sheridan, W. Steele, and Z. Hussain

X-ray Fluorescence Microtomography on a SiC
Nuclear Fuel Shell 233
M. Naghedolfeizi, J-S. Chung, G.E. Ice, W.B. Yun, Z. Cai, and B. Lai

Application of Synchrotron X-ray Fluorescence Microscopy
to the Study of Multi-Metal Oxide Ceramics 241
Dale L. Perry, Scott McHugo, Albert C. Thompson, Joseph C. Farmer,
Bart B. Ebbinghaus, Richard van Konynenburg, William A. Brummond,
Guy Armentrout, Thomas H. Gould, and Nancy Yang

Update on Synchrotron Radiation TXRF: New Results 245
S. Brennan, P. Pianetta, S. Ghosh, N. Takaura, C. Wiemer, A. Fischer-Colbrie,
S. Laderman, A. Shimazaki, A. Waldhauer, and M.A. Zaitz

*Characterizing Trace Metal Impurities in Optical-
Waveguide Materials Using X-ray Absorption 251
P.H. Citrin, P.A. Northrup, R.M. Atkins, P.F. Glodis, L. Niu, M.A. Marcus,
and D.C. Jacobson

PART IV: <u>MATERIALS CHARACTERIZATION WITH X-RAY ABSORPTION</u>

*Anomalous X-ray Scattering Studies of Short-,
Intermediate-, and Extended-Range Order in Glasses 261
David L. Price, Marie-Louise Saboungi, Pascale Armand, and David E. Cox

*Invited Paper

X-ray-Absorption-Fine-Structure (XAFS) Studies of Cobalt
Silicide Thin Films 273
    S.J. Naftel, I. Coulthard, Y. Hu, T.K. Sham, and M. Zinke-Allmang

Electronic Effects at Interfaces in Cu/
(Cr, Mo, W, Ta, Re) Multilayers 279
    A.F. Bello, T. Van Buuren, J.E. Klepeis, and T.W. Barbee, Jr.

Interfacial Electronic-Charge Transfer and Density of
States in Short-Period Cu/Cr Multilayers 285
    A.F. Bello, T. Van Buuren, J.E. Klepeis, and T.W. Barbee, Jr.

Synchrotron X-ray-Absorption-Spectroscopy Studies
of Pt/Si Systems 291
    I. Coulthard, S.J. Naftel, and T.K. Sham

Direct Correlation of Solar Cell Performance With Metal
Impurity Distributions in Polycrystalline Silicon Using
Synchrotron-Based X-ray Analysis 297
    S.A. McHugo, A.C. Thompson, G. Lamble, A. MacDowell,
    R. Celestre, H. Padmore, M. Imaizumi, M. Yamaguchi, I. Perichaud,
    S. Martinuzzi, M. Werner, M. Rinio, H.J. Moller, B. Sopori, H. Hieslmair,
    C. Flink, A. Istratov, and E.R. Weber

EXAFS Characterization of Amorphous GaAs 303
    M.C. Ridgway, C.J. Glover, G.J. Foran, and K.M. Yu

Amorphous Semiconductor Sample Preparation for
Transmission EXAFS Measurements 309
    M.C. Ridgway, C.J. Glover, H.H. Tan, A. Clark, F. Karouta, G.J. Foran,
    T.W. Lee, Y. Moon, E. Yoon, J.L. Hansen, A. Nylandsted-Larsen,
    C. Clerc, and J. Chaumont

Kinetics of the Growth of Copper Clusters on the
Alumina (0001) Surface: Influence of Surface Structure 315
    M. Gautier-Soyer, S. Gota, L. Douillard, P. Le Fèvre, H. Magnan,
    and J.P. Duraud

Temperature-Driven Oxidation Behavior of
Pure Iron Surface Investigated by Time-Resolved
EXAFS Measurements 321
    S.J. Doh, J.M. Lee, D.Y. Noh, and J.H. Je

The Chemical Environment of Er^{3+} in a-Si:Er:H and a-Si:Er:O:H 327
    Leandro R. Tessler, Cinthia Piamonteze, Ana Carola Iñiguez,
    M.C. Martins Alves, and H. Tolentino

An EXAFS Study of Compositional Homogeneity in Sol-Gel
Processed Potassium Tantalum Niobates 333
    A.P. Wilkinson, J. Xu, and S. Pattanaik

X-ray-Absorption-Spectroscopy Studies of
Electrochemically Deposited Thin Oxide Films 339
    M. Balasubramanian, C.A. Melendres, A.N. Mansour, and S. Mini

Synchrotron X-ray-Absorption-Studies of Atomic-Level
Alloying in Immiscible Mixtures 347
    J.H. He, P.J. Schilling, and E. Ma

Characterization of Silver Binding in Cryptomelane by
X-ray Absorption Spectroscopy 353
    R. Ravikumar, D.W. Fuerstenau, and G.A. Waychunas

Electronic- and Spatial-Structure Studies of Cadmium and
Zinc Dialkyldithiocarbamate Molecules in Nonaqueous
Solutions Used in the Processes Spray Pyrolysis 359
    S.B. Erenburg, N.V. Bausk, S.M. Zemskova, S.V. Tkachev,
    and L.N. Mazalov

Surfaces of Semi-Fluorinated Block Copolymers Studied
Using NEXAFS 365
    J. Genzer, E. Sivaniah, E.J. Kramer, J. Wang, H. Körner, M-L. Xiang,
    S. Yang, C.K. Ober, K. Char, M.K. Chaudhury, B.M. DeKoven,
    R.A. Bubeck, D.A. Fischer, and S. Sambasivan

Structural Changes in Iron(II) Polymeric Complexes Upon
Thermally and Optically-Induced Spin Transition
Determined by EXAFS Spectroscopy 371
    S.B. Erenburg, N.V. Bausk, L.G. Lavrenova, Yu.G. Shvedenkov,
    and L.N. Mazalov

Author Index 377

Subject Index 381

PREFACE

This volume contains papers presented at the symposium on "Applications of Synchrotron Radiation Techniques to Materials Science," during the 1998 MRS Spring Meeting held in San Francisco, California, April 13-17. The symposium brought together the materials science community to focus on characterization techniques that use synchrotron radiation—much like the earlier symposia and volumes on this topic:

1988 MRS Fall Meeting	Volume 143, *Synchrotron Radiation in Materials Research* Edited by: R. Clarke, J. Gland, and J.H. Weaver
1993 MRS Spring Meeting	Volume 307, *Applications of Synchrotron Radiation Techniques to Materials Science* Edited by: D.L. Perry, N.D. Shinn, R. Stockbauer, K.L. D'Amico and L.J. Terminello
1994 MRS Fall Meeting	Volume 375, *Applications of Synchrotron Radiation Techniques to Materials Science II* Edited by: L.J. Terminello, N.D. Shinn, G.E. Ice, K.L. D'Amico, and D.L. Perry
1996 MRS Spring Meeting	Volume 437, *Applications of Synchrotron Radiation Techniques to Materials Science III* Edited by: L.J. Terminello, S. Mini, H. Ade, and D.L. Perry

As more synchrotron facilities are constructed and go online both in the United States and in other countries, even more applications of synchrotron radiation will be realized. Both basic and applied research possibilities are manifold, including studies of materials mentioned below and those that are yet to be discovered. Also, the combination of synchrotron-based spectroscopic techniques with ever increasing high-resolution microscopy allows researchers to study very small domains of materials in an attempt to understand their chemical and electronic properties. This is especially important in the areas of composites and other related materials involving material bonding interfaces.

The topics covered in this symposium include surfaces, interfaces, electronic materials, metal oxides, solar cells, thin films, carbides, polymers, alloys, nanoparticles, and graphitic materials. Results reported at this symposium relate recent advances in X-ray absorption and scattering, imaging, tomography, microscopy, and topography methods. It has been an interesting 50 years of synchrotron radiation.

S. M. Mini
S. R. Stock
D. L. Perry
L. J. Terminello

May, 1998

ACKNOWLEDGMENTS

The symposium organizers wish to thank the following for funding used to support this endeavor:

Physical Electronics
Blake Industries
Luxel Corporation
Chemistry and Materials Science Department
    Lawrence Livermore National Laboratory
Advanced Light Source (ALS)
    Lawrence Berkeley National Laboratory
National Synchrotron Light Source (NSLS)
    Brookhaven National Laboratory

The organizers also wish to thank the session chairs (Gene Ice, Patrick Allen, Richard Harlow, David Shuh, Piero Pianetta), who generously gave of their time to make the event so successful, and Marcus Johnson, for his help in preparing the manuscript for publication.

MATERIALS RESEARCH SOCIETY SYMPOSIUM PROCEEDINGS

Volume 481— Phase Transformation and Systems Driven Far From Equilibrium, E. Ma, P. Bellon, M. Atzmon, R. Trivedi, 1998, ISBN: 1-55899-386-X

Volume 482— Nitride Semiconductors, F.A. Ponce, S.P. DenBaars, B.K. Meyer, S. Nakamura, S. Strite, 1998, ISBN: 1-55899-387-8

Volume 483— Power Semiconductor Materials and Devices, S.J. Pearton, R.J. Shul, E. Wolfgang, F. Ren, S. Tenconi, 1998, ISBN: 1-55899-388-6

Volume 484— Infrared Applications of Semiconductors II, S. Sivananthan, M.O. Manasreh, R.H. Miles, D.L. McDaniel, Jr., 1998, ISBN: 1-55899-389-4

Volume 485— Thin-Film Structures for Photovoltaics, E.D. Jones, R. Noufi, B.L. Sopori, J. Kalejs, 1998, ISBN: 1-55899-390-8

Volume 486— Materials and Devices for Silicon-Based Optoelectronics, J.E. Cunningham, S. Coffa, A. Polman, R. Soref, 1998, ISBN: 1-55899-391-6

Volume 487— Semiconductors for Room-Temperature Radiation Detector Applications II, R.B. James, T.E. Schlesinger, P. Siffert, M. Cuzin, M. Squillante, W. Dusi, 1998, ISBN: 1-55899-392-4

Volume 488— Electrical, Optical, and Magnetic Properties of Organic Solid-State Materials IV, J.R. Reynolds, A.K-Y. Jen, L.R. Dalton, M.F. Rubner, L.Y. Chiang, 1998, ISBN: 1-55899-393-2

Volume 489— Materials Science of the Cell, B. Mulder, V. Vogel, C. Schmidt, 1998, ISBN: 1-55899-394-0

Volume 490— Semiconductor Process and Device Performance Modeling, J.S. Nelson, C.D. Wilson, S.T. Dunham, 1998, ISBN: 1-55899-395-9

Volume 491— Tight-Binding Approach to Computational Materials Science, P.E.A. Turchi, A. Gonis, L. Colombo, 1998, ISBN: 1-55899-396-7

Volume 492— Microscopic Simulation of Interfacial Phenomena in Solids and Liquids, S.R. Phillpot, P.D. Bristowe, D.G. Stroud, J.R. Smith, 1998, ISBN: 1-55899-397-5

Volume 493— Ferroelectric Thin Films VI, R.E. Treece, R.E. Jones, S.B. Desu, C.M. Foster, I.K. Yoo, 1998, ISBN: 1-55899-398-3

Volume 494— Science and Technology of Magnetic Oxides, M. Hundley, J. Nickel, R. Ramesh, Y. Tokura, 1998, ISBN: 1-55899-399-1

Volume 495— Chemical Aspects of Electronic Ceramics Processing, P.N. Kumta, A.F. Hepp, D.B. Beach, J.J. Sullivan, B. Arkles, 1998, ISBN: 1-55899-400-9

Volume 496— Materials for Electrochemical Energy Storage and Conversion II—Batteries, Capacitors and Fuel Cells, D.S. Ginley, D.H. Doughty, T. Takamura, Z. Zhang, B. Scrosati, 1998, ISBN: 1-55899-401-7

Volume 497— Recent Advances in Catalytic Materials, N.M. Rodriguez, S.L. Soled, J. Hrbek, 1998, ISBN: 1-55899-402-5

Volume 498— Covalently Bonded Disordered Thin-Film Materials, M.P. Siegal, J.E. Jaskie, W. Milne, D. McKenzie, 1998, ISBN: 1-55899-403-3

Volume 499— High-Pressure Materials Research, R.M. Wentzcovitch, R.J. Hemley, W.J. Nellis, P.Y. Yu, 1998, ISBN: 1-55899-404-1

Volume 500— Electrically Based Microstructural Characterization II, R.A. Gerhardt, M.A. Alim, S.R. Taylor, 1998, ISBN: 1-55899-405-X

Volume 501— Surface-Controlled Nanoscale Materials for High-Added-Value Applications, K.E. Gonsalves, M-I. Baraton, J.X. Chen, J.A. Akkara, 1998, ISBN: 1-55899-406-8

Volume 502— In Situ Process Diagnostics and Intelligent Materials Processing, P.A. Rosenthal, W.M. Duncan, J.A. Woollam, 1998, ISBN: 1-55899-407-6

Volume 503— Nondestructive Characterization of Materials in Aging Systems, R.L. Crane, S.P. Shah, R. Gilmore, J.D. Achenbach, P.T. Khuri-Yakub, T.E. Matikas, 1998, ISBN: 1-55899-408-4

Volume 504— Atomistic Mechanisms in Beam Synthesis and Irradiation of Materials, J.C. Barbour, S. Roorda, D. Ila, 1998, ISBN: 1-55899-409-2

Volume 505— Thin-Films—Stresses and Mechanical Properties VII, R.C. Cammarata, E.P. Busso, M. Nastasi, W.C. Oliver, 1998, ISBN: 1-55899-410-6

Volume 506— Scientific Basis for Nuclear Waste Management XX, I.G. McKinley, C. McCombie, 1998, ISBN: 1-55899-411-4

Volume 507— Amorphous and Microcrystalline Silicon Technology—1998, S. Wagner, M. Hack, H.M. Branz, R. Schropp, I. Shimizu, 1998, ISBN: 1-55899-413-0

MATERIALS RESEARCH SOCIETY SYMPOSIUM PROCEEDINGS

Volume 508— Flat-Panel Display Materials—1998, G. Parsons, T.S. Fahlen, S. Morozumi, C. Seager, C-C. Tsai, 1998, ISBN: 1-55899-414-9

Volume 509— Materials Issues in Vacuum Microelectronics, W. Zhu, L.S. Pan, T.E. Felter, C. Holland, 1998, ISBN: 1-55899-415-7

Volume 510— Defect and Impurity Engineered Semiconductors and Devices II, S. Ashok, J. Chevallier, K. Sumino, B.L. Sopori, W. Goetz, 1998, ISBN: 1-55899-416-5

Volume 511— Low-Dielectric Constant Materials IV, C. Chiang, J.T. Wetzel, T-M. Lu, P.S. Ho, 1998, ISBN: 1-55899-417-3

Volume 512— Wide-Bandgap Semiconductors for High Power, High Frequency and High Temperature, S. DenBaars, M.S. Shur, J. Palmour, M. Spencer, 1998, ISBN: 1-55899-418-1

Volume 513— Hydrogen in Semiconductors and Metals, N.H. Nickel, W.B. Jackson, R.C. Bowman, 1998, ISBN: 1-55899-419-X

Volume 514— Advanced Interconnects and Contact Materials and Processes for Future Integrated Circuits, S.P. Murarka, D.B. Fraser, M. Eizenberg, R. Tung, R. Madar, 1998, ISBN: 1-55899-420-3

Volume 515— Electronic Packaging Materials Science X, D.J. Belton, R. Pearson, M. Gaynes, E.G. Jacobs, 1998, ISBN: 1-55899-421-1

Volume 516— Materials Reliability in Microelectronics VIII, T. Marieb, J. Bravman, M.A. Korhonen, J.R. Lloyd, 1998, ISBN: 1-55899-422-X

Volume 517— High-Density Magnetic Recording and Integrated Magneto-Optics: Materials and Devices, K. Rubin, J.A. Bain, T. Nolan, D. Bogy, B.J.H. Stadler, M. Levy, J.P. Lorenzo, M. Mansuripur, Y. Okamura, R. Wolfe, 1998, ISBN: 1-55899-423-8

Volume 518— Microelectromechanical Structures for Materials Research, S.B. Brown, C. Muhlstein, P. Krulevitch, G.C. Johnston, R.T. Howe, J.R. Gilbert, 1998, ISBN: 1-55899-424-6

Volume 519— Organic/Inorganic Hybrid Materials, R.M. Laine, C. Sanchez, E. Giannelis, C.J. Brinker, 1998, ISBN: 1-55899-425-4

Volume 520— Nanostructured Powders and Their Industrial Application, G. Beaucage, J.E. Mark, G. Burns, H. Duen-Wu, 1998, ISBN: 1-55899-426-2

Volume 521— Porous and Cellular Materials for Structural Applications, D.S. Schwartz, D.S. Shih, H.N.G. Wadley, A.G. Evans, 1998, ISBN: 1-55899-427-0

Volume 522— Fundamentals of Nanoindentation and Nanotribology, N.R. Moody, W.W. Gerberich, S.P. Baker, N. Burnham, 1998, ISBN: 1-55899-428-9

Volume 523— Electron Microscopy of Semiconducting Materials and ULSI Devices, C. Hayzelden, F.M. Ross, C.J.D. Hetherington, 1998, ISBN: 1-55899-429-7

Volume 524— Application of Synchrotron Radiation Techniques to Materials Science IV, S.M. Mini, D.L. Perry, S.R. Stock, L.J. Terminello, 1998, ISBN: 1-55899-430-0

Volume 525— Rapid Thermal and Integrated Processing VII, M.C. Öztürk, F. Roozeboom, P.J. Timans, S.H. Pas, 1998, ISBN: 1-55899-431-9

Volume 526— Advances in Laser Ablation of Materials, R.K. Singh, D.H. Lowndes, D.B. Chrisey, J. Narayan, T. Kawai, E. Fogarassy, 1998, ISBN: 1-55899-432-7

Volume 527— Diffusion Mechanisms in Crystalline Materials, Y. Mishin, N.E.B. Cowern, C.R.A. Catlow, D. Farkas, G. Vogl, 1998, ISBN: 1-55899-433-5

Volume 528— Mechanisms and Principles of Epitaxial Growth in Metallic Systems, L.T. Wille, C.P. Burmester, K. Terakura, G. Comsa, E.D. Williams, 1998, ISBN: 1-55899-434-3

Volume 529— Computational and Mathematical Models of Microstructural Evolution, J.W. Bullard, R. Kalia, M. Stoneham, L-Q. Chen, 1998, ISBN: 1-55899-435-1

Volume 530— Biomaterials Regulating Cell Function and Tissue Development, D. Mooney, A.G. Mikos, K.E. Healy, Y. Ikada, R.C. Thomson, 1998, ISBN: 1-55899-436-X

Volume 531— Reliability of Photonics Materials and Structures, E. Suhir, M. Fukuda, C.R. Kurkjian 1998, ISBN: 1-55899-437-8

Volume 532— Silicon Front-End Technology—Materials Processing and Modelling, N.E.B. Cowern, D. Jacobson, P. Griffin, P. Packan, R. Webb, 1998, ISBN: 1-55899-438-6

Volume 533— Epitaxy and Applications of Si-Based Heterostructures, E.A. Fitzgerald, P.M. Mooney, D.C. Houghton, 1998, ISBN: 1-55899-439-4

Prior Materials Research Society Symposium Proceedings available by contacting Materials Research Society

Part I

Mapping, Microprobes, and Imaging

EXPANSIVE REACTIONS IN CONCRETE OBSERVED BY SOFT X-RAY TRANSMISSION MICROSCOPY

K.E. KURTIS[*], P.J.M. MONTEIRO[*], J.T. BROWN[**], & W. MEYER-ILSE[**]
[*] Department of Civil and Environmental Engineering, University of California, Berkeley, CA 94720
[**] Center for X-ray Optics, Ernest Orlando Lawrence Berkeley National Laboratory, University of California, Berkeley, CA 94720

ABSTRACT

Alkali-silica reaction, sulfate attack, and reinforcing steel corrosion can compromise the long-term durability of concrete structures. The anticipated economic impact of an extensive infrastructure repair scheme has produced a renewed interest in the development of advanced characterization methods to assess the degree of deterioration in the concrete experiencing these deleterious reactions. The products of the alkali silica reaction, sulfate attack, and corrosion as well as the cement hydration products are extremely sensitive to humidity. Consequently, characterization techniques that require high vacuum or drying, as many existing techniques do, are not particularly appropriate for the study of these reactions in concrete as artifacts are introduced.

A high resolution instrument which allows the examination of these reactions and their products without drying and at normal pressures will promote understanding of the reactions and provide further insight into means of mitigating the damage they cause. Only soft x-ray transmission microscopy provides the required high spatial resolution to observe the reaction process in situ. The alkali-silica reaction can be observed over time, in a wet condition, and at normal pressures, features unavailable with most other high resolution techniques. Soft x-rays also reveal information on the internal structure of the sample. This paper reviews published and ongoing applications of soft x-ray transmission microscopy for the study of expansive reactions that occur in concrete.

INTRODUCTION

The existing concrete infrastructure is deteriorating at a rate faster than was anticipated by design engineers. Concrete highways, bridges, dams, spillways, and nuclear power structures are showing the effects of years of neglect and poor maintenance and are in need of repair. The total rehabilitative cost to the concrete infrastructure in the U.S. is estimated to be in excess of $100 billion [1]. The anticipated economic impact of such an extensive repair scheme has produced a renewed interest in the development of advanced characterization methods to assess the degree of deterioration in the concrete infrastructure. By improving characterization methods, the structures expected to experience the greatest levels of distress can be more accurately identified and funds can be reserved for the repair these structures.

Alkali-silica reaction (ASR), sulfate attack, and reinforcing steel corrosion may compromise the long-term durability of concrete structures. Each of these reactions results in expansion which generates tensile strains in the concrete. Since the tensile strength of concrete is typically only 7-11% of its compressive strength, expansive strains of 0.04-0.05% or more are significant enough to crack the concrete [2]. Cracking decreases the elastic modulus and strength of concrete, particularly in flexure — material properties essential for the integrity and performance of the structure under service loads. In addition, as the pressure generated by expansive reactions in concrete increases, cracks will propagate and interconnect, increasing the permeability of the concrete and promoting further deterioration of the structure.

Both the cement hydration products and the deterioration products are extremely sensitive to humidity. Consequently, characterization techniques that require high vacuum or drying, as many existing techniques do, are not particularly appropriate as artifacts are introduced. Traditional characterization methods, such as scanning electron microscopy (SEM) and transmission electron microscopy (TEM), have given limited and misleading information about expansive reactions in concrete which has clouded the fundamental understanding of these reactions.

3

A high resolution instrument which allows the examination of these expansive reactions and their products without drying and at normal pressures will promote understanding of the reactions and provide further insight into means of mitigating the damage they cause. Only soft x-ray transmission microscopy provides the required high spatial resolution to observe the reaction process in situ. The reactions can be observed over time, in a wet condition, and at normal pressures, features unavailable with most other high resolution techniques.

This paper will present an overview of potential applications of soft x-ray transmission microscopy for the study of expansive reactions that occur in concrete. A description of the soft x-ray transmission microscope will be followed by a review of published and ongoing research where this technique has been applied to the study of expansive reactions and their products in concrete.

INSTRUMENT

The Center for X-ray Optics (CXRO) built and operates a high-resolution soft x-ray microscope (XM-1) at the Advanced Light Source in Berkeley, California [3,4]. The microscope is a conventional (full field) x-ray microscope, which uses zone plate lenses to provide high resolution transmission images. The optical setup (Figure 1) is similar to the Göttingen x-ray microscope [5], operated at the BESSY synchrotron radiation facility in Berlin, Germany. A condenser zone plate, fabricated by the Göttingen group [6] illuminates the sample and an objective zone plate, fabricated by E. Anderson [7,8] forms an enlarged image on an x-ray CCD camera. While the optical path of the microscope is in vacuum, the sample is at atmospheric pressure, flushed by helium. The spatial resolution of the microscope is 43 nm, measured as the distance from 10%-90% intensity in the image of a knife-edge [9].

Samples are mounted between two silicon nitride films, each about 100 nm thick. Samples containing water may be prepared about 5-10 μm thick. This thickness is achieved either by using sample material ground to that size or by mixing polystyrene beads of 6 μm diameter with the sample.

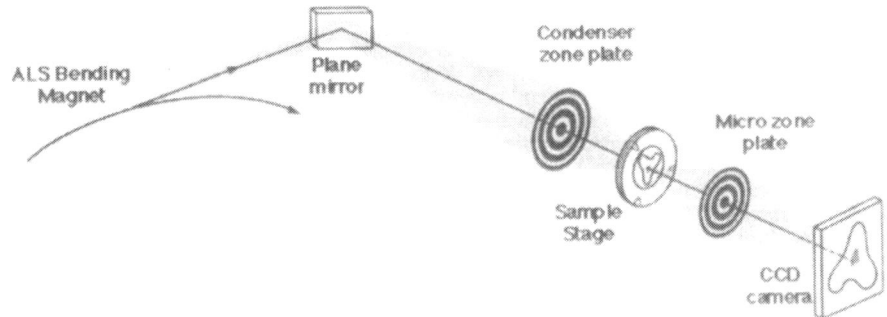

Figure 1
Schematic of the soft x-ray transmission microscope XM-1.

EXPANSIVE REACTIONS IN CONCRETE

Freeze-thaw, corrosion, alkali-silica reaction (ASR), and sulfate attack produce expansion, which generate tensile strains, leading to cracking of the concrete. Damage to the concrete is caused in part by the expansion associated with the reaction itself, as occurs in freezing of water in the concrete, or by expansion of the reaction product, as occurs with corrosion, ASR, and sulfate attack. The focus of this paper will be the use of soft x-ray transmission microscopy for the study of expansive reaction products such as those produced by alkali-silica reaction, sulfate attack, and corrosion of reinforcing steel.

Alkali-Silica Reaction

In concrete, alkali metal ions and hydroxyl ions contributed principally by the cement and reactive silicates present in aggregate can participate in a destructive alkali-silica reaction (ASR). The product of this reaction is a gel that tends to imbibe water found in the concrete pores, leading to swelling of the gel and eventual cracking of the affected concrete member.

Alkali attack of the siliceous aggregate is initiated at aggregate surfaces exposed to the alkaline pore solution in concrete. Ions present in the pore solution, including hydroxyl, sodium, potassium, and calcium ions, can more freely penetrate a poorly crystalline silica or silicate structure as compared to a dense, ordered silica or silicate structure. Thus, some siliceous aggregates are more susceptible than others to this reaction in highly alkaline (pH of 13.5-13.9) pore solution.

The alkali-silica reaction is initiated by hydroxyl ions which rupture Si-O-Si bonds, loosening the network and producing Si-O$^-$ species. Cations abundant in the pore fluid are attracted to these sites of negative charge and become incorporated into the alkali-silica reaction gel. In addition, calcium ions react with silica species in solution to form calcium silicate hydrate (C-S-H*) [10,11]. Because the solubility of silica increases sharply at a pH greater than 10, it is suggested that silica dissolves to a discernible degree in the alkaline concrete environment and that the dissolved species are available for further reaction with species present in solution. The alkali-silica reaction in concrete can be described by four distinct reactions [10,11]:

(1) dissolution of silica in the aggregate by hydroxyl attack

(2) reaction of surface silanol (Si-OH) groups with hydroxyl ions (OH$^-$) in the solution to further promote dissolution

(3) binding of the alkali cations (Na$^+$, K$^+$) and calcium cations (Ca^{++}) at negatively charged sites on the silicate surface

(4) reaction of silica species in solution with calcium ions in solution to form calcium silicate hydrate (C-S-H), a compound of varying composition and morphology which provides strength to concrete.

ASR gel capable of expanding is believed to be produced by the reactions 1-3 listed above [12,13]. The reaction described above as 4 suggests that the alkali-silica reaction may be similar, in some way, to the pozzolanic reaction in concrete [14].

Pozzolans, which are often industrial by-products, may be substituted in part for cement or for fine aggregate in concrete mixtures. Pozzolans are siliceous or siliceous and aluminous material that alone have no cementing properties, but may react chemically with calcium hydroxide in a moist environment at normal temperatures to form calcium silicate hydrates [15]:

$$POZZOLAN + CH + H \rightarrow C\text{-}S\text{-}H$$

Concretes produced with pozzolans typically exhibit enhanced durability because the pozzolanic reaction decreases permeability of the concrete.

Because the product of the alkali silica reaction contains water and the mechanism of expansion is understood to be caused by the imbibtion of water, the removal of the water from the alkali-silica gel product, as required by many existing microscopy techniques, will alter the sample and affect interpretation of the reaction mechanisms. Kurtis *et al.* [16] used soft x-ray transmission microscopy to examine the reaction of ground ASR gel, obtained from a Brazilian dam, and sodium hydroxide, calcium hydroxide, and combined sodium and calcium hydroxide solutions, which is most similar to the composition of concrete pore solution. Reaction of the ASR gel in sodium hydroxide solution resulted in surface dissolution and repolymerization of the ASR gel. Reaction of the ASR gel with calcium hydroxide solution produced spherulitic structures resembling the distinctive 'sheaf of wheat' morphology described in the literature. Reaction of the ASR gel with the combined solution produced dissolution and repolymerization of the gel with some lath-like formations resulting from the reaction of calcium cations and dissolved silica. From

---

* In cement chemistry notation: C=CaO, S=SiO$_2$, A=Al$_2$O$_3$, $=SO$_3$. H=H$_2$O.

the investigation by Kurtis *et al.*[16], it appears that reactive silica combine with alkalis present in pore solution to produce a reaction gel capable of swelling, while the reaction of the silica in the presence of calcium hydroxide and no alkalis results in the formation of a structure resembling C-S-H.

Research is ongoing to further determine if a similarity exists between the pozzolanic reaction and the alkali-silica reaction. The reaction of silica fume, a pozzolan commonly used in concrete, with calcium hydroxide solution has been observed by soft x-ray transmission microscopy. Also, the reaction of chemical grade silica gel, which may be considered pozzolanic but is not typically used in concrete, with calcium hydroxide solution has been observed.

Sulfate Attack

Sulfates present in soils, groundwater, sea water, decaying organic matter, and industrial effluent surrounding a concrete structure may permeate the concrete and react with existing hydration products. In the presence of calcium hydroxide (CH) and water (H), monosulfate hydrate ($C_3A \cdot C\$ \cdot H_{18}$) and calcium aluminate hydrate ($C_3A \cdot C\$ \cdot H_{18}$) react with the sulfate ($\$$) to produce ettringite ($C_3A \cdot 3C\$ \cdot H_{32}$) [1]:

$$C_3A \cdot C\$ \cdot H_{18} + 2CH + 2\$ + 12H \text{ ' } C_3A \cdot 3C\$ \cdot H_{32}$$
$$C_3A \cdot CH \cdot H_{18} + 2CH + 2\$ + 12H \text{ ' } C_3A \cdot 3C\$ \cdot H_{32}$$

In hardened concrete, the formation of ettringite by sulfate attack *can*, but does not always, result in expansion and lead to cracking of the concrete. The conditions under which ettringite formation produces damage in the concrete are uncertain.

It should be noted that ettringite produced by the reactions described above occupies a smaller volume than the reactants occupied. Therefore, the reaction described above must not be responsible for the expansion. It is generally accepted that the expansion caused by sulfate attack is the result of a particular mechanisms associated with the ettringite reaction or is the result of reaction other than the formation of ettringite. Gypsum, in addition to ettringite, can be produced during sulfate attack and is capable of producing expansion.

Two particular mechanisms for expansion associated with the formation of ettringite have been widely published - the topochemical reaction mechanism [17] and the swelling mechanism [18]. According to the topochemical reaction theory of expansion, sulfate and calcium ions in the concrete pore fluid react with dissolving aluminate ions near the surface of the solid phase, and the ettringite produced by this reaction grows perpendicular to the original solid surface [17]. Since the sulfate and calcium ions are in solution, only the volume of the aluminate phase is considered when comparing volume of reactants to the volume occupied by the ettringite produced. In an open system where the concrete is permeable to water present in the environment, water may then occupy the newly formed pore space, producing an expansion.

According to the swelling theory of expansion, poorly crystalline ettringite produces expansion by adsorption of water [18]. In a system containing sufficient concentrations of sulfate, hydroxyl, and calcium ions, small, nearly colloidal ettringite is believed to form. Water from the environment outside the concrete member is adsorbed by the poorly crystalline ettringite, generating an osmotic pressure. If the elastic modulus of the concrete is sufficiently low, a volumetric expansion of the member results.

Thus, the mechanism responsible for expansion caused by sulfate attack in concrete is uncertain. It is possible that both theories described above are valid under certain conditions. An improved understanding of behaviors occurring at the ettringite/pore solution interface is critical in developing a total appreciation for the mechanisms responsible for expansion of concrete from sulfate attack. Because interest in the reaction is focused upon the adsorption of water and the effect of ion concentration on the ettringite produced, the use of an instrument, such as XM-1, that allows the microscopic study of the reaction and the reaction product in a wet environment is integral for understanding the reaction and conditions required for concrete expansion by sulfate attack.

The morphology of the ettringite resulting from the reaction,

$$C_4A_3\$ + 8C\$ + 6C + 96H \longrightarrow 3(C_3A \cdot 3C\$ \cdot 32H)$$

has been examined through soft x-ray transmission microscopy [19]. In addition, O.1M CaCl$_2$ solution was substituted for water (H) in the above equation to study the effect of increased concentrations of Ca^{++} on the product morphology. Variations in the morphology of the products of the calcium sulfoaluminate reaction in water and in 0.1M CaCl$_2$ are readily apparent in Figures 2 and 3. However, an analytical comparison of the reaction products, including the x-ray diffraction pattern and surface charge density of each reaction product, will provide further information as to the expansive nature of these products. These investigations are currently underway.

Figure 2
Calcium sulfoaluminate mixture in water.

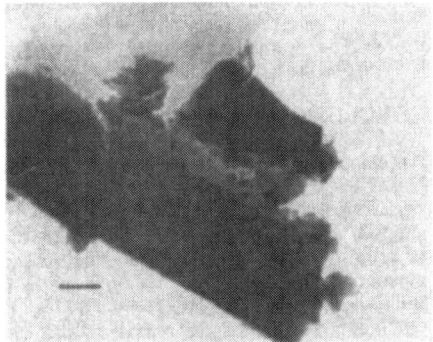

Figure 3
Calcium sulfoaluminate mixture in
0.1M CaCl$_2$ solution.

Reinforcing Steel Corrosion

Corrosion of embedded reinforcing steel is a destructive electrochemical process that ultimately weakens reinforced concrete structures. When the steel reinforcement corrodes, the strength of the member is undermined in several ways. Since corrosion products have a greater volume than the original steel reinforcing bars, internal tensile stresses will develop in the cement mortar at the steel/mortar interface, causing the surrounding concrete to crack and eventually spall away. Formation of such cracks decreases concrete stiffness and tensile strength, while the formation of larger, wider cracks increases concrete permeability. With the steel corroding away, the reinforcing bar cross-section is reduced, decreasing the member's tensile strength. Furthermore, as corrosion advances, the bond between the steel and surrounding concrete is weakened, adversely affecting the load transfer between the two materials.

However, the concrete surrounding the reinforcing steel provides protection from corrosion by several mechanisms. First, good quality concrete is impermeable, physically protecting the steel from the ingress of water and aggressive ions which promote corrosion. Secondly, a thin oxide layer forms a protective, passive film on the steel surface in alkaline environments, such as the concrete pore solution. This passive film must be penetrated or removed for corrosion to occur at a significant rate. Typically, the passive film will be affected by the ingress of chloride ions through the concrete or by the reaction of the hydrated cement products with carbon dioxide in the environment, which decreases the pH of the pore solution, causing the film to be unstable. Thirdly, it has been suggested [20-22] that lime-rich mineral scales formed near the steel/concrete interface may offer an additional means of protection against corrosion by buffering the pore solution, maintaining passivity of the steel.

A critical review of the state-of-the-art on expansion and cracking of reinforced concrete due to corrosion of the embedded steel reveals several areas of uncertainty or ambiguity [22]. Some of these gaps in current understanding stem from the inherent difficulties associated with observing the corrosion of steel within concrete. For instance, the composition and morphology of corrosion products in situ are uncertain. However, it is believed progress in the current understanding of the corrosion process may be promoted through the use of soft x-ray transmission microscopy. In particular, this technique would be advantageous for the study of certain aspects related to the passive layer, mineral scales, and the corrosion product, including:

- The mechanisms of steel passivation in the concrete environment and the mechanisms of depassivation by chloride ions and carbonation.
- The stoichiometry of the passive film on steel and the mechanisms of its formation; The relative significance of the passive film and mineral scales in providing corrosion protection to steel.
- The composition of the mineral scales formed in the vicinity of the steel and the mechanism of protection provided.
- The composition, morphology, and mechanisms of expansion of the steel corrosion products in concrete pore solution.

CONCLUSIONS

Applications of soft x-ray transmission microscopy to the study of expansive reactions in concrete have been discussed. XM-1, a soft x-ray transmission microscopy operated by the Center for X-ray Optics, allows the study of reactions over time with high resolution, but without drying the sample. For each of the reactions discussed -- alkali-silica reaction, sulfate attack, and corrosion of reinforcing steel -- the mechanisms of expansion which result in damage to the concrete are closely associated with humidity. In the alkali-silica reaction, the gel product causes expansion by imbibing water and swelling; with sulfate attack, ettringite in certain conditions may swell; with corrosion, the role of the passive film, the role of mineral scales, and the composition of the corrosion product can only be studied in hydrated condition. Through this examination it is apparent that soft x-ray transmission is beneficial for the study of these deleterious reactions and their products.

ACKNOWLEDGEMENTS

The first-named author wishes to acknowledge financial support by the National Science Foundation graduate research fellowship. Research at XM-1 is supported by the United States Department of Energy, Office of Basic Energy Sciences under contract DE-AC 03-76SF00098.

REFERENCES

1. Federal Highway Administration (1991) 1991 Report to Congress, June.
2. Swamy, R.N. in ACI SP-144, edited by. P.K. Mehta, 105 (1994).
3. Meyer-Ilse, W. *et al., Synchrotron Radiation News*, **8**, 29 (1995).
4. Meyer-Ilse, W.; Medecki, H. ; Brown, J.T.; Heck, J.; Anderson, E.; Magowan, C.; Stead, A.; Ford, T.; Balhorn, R.; Petersen, C.; and Attwood, D.T. in X-ray Microscopy and Spectromicroscopy, edited by. J. Thieme, G Schmahl, E. Umbach, and D. Rudolph, Springer-Verlag, Heidelberg, Germany (1998).
5. Schmahl, G.; Guttmann, P.; Schneider, G.; Niemann, B.; David, C.; Wilhein, T.; Thieme, J.; and Rudolph, D. in X-Ray Microscopy IV, edited by V.V. Aristov and A.I Erko, Chernogolovka, Russia (1994).
6. Hettwer, M. and Rudolph, D. in X-ray Microscopy and Spectromicroscopy, edited by J. Thieme, G. Schmahl, E. Umbach, and D. Rudolph Springer-Verlag, Heidelberg, Germany (1998).
7. Anderson, E.H. and Kern, D. in X-ray Microscopy III, edited by A. Michette, G.R. Morrison, and C.J. Buckley, Springer Heidelberg (1992).
8. Anderson, E.H.; Boegli, V.; Muray, L.P., J. Vac. Sci. Technol. B., **13**, 2529 (1995).

9. Heck, J.M; Meyer-Ilse, W.; Brown, J. T.; Anderson, E.; Medecki, H.; and Attwood, D. T. in X-ray Microscopy and Spectromicroscopy, edited by J. Thieme, G. Schmahl, E. Umbach, and D. Rudolph Springer-Verlag, Heidelberg, Germany (1998).
10. Dent Glasser, L.S. and Kataoka, N., Cem.and Concr. Res., **11**, 1 (1981).
11. Dent Glasser, L.S. and Kataoka, N., Cem.and Concr. Res., **12**, 321 (1982).
12. Powers, T.C. and Steinour, H.H., J. Amer. Concr. Inst., 26, 497 (1955).
13. Powers, T.C. and Steinour, H.H., J. Amer. Concr. Inst., 26, 785 (1955).
14. Taylor, H.F.W., Cement Chemistry, Academic Press, London (1990).
15. Mehta, P.K. and Monteiro, P.J.M. Concrete: Structure, Properties, and Materials, Prentice Hall, New Jersey (1993).
16. Kurtis, K.E.; Monteiro, P.J.M; Brown, J.T.; and Meyer-Ilse, W. Cem. and Concr. Res. **28**, 411 (1998).
17. Odler, I. and Jawed, I., in Materials Science of Concrete II, edited by J. Skalny and S. Mindness, The American Ceramic Society, Westerville, Ohio, 221 (1991).
18. Mehta, P.K., in Materials Science of Concrete III, edited by J. Skalny, The American Ceramic Society, Westerville, OH, 105 (1992).
19. Kurtis, K.E.; Monteiro, P.J.M.; Brown, J.T.; and Meyer-Ilse, W., Advanced Light Source: Compendium of User Abstracts and Technical Reports 1997, in press.
20. Page, C.L. *Nature*, Dec., 514 (1975).
21. Borgard, B.; Warren, C.; Somayaji, S.; and Heidersbach, R., Corrosion Rates of Steel in Concrete, STP1065, ASTM (1990).
22. K.E. Kurtis and P.K. Mehta, CANMET/ACI Fourth International Conference on Durability of Concrete, Sydney (1997).

THE FIRST SYNCHROTRON INFRARED BEAMLINES AT THE ADVANCED LIGHT SOURCE: MICROSCPECTROSCOPY AND FAST TIMING

MICHAEL C. MARTIN and WAYNE R. McKINNEY
Advanced Light Source Division, MS 7-222, Lawrence Berkeley National Laboratory,
1 Cyclotron Road, Berkeley, CA 94720, MCMartin@lbl.gov

ABSTRACT

A set of new infrared (IR) beamlines on the 1.4 bending magnet port at the Advanced Light Source, LBNL, are described. Using a synchrotron as an IR source provides considerable brightness advantages, which manifests itself most beneficially when performing spectroscopy on a microscopic length scale. Beamline (BL) 1.4.3 is a dedicated microspectroscopy beamline, where the much smaller focused spot size using the synchrotron source is utilized. This enables an entirely new set of experiments to be performed where spectroscopy on a truly microscopic scale is now possible. BL 1.4.2 consists of a vacuum FTIR bench with a wide spectral range and step-scan capabilities. The fast timing is demonstrated by observing the synchrotron electron storage pattern at the ALS.

INTRODUCTION

Synchrotron-based infrared beamlines provide considerable brightness advantages over conventional (thermal) IR sources [1]. This brightness advantage manifests itself most beneficially when measuring very small samples. In the commissioning of the first IR beamline at the ALS, BL 1.4.3, we have experimentally measured the small spot-size obtained by our IR microspectroscopy system when using the synchrotron beam as the source and we compare it to the conventional Globar source. We demonstrate the diffraction-limited focus and the corresponding factor of ~100's improvement over conventional sources in measured signal through very small apertures. This new tool enables a host of new scientific pursuits where FTIR spectroscopy (indicating chemical species, phase changes, etc.) can be monitored on a microscopic scale.

Synchrotron-based infrared beamlines also provide a unique opportunity for doing time-resolved IR spectroscopy. Synchrotrons operate with electrons traveling in bunches around the storage ring. Each bunch emits a pulse of synchrotron radiation every time the electrons are forced to turn via a bending magnet. This creates a train of light pulses with a temporal spacing of 2 nanoseconds in the case of the ALS (see Figure 4). Pump-probe type experiments can therefore be performed at speeds up to 2nsec. We provide a demonstration of this fast-timing capability using the electron filling pattern in the ALS storage ring.

EXPERIMENTAL CONFIGURATION

The details of the construction and commission of these IR beamlines is provided in greater detail in a previous publication [2]. Briefly, the synchrotron light is collected from the 1.4 bending magnet through a 10mrad vertical and 40mrad horizontal opening, as schematically drawn in Figure 1. This light is deflected vertically by 0.5 meter, then is refocused outside the shield wall by an ellipsoidal mirror, m2. A 'switchyard' then contains a series of optics to collimate the IR beam and distribute the beam to one of three end stations.

Mat. Res. Soc. Symp. Proc. Vol. 524 © 1998 Materials Research Society

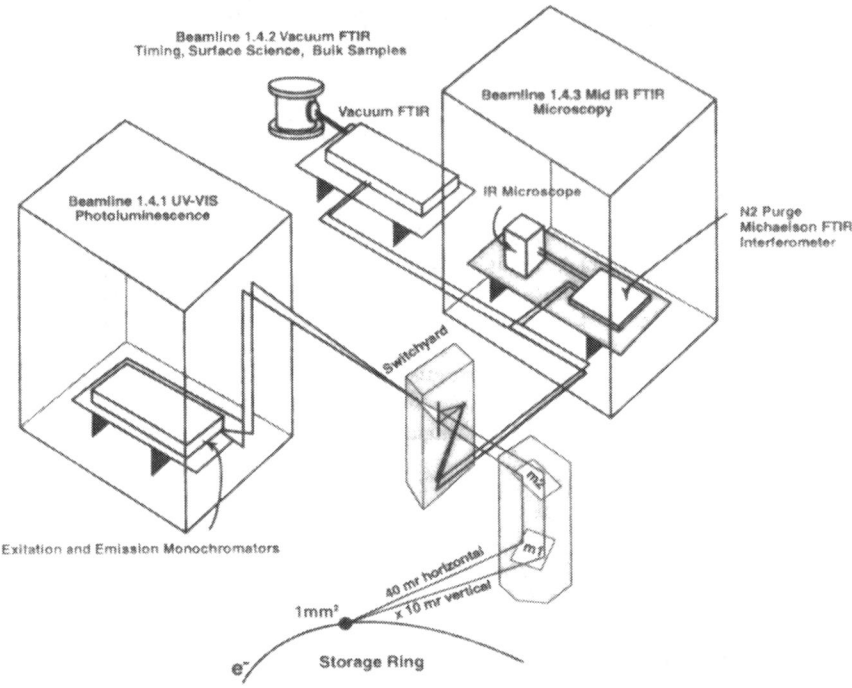

Figure 1. Schematic drawing of ALS beamlines 1.4.1, 1.4.2, and 1.4.3.

BL 1.4.3 inputs the collimated synchrotron light into a Nicolet Magna 760 FTIR bench. The modulated light is then passed through a Nic-Plan IR microscope that can perform both transmission and reflection measurements. The sample stage is computer controlled for motions in the x and y plane enabling automated spectral measurements across a sample with steps as small at 1 μm.

BL 1.4.2 uses the synchrotron light for the input of a Bruker IFS 66v/S vacuum FTIR spectrometer. The light then passes through a standard sample compartment, or out one of three external ports. One external experiment being set up is a UHV surface science chamber that will allow grazing incidence reflectivity measurements. This IFS 66v/S instrument has a wide spectral range, 50 cm^{-1} to 25,000 cm^{-1}, and it has step-scan capabilities in addition to rapid-scan to enable fast timing measurements.

SPOT SIZE

We have made measurements to determine the small spot sizes achievable with the infrared microspectroscopy equipment on beamline 1.4.3. The spectrometer software can obtain an area scan by moving the sample stage with 1 micron spatial precision under the focused IR beam and acquiring FTIR spectra at each point. To determine the actual focused spot size of the synchrotron beam and compare it to the internal Globar IR source, we used a 10 μm pin hole and measured the transmitted spectra as a function of the pin hole position. No other beam-defining apertures were used.

Integrating over the measured mid-IR range, we obtain an intensity number. The intensity as a function of pin hole position is shown in Figure 2. A small spot size readily apparent.

In Figure 3 we plot the x and y profiles of this spot along with fits to a Gaussian line-shape. The data fits very well to a Gaussian with resultant widths

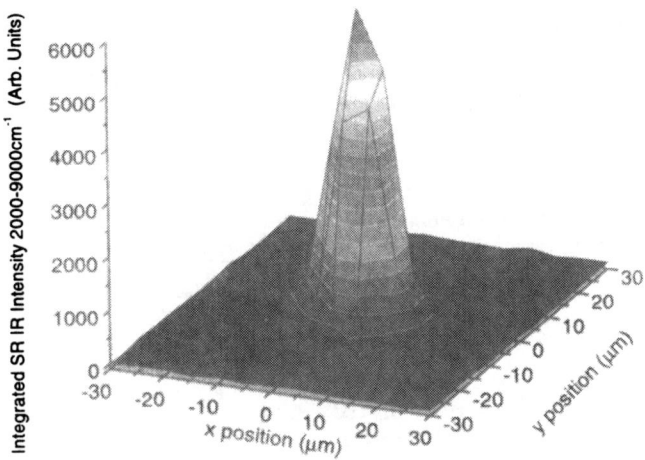

Figure 2. Integrated IR signal intensity from 2000 - 9000cm⁻¹ through a 10μm pin hole being scanned on microscope stage. There are no other apertures in the optical path. This graph demonstrates the small spot size achieved using the synchrotron IR beam

of 10 μm in x and 8 μm in y. This spot size is becoming roughly diffraction limited for the low-energy end of our detection capabilities ($1000 \text{ cm}^{-1} = 10 \text{ μm}$ wave-length).

When we compare the above measurements to analogous measurements made with a conventional Globar thermal IR source, the brightness advantages of the synchrotron become readily apparent. The Globar has a much broader peak profile of around 100 μm in width simply

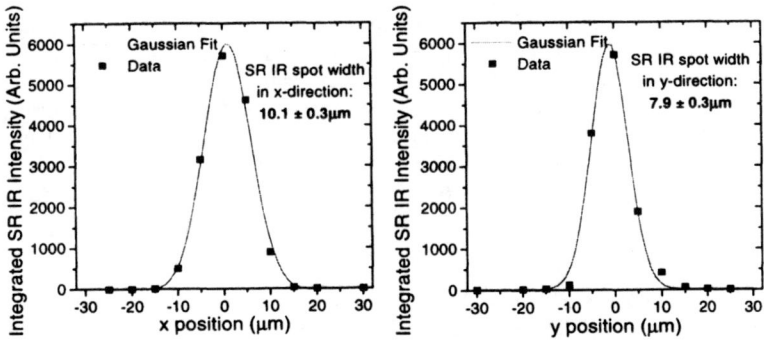

Figure 3. Peak shape profiles in the x (left panel) and y (right panel) directions. We observe a well-defined Gaussian peak shape with widths of 10μm in x and 8μm in y.

due to the large source size of this glowing filament source making a better focus impossible. Therefore, while the overall amount of light passing through the optical system from the Globar and the synchrotron sources is comparable, nearly all of the synchrotron light can be focused onto a 10 μm spot. To achieve a similar spot size using the conventional Globar source, one must simply mask the source and throw away a large factor of signal intensity. We measure improvements of several hundred in intensity through small pin-holes (2 - 10 μm diameter) for the synchrotron source compared to the Globar.

One example of the use of BL 1.4.3's microspectroscopy capabilities is the non-destructive monitoring of the *in situ* inorganic-organic interactions at biological-mineral interfaces [3]. The detailed localization of these bacteriological processes is only possible using a synchrotron based instrument due to the high signal at small spot sizes providing excellent spatial contrast.

FAST TIMING DEMONSTRATION

Standard rapid scan Fourier Transform Infrared Spectroscopy (FTIR) acquires scans on the time scale of one second, so the very fast pulses of a synchrotron source are not noticed. However, using the step-scan capabilities and fast electronics of the Bruker IFS 66v/S FTIR spectrometer on

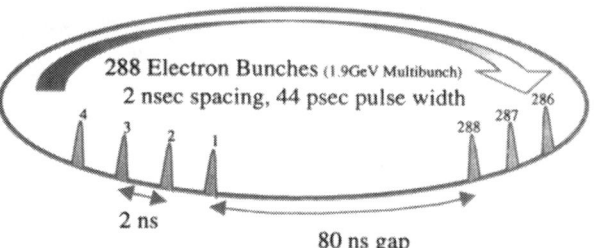

Figure 4. The ALS generates pulses at intervals of 2nsec, except for a single long gap of 80ns. We use this time structure to demonstrate the fast timing capabilities of BL1.4.2.

BL1.4.2, we can measure IR spectra with a time resolution as fast as 5 nsec. This means that processes that occur on the nanosecond all the way up to hour time scales can be monitored with time-resolved spectroscopy using this beamline.

To demonstrate the fast-timing capabilities of this beamline, we synchronized the detection electronics with the ALS ring timing structure (one trigger per complete revolution of the electrons). The ALS typically operates in multibunch mode with 288 bunches spaced 2 nanoseconds apart, followed by an 80 nsec gap. The individual pulse width is 44 psec (FWHM). This electron filling pattern is schematically drawn in Figure 4.

IR spectra were obtained at 5nsec time slices for a total of 650 nsec (a complete revolution of each electron bunch). We plot the measured intensity as a function of time integrated over a region in the visible wavelengths in Figure 5.

Figure 5 clearly demonstrates that we can observe the 80 nsec gap in the synchrotron light pulses. The IR detector and digitization electronics are not yet fast enough to observe the 2 nsec pulse spacing, but forthcoming enhancements should allow sub-nanosecond timing enabling the use of individual synchrotron pulses to serve as the probe in very fast time-resolved experiments.

CONCLUSIONS

BL 1.4.3 utilizes the high brightness aspects of the synchrotron light source to perform IR microspectroscopy at spatial resolutions much higher than is possible with conventional IR sources. Experimental systems where the sample size is quite small, or where features on a sample are on the micron size scale, will gain significantly by using ALS Beamline 1.4.3. A few examples include observing bacteria ingesting specific chemicals in real time [3], localization of

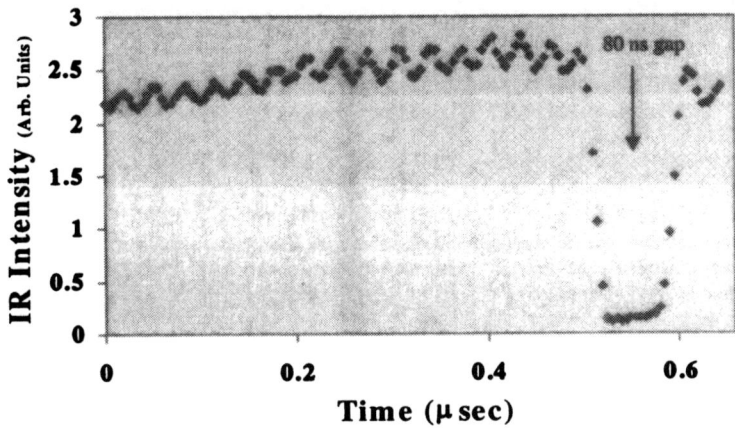

Figure 5. Time-resolved measurement of the IR intensity showing the 80 nsec gap in the ALS synchrotron filling pattern. The time resolution of the current electronics does not yet fully separate the 2 nsec spacing between pulses.

adsorbates on a variety of surfaces, particulate contamination on semiconductors, polymer laminates and composites, and analytical forensic studies of small samples.

The step-scan capabilities of the IFS 66v/S FTIR spectrometer on BL 1.4.2 allow for very fast time-resolved IR spectroscopy. This instrument, in conjunction with the ALS synchrotron's pulsed nature, makes possible a host of new time-resolved and pump-probe spectroscopic experiments. Typical applications for this beamline will include pump-probe measurements (semiconductors, metastable states), environmental science (adsorbates, bacteria, soil chemistry, bioremediation), biological materials (identification of biomolecules, time-resolved chemical reactions), high-pressure systems (materials in diamond anvil cells), and measurements at cryogenic temperatures and/or in high-magnetic fields (far-IR reflectivity from high-T_c and other correlated electronic systems).

ACKNOWLEDGMENTS
This work is supported by the Director, Office of Energy Research, Office of Basic Energy Sciences, Materials Science Division, of the U.S. Department of Energy under Contract No. DE-AC03-76SF00098. Thanks to Carol J. Hirschmugl, Noel Kellogg, Gwyn P. Williams, G. Larry Carr, Howard A. Padmore, T. Lauritzen, N. Andresen, G. Andronaco, R. Patton, and M. Fong.

REFERENCES
1. Hirschmugl, C.J., Ph.D. Thesis Yale University, 1994.
2. W.R. McKinney, C.J. Hirschmugl, H.A. Padmore, T. Lauritzen, N. Andresen, G. Andronaco, R. Patton, and M. Fong, Proceedings of the SPIE, Accelerator-Based Infrared Sources and Applications, Volume 3153, 1997, pp. 59-67.
3. Hoi-Ying N. Holman, Dale L. Perry, Michael C. Martin, and Wayne R. McKinney, "Applications of synchrotron infrared microspectroscopy to the study of inorganic-organic interactions at the bacterial-mineral interface," this proceedings volume.

APPLICATIONS OF SYNCHROTRON INFRARED MICROSPECTROSCOPY TO THE STUDY OF INORGANIC-ORGANIC INTERACTIONS AT THE BACTERIAL- MINERAL INTERFACE

HOI-YING N. HOLMAN*, DALE L. PERRY, MICHAEL C. MARTIN, AND WAYNE R. McKINNEY, Lawrence Berkeley National Laboratory, University of California, Berkeley, CA 94720, hyholman@lbl.gov

ABSTRACT

Synchrotron microspectroscopy has been used to study the inorganic-organic interactions in the mid-infrared region (4000-400 cm^{-1}) as *Arthrobacter oxydans* attach themselves to magnetite surfaces. Relative band intensities and band intensity ratios for functional groups of organically-derived biological molecules that are inherent to the experimental system are discussed. The molecular components as they are perturbed by interactions with water, dichromate and chromate metal ions on the mineral surfaces are investigated. Mapping of the spectral markers for the inorganic-organic interactions at the biological-mineral interfaces is presented and discussed. Comparative analyses of the synchrotron infrared microspectra suggest that the bacterial-chromium interactions depend on the solubility and toxicity of the chromium compounds.

INTRODUCTION

Pollution of subsurface geologic zones by heavy metals and the possibility of using intrinsic endolithic (rock/mineral-inhabiting) bacteria to either detoxify or immobilize the pollutants have stimulated new interests in the exploration of the bacterial responses to metal ions in geologic environments [1-2]. It is widely believed that the inorganic-organic interactions at the biological-mineral interfaces, which mostly arise from bacterial defense mechanisms, can change the effectiveness of bacteria to remediate metal ions in the immediate vicinity [3]. In the case of chromium, it is generally accepted that the enzymatic reduction of hexavalent chromium Cr(VI) (chromate) is one defense mechanism employed by bacteria living in the chromate-contaminated environment [4]. Most of the previous studies on the interactions between bacteria and metal ions were conducted by means of batch or column experiments [4-5], which can only globally address the questions of whether or not the bacteria are interacting with the metal ions. To evaluate if the mechanisms of the interaction exist at the biological-mineral interfaces, one needs to examine and monitor, nondestructively, the structural changes of molecules as the reactions take place *in situ*.

In this study we present the use of synchrotron-infrared absorption-reflectance microspectroscopy as a chemical/biological microprobe to study — *in situ* and nondestructively — the responses of bacteria to Cr(VI) at the biological-mineral interfaces. Specifically, we investigated if the synchrotron microprobe could successfully (1) monitor the inorganic-organic interactions and (2) illustrate the highly localized differences in the progress of the interactions and the structural details of clusters of living-bacteria in minerals that inherently have low infrared reflective surfaces. *Arthrobacter oxydans* and magnetite, the bacteria and mineral used in this study, are representative of those commonly encountered in the subsurface [6,7]. Results from the synchrotron infrared microspectroscopy analysis provide new insights into factors that control bacterial defense mechanisms and further our understanding of the bacteria's capability to remediate metal contaminants in geologic environments.

EXPERIMENTAL

Chemicals

All chemicals were purchased from Aldrich (USA). They were of A.C.S. reagent grades with a purity of at least 99%. Cr(VI) in this study was derived from two different chromium compounds:

potassium chromate and dichromate (from Aldrich, USA). Their aqueous solutions were prepared at pH 7.4 (physiologic pH) in phosphate buffer solutions. Potassium chromate and dichromate were selected as our model compounds to represent a high solubility/toxicity and a low solubility/toxicity chromate compound at the physiologic pH. Toluene vapor was used as a carbon source for the bacteria when needed.

Bacteria and Minerals

A. oxydans were isolated from rock cores collected from a subsurface rock vadose zone at a depth of 66-70 m at a site within the extensive Columbia basalt flow in southeastern Idaho as previously described [8]. *A. oxydans* were maintained in a filter-sterile liquid growth medium that consisted of 40% of basalt extract solution and 60% of distilled and deionized water. The basalt extract solution was prepared by mixing vigorously and heating (without boiling) for one-hour 500 g of basalt grains in 1 liter of distilled, deionized water. The liquid was strained through a double-layer cheesecloth and filter sterilized with 0.2 μm cellulose acetate/cellulose nitrate mixed esters (CA/CN) filters.

Magnetite mineral samples were obtained commercially from Minerals Unlimited (Ridgecrest, CA). Thin magnetite specimens were prepared by cleaving fragments off the micro-fissiles. The specimen surfaces were cleaned by sonification in deionized and organic-free water for fifteen minutes and sterilized by UV-irradiation for twenty minutes.

Inorganic-Organic Interaction Experiments

To generate the experimental conditions for studying the inorganic-organic interactions at the bacterial-mineral interfaces, *A. oxydans* were introduced onto the magnetite surfaces. This procedure involved first capturing the bacteria from the growth liquid medium onto the 0.2 μm CA/CN filters and printing onto the magnetite surfaces using a technique similar to the agar printing-off procedure described in [9]. Five hours after the bacterial attachment to the magnetite surfaces, they were exposed to 10 ppm of either chromate or dichromate solutions. The temporal variation and the spatial distribution of the inorganic-organic interactions on the biological-mineral surfaces were measured spectroscopically with the infrared microprobe at the Advanced Light Source (ALS) at Lawrence Berkeley National Laboratory. For each measurement at a given location, 128 spectra were co-added at a spectral resolution of 4 cm^{-1}. All infrared spectral characteristics were interpreted according to well-documented spectral research literature [12-15].

Instrumentation

The progress and the spatial distribution of chromium-bacteria interactions at the biological-magnetite interface was monitored using the infrared microspectroscopy Beamline 1.4.3 at the Advanced Light Source (ALS), Lawrence Berkeley National Laboratory as previously described [10,11]. The infrared microprobe uses a synchrotron source that has much higher brightness than a conventional thermal IR source. The synchrotron light is input into a Nicolet Magna 760 FTIR bench, then passed through a Nic-Plan IR microscope. As detailed elsewhere [10], the experimental spot size of the unmasked synchrotron beam focused through an infrared microscope is 10 microns, a 100 fold smaller than is possible with an internal thermal source. This experimental procedure is non-destructive to the biological materials being studied.

RESULTS AND DISCUSSION

Dichromate-Bacteria Interactions

In the first study *A. oxydans* was exposed to a 10 ppm dichromate $(Cr_2O_7)^-$ solution. The dichromate-bacteria interactions on the magnetite surfaces were monitored for two days. Figure 1 shows the synchrotron FTIR absorption microspectra in the region 1750-650 cm^{-1} for the living *A. oxydans* cells before their exposure to the 10-ppm $(Cr_2O_7)^=$ solution. The global infrared spectral features for the living cells prior to the bacterial exposure to $(Cr_2O_7)^=$ are consistent with those reported in the literature [12]. All absorption bands are related to the vibrations of the functional groups of the biomolecules of bacteria cell walls. The prominent absorption envelopes in the 1695-1620 cm^{-1} and 1515-1570 cm^{-1} regions arise from the C=O amide vibrations of the peptide linkage in the bacteria protein Amide I and II systems. The absorption band between 1250 and 1200 cm^{-1} arises from the P=O double-bond asymmetric stretching mode of phosphodiester, free phosphate, and monoester phosphate functional groups of the polysaccharide backbone structures. The complex sequence of peaks in the 1200-900 cm^{-1} region are essentially due to C-O-C and C-O-P stretching vibrations of mostly oligo- and polysaccharidic components.

Two additional absorption peaks at 906 and 884 cm^{-1} were detected after the $(Cr_2O_7)^=$ solution was introduced into the biological-mineral system. The vibrational frequencies of these peaks are the same as those reported for $(Cr_2O_7)^=$ in the literature [13], this confirms the presence of $(Cr_2O_7)^=$ in our system. The shifting of the 1640 cm^{-1} band to a higher frequency of 1650 cm^{-1} after the bacterial exposure to $(Cr_2O_7)^=$ implies that the bacterial protein Amide I structure increased its "randomness" or became more disordered. The width of the Amide I vibration band, which is proportional to the amount of water bound to the albumin molecules [12], remained constant and did not accompany the structural change of the bacterial protein. The occurrence of a prominent new peak at 935 cm^{-1} shortly after the exposure to $(Cr_2O_7)^=$ is probably due to the coupling of C-O-Cr and ring C-O-C [14]. The shifting of the 935 cm^{-1} band to 950 cm^{-1} with time in a biological system has not been reported in the literature. We speculate that it is possibly due to the weakening in the strength of hydrogen bonding in the presence of $(Cr_2O_7)^=$. This needs to be investigated. The increase in its relative absorption intensity with time implies the presence of the $(Cr_2O_7)^=$ stimulated the production of polysaccharide. Similar phenomena have been reported for other metal ions in the literature [14].

Chromate-Bacteria Interactions

This study was carried out by exposing *A. oxydans* to a 10 ppm chromate $(CrO_4)^=$ solution. The chromate-bacteria interactions on the magnetite surfaces were also monitored for two days. Figure 2(a,b) shows the synchrotron FTIR absorption microspectra in the region 1750-650 cm^{-1} for the living *A. oxydans* cells before their exposure to the 10-ppm $(CrO_4)^=$ solution. Their global infrared spectral features prior to the exposure in this experiment again are consistent with those reported in the literature [12].

An additional absorption peak at 846 cm^{-1} (Figures 2b) was detected after the $(CrO_4)^=$ solution was introduced into the biological-mineral system. The vibrational frequency 846 cm^{-1} has been assigned to $(CrO_4)^=$ in the literature [13], thus confirming the presence of chromate $(CrO_4)^=$ in the biological-mineral system. After contact with *A. oxydans* cells, the $(CrO_4)^=$ peak shifted to a lower frequency with time (Figure 2b), which arose from a decrease in the oxidation state of the chromium [15], in the presence of the bacteria.

The spectral characteristics in Figure 2a illustrate that the biomolecules of *A. oxydans* interacted with $(CrO_4)^=$ differently. In addition to a shift of the Amide I vibrational band from 1640 cm^{-1} to a higher frequency, a narrowing of the bandwidth of the Amide I vibration band was also detected. This implies that the interaction between $(CrO_4)^=$ and biomolecules could decrease the amount of water bound to the albumin molecules and resulted in an increase in the secondary protein structure to unordered structure. The appearance of multiplets in the Amide I and Amide II absorption envelopes was probably caused by the degradation and denaturing of the proteins, most likely caused by the decrease in water in the cells. The progressive decrease in the relative intensity of absorption bands for functional groups of other biomolecules in region below 1460 cm^{-1} was associated with the observed microscopic change of *A. oxydans* cells from the initial rods to their stressed form cystites after their exposure to $(CrO_4)^=$.

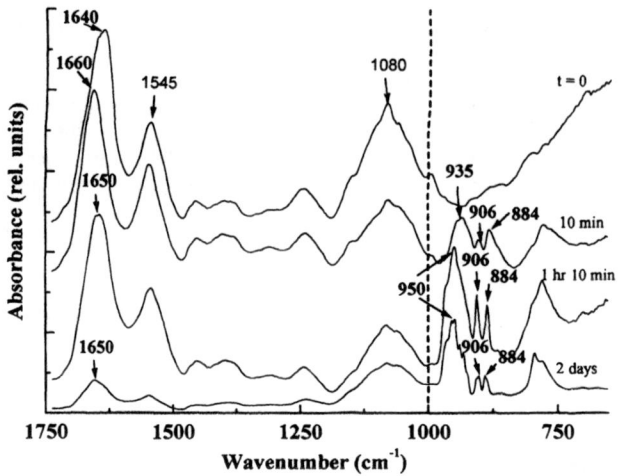

Fig. 1. Synchrotron FTIR absorption microspectra in the region 1750-650 cm^{-1} for the living *A. oxydans* cells before and after their exposure to the 10-ppm $(Cr_2O_7)^=$ solution.

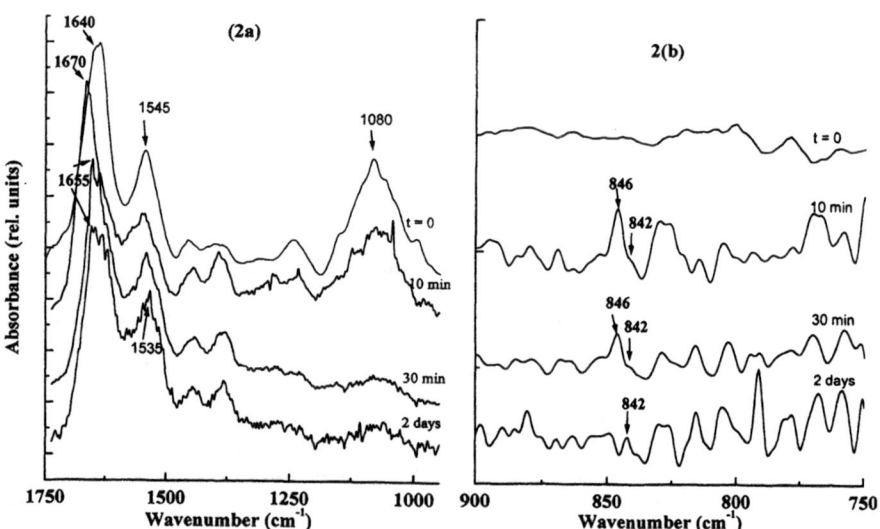

Fig. 2. Synchrotron FTIR absorption microspectra in the region (a) 1750-1000 cm^{-1} (b) 1000-650 cm^{-1} for the living *A. oxydans* cells before and after their exposure to the 10-ppm $(CrO_4)^=$ solution.

Mappings of Chromate-Bacteria Interactions on Magnetite

This study was carried out by first exposing the magnetite surfaces to a 80-ppm toluene vapor for five hours prior to the introduction of *A. oxydans*. This additional experimental step was to provide *A. oxydans*, a toluene degrader [16], a carbon source so that the experiment could last for a longer period of time. Figure 3 shows a representative synchrotron FTIR absorption microspectrum in the region 1750-650 cm[-1] for the surface of our biological-mineral system after exposure to a 10-ppm chromate solution in the presence of toluene as a carbon source.

The spatial variation of chromate-bacteria interactions on the magnetite surfaces were monitored five days after the initial exposure to the 80-ppm toluene vapor and a subsequent exposure to a 10-ppm chromate solution. Figure 4 shows the 3-D maps of the spatial distribution of the vibrational modes associated with *A. oxydans* (1665 cm[-1]), chromate $(CrO_4)^-$ (846 cm[-1]), and the toluene (728 cm[-1]) on a magnetite surface. We observed that at the location of the *A. oxydans* (peak in the leftmost plot) there is a significant decrease in the amount of both chromate and toluene (right two plots).

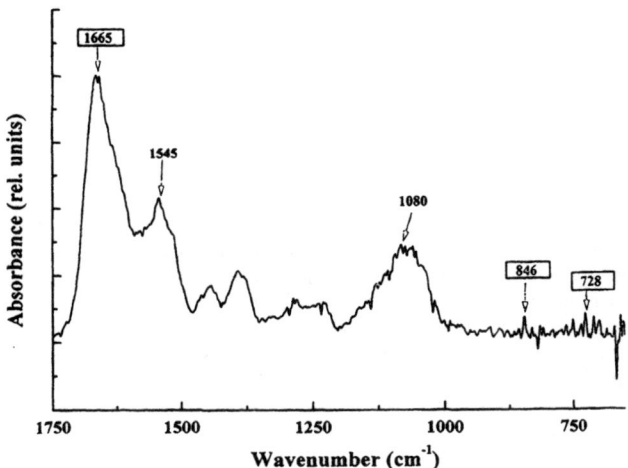

Figure 3. A synchrotron FTIR absorption microspectrum in the region 1750-650 cm[-1] for the surface of our biological-mineral system after exposure to a 10-ppm chromate solution in the presence of toluene vapor.

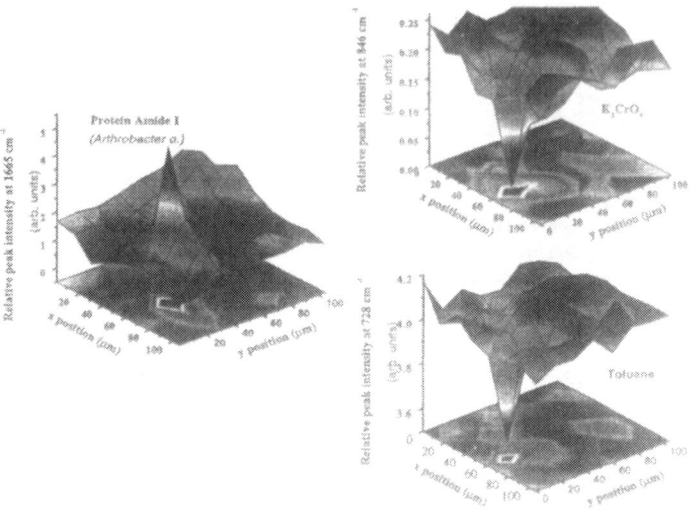

Figure 4. 3-D maps of the spatial distribution of the vibrational modes associated with
Arthrobacter o. (1665 cm^{-1}), chromate $(CrO_4)^=$ (846 cm^{-1}), and the toluene (728 cm^{-1})
on a magnetite surface.

CONCLUSIONS

The above results demonstrate that the synchrotron microprobe could nondestructively monitor
the *in situ* inorganic-organic interactions at the biological-mineral interfaces. It also illustrates the highly
localized differences in the progress of the interactions and the structural details of clusters of living-
bacteria in minerals. Comparative analyses of the synchrotron infrared microspectra suggest that the
mode of bacterial transformation of the hexavalent chromium Cr(VI) depends on the solubility and
toxicity of the chromium compounds.

ACKNOWLEDGMENTS

We thank Dr. S. Goldman at the University of California for reviewing the manuscript. This
work was performed with support by the Directors, Office of Energy Research, Offices of Health and
Environmental Sciences, Biological and Environmental Research Program and Basic Energy Sciences,
Materials Science Division, of the United States Department of Energy under Contract No. DE-AC03-
76SF00098.

REFERENCES

1. R.J. Lenhard, R.S. Skeen and T.M. Brouns in Bioremediation: Science and Applications, edited by H.D. Skipper and R.F. Turco (SSSA Special Publication #43, Madison, Wis., 1995), p.157-172.
2. W.T. Frankenberger, Jr. and M.E. Losi in Bioremediation: Science and Applications, edited by H.D. Skipper and R.F. Turco (SSSA Special Publication #43, Madison, Wis., 1995), p.173-210.
3. R. Margesin and F. Schinner, J. Basic Microbiol., 4, 269(1996).
4. E.M.N. Chirwa and Y.T. Wang, Environ. Sci. Technol., 31, 1446(1997).
5. Y.T. Wang and H. Shen, J. Ind. Microbiol., 14, 154(1995)
6. D.L. Balkwill and D.R. Boone, in the Microbiology of the Terrestrial Deep Subsurface, edited by P.S. Amy and D.L. Haldeman (CRC, Lewis Publishers, N.Y., 1997), p. 105-118.
7. J.E. Kostka and K.H. Nealson, Environ. Sci. Technol., 29, 2535(1995)
8. H.-Y.N. Holman, D.L. Perry and J.C. Hunter-Cevera, submitted to J. Microbiol. Methods, 1998.
9. P. Hirsch, F.E.W. Eckhardt and R.J. Jr. Palmer, J. Microbiol. Methods 23, 143(1995).
10. W.R. McKinney, C.J. Hirschmugl, H.A. Padmore, T. Lauritzen, N. Andresen, G. Andronaco, R. Patton, and M. Fong, Proceedings of the SPIE, Accelerator-Based Infrared Sources and Applications, 3153, pp 59-67.
11. Michael C. Martin and Wayne R. McKinney, this proceeding volume.
12. K. Brandenburg and U. Seydel in Infrared Spectroscopy of Biomolecules, edited by H.H Henry and D. Chapman (Wiley-Liss, N.Y., 1996), p. 203-238.
13. J.A. Campbell, Spectrochimica Acta, 21, 1333(1965).
14. D. Grant, W.F. Long, and F.B. Williamson, Biochem. Soc. Trans. 18, 1281(1990)
15. J.A. Campbell, Spectrochimica Acta, 21, 851(1965).
16. H.-Y.N. Holman and J.C. Hunter-Cevera, manuscript in preparation.

HIGH-RESOLUTION X-RAY PHOTOEMISSION ELECTRON MICROSCOPY AT THE ADVANCED LIGHT SOURCE

THOMAS STAMMLER[*], SIMONE ANDERS[*], HOWARD A. PADMORE[*],
JOACHIM STÖHR[**], MICHAEL SCHEINFEIN[***], and HARALD ADE[****]

[*] *Lawrence Berkeley National Laboratory, 1 Cyclotron Road, Berkeley, CA 94720*

[**] *IBM Almaden Research Center, 650 Harry Road, San Jose, CA 95120*

[***] *Department of Physics and Astronomy, Arizona State University, Tempe, AZ 85287*

[****] Department of Physics, North Carolina State University, Raleigh, NC 27695

ABSTRACT

X-ray Photoemission Electron Microscopy (X-PEEM) is a full-field imaging technique where the sample is illuminated by an x-ray beam and the photoemitted electrons are imaged on a screen by means of an electron optics. It therefore combines two well-established materials analysis techniques - photoemission electron microscopy (PEEM) and x-ray spectroscopy such as near edge x-ray absorption fine structure (NEXAFS) spectroscopy. This combination opens a wide field of new applications in materials research and has proven to be a powerful tool to investigate simultaneously topological, elemental, chemical state, and magnetic properties of surfaces, thin films, and multilayers at high spatial resolution. A new X-PEEM installed at the bend magnet beamline 7.3.1.1 at the Advanced Light Source (ALS) is designed for a spatial resolution of 20 nm and is currently under commissioning. An overview of the ongoing experimental program using X-PEEM in the field of materials research at the ALS is given by elemental and chemical bonding contrast imaging of hard disk coatings and sliders, field emission studies on diamond films as possible candidates for field-emission flat-panel displays, and the study of dewetting and decomposition phenomena of thin polymer blends and bilayers.

INTRODUCTION

Near Edge X-ray Absorption Fine Structure (NEXAFS) spectroscopy is an established technique to study materials properties such as elemental composition, chemical bonding, and molecular orientation [1]. It is based on the availability of X-ray radiation of tunable wavelength produced by a synchrotron. Utilizing the polarization of the synchrotron radiation, one can perform x-ray magnetic dichroism (XMD) spectroscopy to investigate the magnetic structure and atomic magnetic properties of magnetically ordered materials [2]. Third-generation sources of high-brilliance synchrotron radiation make it possible to combine effectively spectroscopic methods like NEXAFS and high spatial resolution microscopy techniques. The latter can be achieved either by scanning techniques or by parallel image acquisition. In scanning x-ray microscopy, x-ray optics focus the beam, and a detector senses photons or electrons as the x-ray spot moves across the sample surface (or as the sample rasters through a fixed x-ray spot). The lateral resolution is therefore determined by the size of the x-ray spot. In contrast, x-ray photoemission electron microscopy (X-PEEM) is a full-field imaging technique where the sample is illuminated by an x-ray spot focused to a size of the largest field of view considered. In this case the resolution is determined by the aberrations of the electron-optical imaging system consisting of two or more electrostatic lenses. In addition, due to a parallel collection of information an imaging technique is inherently fast and allows the study of time dependent processes at video rates. At present, there are several spectromicroscopes in operation at synchrotron radiation facilities. The highest spatial resolution so far reported for such an instrument is 40 nm and has been achieved by the Clausthal group at BESSY I [3]. The new X-PEEM installed at the bend magnet beamline 7.3.1.1 at the ALS is designed for a spatial resolution of 20 nm and is currently under commissioning. An even more advanced instrument equipped with an electrostatic hyperbolic-field mirror that allows correction of spherical and chromatic aberrations simultaneously is in the design phase. The use of an aberration–correcting mirror has the potential for pushing the spatial-resolution limit of the X-PEEM to a few nm. Another UHV spectromicroscope called SMART with a calculated spatial resolution of 2 nm is under construction for a soft x-ray undulator beamline at BESSY II [4].

EXPERIMENTAL

Figure 1 shows the schematic layout of the spectromicroscopy beamline 7.3.1.1 at the ALS. The spherical grating monochromator provides soft x-rays in a spectral range from 260 eV to 1500 eV with a resolving power of $E/\Delta E = 1800$ and 3×10^{12} photons/s/0.1%BW at 800 eV. The use of a bend magnet gives the choice of linearly or circularly polarized radiation. The x-ray beam is focused on the sample with a spot size of 30 μm x 30 μm. The objective lens is of conical shape allowing the sample to be illuminated by the x-ray beam at an angle of about 30 degrees (c.f., Fig. 2). The photoemitted electrons are accelerated by the immersion objective lens and form an intermediate image with a magnification of $m = 10$. In order to reduce spherical and chromatic aberrations the microscope is designed for a rather high accelerating voltage of 30 kV. Three corrector elements are incorporate into the microscope column to accommodate for lens imperfections and misalignment of the individual components. In order to correct for astigmatism an octopole stigmator is located in the back focal plane of the objective lens. Two hexapole deflectors, one located right behind the stigmator, the second one at the back focal plane of the intermediate lens (cf., Fig. 2), correct for beam deflection caused by misalignment of the lens elements.

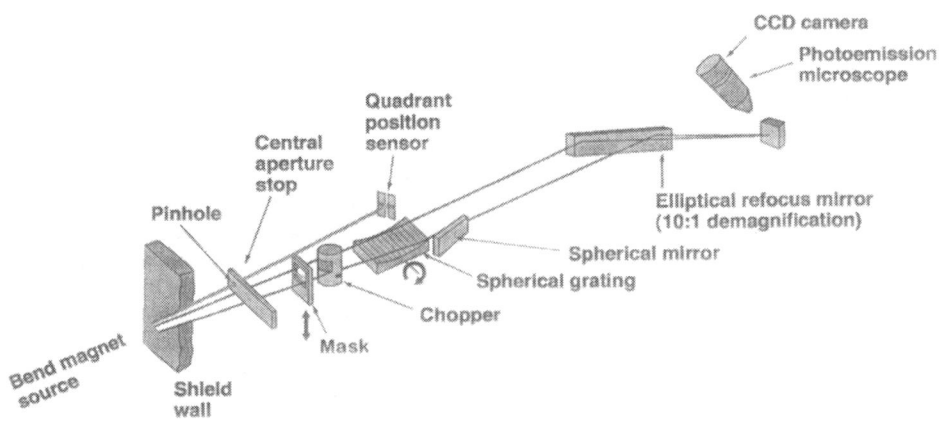

Figure 1: Schematic layout of the spectromicroscopy beamline 7.3.1.1 at the ALS.

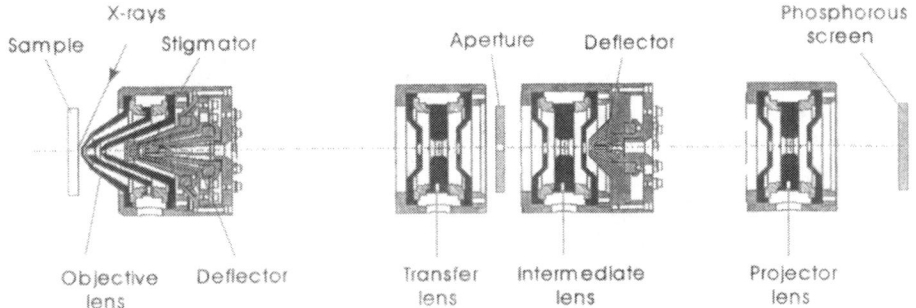

Figure 2: Schematic layout of the new X-PEEM electron optics. Part of the electron optics design was adapted from an existing x-ray transmission microscope [5].

The back focal plane of the objective lens would also be a well-suited place for an angle-limiting aperture which has an electron energy filtering effect and therefore reduces the chromatic aberrations. Since the stigmator/deflector arrangement does not leave enough space, a transfer lens with a magnification of unity is added to form another (conjugate) back focal plane where the aperture can be located. However, the insertion of an aperture reduces the chromatic aberrations. Since the stigmator/deflector arrangement does not leave enough space, a transfer lens with a magnification of unity is added to form another (conjugate) back focal plane where the aperture can be located. However, the insertion of an aperture reduces the transmittance of the microscope to typically 5% (for an aperture diameter of 12 μm). The intermediate and projective lenses form a magnified image on a phosphorous screen from where the image is transferred to a slow-scan CCD camera using a fiber-optics coupling with a 1:2 taper. The calculated resolution limit is about 20 nm. The new X-PEEM at beamline 7.3.1.1 is not only equipped with a surface science preparation chamber and load-lock system for fast sample transfers, it has also incorporated an electron-beam evaporator for *in situ* deposition. The sample manipulator is equipped with an electron-beam heating system for temperatures up to 1500 °C.

APPLICATIONS

A number of studies using X-PEEM in the field of materials science have been performed at the ALS during the last year using a previously built PEEM installed at beamline 8.0. This microscope is a two-lens system operating at a nominal voltage of 10 kV with a spatial resolution of about 200 nm and is described in detail elsewhere [6].

One example is the study of the tribology of the head/disk interface of magnetic storage devices. By decreasing the spacing between slider and disk approaching pseudo-contact and contact recording, the tribochemical properties of the system consisting of the carbon overcoat of the disk, the lubricant, and the slider surface become more and more important. Due to its elemental and chemical bonding sensitivity, X-PEEM can be used to study wear tracks on disks and slider surfaces as discussed in detail in Anders *et al.* [7]. As an example, Figure 3a shows an X-PEEM image of the edge of a wear track on an amorphous hard carbon coated (7 nm sputter deposited) and lubricated (0.85 nm perfluoropolyether [Z-dol]) hard disk. Local NEXAFS spectra were taken in the undamaged area of the disk and at the edge of the wear track caused by the rail. Fig. 3b shows the carbon K-edge spectra as observed in the different regions indicated in Fig. 3a. The spectra show that in the scratch caused by one of the rails, the carbon spectrum shows a distinguished peak at 288.5 eV which can be attributed to the formation of carboxylic bonds [1]. This indicates that the lubricant has been chemically modified by the wear. These studies demonstrate the sensitivity of X-PEEM to elemental and chemical contrast.

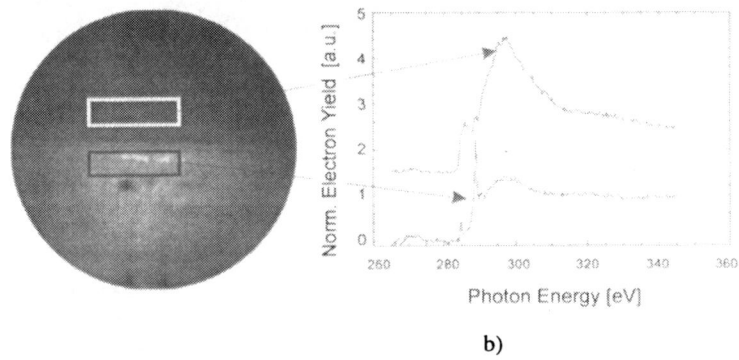

a)   b)

Figure 3: (a) X-PEEM image of the edge of a wear track on an amorphous hard carbon coated and lubricated hard disk taken at a photon energy of 290 eV. The field of view is 100 μm. (b) Local carbon K-edge NEXAFS spectra as observed in the different regions indicated in (a).

With a field strength of the extraction field of the objective lens in the microscope higher than the onset field strength of high field emission yield samples, X-PEEM can be used to study the nature of the emission site of possible field emission display materials. A field emission display is a flat panel display in which electrons are emitted from an array of field emission cathodes, accelerated across a vacuum gap, and form an image on a flat phosphor screen. Diamond is a strong candidate for field emission microcathodes for field emission displays because of its low electron affinity and chemical inertness. The films were produced by plasma enhanced CVD. Electron emission at fields as low as 3 V/μm was observed for some of the samples. A characterization of various field emission sites can be done in two steps. First, at a field strength of $E \approx 5\text{-}7$ V/μm, one can easily locate and image active emission sites and differentiate between photoemission and field emission by switching the x-rays on and off. In a second step, the field strength is reduced by decreasing the sample voltage to a value below the field emission threshold and reducing all lens voltages proportional to the sample voltage in order to keep imaging conditions. The X-PEEM can then be used to perform local NEXAFS spectroscopy to investigate the nature of the emission site. In particular, the elemental and chemical bonding specificity can be utilized to obtain information to find out if the field emission is due to a difference in the chemical bonding situation such as a locally different amount of sp^3/sp^2 or contamination. The absence of any elemental or chemical contrast at the emission sites of the samples investigated indicates that the difference between an emitting and a non-emitting area might just be a difference in the local topology. However, this study was performed with a rather low spatial resolution of 0.2 μm, and further investigations will be performed as soon as the commissioning of the new X-PEEM is finished and the design resolution of 20 nm is reached.

Dewetting and decomposition phenomena of thin films of polymer blends and bilayers are crucial for the use of these polymers in, e.g., colloidal paint systems where wetting controls the dispersion of color pigments. Blends and bilayers of polystyrene (PS) and brominated polystyrene (PBrS) have recently been studied using a scanning transmission x-ray microscope (STXM) at the National Synchrotron Light Source [8]. STXM provides quantitative composition maps integrated along the direction of the surface normal. The different energy position of the π^* resonance of PS and PBrS can be utilized as a contrast mechanism. These studies reveal the morphology formed as the PS and PBrS phase separates. However, since STXM is rather bulk sensitive, it cannot distinguish the surface from the subsurface polymer. Thus X-PEEM has been utilized to determine the composition of the surface (probing depth about 10 nm) of various PS/PBrS thin films. Starting out from bilayers consisting of a 30 nm thick PBrS layer on top of a 30 nm thick PS layer produced by spin casting on silicon substrates, the samples were annealed for different duration at 180°C in a vacuum oven. Figure 4 shows an X-PEEM image of a PS/PBrS bilayer which was annealed for 4 days.

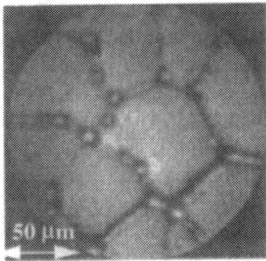

Figure 4: X-PEEM image of a PS/PBrS bilayer which was annealed for 4 days. The image was taken at the PBrS π^* resonance at a photon energy of $E = 286.3\text{eV}$.

Whereas after 1 hour of annealing there was no indication of any dewetting, the 2 days annealed sample showed small areas (about 20 μm in size) where the dewetting starts. For the 4 days annealed sample, a distinguished spine pattern could be observed. STXM measurements [7] revealed that the spines are composed of PBrS, and the X-PEEM experiments showed clearly that these PBrS spines are covered by a thin PS layer [9]. As shown, the combination of STXM and X-PEEM allows the study of complex polymer systems, and both methods yield complementary information which, if combined, give a complete understanding of the processes taking place during dewetting of the polymer systems.

CONCLUSIONS

It was shown that X-PEEM is a powerful tool to investigate elemental and chemical bonding contrast imaging of hard disk coatings and sliders, field emission of, e.g., diamond films, and the dewettir.g and decomposition phenomena of thin polymer blends and bilayers. The combination of x-ray spectroscopy and high spatial resolution photoemission electron microscopy, together with surface science sample preparation capabilities, will make the new X-PEEM at the ALS a very versatile instrument in the field of materials science.

ACKNOWLEDGEMENTS

The authors are grateful to W. Tong for putting the field-emission sample to our disposal. We thank W. Fong, C.S. Bhatia, and D.B. Bogy for preparing and providing the hard disk samples. The support by the technical staff of the ALS is gratefully acknowledged. This work was supported by the Director, Office of Energy Research, Office of Basic Energy Science, Materials Science Division, of the U.S. Department of Energy under Contract No. DE-AC03-76SF00098.

REFERENCES

[1]   J. Stöhr, *NEXAFS Spectroscopy*, New York: Springer, 1992.
[2]   G. Schütz, W. Wagner, W. Wilhelm, P. Kienle, R. Zeller, R. Frahm, and G. Materlik, Phys. Rev. Lett. **58**, 737 (1987)
[3]   E. Bauer, T. Franz, C. Koziol, G. Lilienkamp, T. Schmidt, in: *Chemical, Structural and Electronic Analysis of Heterogeneous Surfaces on the Nanometer Scale*, edited by R. Rosei, Kluwer Academic, Dordrecht, in press.
[4]   R. Fink, M.R. Weiss, E. Umbach, D. Preikszas, H. Rose, R. Spehr, P. Hartel, W. Engel, R. Degenhardt, R. Wichendahl, H. Kuhlenbeck, W. Erlebach, K. Ihmann, R. Schlögl, H.-J. Freund, A.M. Bradshaw, G. Lilienkamp, Th. Schmidt, E. Bauer, G. Benner, Journal of Electron Spectroscopy and Related Phenomena **84** (1997) 231-250.
[5]   R.N. Watts, S. Liang, Z.H. Levine, T.B. Lucatorto, F. Polack, and M.R. Scheinfein, Rev. Sci. Instrum. **68** (1997) 3464-3476.
[6]   B.P. Tonner, D. Dunham, T. Droubay, and M. Pauli, Journal of Electron Spectroscopy and Related Phenomena **84** (1997) 211.
[7]   S. Anders, Th. Stammler, C. Singh Bhatia, J. Stöhr, W. Fong, and D.B. Bogy, Spring Meeting of the Material Research Society, San Francisco, 1998, to be published.
[8]   D. A. Winesett, H. Ade, A. P. Smith, M. Rafailovich, S. Sokolov, and D. Slep, Microscopy and Microanalysis, to be published (1998).
[9]   S. Anders, Th. Stammler, H. Ade, M. Rafailovich, J. Sokolov, D. Slep, C. Heske, and J. Stöhr, to be published.

SYNCHROTRON RADIATION HARD X-RAY MICROPROBE BY MULTILAYER FRESNEL ZONE PLATE

S. TAMURA * , K. OHTANI * , M. YASUMOTO * , K. MURAI * , N. KAMIJO ** ,
H. KIHARA ** , K. YOSHIDA *** and Y.SUZUKI ****
*Osaka National Research Institute, AIST, Ikeda, Osaka, Japan.
**Kansai Medical University, Hirakata, Osaka, Japan.
***Osaka Institute of Technology, Ohmiya, Asahi-ku, Osaka, Japan.
****Japan Synchrotron Radiation Research Institute, Kamigori, Ako-gun, Hyogo, Japan.

ABSTRACT

A hard X-ray microbeam with submicrometer spot size from synchrotron radiation (SR) sources is expected to add a new dimension to various X-ray analysis methods. A Fresnel zone plate (FZP) is one of the promising focusing elements for X-rays. In order to develop high performance multilayer FZP for use in the hard X-ray region, Cu/Al concentric multilayers were fabricated by use of a DC sputtering deposition process. Lower Ar gas pressure or higher rotating speed of a wire substrate has been effective in forming smoother multilayer interfaces. From a focusing test of the Cu/Al FZP (100-zones) by the SR (λ=0.154nm), microbeams of 1.5 μm ϕ and 0.8 μm ϕ have been achieved for the first- and third-order focal beams, respectively.

INTRODUCTION

A hard X-ray microbeam is considered to be a key technology for third-generation high-brilliance synchrotron radiation (SR) sources such as the APS (Advanced Photon Source, USA), the ESRF (European Synchrotron Facility, France) and the SPring-8 (Super Photon ring-8 GeV, Japan). The hard X-ray microbeam with submicrometer spot size is expected to add a new dimension to various X-ray analysis methods. In the hard X-ray region, a focused beam of 0.7 μm ϕ has been obtained by a circular Bragg Fresnel lens[1], and focused beams of 0.6 $\sim$ 0.7 μm ϕ and 0.46 $\times$ 1.2 μm have been obtained by a Fresnel zone plates (FZP) fabricated by a lithography-based technique[2,3]. The FZP, a circular diffraction grating of alternate transparent and opaque zones to X-ray, is one of promising focusing elements for the hard X-ray. We have developed multilayer (sputtered-sliced) FZPs, Ag/C and Cu/Al. For the Ag/C FZP (50-zones, outermost zone width: 0.25 μm), a microbeam of 1.3 μm ϕ was obtained for the first-order focal beam using 8 KeV (λ=0.154nm) X-ray from a bending magnet beamline (BL-8C) at the National High Energy Physics (KEK, Japan), and a microbeam of 0.5 μm ϕ was obtained using 8.54 KeV (λ=0.145nm) X-ray from an undulator beam line (MR-BW-TL) at the KEK[4,5]. For the Cu/Al FZP, a microbeam of 1.3 μm ϕ was obtained for the first-order focal beam using 8 KeV X-ray from the bending magnet beamline (BL-8C) [6]. In order to attain better focusing characteristic, it is necessary to reduce the roughness of the multilayer zone boundaries as well as to reduce thickness error. In the previous paper, we reported the experimental results on the dependence of the zone boundary structure on the sputtering Ar gas pressure of the Ag/C or the Cu/Al concentric multilayers[6,7].

In the following, firstly, a new experimental dependence of the zone boundary structure on the deposition condition of the Cu/Al multilayer is shown. Next, the FZP focusing test is described.

31

EXPERIMENT

The outline of the fabrication process of the multilayer FZP is shown in Fig.1, where "t" is the thickness and "Δ r" is the outermost zone width. The zone width decreases gradually from the center to the outer edge. The focused beam size is comparable to its outermost (minimum) zone width.

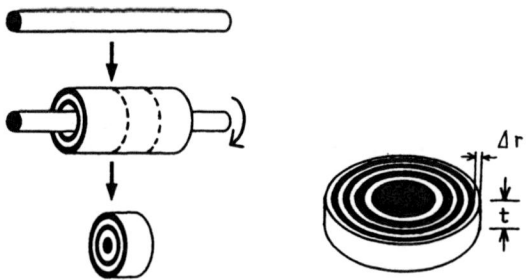

Fig.1   Outline of fabrication process of multilayer FZP.

The concentric multilayer samples (30 ~ 50 layers) were deposited onto rotating Au wire substrate (47 μm φ) by use of a DC magnetron sputtering technique. The deposition parameters of the multilayers are shown in Table I.   The base pressure of the deposition system was 1×10^{-4} Pa. The substrate temperature was not controlled. The transparent (low-Z) material for X-ray is Al, and the opaque (high-Z) materials is Cu. Each film thickness (zone width) was 0.25 ~ 0.40 μm. After the deposition, the multilayer sample was fixed into a low melting point alloy, sliced into a plate of 1 mm thickness perpendicular to the wire axis. One sliced surface was polished mechanically and the layer structure was observed by a scanning electron microscopy (SEM). The FZP for hard X-ray was obtained by thinning the sample to less than 50 μm. A Fresnel phase zone plate (FPZP), with high focusing efficiency and low background at the focus, was obtained by thinning the sample to less than 10 μm.

Table I   Deposition parameters of concentric multilayer samples.

Sample number	Layer material	Growth rate (nm/s)	Ar Pressure (Pa)	Substrate rotating (rpm)
(1)	Cu/Al	1	0.20	15
(2)	Cu/Al	1	0.67	15
(3)	Cu/Al	1	0.20	50
(4)	Cu/Al	1	0.67	50

RESULTS

Cu/Al multilayers

The SEM micrographs of the concentric multilayers are shown in Fig.2. The zone roughness is amplified toward the top of the layer stack. The roughness of the sample (2) is greatly enhanced. At the rotating speed of 15 rpm, the multilayer prepared at lower Ar gas pressure had smoother zones. This result is similar to that of the Ag/C multilayer in refs. 6 and 7. At the rotating speed of 50 rpm, the zone roughness of the multilayer prepared at the Ar gas pressure of 0.67 Pa was improved.

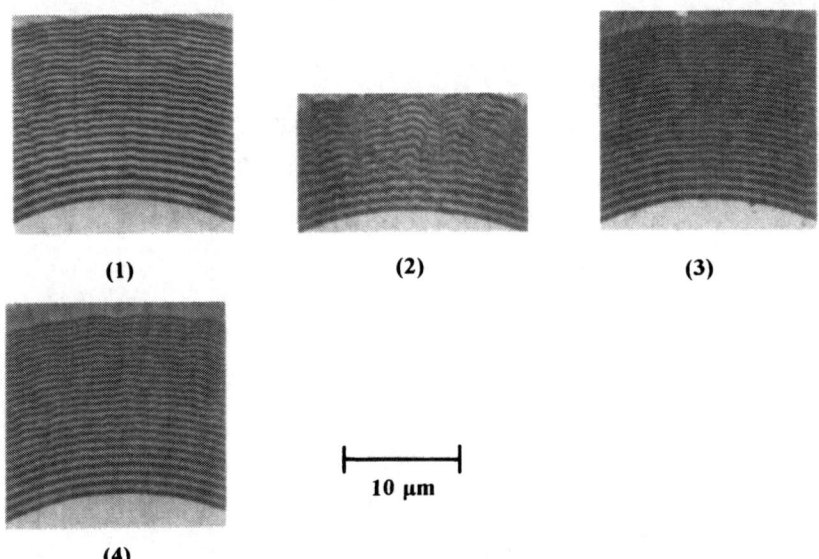

(1)   (2)   (3)

10 μm

(4)

Fig.2   SEM micrographs of Cu/Al concentric multilayers on Au wire substrate. Black rings are Al layers, and white rings are Cu layers. Each sample number coincides with those in Table I.

No visible difference of the shape is observed at the zone roughness among the sample (1), (3) and (4). It is well known that the microstructure and the stress of the sputtered thin films are strongly correlated with the sputtering gas pressure[8]. The difference of the surface mobility of adatoms results in the difference characteristic of the films[9]. In addition, the difference of the substrate rotating speed will also influence in the characteristic of the films in our experiment. In order to improve the zone configuration, many trial and error experiments under various deposition parameters are required.

Zone plate focusing test

A Cu/Al concentric multilayer (100-layers) was deposited by the parameter of Table I-(1). The width of the first inner zone (Al) was 0.40 μm and that of the outermost zone (Cu) was 0.19 μm. Each zone width (film thickness) was determined so that the focal length might be 124 mm for 8 KeV X-ray. The SEM micrographs of the multilayer are shown in Fig.3. A FPZP was fabricated by thinning the sample. The thickness of the FPZP is $\sim$ 8 μm. The laser microscopy image of the FPZP is shown in Fig.4. The circular form of the FPZP has not changed by the thinning process.

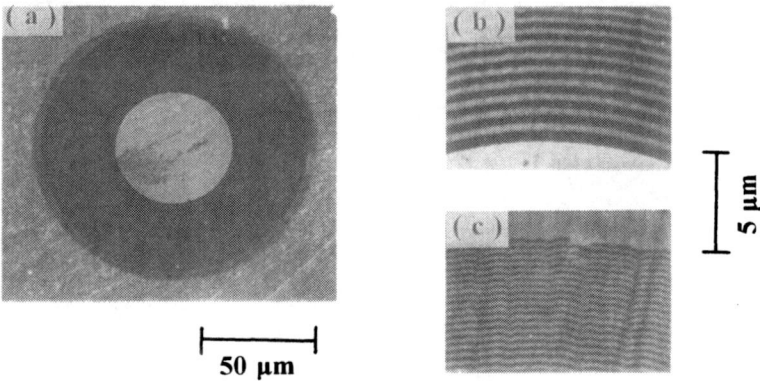

Fig.3   SEM micrographs of Cu/Al concentric multilayer on Au wire substrate. (a) Full view, (b) close-up view (inner part), (c) close-up view (outer part). Black rings are Al layers, and white rings are Cu layers.

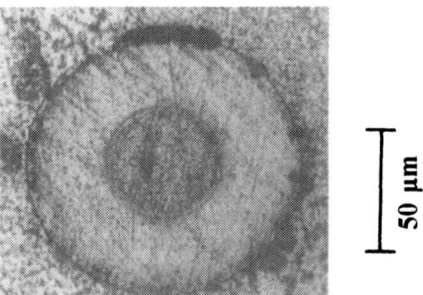

Fig.4   Laser microscopy image of FPZP.

A focusing test of the FZP was performed at the BL-8C. The schematic diagram of the optical system is shown in Fig.5. The focused beam size [full width at half-maximum (FWHM)] and the focusing efficiency for the first-order focal beam were 1.5 μm ϕ and 9 %, respectively. These for the third-order focal beam were 0.8 μm ϕ and 2 %, respectively. The intrinsic focusing efficiencies achieved by the zoned area (excepting the large central stop of the gold core) for the first- and third-order focal beams were 12 % and 3 %, respectively. The focused beam profiles measured by a knife-edge scanning are shown in Fig.6. In Table II, the abilities of our FZPs are summarized.

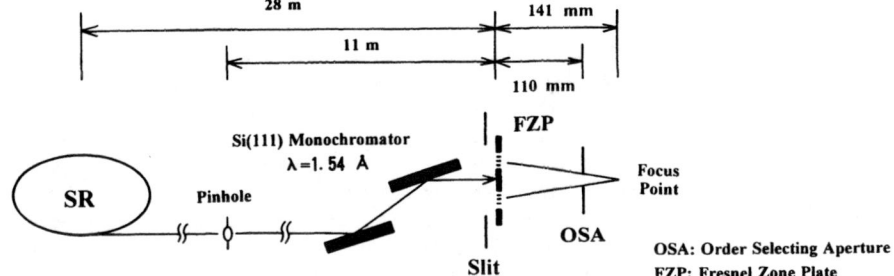

Fig.5 Schematic diagram of optical system.

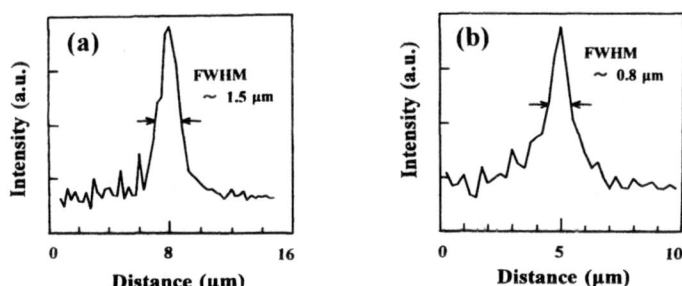

Fig.6 Focused beam profiles of FPZP. (a) first-order focal beam,
(b) third-order focal beam.

Table II Multilayer Fresnel Zone Plate at ONRI

Sample number	Zone number	Thickness (μm)	Δr (μm)	Beam size[1] (μm φ)	Efficiency [3] (%)	SR Beamline[4]	Ref. No.
Ag/C-1	50	50	0.25	2.3	4	BL-8C	10
Ag/C-2	50	8 ～ 9	0.25	1.3	16	BL-8C	4,5
				0.5	16	MR-BW-TL	
Ag/C-3	28	30	0.29	0.8 ～ 0.9	4	BL-8C	7
Cu/Al-1	50	20	0.25	1.3 ～ 1.5	10	BL-8C	6
Cu/Al-2	100	～ 8	0.19	1.5	12	BL-8C	
				0.8 [2]	3		

[1] First-order focal beam.   [2] Third-order focal beam.
[3] Intrinsic focusing efficiencies achieved by the zoned area only.
[4] BL-8C : bending magnet beamline (2.5 GeV).
　　MR-BW-TL : undulator beam line (8 GeV).

CONCLUSIONS

A Cu/Al multilayer FZP was fabricated by use of a DC sputtering deposition process and mechanical thinning (polishing) process. During the deposition process, lower Ar gas pressure or higher rotating speed of a wire substrate has been effective in forming smoother multilayer interfaces (zones). From a focusing test of the FZP (100-zones) by the SR (8 KeV), microbeams of 1.5 μm φ and 0.8 μm φ have been achieved for the first- and third-order focal beams, respectively. The intrinsic focusing efficiencies achieved by the zoned area for the first- and third-order focal beams have been 12 % and 3 %, respectively. The focusing efficiency of the third-order focal beam has been somewhat larger than the theoretical value. Theoretically, it is one-ninth of the efficiency of the first-order focal beam. The measured focal length (141 mm, see Fig.5) was longer than the designed one. We are now investigating the reasons.

ACKNOWLEDGEMENTS

This work was performed with the approval of Photon Factory Advisory Committee (Proposal No.95G322).

REFERENCES

1. A.Snigirev, I.Snigireva, P.Engstom, S.Lequien, A.Suvorov, Ya.Hartman, P.Suvorov, P.Chevallier, M.Idir, F.Legrand, G.Soullie and S.Engrand, Rev. Sci. Instrum., **66**, 1461 (1995).
2. B.Lai, W.B.Yun, D.Legnini, Y.Xiao, J.Crzas, P.J.Viccaro, V.White, S.Bajikar, D.Denton, F.Cerrina, E.Di.Fabrizio, M.Gentili, L.Grella and M.Baciocchi, Appl. Phys. Lett., **61**, 1877 (1992).
3. D.M.Mills, J. Synchrotron Radiation, **4**, 117 (1997).
4. N.Kamijo, S.Tamura, Y.Suzuki, K.Handa, A.Takeuchi, S.Yamamoto, M.Ando, K.Ohsumi and H.Kihara, Rev. Sci. Instrum., **68**, 14 (1997).
5. Y.Suzuki, N.Kamijo, S.Tamura, K.Handa, A.Takeuchi, S.Yamamoto, H.Sugiyama, K.Ohsumi and M.Ando, J. Synchrotron Radiation, **4**, 60 (1997).
6. S.Tamura, K.Mori, T.Maruhashi, K.Yoshida, K.Ohtani, N.Kamijo, Y.Suzuki and H.Kihara in Thin Films --- Structure and Morphology, edited by S.C.Moss, D.Ila, R.C.Cammarata, E.H.Chason, T.L.Einstein and E.D.Williams (Mater. Res. Soc. Proc. **441**, Boston, MA, 1996) pp.779-784.
7. S.Tamura, K.Ohtani, N.Kamijo, T.Suzuki and H.Kihara, Thin Solid Films, **281/282**, 243 (1996).
8. J.A.Thornton, J. Vac. Sci. Technol., **A4**, 3059 (1986).
9. D.L.Windt, W.L.Brown, C.A.Volkert and W.K.Waskiewicz, J. Appl. Phys., **78**, 2423 (1995).
10. N.Kamijo, S.Tamura, Y.Suzuki and H.Kihara, Rev. Sci. Instrum., **66**, 2132 (1995).

MICRO-, MESO- AND MACRO-TEXTURE AND
FATIGUE CRACK ROUGHNESS IN Al-Li 2090 T8E41

J.D. HAASE, A. GUVENILIR, J.R. WITT and S. R. STOCK
School of Materials Science and Engineering and Mechanical Properties Research Laboratory
Georgia Institute of Technology, Atlanta, GA 30332-0245

ABSTRACT

The use of synchrotron polychromatic x-ray microbeams in the transmission geometry is described for mapping grain orientation as a function of position and for relating this microtexture to the formation of large asperities on fatigue crack surfaces in Al-Li 2090 T8E41. In common with the centers of rolled plates of many aluminum alloys, Al-Li 2090 T8E41 has a sharp average texture or macrotexture different from that in the outer portions of the plate. The geometry of large asperities in Al-Li 2090 has been related to this macrotexture, and the resulting roughness-induced crack closure is recognized to be responsible for the very low crack propagation rates in certain plate orientations. This report focuses on why asperities form at certain positions and why the crack remains relatively planar elsewhere. The microtexture (i.e., the grain-to-grain orientation variation) seems to be organized into a specific type of mesotexture: multiple adjacent grains have nearly identical orientations and form substantial volumes of near-single-crystal material. Transitions between differently oriented near-single-crystal volumes or between a near-single- crystal region and more randomly oriented grains appear to bound asperities.

INTRODUCTION:

In Al-Li 2090 T8E41 fatigue crack growth rates along certain plate orientations are unusually low compared to those of other Al alloys [1,2]; very rough cracks are formed and the faces of asperities have been correlated with the strong macrotexture (i.e., the sample's average texture) in the center of the plates [3]. This crack face roughness produces crack closure, the phenomenon where the crack faces come into contact prematurely during unloading of the sample (i.e., before the minimum stress of a fatigue cycle is reached) or where the crack faces remain in contact much longer than expected during loading [4], and leads to a reduced driving "force" for crack propagation and lower fatigue crack growth rates [5]. X-ray microtomography demonstrates complex patterns of crack face contact as a function of applied load [6-10] and reinforces the view that what determines the particular crack path is the three-dimensional distribution of grain orientations, not only immediately adjacent to the crack but also in the volumes where the crack could have propagated. Recording transmission Laue patterns with synchrotron microbeams avoids the enormous effort of serial sectioning required with electron-based techniques, but very few reports of microbeam diffraction have appeared for polycrystalline materials [11-16] despite the efficiency of this approach for studying three-dimensional microtexture and its relationship to macroscopic phenomena such as fatigue crack propagation.

37

EXPERIMENTS:

Fatigue cracks were grown in L-T oriented compact tension samples of Al-Li 2090 T8E41 (i.e., loading along the L or rolling direction and crack propagation along the plate's T or transverse direction) with R = 0 .1 (i.e., $\sigma_{min}/\sigma_{max}$), 5 Hz frequency and haversine wave form. The 2.7 mm thick compact tension samples were machined from the center of plates with scaling in accordance with ASTM E-399-83 [8, 9, 17]. After fracture, the volume of material adjacent to the crack was cut from the sample so that the specimen could be examined in the transmission parallel to the rolling direction (Fig. 1). The sample thicknesses (along L) was kept less than 3 mm because examining material within 3 mm above and below the nominal crack plane reveals the grain orientations from which the crack selected its path. One side of sample CT21 and both sides of CT11 were studied.

Polychromatic bending magnet radiation at Stanford Synchrotron Radiation Laboratory (SSRL) beamline 2-2 (3.0 GeV, beam currents between 20 and 100 mA) was used with a 10 μm diameter pinhole collimator to record the transmission Laue patterns on image storage plates [18, 19]. Data was collected on image plates, and exposures were on the order of 1×10^4 mA.sec. The collimator was placed 55 cm from the sample, and the beam broadened to about 80 μm vertically at the sample position (which is well-suited for mapping the ~40-50 μm thick, pancake-shaped grains). The image plates for the Fuji BAS 2500 system were read with 50 μm pixels and 256 levels of contrast, and an absorption edge filter [20] was used in some exposures to help index the diffraction streaks.

Mapping of grain orientations involved recording transmission Laue patterns with the x-ray beam along the sample's L direction (i.e., in the parallel geometry). Data collection centered on large asperities and their surroundings and, where available, the matching holes on the other crack face. The separation between sampling positions along the plate's short or S axis was chosen so that each grain would contribute to more than one diffraction pattern and ranged from 20 to 100 μm.

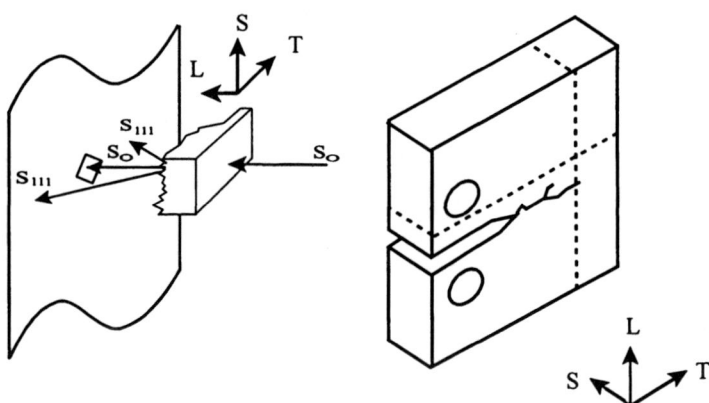

Figure 1. Geometry for recording the microbeam Laue patterns is shown on the left. The incident beam S_0 passes through the sample and strikes the beam stop while the diffracted beams S_{hkl} reach the image plate. The dashed lines (right) indicate cuts required to remove sample volumes from compact tension specimens, and the axes show the plate orientations.

RESULTS AND DISCUSSION

Microbeam diffraction reveals that the boundary between planar regions of the crack face and asperities corresponds to a specific change in orientation of grains within and outside the asperities. Figure 2 shows a typical example of this observation for sample CT21. The top pair of images show large parts of the Laue patterns from positions inside an asperity (right) and outside the same asperity (left) while the bottom pair of images show enlargements of the central part of the two patterns. The white area near the center of the pattern is the beam stop, and the dark streaks on the right-hand side are 111 "spots" from the several grains intercepted by the beam (the darker the pixel the greater the diffracted intensity). The abrupt change of texture at the edge of the asperity is seen by the quite pronounced change in 111 streak orientations. Within the asperity volume, the 111 streaks showed only minor changes from the positions shown in Fig. 2 (right), which is consistent with what is seen for other asperities. This means that multiple adjacent grains have nearly identical orientation and asperities consist of near-single-crystal material and that the presence of this type of mesotexture is important for asperity formation in Al-Li 2090 T8E41.

Figure 3 shows Laue patterns from a second orientation of near-single-crystal material: well-defined ellipses of diffraction streaks are seen, characteristic of transmission Laue patterns of single crystals. The two patterns were recorded from positions over 1 mm apart, and similar patterns appear throughout the central 3 mm thickness of plates of Al-Li 2090. Plotting the diffracted beam directions of the ellipses on a stereographic projection, and comparing the zone axis with 111, 200 and 220 pole figures reveals the zone axes to be <110>. Indices for each "spot" in a typical ellipse, consistent with the measured Bragg angle and the range of wavelengths in the polychromatic beam, are shown in Fig. 4 which

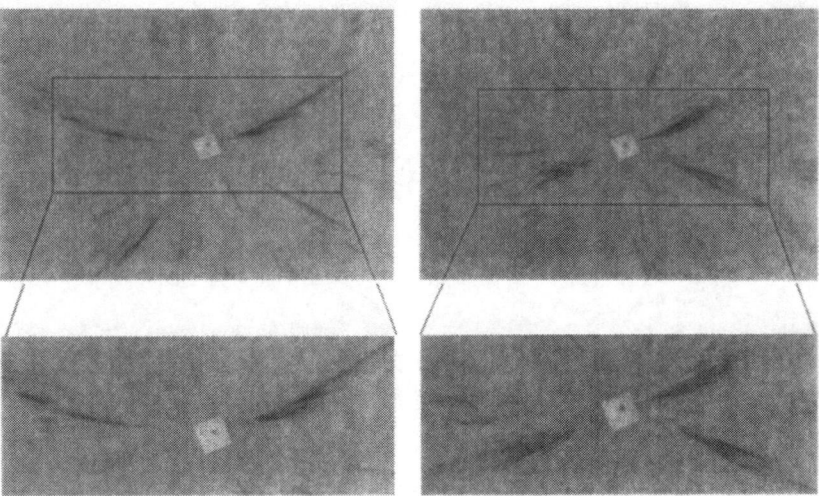

Figure 2. Transmission Laue patterns within (right) and outside (left) of an asperity. Enlargements appear below, 111 diffraction streaks are horizontal near the shadow of the square beam stop and darker pixels correspond to greater diffracted intensities.

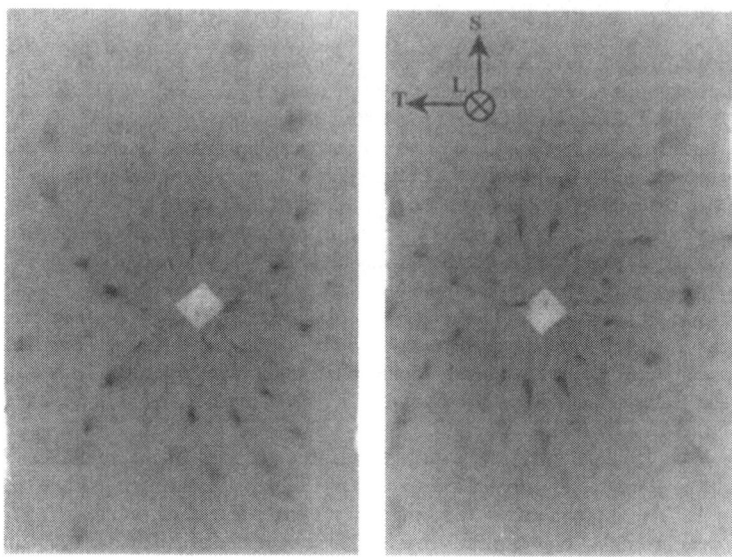

Figure 3. Diffraction patterns showing near-single-crystal volumes from two positions over 1 mm apart. The elliptical pattern of streaks are from a <110> zone, and increasing diffracted intensity is indicated by darker pixels.

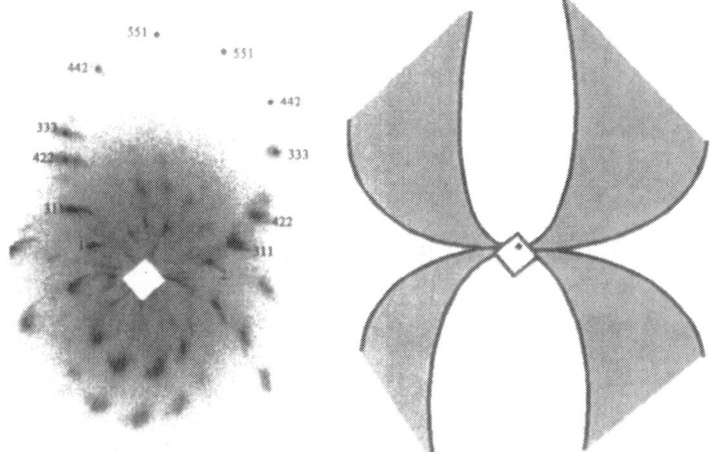

Figure 4. Indexed diffraction streaks comprising a single ellipse (left) and a schematic illustrating the range of ellipse orientations (right).

also shows the limits of the ellipses schematically. The range of ellipse orientations is consistent with the range of orientations in the 220 pole figure [3].

A consistent picture emerges of the relationship between microtexture, mesotexture, macrotexture, asperity formation and "choice" of crack path of the central portion of plates of Al-Li 2090 T8E41. Groups of adjacent, highly-oriented, pancake-shaped grains form near-single-crystal regions within the plate, approximately 45% of the plates' volumes consists of near-single-crystal regions of one orientation type or another, and this picture is consistent with the distribution of grain boundary disorientation angles measured by others [21]. Slip in Al-Li alloys is extremely planar and leads to pronounced crystallographic faceting of fracture surfaces, cracks would be expected to propagate crystallographically within or adjacent to these highly-oriented volumes and crack deflection is likely at the boundary of such a region. Normally one expects fracture feature sizes and aspect ratios similar to the grain size, but, with the pancake-shaped grains being correlated spatially and orientationally, it is not surprising that large asperities are often seen. The orientation of these near-single crystal mesotexture regions has been correlated to the macrotexture of the material and the location of these regions has also been correlated to fatigue surface features. Thus, the spatial distribution of these highly-oriented regions appears to govern whether or not asperities are present on fatigue crack surfaces in this alloy and where on the surfaces these asperities form.

CONCLUSIONS:

A characteristic mesotexture exists in the center of plates of Al-Li 2090 T8E41 and consists of sets of adjacent grains are so highly aligned that they may be regarded as nearly single crystalline volumes. Several different orientations of near-single-crystal volumes are observed, and these variants and their range of orientations are consistent with the macrotexture revealed in pole figures. It appears that a single orientation of near-single-crystal material comprises each large asperity and that large crack deflections occur in response to changes from one near-single-crystal volume to another or from a single-crystal volume to more randomly oriented material. Thus, this characteristic mesotexture, a scale of texture between micro- and macro-texture, appears as the origin of roughness-induced crack closure and extraordinarily low fatigue crack growth rates in Al-Li 2090 T8E41. It will be very interesting to see whether this mesotexture occurs in other alloy which exhibit the same macrotexture.

ACKNOWLEDGMENTS:

We gratefully acknowledge the support of the Office of Naval Research through grants N00014-94-1-0726 and -0306 and the encouragement of Dr. George Yoder, Office of Naval Research. Experiments were performed at SSRL, which is operated by the Department of Energy, Office of Basic Energy Sciences. We also thank Dr. Zofia Rek of SSRL for her invaluable assistance with the microbeam experiments and Mr. R Brown for his help with fatigue crack propagation. The authors would also like to acknowledge Ms. S. Spragg and Mr. T. Watt for their help in outputting the gigabytes of diffraction patterns leading to the observations reported in this paper.

REFERENCES:

1. K.T. Venkateswara Rao, W. Yu and R. O. Ritchie, Met Trans **19A**, 549 and 563 (1988).
2. P.S. Pao, L.A Cooley, M.A. Imam and G.R. Yoder, *Scr. Met.* **23**, 1455 (1989).
3. G.R. Yoder, P.S. Pao, M.A. Imam and L.A. Cooley, in *Proc. Fifth Int. Aluminum-Lithium Conf.*, T.H. Sanders, Jr. and E.A Starke, Jr., Eds, Mat. and Comp. Eng. Publ., Birmingham, UK p. 1033 (1989).
4. R.O.Ritchie, in *Fatigue Thresholds*, J. Backland, A. Blom and C.J. Beevers, eds., Eng Advisory Services, Warley, UK, p. 503 (1981).
5. R.O.Ritchie, *Mater. Sci. and Eng.* **103**, 15 (1988).
6. T.M. Breunig, S.R. Stock, S.D. Antolovich, J.H. Kinney, W.N. Massey and M.C. Nichols, *ASTM STP* **1131**, 749 (1992).
7. A. Guvenilir, T.M. Bruenig, J.H. Kinney and S.R. Stock, *Acta. Mat.* **45**, 1977 (1997).
8. A. Guvenilir, *Investigation into Asperity Induced Closure in an Al-Li Alloy Using X-Ray Tomography*, PhD Thesis, Georgia Inst. of Technology. December 1995.
9. A. Guvenilir, S.R. Stock, M.D. Barker, and R.A. Betz, in *4th Int. Conf. On Aluminum Alloys,* T.H. Sanders, Jr. and E.A. Starke, Jr., Eds., Georgia Institute of Technology, Atlanta, p. 413 (1994).
10. A. Guvenilir and S.R. Stock, *Fract. Fatigue Eng. Mater. Struct.*, in press May 1998.
11. S.R. Stock, A. Guvenilir, D.P. Piotrowski, and Z.U. Rek, *MRS Symp. Proc.* **375**, 275 (1995).
12. D.P. Piotrowski, *Synchrotron Polychromatic X-ray Diffraction Tomography of Large Grained Polycrystalline Samples*, MS Thesis, Georgia Institute of Technology. March 1996.
13. D.P. Piotrowski, S.R. Stock, A. Guvenilir, J.D. Haase, and Z.U. Rek, *MRS Symp. Proc.* **437**, 125 (1996).
14. H.F Poulsen, S. Garbe, T. Lorentzen, D. Juul Jensen, F.W. Poulsen, N.H. Anderson, T. Frello, R. Feidenhans'l and H. Graafsma, *J. Synchrotron Rad.* **4**, 147 (1997).
15. P.C Wang, G.S. Cargill III and I.C. Noyan, *MRS Symp. Proc.* **375**, p.247 (1995).
16. J.D. Haase, A. Guvenilir, J.R. Witt and S.R. Stock, *Acta Mat.* in press (1998).
17. ASTM Standard E-399-83.
18. Y. Ameniya, T. Matsushita, A. Nakagawa, Y. Satow, J. Miyahora and J. Chikawa, *Nucl. Instrum. Meth.* **A266**, 645 (1988).
19. B.R. Whiting, J.F. Owen and B.R. Rubin, *Nucl. Instrum Meth.* **A266**, 628 (1988).
20. S.R. Stock, Z.U. Rek, Y.H. Chung, P.C. Huang, and B.M. Ditchek, *J Appl Phys*, **73**, 1 (1993).
21. F. Barlat, S.M. Miyasota, J. Liu and J.C. Brem, in *4th Int. Conf. on Aluminum Alloys*, T.H. Sanders, Jr. and E.A. Starke, Jr., Eds., Georgia Inst. Tech., Atlanta, p.389 (1994).

X-RAY MICROBEAM QUANTIFICATION OF GRAIN SUBDIVISION ACCOMPANYING LARGE DEFORMATIONS OF COPPER

G.C. Butler[1], A. Guvenilir[2], D.L. McDowell[1,2] and S.R. Stock[2]
Mechanical Properties Research Laboratory and [1]George W. Woodruff School of Mechanical Engineering or [2]School of Materials Science and Engineering, Georgia Institute of Technology, Atlanta, Georgia 30332-0245

ABSTRACT

Polychromatic synchrotron x-ray microbeams offer a very efficient alternative to electron beam methods for quantifying the amount and character of grain subdivision accompanying large deformations. With a 0.01 mm diameter collimator, bending magnet radiation from a 3.0 GeV source and image storage plates, samples of copper with thicknesses greater than 0.1 mm have been studied. Results from an as-received sample and a sample deformed to 100% torsion are compared and illustrate how efficiently grain subdivision can be quantified with polychromatic microbeam diffraction.

INTRODUCTION

Metal forming operations are often designed to produce specific types of anisotropy in grain shape, size and orientation. The pattern of preferred orientation of the grains is termed texture and can be considered to consists of three size scales. Macrotexture refers to the average texture of a sample, microtexture describes the orientation of individual grains and mesotexture denotes the orientation of ensembles of adjacent grains [1], i.e., a local average texture. Despite experimental estimates showing that grain subdivision contributes levels of anisotropy comparable to that of grain reorientation at low to moderate strains [2], texture modeling has largely ignored this complication.

The Taylor model [3, 4] is widely used to average the response of individual grains and assumes each subunit (grain) experiences the same deformation as the aggregate. Texture prediction is often based on the Taylor model [5-7]; while this approach represents the qualitative aspects of texture distribution reasonably well, experiment shows a less sharp, more slowly developing texture than predicted [8, 9]. Treatment of grains as discrete entities largely ignores the development of substructure observed within the grains. Disorientations greater than 10° between adjacent (~1 μm) domains are common in fcc materials after equivalent von Mises strains $\epsilon_{vM} > 1$ [10]. If the resulting spread of orientations within each grain does not retard its net rotation, it most certainly produces less sharply defined macrotexture than would be the case if grains were regarded as monolithic entities. Thus, macrotexture does not sharpen as rapidly as predicted by grain rotation models, and it is difficult to see how models could match experiment without incorporating this effect.

Quantification of preferred orientation or macro-texture began with Wever's x-ray diffraction work [11] and still receives significant coverage in basic x-ray diffraction textbooks [12, 13]. Recently, electron diffraction in transmission electron microscopy (TEM) and electron channeling or backscattering (i.e., orientation imaging microscopy or OIM [14, 15]) in scanning electron microscopy (SEM) has been used in this application. Recent texture quantification emphasizes OIM for microtexture measurements used to construct macrotexture pole figures [16] or for quantification of disorientations between adjacent grains [17] or TEM

43

for quantification of subgrain domain disorientations [10, 18]. Microbeam x-ray diffraction was relatively neglected, insofar as it was applied to mapping strain, microtexture, etc in polycrystalline sample, until dedicated storage rings for production of synchrotron radiation became routinely available [1, 19-22]. Mapping with 1 μm diameter or smaller microbeams is being devedloped by several groups [23, 24].

EXPERIMENTS

The samples were OFHC (oxygen-free high conductivity) Cu with an initial grain size of approximately 60 μm [25]. One sample was cut from the as-received material, a second sample was from material subjected to 50% compression, the third was from a sample subjected to 50% compression followed by 50% torsion and the fourth was from specimens strain in torsion to 100%. The thin-walled torsion specimen design, adapted from that of Lindholm et al [26], limits deformation to the gage section, even at large strains, and maintains a high degree of shear strain uniformity in this volume. The free-end torsion tests were conducted at room temperature at an effective strain rate of 4×10^{-4} s^{-1}, and 111, 200, and 220 pole figures from the samples are reported elsewhere [27] and are consistent with those reported in the literature. Planar sections of the samples were cut with a slow speed diamond wheel, were hand ground from 1 mm thick sections into wedges, were polished on SiC paper and were etched to remove polishing damage. The portion of the samples examined with the x-ray microbeam was about 100 μm and no more than 200 μm thick.

Diffraction data were collected with polychromatic bending magnet radiation at the SSRL beamline 2-2 (3.0 GeV, beam currents between 100 and 50 mA). Microbeam diffraction was performed initially with a 10 μm diameter pinhole collimator placed 55 cm from the sample; subsequently the separation was reduced to approximately 8 cm. In the former case the ⁻20 arcseconds of vertical divergence of the beam was enough to broaden the beam from the 10 μm diameter collimator to 80 μm vertically at the sample position [28]. Image storage plates [29,30] record the transmission Laue patterns, and both the image storage plates and the wedge-shaped samples were normal to the incident beam. Initially, 20 x 25 cm plates were read with 100 μm pixel size and 1024 levels of contrast in a Fuji BAS-2000 Imaging Plate Scanner. Once a Fuji BAS-2500 Imaging Plate Scanner became available, all data were collected on 20 x 40 cm plates and read with 50 μm pixel resolution and 256 levels of contrast. Additional levels of contrast could be obtained in the BAS-2500 system at the cost of much larger data file sizes, but 256 levels of contrast provided adequate range for these experiments. Exposures were on the order of 4×10^4 mA·sec.

Beams diffracted from the specimen passed through 3 mm of Al on the front of the holder for the image plates. A diamond-shaped, Pb beam stop was attached to the front of the plate holder, and its shadow is a prominent feature in all of the diffraction patterns. For some of the diffraction patterns, a filter with and absorption edge in the wavelength range of interest (i.e., Pd, Zr, or Mo) was placed in the incident beam to produce a sharp change in contrast in the polycrystalline diffraction pattern and to simplify indexing the diffraction pattern [31]. Wavelengths below the absorption edge were heavily attenuated while those above the absorption edge are lightly attenuated. The image plate to sample distance and the separation between the edge position and the incident beam position (i.e., where it was recorded on the image plate) were used to compute the diffraction angle for that absorption edge, and, with the wavelength and diffraction angle defined, Bragg's law was used to determine the corresponding d-spacing and hkl indices. Microtexture was mapped in the specimens by translating the sample along the two orthogonal axes perpendicular to the beam by fixed increments (generally 10 μm) and recording the resulting transmission Laue pattern.

RESULTS AND DISCUSSION

Figure 1 compares transmission Laue patterns from the as-received copper (top left), the 50% compression sample (top right), the 50% compression-50% torsion sample (bottom left) and the 100% torsion sample (bottom right); this data was recorded with the smaller sample-collimator separation. The black circle marks the position where the azimuthal variation of diffracted intensity was measured, and wider streaks indicate greater spread of orientations associated with the irradiated volume. The background is white, with increasing diffraction intensity shown by darker pixels. Little broadening of the diffraction streaks is seen in the as-received material, more broadening is evident after 50% compression, and the two samples with 100% effective strain show the greatest azimuthal widths. Figure 2 shows the azimuthal variation of diffracted intensity, but numerical values of peak widths, etc. are not yet available. Data with the larger collimator-sample separation shows azimuthal full-widths at half maximum intensity between 2° and 5° while this quantity was between 25° and 30° after 100% torsion. Furthermore, three or more sharply-defined substreaks were within the envelope of a single streak after 100% torsion.

Figure 3 is of the 50% compression, 50% torsion sample and shows diffraction patterns recorded from four positions separated by 10 μm translations. The horizontal diffraction

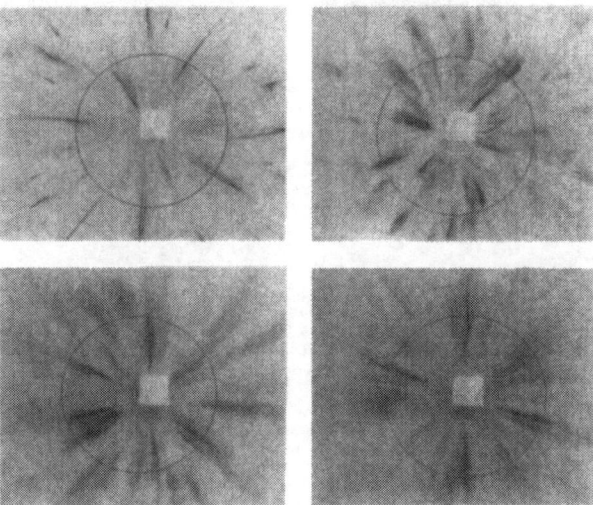

Figure 1. Transmission Laue patterns recorded with a 10 μm diameter pinhole collimator for as-received OFHC copper, after 50% compression, after 100% torsion, and after 50% compression followed by 50% torsion (clockwise from upper left). Lowest intensities are white with increasing diffraction intensity shown by darker pixels. The dark square in each is the beam stop.

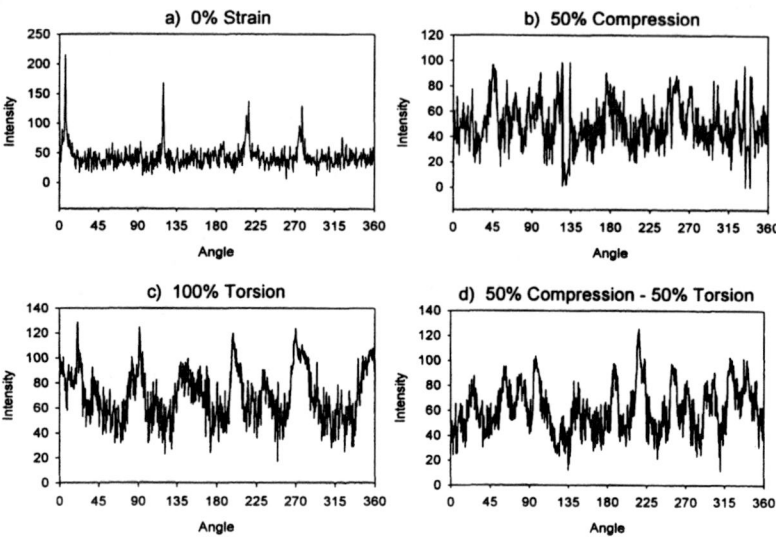

Figure 2. Azimuthal variation of diffracted intensity along the black circles shown in Fig. 1

Figure 3. Laue patterns of the 50% compression, 50% torsion sample at four positions separated by 10 μm translations. The gray scale is the same as Fig. 1.

streaks just to the left of the beam stop persist between 10-30 μm; since the maximum beam dimension is between 15 and 20 μm, this demonstrates that the structure observed occurs on a length scale substantially smaller than the original ⁻60 μm grain size. Further analysis is underway and should provide quantitative and statistically representative data on grain subdivision processes in copper.

CONCLUSIONS

Microbeam diffraction using polychromatic synchrotron x-ray radiation was demonstrated for mapping the amount and spatial distribution of deformation at length scales substantially below that of the grains in the as-received OFHC copper material. The preliminary results for samples deformed 50% in compression, 50% in compression followed by 50% torsion and 100% torsion illustrate how grain subdivision can be quantified in relative thick samples requiring little in the way of sample preparation. Even at this early stage of analysis, the results show that the spread in macrotexture observed in pole figures is due to grain subdivision processes. It is not surprising that model based on grains as the fundamental unit of crystallographic orientation predict too rapid and too sharp texture for heavily deformed samples.

ACKNOWLEDGMENTS

We thank Mr. S. Graham, Jr. of the MPRL for conducting the deformation experiments and Dr. Z.U. Rek of SSRL for her assistance. The data were recorded at SSRL which is supported by the Department of Energy. We are also grateful for the support from the Office of Naval Research (through grants N00014-94-1-0306 and -0726) and from the Army Research Office (through grants DAAH0493G00138 and DAAH049510177).

REFERENCES

1. J.D. Haase, A. Guvenilir, J.R. Witt and S.R. Stock, Adv X-Ray Anal **41** (1998) in press.
2. N. Hansen and D. Juul Jensen, Acta Mater **40** (1992) 3265.
3. G.I. Taylor, J Inst Metals **62** (1938) 307.
4. G.I. Taylor, in Steven Timoshenko 60ᵗʰ Anniv. Vol., Ed. J.M. Lessels, 1938, p.218.
5. R.J. Asaro and A. Needleman, Acta Metall **33** (1985) 923.
6. K.K. Mathur and P.R. Dawson, Int J Plasticity **5** (1989) 67.
7. S.R. Kalidindi and L Anand, in Adv. In Finite Deformation Prob. In Mater. Processing Structures, AMD **125** N. Chandra, J.N. Reedy, eds, ASME, 1991, p.3
8. N. Hansen and D. Kuhlman-Wilsdorf, Mater Sci Eng **81** (1986) pp. 141-161.
9. M.G. Stout, J.S. Kallend, U.F. Kocks, M.A. Przystupa and A.D. Rollet, in Proc. 8ᵗʰ Conf. On Textures of Mater. (ICOTOM 8), eds J.S. Kallend and G. Gottstein, TMS, 1988, p. 479.
10. D.A Hughes and N. Hansen, Acta Mater **45** (1997) 3871.
11. F. Wever, Z Phys **28** (1924) 69.
12. B.D. Cullity, Elements of X-ray Diffraction, Second Ed. Addison-Wesly, 1978.
13. C.S. Barrett and T.B. Massalski, Structure of Metals, Third Ed., Pergamon, 1986.
14. D.J. Dingley, Scanning Electron Microscopy **2** (1984) 569.

15. V. Randall, <u>Microtexture Determination and its Applications</u>. Inst. Metals, 1992.
16. B.L. Adams, S.I. Wright and K. Kunze, Met Trans. **24A** (1993) 819.
17. K. Matsumoto, T. Shibayanagi and Y. Umakoshi, Acta Mater **45** (1997) 439
18. D.A. Hughes, Q. Liu, D.C. Chrzan and N. Hansen, Acta Mater **45** (1997) 105
19. S. Stock, G.E. Ice, A. Habenschuss and C.J. Sparks, Jr., National Synchrotron Light Source Annual Report 1986, BNL 52045, p 354.
20. R. Rebonato, G.E. Ice, A. Habenschuss, and J.C. Billelo, Phil Mag **A60** (1989) 571.
21. G.E. Ice, Summarizing [19 and 20] in Nucl Instrum Meth **B24-25** pt. 1 (1987) 397.
22. J.D. Haase, A. Guvenilir, J.R. Witt and S.R Stock, Acta Mater in press (1998).
23. P.C. Wang, G.S. Cargill, III, and I.C. Noyan, Mat Res Soc Proc **375** (1995) 247.
24. H.F. Poulsen, S. Garbe, T. Lorentzen, D. Juul Jensen, F.W. Poulsen, N.H. Andersen, T. Frello, R. Feidenhans'l and H. Graafsma, J Synch Rad **4** (1997) 147.
25. S. Graham Jr., "The Stress State Dependence of Finite Inelastic Deformation Behavior of F.C.C. Polycrystalline Materials," M.S. Thesis, Georgia Inst. Technology, 1995.
26. U.S. Lindholm, A. Nagy, G.R. Johnson and J.M. Hoegfeldt, ASME J Eng Mater Tech **102** (1980) 376.
27. G.C Bulter, S. Graham, D.L. McDowell, S.R. Stock and V.C. Ferney, ASME J Eng Mater Tech in press (1998).
28. S.R. Stock, M.A. Langoy and R. Morano, unpublished data, December 1997.
29. Y. Ameniya, T. Matsushita, A. Nakagawa, Y Satow, J. Miyahora, and J. Chikawa, Nucl Instrum Meth. **A266** (1988) 645.
30. B.R. Whiting, J.F. Owen and B.R. Rubin Nucl Instrum Meth A266 (1988) 628.
31. S.R. Stock, Z.U. Rek, Y.H. Chung, P.C. Huang, and B.M. Ditchek, J Appl Phy **73** (1993) 1.

AUTOMATED INDEXING OF LAUE IMAGES FROM POLYCRYSTALLINE MATERIALS

Jin-Seok Chung and Gene E. Ice
Oak Ridge National Laboratory, Metals & Ceramics Division, Oak Ridge, TN 37830-6118

ABSTRACT

Third generation hard x-ray synchrotron sources and new x-ray optics have revolutionized x-ray microbeams. Now intense sub-micron x-ray beams are routinely available for x-ray diffraction measurement. An important application of sub-micron x-ray beams is analyzing polycrystalline material by measuring the diffraction of individual grains. For these measurements, conventional analysis methods will not work. The most suitable method for microdiffraction on polycrystalline samples is taking broad-bandpass or white-beam Laue images. With this method, the crystal orientation and non-isostatic strain can be measured rapidly without rotation of sample or detector. The essential step is indexing the reflections from more than one grain. An algorithm has recently been developed to index broad bandpass Laue images from multi-grain samples. For a single grain, a unique set of indices is found by comparing measured angles between Laue reflections and angles between possible indices derived from the x-ray energy bandpass and the scattering angle 2 theta. This method has been extended to multigrain diffraction by successively indexing points not recognized in preceding indexing iterations. This automated indexing method can be used in a wide range of applications.

INTRODUCTION

In recent years, third generation hard x-ray synchrotron-radiation sources have improved source brilliance by orders of magnitude, and revived interests in x-ray microprobes. Despite their superior chemical and crystallographic sensitivities, x-ray microprobes had been restricted by low flux and superseded by electron probes. But with these new x-ray sources and much improved x-ray optical components like hard x-ray zone plates[1,2], elliptic Kirkparick-Baez mirrors[3,4], and glass capillaries[5], the flux in submicron x-ray microbeam can exceed the flux in a mm^2 beam from a conventional x-ray source with fixed energy. These advances make x-ray microdiffraction analysis a powerful tool for the nondestructive measurement of crystallographic distributions in materials[6-8]. Unlike electron probes, x-ray microprobes can nondestructively analyze grains or structures even when covered by other materials. For example with x-ray microdiffraction, the strain of Al interconnections under a passivated layer of ultra large scale integrated circuits(ULSI) can be measured[9]. It is also possible to generate 3 dimensional maps of crystallographic information in polycrystalline materials.

Previous local microtexture analysis has relied on electron back-scatter diffraction technique(EBSD)[10], which uses electron beam microprobes. This technique has demonstrated the potential of microprobe measurements; new material science opportunities are possible with individual grain analysis. But EBSD has the limitation of electron probes, which are highly surface sensitive. X-ray microprobes overcome this limitation because x-rays are a penetrating probe. Also x-rays have much better signal-to-background ratio compared to electron probes and have strain sensitivity to ~10^{-5}. With these advantages, x-ray microprobes will benefit many material studies including stress induced cracking, electromigration, recrystallization, etc.

Despite the promise of x-ray microdiffraction, progress has been limited by the intrinsic difficulty of using conventional diffraction methods. For example, the motivation for using x-ray microprobes is to perform diffraction with sub-micron spatial resolution, but it is very hard to rotate a sample with micron accuracy using conventional diffractometers. As shown in Fig. 1, even if it is possible, the grains illuminated by an x-ray probe change with rotation due the penetration of x-rays. Also the excited reflections from many different grains make the analysis very difficult. Another problem with conventional diffraction is the time scale of getting two or three dimensional distributions of orientation or strain. Since the time required to obtain and analyze data at each point is not trivial, it would take unreasonably long time to get a map of diffraction information. Especially for polycrystalline samples with unknown crystal orientations, a new method of measuring diffraction information is required to allow measurements at tens of thousands of points.

One of the possible solutions in microdiffraction of polycrystalline samples is an automated analysis of wide bandpass or white beam Laue images. Laue patterns can be digitized by x-ray sensitive CCD area detectors, image plates, or 2D wire proportional counters. A single Laue image from polycrystalline samples has all the diffraction data from many grains excited by an x-ray probe without rotating the sample. Since it takes only a few seconds at most to take a CCD image, this method is favored, when many points need to be scanned.

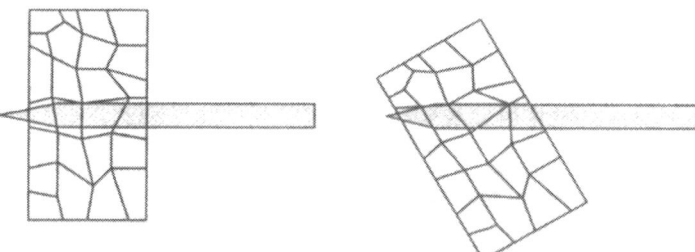

Figure 1. Rotation of a polycrystalline sample. The grains illuminated by x-ray microprobe are different, when the crystal is rotated.

Compared to conventional single crystal diffraction, however, the analysis of Laue images is quite different. The most crucial step is indexing reflections. There has been previous work on indexing Laue images for biological samples[11], which usually have thousands of reflections on one Laue image. Because there are so many reflections, zone axes can be identified and it is relatively easy to derive indices of each reflection from zone axes The method has recently been applied to synchrotron radiation microdiffraction. In this paper, a newly developed automated indexing algorithm, designed for Laue images taken at wide bandpass undulator synchrotron radiation sources, is introduced and its application is discussed. This computing algorithm is capable of indexing Laue patterns from single crystals or polycrystalline samples with known crystal structures, but where only a few reflections (>5) per grain are recorded in the Laue image.

METHOD

Most modern x-ray microprobes use synchrotron radiation from either bending magnets or hard x-ray undulators. Since bending magnet sources have wider bandpass and closer to ideal white source, their analysis can be done in somewhat similar way as the approach for the biological samples[12]. However, undulator sources like the Advanced Photon Source (APS), are many orders of magnitude brighter and have important advantages. Typically undulators have ~ 1% bandpass, but 10% bandpass can be obtained by tapering an undulator, going off axis, or scanning energy. Around 20 keV and with a practical solid angle of ~0.4, a 10% bandpass will give 5 ~ 10 reflections for fcc single crystals. This is about the right number of reflections to identify unique indices in our algorithm. The number of reflections increases as the bandpass become wider, but indexing reflections become more difficult. The minimum requirement to get unique sets of indices is more than 3 non-coplanar reflections for each grain.

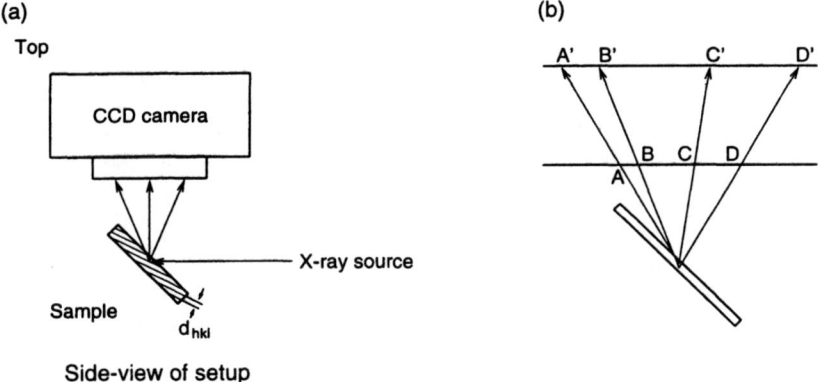

Figure 2. (a) A side view of an experimental setup for microdiffraction. (b) Ray tracing can find the relative location of the CCD with respect to the beam.

Figure 2 shows a typical experimental setup to take wide bandpass Laue images. Because synchrotron radiation is polarized in the plane of the storage ring, a CCD above the sample is more efficient than one in the ring plane. The sample can be scanned along 3 axes and the CCD can be moved up and down. This motion of the CCD is important since by translating vertically as shown in Fig. 2 (b), the relative location of the CCD to the location where the beam hits the sample can be determined by a ray-tracing method. The accuracy of this relative position is directly connected to measurement errors in angles. The best way of fine-tuning the measured relative position is to take a Laue image of a single crystal with known orientation and to find the position with the minimum error in reflection angles by iteration. The accuracy of angles is especially crucial for strain measurements.

Once the angles of observed reflections are measured from the CCD, the directional vector of the scattering vector $\hat{q}$ for each Laue reflection are determined from the direction of the incident beam, $\hat{k}_{in}$ and the diffracted beam, $\hat{k}_{out}$:

$$\vec{q} = \vec{k}_{out} - \vec{k}_{in} , \quad |\vec{k}_{out}| = |\vec{k}_{in}| \tag{1}$$

$$\hat{q} \, // \, \hat{k}_{out} - \hat{k}_{in} . \tag{2}$$

These directional vectors give the measured angles between reflections. For a given bandpass and a known crystal structure, there are only a limited number of possible pairs of indices, h, k, and l's. The upper bound of h, k, and l is set by the energy bandpass. Also, for larger indices, reflections are weaker because atomic scattering factors diminish for large scattering vectors. Typically indices above 20 can be ignored. The symmetry of the crystal structure further eliminates certain combinations of indices by selection rules. With a reasonable selection of bandpass, a manageable number of possible indices can be obtained. For example, with a 10 % bandpass at 20 keV, the number of possible indices per each reflection become 200 ~ 500 for a Si crystal. As the bandpass increases, this algorithm becomes increasingly inefficient. The algorithm makes a list of possible indices for each reflection and simply compares the angles between reflections with potential indices, to the measured angles between diffracted beams. It discards the pair if it does not satisfy the angle requirements. All the possible combinations of indices are checked, until only pairs that are consistent with the known crystal structure are found. When the angles between more than three reflections are specified, the number of possible pairs drops quickly. With more than three non-coplanar reflections, unique pairs can be found by this algorithm. Although this process may seem tedious, it is very fast with modern computers; a typical Laue image of Si with an 18 ~ 20 keV bandpass, takes only a few seconds to index on a Pentium Pro 200 MHz PC.

For polycrystalline samples, some reflections may not find any index, which satisfies the angle requirements because they are from different grains. These points are set aside and the program continues to index as many points as possible. Later the algorithm is re-applied to the unindexed reflections to index and identify separate grains.

RESULTS

In tests of our algorithm for Si crystals, a conventional x-ray tube with W target was used because our third-generation x-ray microbeam setup was not ready. To simulate 10 % bandpass around 20 keV, a balanced pair of Mo and Zr filters was used. First, our algorithm was tested for Si single crystals in random orientations and compared with results from an independent Laue simulation program, called *laueX* [13]. For the simulation, the orientation was guessed from white x-ray Laue images that had a lot more reflections. Our algorithm showed the same results as those from the simulation program. Next a Si (001) wafer was fractured into small pieces ~ 1 mm^2, which were slightly larger than our x-ray beam size. Several pieces were mounted on a translation stage. Fig. 3 (a) shows a Laue image taken near the boundary of two Si crystal pieces. The algorithm found two grains from this image, which implied that two pieces were excited simultaneously. Two more Laue images were taken after the stage was translated until the x-ray beam hit the centers of the two grains. Since the beam was smaller than each piece, only one grain was excited in these Laue images. Those images were used to index reflections with the simulation program. The results of simulation for these two images were identical to the result of our algorithm applied to the image that had reflections from two grains at the same time. For further evaluation, another Laue image was taken for the third grain and an artificial data file was made with locations of reflections from the three independent "grains" of Si. Our algorithm was applied to this file and found three grains. The Fig. 3 (b) shows the result of automated indexing for three grains. This result was also in a good agreement with the results from simulations of each grain.

(a)   (b)

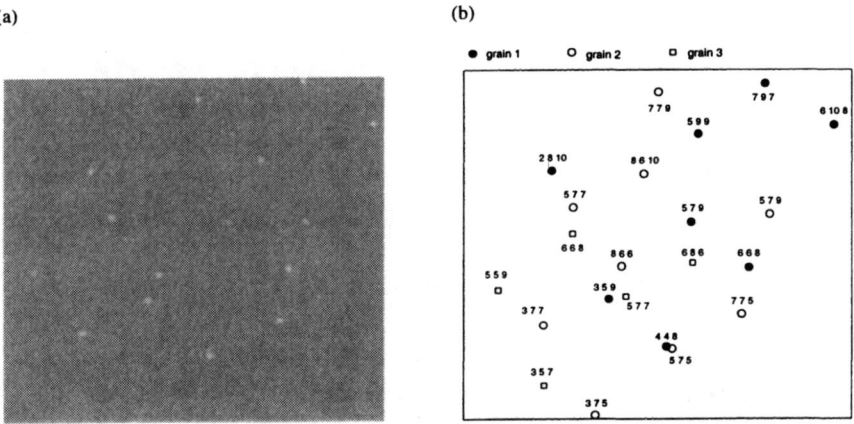

Figure 3. (a) Laue image of two Si grains (b) Result of an automated indexing of Laue reflections from 3 grains of Si.

CONCLUSIONS

Our automated indexing algorithm has proved to work efficiently for simulated polycrystalline samples. Time required for indexing was fast enough to index tens of thousands of Laue images in a reasonable time. This is a key step towards scanning x-ray microdiffraction of polycrystalline samples. The bandpass of the incident x-ray beam is critical to the efficiency of this algorithm. It must be wide enough to excite more than three Laue reflections from single grains, but narrow enough to have a manageable number of possible indices. The alignment of the CCD relative to the position where the x-ray beam hits the sample is critical for successful indexing. Strain in each grain can be measured by analyzing the difference between the theoretical angles from indices and the measured angles.

ACKNOWLEDGEMENTS

Research sponsored by the Laboratory Directed Research and Development Program of the Oak Ridge National Laboratory and the Division of Material Sciences, U.S. Department of Energy under contract DE-AC05-96OR22464 with Lockheed Martin Energy Research Corporation.

REFERENCES

1. W. B. Yun, B. Lai, D. Legnini, Y. H. Xiao, J. Chrzas, K. M. Skulina, R. M. Bionta, V. White and F. Cerrina, SPIE **1740**, pp. 117-129 (1992).

2. R. M. Bionta, E. Ables, O. Clamp, O. D. Edwards, P. C. Gabriele, K. Miller, L. L. Ott, K. M. Skulina, R. Tilley, T. Viada, Opt. Eng. **29**, p. 576 (1990).

3. J. H. Underwood, A. C. Thompson, J. B. Kortright, K. C. Chapman, and D. Lunt, Rev. Sci. Inst. Abstract **67**, p. 3359 (1996).

4. B. X. Yang, M. Rivers, W. Schildkamp, P. J. Eng, Rev. Sci. Inst. **66**, p. 2278 (1995).

5. D. H. Bilderback, S. A. Hoffman, and D. J. Thiel, Science **263**, p. 201 (1994).

6. P. C. Wang, G. S. Cargill III, and I. C. Noyan, MRS Symp. Proc. **375**, p. 247 (1995)

7. D. P. Piotrowski, S. R. Stock, A. Guvenilir, J. D. Haase and Z. U. Rek, MRS Symp. Proc. **437**, p. 125 (1996)

8. S. R. Stock, A. Guvenilir, D. P. Piotrowski and Z. U. Rek, MRS Symp. Proc. **375**, p. 275 (1995)

9. N. Yamamoto and S. Sakata, Jpn. J. Appl. Phys. **34**, p. L664 (1995).

10. D. J. Dingley and V. Randle, J. Mater. Sci. **27**, p. 4545 (1992).

11. R. B. G. Ravelli, A. M. F. Hezemans, H. Krabbendam and J. Kroon, J. Appl. Cryst. **29**, pp. 270-278 (1996)

12. H. R. Wenk, F. Heidelbach, D. Chateigner and F. Zontone, J. Synchrotron Rad. **4**, pp. 95-101 (1997)

13. Alain Soyer, *laueX*, This UNIX-based Laue simulation program is available from `ftp.lmcp.jussieu.fr` in `/pub/sincris/software/general/laueX/`.

Grain Orientation Mapping of Passivated Aluminum Interconnect Lines with X-ray Micro-Diffraction

A.A. MacDowell[1], C.H. Chang[1,2], H.A. Padmore[1], J.R. Patel[1,2], A.C.Thompson[3]

[1] Advanced Light Source, Lawrence Berkeley National Laboratory, Berkeley, CA 94720
[2] SSRL/SLAC, Stanford University, Stanford, CA 94309
[3] Center for X-Ray Optics, Lawrence Berkeley National Laboratory, Berkeley, CA 94720

ABSTRACT

A micro x-ray diffraction facility is under development at the Advanced Light Source. Spot sizes are typically about 1-μm size generated by means of grazing incidence Kirkpatrick-Baez focusing mirrors. Photon energy is either white of energy range 6-14 keV or monochromatic generated from a pair of channel cut crystals. A Laue diffraction pattern from a single grain in a passivated 2-μm wide bamboo structured Aluminum interconnect line has been recorded. Acquisition times are of the order of a few seconds. The Laue pattern has allowed the determination of the crystallographic orientation of individual grains along the line length. The experimental and analysis procedures used are described, as is a grain orientation result. The future direction of this program is discussed in the context of strain measurements in the area of electromigration.

INTRODUCTION

Electromigration is the physical movements of atoms in metallic interconnect lines passing current at high electron density (typically in the range of 10^5 amp/cm^2). Significant material movement results in voids that consequently lead to breakage and circuit failure in the metal lines. This problem gets more severe as the line dimensions continue to shrink on integrated circuits. In spite of much effort in this field, (1,2,3) electromigration is not understood in any depth or detail, but is strongly associated with the physical material properties (stress and strain) within the interconnect material. Throughout this century x-rays have been a powerful tool to measure such material properties, but the ability to make such measurements on the micron scale required by the semiconductor industry has only come into realization with the advent of the latest generation of high brightness synchrotron sources (see for example ref. 4). In this paper we describe the beginnings of a program to carry out various x-ray diffraction measurements on the micron scale. It is presumed that the electromigration properties of a metal line will be dependent to some extent on the grain orientation of adjacent grains in the line. This paper describes the experimental and analysis techniques that allow the grain orientation and indexing of individual micron sized grains along the length of aluminum interconnect line.

X-rays are quite well suited to such measurements as they are able to penetrate several microns into matter. In general, interconnect lines are encased in the insulator silicon dioxide (passivation). X-rays are able to penetrate and study such buried samples in the environment that they will be used.

EXPERIMENTAL

Figure 1 shows the experimental setup. The synchrotron source of size typically 300 × 30 μm FWHM (horizontal and vertical) is imaged with demagnifications of 300 and 60 respectively by a set of grazing incidence platinum-coated elliptically bent Kirkpatrick-Baez (K-B) focusing

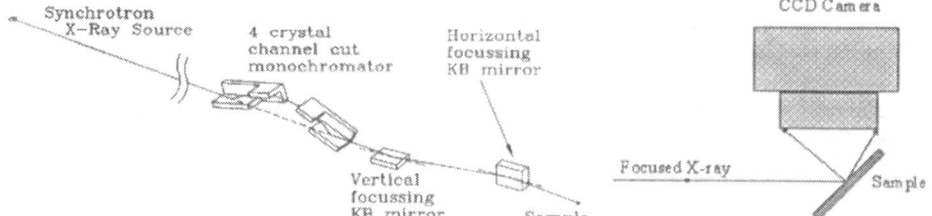

Figure 1. Schematic layout of the K-B mirrors and the four crystal channel-cut monochromator

Figure 2. Schematic layout of the arrangement around the sample

mirrors (5). Imaged spot sizes on the sample are about a micron in size. Photon energy is either white of energy range 6-14 keV or monochromatic generated by inserting a pair of Si(111) channel-cut monochromator crystals into the beam path. A property of the four crystal monochromator is its ability to direct the monochromatic primary beam along the same direction as the white radiation. Thus, the sample can be irradiated with either white or monochromatic radiation. The x-ray probe motion on the sample between white and monochromatic modes has been measured to be less than 0.5-µm. White radiation is chosen for Laue experiments which allow for crystal grain orientation determination. Monochromatic radiation is to be used for d-spacing measurements to determine stress/strain determination of single grains in the metal line.

The sample was a 2-µm wide aluminum line deposited to a thickness of 0.5-µm onto an oxidized silicon substrate. The line was passivated with a plasma-enhanced chemical vapor deposition (PECVD) nitride at 300°C to 0.3-µm thickness. Laue patterns were collected using white radiation and a x-ray CCD camera. The exposure time was 0.5 sec and sample-to-CCD distance was 16.4 mm. Fig. 2 shows the arrangement of the sample and CCD detector.

RESULTS

Figure 3 shows the Laue pattern from the silicon substrate and figure 4 shows the Laue patterns from the silicon substrate with the fainter diffraction spots from a single grain in the aluminum line. Digital subtraction of the silicon pattern (Fig. 3) from the silicon and aluminum pattern (Fig.4) results in the Laue pattern of the single aluminum grain under observation. (Fig.5)

The origin of the Laue pattern (the sample) with respect to the CCD detector was determined by moving the CCD camera radially from the sample and recording the silicon Laue patterns at various distances from the sample. The Laue pattern origin was determined from the intersection of lines drawn through a succession of the same Laue spots for different radial locations of the CCD camera. All aluminum spot positions were coordinated to the origin and indexed using the indexing software package - LaueX (6). Figure 6 shows the simulated pattern with reflections indexed indicating good agreement for the spot positions. Confirmation of the indexation was also achieved by inserting the 4 crystal monochromator into the beam, thus illuminating the aluminum grain with monochromatic light. The photon energy was scanned and the rocking curve (of the central aluminum spot in Figure 5) mapped out. From the diffracted photon energy (7323eV) and the angular direction of the diffracted spot, the d-spacing of this spot was measured, which confirmed its indexation as Al(111). The accuracy of this d spacing measurement was around 1 part in 1000, as limited by the angular resolution of the CCD camera to about 1-mrad (determined from the CCD pixel spacing (23.5-µm) and the CCD radial scan range of ~20cm). Detailed strain measurements will require an improved d spacing measurement

Figure 3. Laue pattern of silicon substrate

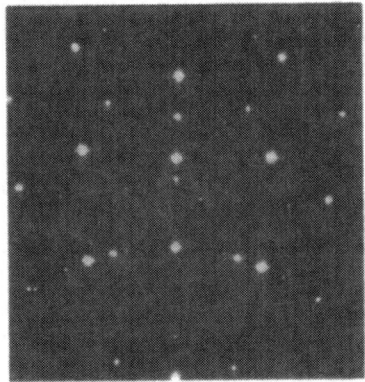

Figure 4. Laue pattern of a single grain in aluminum line as well as the silicon substrate.

Figure 5. Laue pattern of a single aluminum grain obtained by substracting the silicon pattern (Fig.3) from Fig. 4

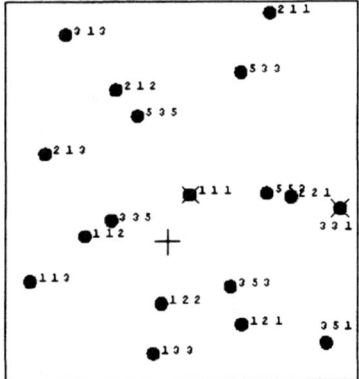

Figure 6. Simulated Laue pattern with the aluminum spots indexed

some 10-100 times better. It is proposed to increase the radial scan range of the detector in order to improve the angular accuracy required for detailed strain measurements.

The aluminum grain orientation can be referenced to the silicon substrate based on the orientation matrix $\mathbf{R}_{Si}$ and $\mathbf{R}_{Al}$ of the silicon substrate and aluminum grain, respectively. The matrix $\mathbf{R}$ relates the crystal system S with axes parallel to the basic crystallographic axes in the crystal to the reference system S^R related to the primary beam direction:

$$S^R = \mathbf{R}S$$

57

The aluminum grain orientation measured is then referenced to the silicon substrate as the following orientation matrix:

$$\mathbf{M} = \mathbf{R}_{Si}^{-1}\mathbf{R}_{Al}$$

$$= \begin{pmatrix} 0.707 & 0.475 & -0.524 \\ 0.003 & 0.737 & 0.676 \\ 0.707 & -0.481 & 0.518 \end{pmatrix}^{-1} \begin{pmatrix} 0.916 & 0.039 & 0.399 \\ 0.308 & 0.571 & -0.761 \\ -0.257 & 0.822 & 0.509 \end{pmatrix}$$

$$= \begin{pmatrix} 0.468 & 0.612 & 0.637 \\ 0.787 & 0.045 & -0.615 \\ -0.404 & 0.792 & -0.458 \end{pmatrix}.$$

The Laue diffraction patterns from aluminum grains are always accompanied by the Laue pattern from the silicon substrate. We use the silicon substrate as the reference for the aluminum grain orientation.

The experimental accuracy in the determination of the orientation matrix $\mathbf{M}$ depends mainly on the angular resolution of the Laue camera system. In the present case, the Laue patterns were recorded with a sample-to-CCD distance of 16.4mm. This results in a misorientation angle determined to a precision of several minutes of arc.

CONCLUSION AND FUTURE DEVELOPMENT

We have demonstrated that x-ray micro-diffraction is capable of determining the crystallographic orientation of individual grains in passivated interconnect lines. The orientation mapping can be done by collecting the Laue patterns from individual grains along the length of the lines. A computerized indexing code to automate this is under development. Beyond this the requirement is to measure the d-spacing of various aluminum planes to high accuracy to determine the stress and strain state of individual grains along the length of the aluminum interconnect line whilst passing a high current as in a typical electromigration experiment. This work is underway with the construction of a new micro-diffractometer.

ACKNOWLEDGEMENTS

This work was supported by the Director, Office of Basic Energy Sciences, Materials Sciences Division of the US Department of Energy, under Contract no. DE-AC03-76SF00098. Samples from T. Marieb and equipment support from Intel Corporation, Santa Clara, CA.

REFERENCES

1. T.Marieb, P.Flinn, J.C.Bravman, D.Gardner, and et al. *J. App Phys.* **78**, 1026-1032 (1995).
2. P.C.Wang, G.S.Cargill, I.C.Noyan, E.G.Liniger, C.K.Hu and K.Y.Lee, *MRS Proceedings* **473**, 273-278 (1997).
3. J.S.Chung, G.E.Ice, *These proceedings*.
4. A.Rindby, P.Engstrom and K.Janssens, J.Synchrotron Rad. **4**, 228-235 (1997)
5. A.A.MacDowell, R.Celestre, C.H.Chang, K.Franck, M.R.Howells, S.Locklin, H.A.Padmore, J.R.Patel, and R.Sandler, *SPIE Proceedings* **3152**, 126-135 (1998).
6. A.Soyer, *J. App. Cryst.* **29**, 509 (1996); http://www.lmcp.jussieu.fr/sincris/logiciel/laueX/en/laueX_en.html.

SYNCHROTRON RADIATION DIFFRACTION IMAGING STUDY OF THE MAGNETOACOUSTICALLY INDUCED X-RAY FOCUSING EFFECT IN FeBO3

I. MATSOULI††, V. V. KVARDAKOV*, J. I. ESPESO†§, L.CHABERT†, J. BARUCHEL†
† ESRF, BP 220, Grenoble 38043, France, matsouli@esrf.fr
‡ University of Warwick, Coventry, CV4 7AL, UK
* Kurchatov Institute of Atomic Energy, Moscow 123182, Russia
§ Universidad de Cantabria, Santander, Spain

ABSTRACT

Synchrotron radiation diffraction imaging ('topography') was used at the ESRF to visualize ultrasonic standing waves in a magnetoacoustically excited FeBO3 crystal. Images were recorded at long sample-to-film distances (up to 1.5 m) without substantial loss of resolution. The resonant patterns reveal that the crystal acts as a pulsed X-ray focusing lens and strongly depend on both the amplitude of the magnetic field and the sample-to-detector distance. Different resonant frequencies can also lead to more complicated images due to interference effects of several vibrating modes present in the crystal. A model is proposed followed by a numerical integration which predicts the focusing of the X-ray beam and demonstrates the agreement between theory and experiment. This effect could be used to simultaneously monochromatize and focus the X-ray beam.

INTRODUCTION

The application of an alternating magnetic field provides a contactless method for the excitation of ultrasonic waves in weak ferromagnets. The oscillation of the magnetic moments can lead, through magnetoelastic coupling, to the propagation of lattice waves and the development of standing elastic waves. Iron borate, FeBO3, is a good candidate as it grows in thin platelets that readily support membrane type vibrations and it is characterized by a magnetically easy plane with a vanishingly small in-plane anisotropy [1,2].

In the first neutron diffraction investigation performed on FeBO3 [3,4] it was found that at given frequencies, when under the influence of a high frequency alternating magnetic field, the scattered intensity was enhanced by up to a factor of four and standing elastic waves were visualized using neutron topography. In the case of synchrotron radiation (SR) sources, the highly parallel beam allows the observation of effects associated with vibrations of the crystal planes. The small angular source size viewed from the sample in combination with the ability to vary the sample-to-film distance as an additional parameter can lead to the focusing of the X-ray beam, already observed on surface acoustic waves [5]. We, therefore, combined the new possibilities of the ESRF third generation machine together with the non-contact magnetoacoustic generation of ultrasound to investigate the influence of volume acoustic waves on the diffracted X-ray beam from an FeBO3 crystal. A suitable theoretical model and simulations that follow predict the X-ray focusing as well as interpreting the additional features that appear on the topographs.

EXPERIMENT

The FeBO3 crystals were grown at the Physical Institute (Prague) of the Czech Academy of Sciences [6]. FeBO3 is a weak ferromagnet, usually 50-100μm thick that takes the

Mat. Res. Soc. Symp. Proc. Vol. 524 © 1998 Materials Research Society

form of good quality, thin hexagonal platelets parallel to the (111) plane, the plane of easy magnetization. A moderate magnetic field H_0, of 9 A cm^{-1}, was sufficient for the alignment of the nett magnetic moments. An additional alternating magnetic field, H_{ac}, was then applied at right angles, at frequencies ranging from 1-1.5 MHz. Ultrasonic waves were this way introduced into the crystal, through the strong magnetoelastic coupling, and the resulting standing-wave-related contrast was visualized using SR diffraction topography and X-ray films as a detector. White beam experiments were performed on the X-ray imaging and high resolution diffraction beam line, ID19 at the ESRF, the European Synchrotron Radiation Facility in Grenoble, France. Diffraction imaging is a non-destructive technique, sensitive to very weak distortions (10^{-7}). It is based on Bragg diffraction from single crystals where every Bragg spot is a direct space image (topograph) of the given sample [7].

RESULTS

During resonance, intensity maxima were observed on the topographs due to standing elastic waves in the crystal, where extra blackening on the film represents extra X-ray intensity. Figure 1 shows a set of topographs of the $\bar{1}2\bar{1}$ reflection.

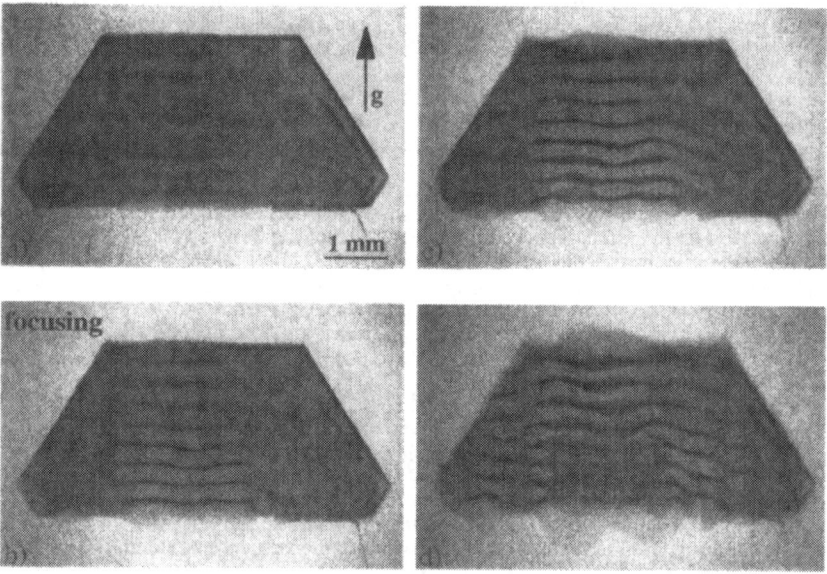

Figure 1: White beam topographs of the $\bar{1}2\bar{1}$ reflection in FeBO$_3$ as a function of the sample-to-film distance; (a) 19.5 cm, (b) 41 cm, (c) 55 cm, (d) 114 cm. The resonance pattern corresponds to a frequency of 1.3 MHz and clearly reveals a focal distance at 41 cm beyond the sample.

The images were recorded for increasing sample-to-film distances ranging from 19 to 114 cm beyond the sample without substantial loss of resolution due to the highly parallel synchrotron beam and the small source size at the ESRF. The contrast appeared to be hardly visible for the shortest distance and very sharp at 41 cm. At distances larger than this focusing distance, an intensity redistribution took place due to the defocusing of the X-ray beam. The contrast associated with the intensity maxima now appeared deteriorated with dark lines superimposed on the image that coincided with the position of the in-focus sharp maxima. At the border of the defocused image there was a sharp drop of intensity and, when the distances exceeded 55 cm additional overlapping effects of neighbouring 'defocused' maxima appeared. Similar results were recorded as a function of the amplitude of the a.c. magnetic field.

Simultaneously imaging several reflections provided us with complementary information. Different vibration modes were present in the crystal leading to interference effects. In fig.1, the dominant mode is the one propagating parallel to the diffraction vector. Figure 2 shows, images of the same $\bar{1}2\bar{1}$ reflection recorded at a resonant frequency of 1.42 MHz. In this case, there exist two dominant interfering modes, nearly perpendicular the one to the other resulting in more complex resonance patterns.

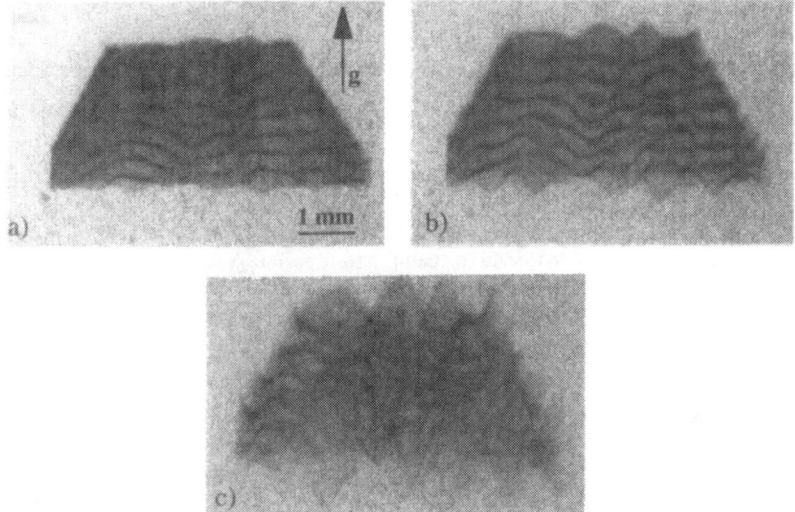

Figure 2: White beam topographs of the $\bar{1}2\bar{1}$ reflection in FeBO₃ as a function of the sample-to-film distance; (a) 45 cm, (b) 100 cm, (c) 261 cm. The resonance pattern reveals strong interference effects and corresponds to a frequency of 1.42 MHz; the images are inversed compared to the ones of fig.1.

DISCUSSION

Figure 3, is a schematic representation of the membrane type vibration of FeBO₃. In a non-vibrating crystal the X-rays are diffracted in the same way from all the diffracting planes that fulfill the same Bragg condition. When under a magnetically induced resonance, however, the planes rotate with respect to each other and the crystal exhibits a focusing/defocusing behaviour. During defocusing the intensity arising from the segment S is spread out over the region BD whereas during the focusing phase the X-rays converge on C. In other words, the crystal splits into several regions equal to the segment S that act as focusing lenses for the X-ray beam periodically in time.

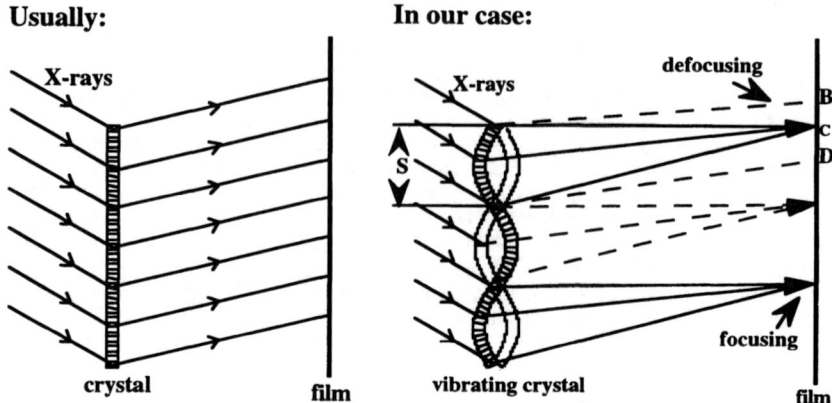

Figure 3: An illustration of the membrane type vibration in FeBO3; each segment S acts as a focusing lens periodically in time.

The observed images of the standing elastic waves for a given diffraction vector strongly depend both on the sample-to-film distance and the amplitude of the vibrations. A model was produced which accounts for all the observed changes in contrast, predicts the focusing of the X-ray beam as well as confirming an agreement between theory and experiment. These calculations involve spatial and time integrations of the diffracted beam intensity as a function of the position on the detector [8]. The diffracted intensity on a point x of the film from an arbitrary point ξ on the crystal at a time t is represented by $i(x,\xi,t)$. The time integrated intensity $i(x,\xi)$ was found to be:

$$i(x,\xi) \propto \frac{1}{\sqrt{(A\sin\xi)^2 - (x-\xi)^2}} \quad \text{if } |x - \xi| \leq |A\sin\xi|$$

$$i(x,\xi) = 0 \text{ otherwise}$$

(1)

where A is a dimensionless parameter defined as $A = 2\phi_o L$, L being the dimensionless sample-to-film distance. Since in white beam we can only sense plane rotations, ϕ is an angle describing the plane vibration and has a maximum amplitude ϕ_o corresponding to the amplitude of the applied a.c. field. According to this model, the exact focal position takes place at $A = 1$. By measuring the focusing distance one can, therefore, indirectly calculate that in our case the maximal rotational motion of the vibrating planes ϕ_o is 50".

In order to find the total intensity distribution $I(x)$ on the film one has to integrate $i(x,\xi)$ over the whole crystal:

$$I(x) \propto \int i(x,\xi)d\xi$$

(2)

This was done using computer simulations based on Simpson's rule of numerical integration. Figure 4 is a contour map of the intensity distribution plotted as a function of A and the x-position on the film. It graphically represents all the intensity changes that occur at different values of the parameter A (~ distance x vibration amplitude) along the film. At $A = 1$ sharp maxima appear with maximum intensity corresponding to the focal position. As A increases defocusing takes place and both the sharp drop of intensity at the border of the defocused image as well as the high intensity lines always present in the middle are in agreement with the experimental results. When A approaches 3, the defocused images overlap giving rise to additional intensity peaks clearly seen as an intensity jump on the contour map. The misleading impression that the intensity increases as A is closer to 5 is due to a strong optical effect. Figure 5 shows two sections of the contour map for $A = 1$ and $A = 2.5$ respectively.

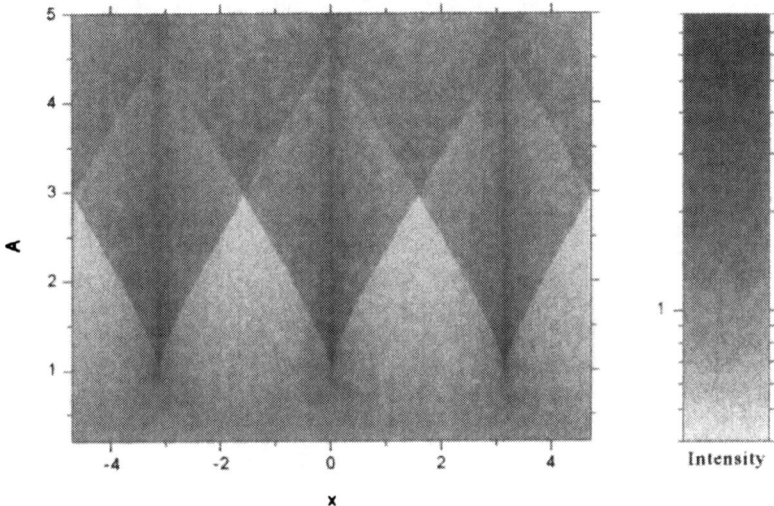

Figure 4: Contour map of the diffracted intensity distribution plotted as a function of A and the x-position on the film.

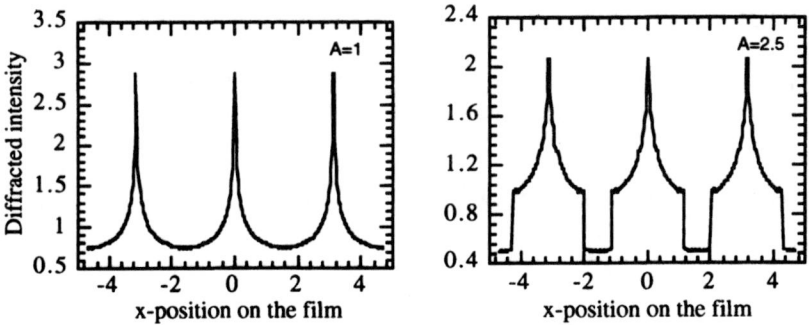

Figure 5: Contour map sections for $A = 1$ and $A = 2.5$ respectively; the diffracted intensity is plotted as a function of the position x on the detector.

CONCLUSIONS

The numerical simulations of the presented theoretical model are in very good agreement with the experimental results. They predict the focusing of the diffracted X-ray beam and demonstrate the additional features that appear on the topographs. Using the present set-up one can focus the X-ray beam over a range of distances from the $FeBO_3$ plate by simply modulating the amplitude of the applied a.c. field. This provides a promising application where the crystal could be used to focus X-rays on the micrometer scale, a field where other methods are being actively pursued such as capillary or Bragg-Fresnel optics [9,10], bent crystals [11], refractive lenses [12]. This would involve the appropriate collimation of the diffracted beam and the use of the pulsed structure of the synchrotron source through the implementation of a stroboscopic method on the microsecond scale. The advantages and disadvantages of the current method are under further investigation.

ACKNOWLEDGMENTS

The author gratefully aknowledges the help of Frank Heyroth and the support of the European Training and Mobility of Researchers (TMR) grant under the reference number ERB4001GT957285.

REFERENCES

[1] V I Ozhogin, V L Preobrazhenskií, *Sov. Phys. Usp.* 31(8), p. 713-729, 1988.

[2] R Diehl, W Jantz, B I Noläng, W Wettling in Current Topics in Material Science, Vol. 11, Ch. 3, 1984.

[3] V V Kvardakov, V A Somenkov, *J. Moscow Phys. Soc.* 1, p. 33-57, 1991.

[4] V V Kvardakov, V A Somenkov, A B Tyugin, *JETP Lett.* 48(7), p. 437-439, 1991.

[5] H Cerva, W Graeff, Phys. Stat. Sol. (a) 82, p. 35-45, 1984.

[6] M Kotrbová, S Kadecková, J Novák, J Brádler, *J. Cryst. Growth* 71, p.607-614, 1985.

[7] A Authier, S Lagomarsimo, B K Tanner (editors), X-ray and neutron dynamical diffraction: theory and applications (NATO ASI series, Plenum Press, NY), 1996.

[8] I Matsouli, V V Kvardakov, I E Espeso, L Chabert, J Baruchel, submitted to *J. Phys.: Appl. Phys. D*, 1998.

[9] P Engströem P, C Riekel C, *J. Synchrotron Rad.* 3, p. 97-100, 1996.

[10] A Snigirev, *Rev. Sci. Instrum.* 66(2), p. 2053-2058, 1995.

[11] C Schulze, U Lienert, M Hanfland, M Lorenzen, F Zontone, *J. Synchrotron Rad.* 5, p. 77-81, 1998.

[12] A Snigirev, V Kohn, I Snigireva, B Lengeler, *Nature* 384, p.49-51, 1996.

THE MECHANISM OF TWINNING IN ZINCBLENDE STRUCTURE CRYSTALS: NEW INSIGHTS ON POLARITY EFFECTS FROM A STUDY OF MAGNETIC LIQUID ENCAPSULATED CZOCHRALSKI GROWN InP SINGLE CRYSTALS

M. DUDLEY*, B. RAGHOTHAMACHAR*, Y. GUO*, X.R. HUANG*, H. CHUNG*, D.J. LARSON*, JR., D. T. J. HURLE**, D. F. BLISS***, V. PRASAD****, and Z. HUANG*****
*Dept. of Materials Science & Engineering, SUNY at Stony Brook, NY 11794-2275, USA
**H. H. Wills Physics Laboratory, University of Bristol, Tyndall Avenue, Bristol BS8 1TL, UK
***USAF Research Laboratory, Hanscom AFB, MA 01731
****Dept. of Mechanical Engineering, SUNY at Stony Brook, NY 11794-2300, USA
*****James Franck Institute, University of Chicago, Chicago IL 60637

ABSTRACT

Synchrotron White Beam X-ray Topography (SWBXT) and synchrotron X-ray anomalous scattering have been employed to determine the polarity of $\{111\}$ edge facets, anchored to the three phase boundary (TPB) on which twinning is observed to nucleate in Magnetic Liquid Encapsulated Czochralski (MLEC) grown sulfur doped, <001> InP single crystals. Analysis of the results indicates that both the formation of edge facets and the nucleation of twins occur preferentially on $\{\bar{1}\bar{1}\bar{1}\}_p$ faces. Of the four possible sets of edge facets, belonging to the $\{\bar{1}\bar{1}\bar{1}\}_p$ form, which are oriented so as to be thermodynamically favored to be anchored to the TPB, two can give rise to a $\{115\}$ to $\{\bar{1}\bar{1}\bar{1}\}_p$ external shoulder facet conversion upon twinning, while the other two can give rise to a $\{114\}$ to $\{110\}$ conversion. For these cases, twinning is only observed when the $\{\bar{1}\bar{1}\bar{1}\}_p$ edge facets are anchored to the TPB in a region where the shoulder angle is close to $74.21°$ or $70.53°$, facilitating the production of the $\{115\}$ and $\{114\}$ external shoulder facets, respectively, prior to twinning. These observations are discussed in light of calculated surface energies of the various internal and external facets.

INTRODUCTION

The phenomenon of growth-twinning during the bulk crystal growth of zincblende structure crystals (especially for Indium-containing III-V compounds such as InP and InSb) has, for many years, remained a persistent problem. Despite much experimental work in this area (e.g. [1-3]) it was not until 1991 that Hurle proposed comprehensive model to explain and predict the incidence of twinning in the growth of III-V compound semiconductors [4] (also reported in more complete form in [5]). This model was based on the calculation of the thermodynamic conditions under which edge facets could be anchored to the three-phase-boundary (TPB), which were translated into a range of external-shoulder or grow-out angles. If the range of shoulder angles was such that it became possible for twinning to replace a high index, high surface energy external facet with a low index, low surface energy facet, provided the local undercooling was sufficient, then twin nucleation would be favored. Hurle used this approach to prescribe the most dangerous grow-out angles for <001> and <111> growth of several zincblende crystals, including InP. Unfortunately, there has been very little in the way of experimental support for this model. However, in a recent paper [6], modifications to the Hurle theory were reported. These modifications were based on direct observations of the nucleation of twins on edge facets anchored to the TPB in S-doped, MLEC grown, <001> InP single crystals using SWBXT and optical microscopy [6,7]. Twins were directly observed to induce {115} to {111}, and {114} to {110} external shoulder facet conversions. The most significant modifications to the Hurle model for twinning in InP included: a change in the most dangerous grow-out angle for <001> growth, from the point of view of twin nucleation, from 35.5° to 74.21°; an expansion of the range of shoulder angles over which edge facets are thermodynamically favored to be anchored to the TPB from $31° < v < 86.5°$ to $31° < v < 112°$, where v is the angle between the edge facet and the extension of the crystal surface; and a significant reduction in the estimated undercooling required to promote twinning from ~15°C to ~2°C.

In addition to the factors considered above, another important factor which can influence twinning in zincblende structure crystals is the polarity along the <111> directions. The influence of polarity on facet formation and twinning in InSb was discussed by Hurle [4,5], and studies of its effects have also been reported by Hulme and Mullin for InSb [2] and by Steinemann and Zimmerli for GaAs [8]. However, an understanding of the influence of polarity on facet formation and twinning is not available for all zincblende structure crystals. In particular such understanding is lacking for InP, although the influence of the polarity of the seed face which is in contact with the melt on the preponderance of twinning in <111> grown InP has long been recognized [3].

In this paper, SWBXT is employed to identify the presence of anchored {111} edge facets which give rise to twin nucleation, and synchrotron white-beam X-ray anomalous scattering is employed to explicitly determine the polarity of those facets. Knowledge of the polarity enables us to extend our previous study [6] by revealing further details regarding the influence of crystal structure orientation on twinning configurations in S-doped, [001] MLEC grown InP, with particular reference to the influence of the polarity of the facets on which twinning occurs. In combination with our previous results, the work reported here provides a comprehensive understanding of twinning and faceting in InP single crystals.

EXPERIMENTAL

The sulfur-doped [001] InP crystal studied here was grown at Rome Laboratory, Hanscom AFB using the MLEC technique with a cusp magnetic field. To facilitate the study of the overall defect distribution in the crystal, the as-grown [001] boule was cut longitudinally into twenty-four (110) wafers. The boule could be reconstructed, for the purposes of relating internal features with the growth surface morphology, by simply re-assembling these wafers side-by-side. Results from one of these wafers are presented here.

SWBXT and synchrotron white-beam anomalous-scattering experiments were carried out at the Stony Brook Synchrotron Topography Station, Beamline X-19C, at the National Synchrotron Light Source (NSLS), Brookhaven National Laboratory. Transmission topographic images were recorded using wavelengths of around 0.45 Å, which is just above the In K absorption edge (0.444 Å), in order to minimize photoelectric absorption. All images were recorded on 8" × 10" Kodak SR-1 high-resolution X-ray film.

The polarity of the planes on which both faceting and twinning was observed to occur was determined, using anomalous scattering, by measuring and comparing the diffracted intensities from the $(1\bar{1}1)$ and $(\bar{1}1\bar{1})$ symmetric reflections in the vicinity of the In absorption edge (λ_K=0.444 Å, θ_B = 3.75°), in the transmission geometry. These experiments were carried out using filtered synchrotron white radiation (using a 1.9 cm thick aluminum filter) and a θ-2θ diffractometer in the vertical configuration. The filtering removed most of the long wavelength components from the incident beam spectrum, significantly reducing the background noise.

RESULTS AND DISCUSSION

Figure 1 shows a detail from a scanned SWBXT image, recorded, in transmission Laue geometry, from a (110) wafer cut from a S-doped, [001] grown InP single crystal which had undergone twinning in a region wherein $(1\bar{1}1)$ edge facets intersected the external shoulder of the crystal (i.e. were anchored at the TPB during crystal growth). Examination of the right-hand side of this image illustrates that twinning nucleates in a region containing anchored edge facets, when the shoulder angle reaches 74.21°, which is the critical value enabling the creation of a $(1\bar{1}5)$ external shoulder facet. Upon twinning, this $(1\bar{1}5)$ facet is replaced by a $(\bar{1}11)$ one. In our experiments, the actual external shoulder facets present before and after twinning were identified using synchrotron white-beam back-reflection spot patterns [6,7].

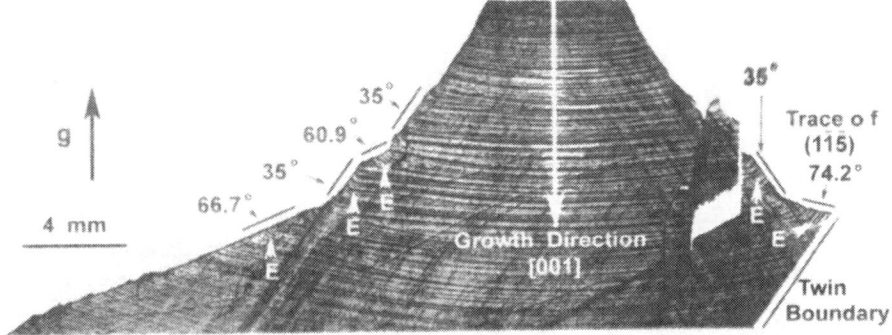

Figure 1. SWBXT image recorded in transmission Laue geometry (g=004, λ=0.45 Å) showing the shoulder region of crystal. E indicates anchored edge facets. Local shoulder angles are indicated.

On the left-hand side of this image, anchored edge facets can be observed in several regions but never when the shoulder angle is close enough to 74.21° to enable the creation of the $(\bar{1}1\bar{5})$ external shoulder facet required for the nucleation of a twin. Consequently no twinning is observed in this region, emphasizing the importance of shoulder geometry. In the same boule, twinning was also observed to occur on the $(\bar{1}\bar{1}\bar{1})$ planes, leading to the conversion of a $(\bar{1}\bar{1}4)$ external shoulder facet to a $(\bar{1}\bar{1}0)$ one. In this case the twin, although nucleated in the shoulder region, grows out of the crystal. The local shoulder angle, in this case, must become equal to 70.53°, i.e. the shoulder must be parallel to $(\bar{1}\bar{1}4)$. Again the presence of the various shoulder facets, in the actual crystal, was confirmed using synchrotron white-beam back-reflection spot patterns.

The index system used to this point does not explicitly indicate the crystallographic polarity. In order to determine the polarity of the edge facets and subsequent twin plane for the twin shown in figure 1, anomalous scattering was utilized. Reproducible plots of diffracted intensity as a function of Bragg angle corresponding to the $(1\bar{1}1)$ and $(\bar{1}1\bar{1})$ reflections as well as a plot of the ratio of the intensities of these reflections are presented in figure 2.

It can be seen that for both reflections, the diffracted intensity changes abruptly at the absorption edge due to the drastic variation of the absorption coefficient. More importantly, the intensity values of the two reflections are significantly different in the vicinity of the absorption

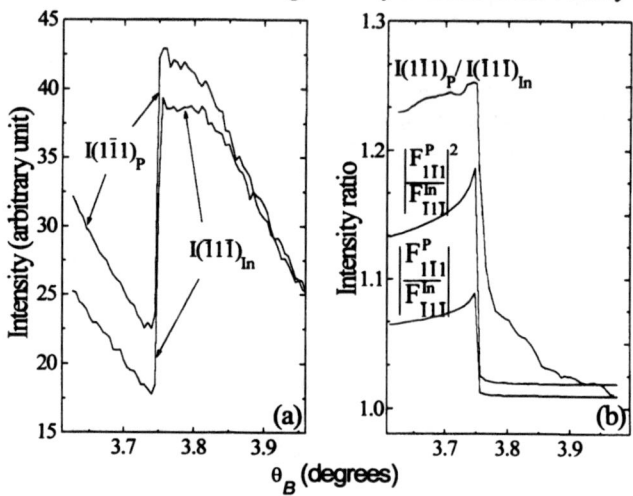

Figure 2 (a) Observed intensity versus Bragg angle plots for the two reflections, and (b). Plots of observed ratios of diffracted intensity as a function of Bragg angle from the $(1\bar{1}1)_p$ and $(\bar{1}1\bar{1})_{In}$ planes, and calculated ratios of diffracted intensity according to kinematical theory, i.e., $|F_{1\bar{1}1}^P / F_{\bar{1}1\bar{1}}^{In}|^2$ and dynamical theory, i.e., $|F_{1\bar{1}1}^P / F_{\bar{1}1\bar{1}}^{In}|$.

edge. This difference is due to the breakdown of Friedel's law (i.e., $|F_{1\bar{1}1}|^2 \neq |F_{\bar{1}1\bar{1}}|^2$) as the scattering factor of the In atoms becomes complex near the absorption edge (i.e. $f_{In} = f_{In}^0 + f_{In}' + if_{In}''$, where f_{In}^0 is the normal scattering factor and $f_{In}' + if_{In}''$ is the wavelength dependent anomalous-dispersion correction). Based on a computer program compiled by Brennan and Cowan [9], we have calculated the wavelength dependent structure factors, $F_{1\bar{1}1}^P$ and F_{111}^{In}, where $[1\bar{1}1]$ is defined as the direction pointing from the In face to the P face, along the direction of the P-In chemical bond. The calculated ratios $|F_{1\bar{1}1}^P / F_{\bar{1}1\bar{1}}^{In}|$ and $|F_{1\bar{1}1}^P / F_{\bar{1}1\bar{1}}^{In}|^2$ are also presented in figure 2. These two curves represent the theoretical diffracted intensity ratios according to the kinematical and dynamical X-ray diffraction models, respectively. It should be noted that for a single crystal containing defects, the diffraction generally contains both dynamical and kinematical contributions. It can clearly be seen that the profile of the ratios of the measured intensities is similar in shape to the theoretical predictions. Therefore, synchrotron X-ray anomalous scattering shows unambiguously that $(1\bar{1}1)$ is a P face while $(\bar{1}1\bar{1})$ is an In face. The discrepancies between the theoretical and experimental anomalous scattering data mainly arise due to slight inaccuracies in the theoretical models which are currently being addressed but which are beyond the scope of this article. More detailed demonstrations and descriptions of the anomalous scattering behavior of InP will be presented in a separate paper [10]. The polarity results obtained using anomalous scattering also confirm those obtained using etching [11].

With the polarity of $(1\bar{1}1)$ and $(\bar{1}1\bar{1})$ being unambiguously defined, we can now define the polarity in this crystal according to the (universal) convention adopted by Brown et al [12] for CdTe, i.e. planes belonging to the {111} form are In faces while those belonging to the $\{\bar{1}\bar{1}\bar{1}\}$ form (which, for the zincblende structure is not equivalent to {111}) are P faces. Therefore, in this crystal, twinning is observed to nucleate on $\{\bar{1}\bar{1}\bar{1}\}_P$ facets.

As briefly mentioned earlier, polarity effects on facet formation and twinning have been reported for various III-V crystals. For the case of InSb, Hulme and Mullin [2] used autoradiography of radio-Te-doped crystals to reveal the production of edge facets during the growth of crystals in the $[111]_{In}$ direction and also the non-appearance of edge facets during growth in the opposite $[\bar{1}\bar{1}\bar{1}]_{Sb}$ direction (according to the indexing convention adopted here), clearly indicating a preference on the part of the crystal to the production of $\{\bar{1}\bar{1}\bar{1}\}_{Sb}$ edge facets. This was used to explain the high incidence of twinning encountered when growth of $[111]_{In}$ is attempted and the reduced incidence of twinning encountered in $[\bar{1}\bar{1}\bar{1}]_{Sb}$ crystals. These results are analogous to those of Bonner in InP [3], who reported that the normally preferred growth direction for <111> InP growth is that for which a $\{\bar{1}\bar{1}\bar{1}\}_P$ face is in contact with the melt. In this orientation, twin nucleation on edge facets is avoided. However, when the shoulder angle exceeds ~19.68°, patch twins are nucleated on the three symmetrically related external shoulder facets on planes belonging to the $\{\bar{1}\bar{1}\bar{1}\}_P$ form indicating a preference for both faceting and the nucleation of twins (patch type in this case) on $\{\bar{1}\bar{1}\bar{1}\}_P$ faces. Similar results were reported by Steinemann and Zimmerli [8] for GaAs which exhibited a preference for twinning on $\{\bar{1}\bar{1}\bar{1}\}_{As}$ planes. Hurle [5], argued that the marked effect of <111> InSb seed polarity on twinning incidence could be accounted for if the surface energy of the In-terminating facet were of the order of 10% higher than that of the Sb-terminating one. He cited the results of the model calculations of Oshcherin [13] who obtained, for InSb, a value ~7% higher. It is interesting to note that, for the case of GaAs, Oscherin reports a value for the surface energy of the Ga-terminating facet which is ~12% higher than that of the As-terminating one. This may again account for the preference for twinning on $\{\bar{1}\bar{1}\bar{1}\}_{As}$ planes reported by Steinemann and Zimmerli. Hence, it appears that in the general case, the occurrence of twinning is strongly dependent on the polarity of the zincblende structure due to differences in the surface energy of the two polar faces. This is also borne out by surface energy values reported by Oscherin for InP. Calculated values of the surface energy of $\{111\}_{In}$ faces were reported to be more than 30% larger than those for $\{\bar{1}\bar{1}\bar{1}\}_P$ faces. Interestingly, he also reports calculated values for {110} faces which are slightly less than 1% lower than those for $\{111\}_{In}$.

CONCLUSIONS

1. The polarity of {111} edge facets, anchored to the TPB, which give rise to growth-twin nucleation in the shoulder region of a [001], S-doped, MLEC grown, InP boule, has been studied by synchrotron X-ray anomalous scattering. Analysis of the results indicates that both the formation of edge facets and the nucleation of twins occur preferentially on $\{\bar{1}\bar{1}\bar{1}\}_P$ faces.

2. Of the four sets of edge facets, belonging to the $\{\bar{1}\,\bar{1}\,\bar{1}\}_p$ form, which are oriented so as to be thermodynamically favored to be anchored to the TPB, two of them can give rise to the conversion of a $\{115\}$ external shoulder facet to a $\{\bar{1}\,\bar{1}\,\bar{1}\}_p$ one upon twinning, while the other two can give rise to the conversion of a $\{114\}$ external shoulder facet to a $\{110\}$ one. In both cases, twinning is only observed when the $\{\bar{1}\,\bar{1}\,\bar{1}\}_p$ edge facets are anchored to the TPB in a region where the shoulder angle is close to $74.21°$ or $70.53°$, facilitating the production of the $\{115\}$ and $\{114\}$ external shoulder facets, respectively, prior to twinning.
3. The preference for faceting and twinning on $\{\bar{1}\,\bar{1}\,\bar{1}\}_p$ faces appears to be accounted for by surface energy values calculated by Oscherin [13]. Calculated values indicate that $\{\bar{1}\,\bar{1}\,\bar{1}\}_p$ surface energies are 30% lower than those of $\{111\}_{In}$, while $\{110\}$ surface energies are also lower than $\{111\}_{In}$ by 1%.

ACKNOWLEDGMENTS

Research supported by the DARPA/AFOSR Consortium for Crystal Growth Research under contract No. F496209510407, and in part by NASA under contracts NAS8-38147, NAG8-891, NCC8-53 and NCC8-48, and the Universities Space Research Association under contracts 3536-03 and 3537-04. Topography was carried out at the Stony Brook Synchrotron Topography Facility, beamline X-19C, at the NSLS, at Brookhaven National Laboratory, which is supported by the U.S. Department of Energy, under contract No. DE-AC02-76CH00016.

REFERENCES

[1] E. Billig, J. Inst. Metals, **83**, 53, (1954/5).

[2] K. F. Hulme and J. B. Mullin, Solid State Electron., **5**, 211, (1962).

[3] W. A. Bonner, J. Cryst. Growth, **54**, 21, (1981).

[4] D. T. J. Hurle, in: *Sir Charles Frank, OBE, FRS: An 80th Birthday Tribute*, Eds. R. G. Chambers, J. E. Enderby, A. Keller, A. R. Lang and J. W. Steeds, (Adam Hilger, 1991), p. 188

[5] D. T. J. Hurle, J. Cryst. Growth, **147**, 239, (1995).

[6] H. Chung, M. Dudley, D. Larson, D.T.J. Hurle, D. Bliss, and V. Prasad, J. Cryst. Growth, **187**, 9, (1998).

[7] H. Chung, Ph.D. Thesis, Stony Brook, (1997).

[8] A. Steinemann and U. Zimmerli, Solid State Electron., **6**, 597, (1963).

[9] S. Brennan and P.L. Cowan, Rev. Sci. Instrum., **63**, 850 (1992). The latest version of the program is available on the Internet: ftp://ftp.minerals.csiro.au/pub/xtallography /absorb.

[10]X.R. Huang, M. Dudley, B. Raghothamachar, in preparation.

[11]M. Dudley, B Raghothamachar, Y. Guo, X.R. Huang, H. Chung, M. Dudley, D.T.J. Hurle, and D. Bliss, J. Cryst. Growth, (1998) (In Press).

[12]P.D. Brown, K. Durose, G.J. Russell and J. Woods, J. Cryst. Growth, **101**, 211, (1990).

[13]B.N. Oshcherin, Phys. Status Solidi (a) **34**, K181, (1976).

CONTRAST MECHANISM IN SUPERSCREW DISLOCATION IMAGES ON SYNCHROTRON BACK-REFLECTION TOPOGRAPHS

X.R. Huang*, M. Dudley*, W.M. Vetter*, W. Huang*, S. Wang*a, C.H. Carter, Jr.**
*Dept. of Materials Science & Engineering, SUNY at Stony Brook, NY 11794-2275
**Cree Research, Inc., 4600 Silicon Drive, Durham, NC 27703

ABSTRACT

The topographic contrast of superscrew dislocations in 6H-SiC crystals has been studied by synchrotron white-beam x-ray topography in the Bragg reflection geometry. The diffraction images of these dislocations are simulated using a ray-tracing method. Systematical simulations, which coincide with the dislocation images taken by back- and grazing-reflection topography, clearly reveal the kinematic diffraction mechanisms of the superscrew dislocation, and illustrate that synchrotron reflection topography is capable of providing accurate descriptions of the strain fields, the Burgers vector magnitudes, and the senses of these dislocations. In addition, our experiments and simulations demonstrate straightforwardly the relation between the topographic contrast and the lattice distortions, and therefore the general mechanisms underlying contrast formation of defect images in synchrotron reflection topographs are provided.

INTRODUCTION

Synchrotron white-beam x-ray topography (SWBXT) is a powerful and efficient diffraction imaging technique for revealing distribution and structure of crystallographic defects in single-crystal materials [1]. The most commonly used geometry of this technique is the Laue transmission geometry, which is analogous to conventional Lang topography, and has been well developed both experimentally and theoretically [2]. For the Bragg reflection geometry, although pseudo-plane-wave reflection topography is frequently applied to study heteroepitaxial systems [3,4], back-reflection SWBXT has been seldom used, and consequently the contrast formation mechanisms are not well understood.

In recent years, it has been shown that, in some cases, back-reflection SWBXT is an especially useful tool complementary to transmission topography. An example is that it can be conveniently applied to image superscrew dislocations (*i.e.* micropipes) running parallel to [0001] in SiC bulk crystals and heterostructures [5-7]. For instance, Figure 1(a) shows a synchrotron back-reflection topograph taken from a (0001) 6H-SiC wafer with the diffraction geometry plotted in Fig. 1(b). In such a topograph, each superscrew dislocation, appearing as a black ring encircling a white center, can be clearly distinguished and analyzed.

Based on the diffraction geometry, however, it is not obvious why a screw dislocation can produce a large black ring with the diameter in the order of a few tens of microns. In order to understand the contrast mechanism, one has to resort to contrast simulation. Back-reflection topographs may be simulated, in principle, in a way similar to the simulation of transmission topographs on the basis of Takagi-Taupin equations [8], with concomitant complexity in the computing process. In view of the large lattice deformation associated with the superscrew dislocations as well as the low x-ray absorption of SiC crystals, however, an alternative technique based on kinematic diffraction principles is applied, in this paper, to simulate the direct images of the super dislocations. For this simplified treatment, the traces of x-rays diffracted from any point of the deformed lattice may be calculated precisely, and then the topographic contrast is regarded as being produced by the overlap and separation of the diffracted rays on the recording plate. As

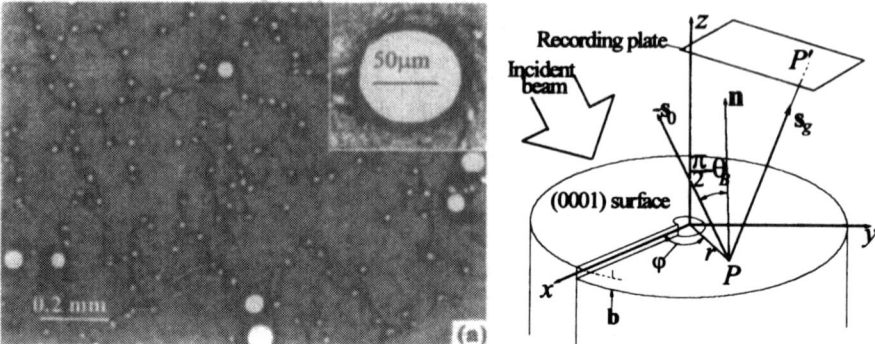

Fig. 1. (a) Synchrotron back-reflection topograph of superscrew dislocations in 6H-SiC (g = 00024 and $\theta_B = 81.4°$). Inset is an enlarged image of a micropipe with an 8c Burgers vector. (b) Corresponding diffraction geometry .

will be demonstrated in the following, such a diffraction model can give a good agreement between the recorded and simulated images of superscrew dislocations in synchrotron back- and grazing-reflection topographs.

PROCESS OF CONTRAST SIMULATION

The kinematic diffraction geometry of back-reflection SWBXT, as plotted in Fig. 1(b), can be simply described by

$$s_g = s_0 + 2\sin\theta_B \frac{g}{|g|}, \qquad (1)$$

where s_g and s_0 are unit vectors in the incident and diffracted beam directions, respectively, g is the diffraction vector, and θ_B [$= -\cos^{-1}(-s_0, g) + \pi/2$] is the Bragg angle. In simulating the kinematic defect contrast, it is usually precise enough to neglect the slight divergence of the incident synchrotron radiation and to consider s_0 fixed. For the distorted lattice around a defect, g varies with space coordinates by

$$g = g_0 - \text{grad}(g_0 \cdot u), \qquad (2)$$

where g_0 is the principle diffraction vector corresponding to the perfect matrix [2]. Based on the approximation of isotropic elastic media, the displacement of the helically deformed lattice around the screw dislocation is expressed in the polar coordinate system as

$$u = (u_r, u_\varphi, u_z) = \left(0, -\frac{br}{2\pi(\sqrt{r^2 + z^2} - z)}, \frac{b\varphi}{2\pi} \right), \qquad (3)$$

where b is the magnitude of the Burgers vector b and u_φ is the displacement correction due to the surface effect [9]. Eqs. (1), (2), and (3) then enable us to calculate the trace of the diffracted beam emitted from an arbitrary point P of the deformed lattice.

From our experiments it is found that the intensity recorded by the photographic plate (Kodak Industrex SR-5 films) mainly comes from the diffraction of a thin layer near the crystal surface. In the simulations, therefore, we consider only the surface layer. In our computing process, both the crystal surface and the recording plate are divided into a set of small squares of constant area, and the diffraction intensity from each square is then projected onto the corresponding square of the recording plate according to the calculated ray path. Eventually, the added intensity values of all the squares on the recording plate are transferred to a plotting program, enabling them to be plotted as a gray scale "topograph" with at most 256 gray levels.

SIMULATIONS OF BACK- AND GRAZING-REFLECTION TOPOGRAPHS

Figure 2(a) is the simulation of a pure screw dislocation computed under the actual diffraction conditions of Fig. 1(a). It is interesting that the simulated image is, indeed, a circular ring of intensified black contrast surrounding a white center where diffraction intensity is completely absent. This contrast feature is in good agreement with the real topographic images in Fig. 1(a), confirming clearly the existence of superscrew dislocations in SiC crystals.

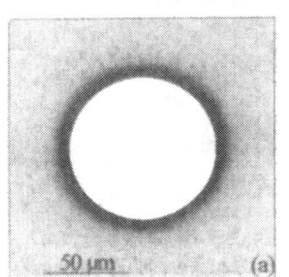
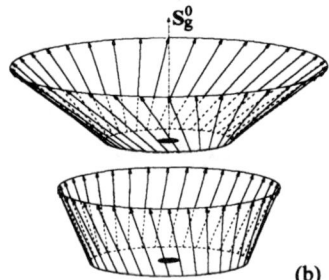

Fig. 2. (a) Simulation of an $8c$ superscrew dislocation contrast. (b) Schematic representation of diffraction cones.

The contrast feature of Fig. 2(a) shows that x rays incident on the dislocation-core region are scattered away from the principle diffraction s_g^0 by the deformed lattice. In fact, the simulation reveals that x rays diffracted from different circles around the dislocation core actually form a set of diffraction cones. Two of these cones are plotted separately in Fig. 2(b), from which it can be seen that the semiapex angles of the cones decrease continuously in the radial direction. Consequently, these cones intersect each other in space, giving rise to a circular contrast feature on the recording film. Our calculations also show that under a fixed diffraction condition, the diameter of the white center is approximately proportional to the magnitude of the Burgers vector. Based on this principle, we can determine that most of the small white spots in Fig. 1(a) correspond to $1c$ superscrew dislocations while the larger spots are images of superscrew dislocations with Burgers vectors ranging from $2c$ to $8c$ ($c = 1.517$ nm is the lattice constant along [0001] of 6H-SiC).

The other interesting feature of Fig. 2(b) is that the diffracted x rays are twisted around s_g^0. This feature may be easily understood from the distorted lattice of the superscrew dislocation in Fig. 1(b): the helical (0001) basal plane make its normal inclined toward the direction of -φ, and then the negative value of n_φ leads to the twist of the diffracted rays. Obviously, the ray twist of a right-handed superscrew dislocation is left-handed, as shown in Fig. 2(b), while the diffracted rays from a left-handed superscrew dislocation are twisted in the right-handed direction. Therefore, it is possible to discern the dislocation senses from the ray-twisting directions. Because the ray-twisting feature can not be observed from the overall topographic images of superscrew dislocations, we use the narrow-beam imaging technique, namely back-reflection section topography, and the corresponding simulation to show the senses of superscrew dislocations in 6H-SiC.

Figure 3(a) shows a simulation of a right-handed superscrew dislocation illuminated by an incident beam restricted to a narrow strip (20 μm in width) in the section topographic geometry. In comparison with Fig. 2(a), only two "tails" of the whole dislocation image remain in the section topograph. The simulation also shows that with respect to the dislocation center, the upper tail is the image of the left strained region illuminated by the beam while the lower tail corresponds to the right region. The left-handed twist of the diffracted rays is clearly indicated by the shapes of the tails. It is apparent that for a left-handed superscrew dislocation, the simulation is a mirror image of Fig. 3(a). Figures 3(b) and 3(c) are real section topographs of left-handed and right-handed superscrew dislocations, respectively. It can be seen that the contrast features are consistent with the simulation results. This verifies the above contrast formation mechanism for synchrotron back-reflection topography of superscrew dislocations. Meanwhile, the section imaging technique is demonstrated to be capable of discerning the senses of superscrew dislocations.

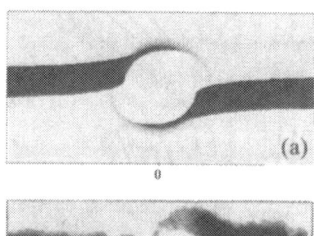

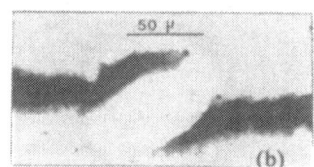

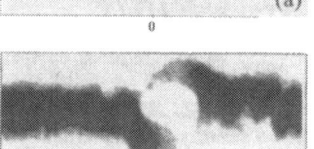

Fig. 3. Section topographs of super screw dislocations. (a) Simulation. (b) and (c) are real topographs of two su perscrew dislocations with opposite senses.

In addition to back-reflection topography described above, another useful diffraction imaging technique for observation of superscrew dislocations is the grazing-reflection topography. This technique is especially suited to investigations of dislocations in SiC epitaxial films and heterostructures. For simplicity, we still use the 6H-SiC bulk crystal to demonstrate the diffraction behavior of superscrew dislocations with the $11\bar{2}6$ grazing reflection [10]. For this reflection, the recording film must be arranged parallel to the (0001) surface to avoid geometric distortion of the projection topograph. Figure 4(a) is a topograph recorded in this geometry. Compared with the circular contrast in the back-reflection topographs, the dislocation images in the grazing-reflection topograph appear as oval-shaped white spots surrounded by complicated black contrast patterns.

The grazing-reflection image of a superscrew dislocation simulated on the basis of Eqs. (1)-(3) is plotted in Fig. 4(b), where most of the scattered x rays overlap at the right edge of the white contrast region and a small part at the left edge. The shape of the simulated images coincides well with that of the recorded ones except that the black contrast of the latter images is intensified by contrast due to the presence of basal plane dislocations. Transmission topography has shown that basal plane dislocations usually originate from superscrew dislocations and form network structures on the (0001) lattice planes [9]. In the $11\bar{2}6$ grazing reflection, these dislocations show strong contrast because of the condition $\mathbf{g} \cdot \mathbf{b} \neq 0$. However, because the strain fields of the basal plane dislocations are much smaller than those of superscrew dislocations, the influence of the basal plane dislocations on the shapes of the white contrast associated with the superscrew dislocations is negligible although they change the intensity distribution outside the white contrast regions.

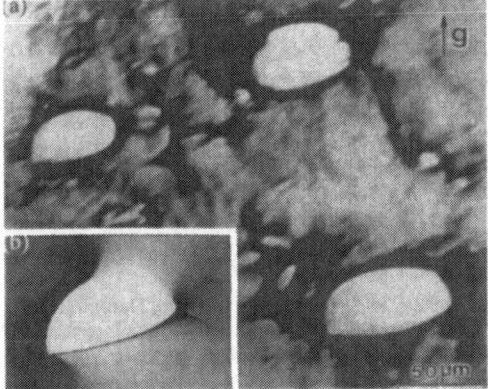

Fig. 4. (a) Grazing-reflection topograph of superscrew dislocations ($g = 11\bar{2}6$, $\lambda = 0.07$ nm). (b) Simulated superscrew dislocation image.

It is worth noting that for the symmetric back reflection, as described earlier, the horizontal lattice displacement u_φ arising from the surface-relaxation effect has no influence on the topographic contrast. The $11\bar{2}6$ asymmetric reflection, however, is sensitive to both u_z and u_r. Consequently, if the simulation is performed without taking into account u_φ (corresponding to the case that the dislocation is in an infinite medium), we can obtain a much narrower dislocation image which is quite different from the actual topographic contrast feature. This difference further proves that the displacement fields of the superscrew dislocation normally outcropping on the free crystal surface are, indeed, represented by Eq. (3).

DISCUSSION AND CONCLUSION

Based on ray-tracing diffraction principles, we have simulated the diffraction images of superscrew dislocations in synchrotron back- and grazing-reflection topographs. The good agreement between the simulations and the recorded topographs indicates that the contrast formation mechanism of superscrew dislocations in 6H-SiC can be described precisely by the kinematic diffraction model. This model is justified because of the low absorption coefficient of SiC crystals and the large Burgers vectors of superscrew dislocations. In particular, the large Burgers vectors mean that the crystal lattice around the superscrew dislocations is significantly distorted. In white-beam diffraction, both the directions and the wavelengths of the diffracted x rays vary quickly from point to point in such deformed lattice. Therefore, the diffraction around superscrew dislocations is similar to that of "mosaic crystals" and has negligible dynamic effects.

Topographic images of crystal defects described by the kinematic diffraction model are generally called the "direct images" [11]. In fact, this kind of images are the most commonly observed contrast features in synchrotron reflection topographs, even for defects with small strain fields. Therefore, it is generally adequate to use Eqs. (1) and (2) to correlate the contrast features with the actual lattice displacement u associated with an arbitrary defect.

In conclusion, back- and grazing-reflection SWBXT techniques have been used to study the diffraction contrast of superscrew dislocations in 6H-SiC crystals. With the help of numerical simulations which are based on the kinematic diffraction principles, these techniques are demonstrated to be capable of quantitatively analyzing the structure of superscrew dislocations, including their strain fields, their senses, and even their surface-relaxation effects. Meanwhile, the simulation process presented in this paper provides a general method for interpretation of defect contrast in synchrotron reflection topographs.

ACKNOWLEDGEMENTS

This work was supported by DARPA/AFWL under Grant No. F33615-95-C-5426 and by the US Army Research Office under Grants No. DAAH04-94-G-0091 and DAAH04-94-G--121. The experimental work was performed at the Stony Brook Synchrotron Topography Facility (Beamline X-19C), National Synchrotron Light Source, Brookhaven National Laboratory. The authors are indebted for Prof. Y. Epelboin for very helpful discussions about the simulation of synchrotron topographs.

REFERENCES

[a]Present address:Advanced Technology Materials, 7 Commerce Drive, Danbury, CT 06810.
[1] M. Sauvage and J. F. Petroff, in: Synchrotron Radiation Research, Eds. H. Winick and S. Doniach, Plenum Press, New York, 1980.
[2] B. K. Tanner, X-ray Diffraction Topography, Pergamon Press, Oxford, 1976.
[3] M. Riglet, M. Sauvage, J. F. Petroff, and Y. Epelboin, Philos. Mag. A **42**, p. 339 (1980).
[4] S. J. Barnett, A. M. Keir, A. G. Gullis, A. D. Johnson, J. Jefferson, G. W. Smith, T. Martin, C. R. Whitehouse, G. Lacey, G. F. Clark, B. K. Tanner, W. Spirkl, B. Lunn, J. C. Hogg, P. Ashu, W. E. Hagston, C. M. Castelli, J. Phys. D: Apply. Phys. **28**, p. A17 (1995).
[5] M. Dudley, S. Wang, W. Huang, C. H. Carter Jr, V. F. Tsvetkov, and C. Fazi, J. Phys. D: Appl. Phys. **28**, p. A63 (1995).
[6] S. Wang, M. Dudley, C. H. Carter Jr, and H. S. Kong, Mat. Res. Soc. Symp. Proc. **339**, p. 735 (1994).
[7] M. Dudley, W. Si, S. Wang, C. H. Carter Jr, R. Glass, V. Tsvetkov, Il Nuovo Cimento **19D**, p. 153 (1997).
[8] M. Epelboin, Materials Science and Engineering **73**, 1 (1985); C. A. M. Carvalho, and Y. Epelboin, Acta Crystallogr. A **49**, p. 467 (1993).
[9] J. D. Eshelby and A. N. Stroth, Philos. Mag. **42**, p. 1401 (1951).
[10] S. Wang, Ph. D. Thesis, State University of New York at Stony Brook, 1995.
[11] H. Klapper, Crystals **13**, p. 109 (1991)

TWIN AND GRAIN BOUNDARY IN InP :A SYNCHROTRON RADIATION STUDY

YUJIE HAN*, JIANHUA JIANG**, ZHOUGUANG WANG**, XUNLANG LIU*,
JINGHUA JIAO*, YULIAN TIAN**, LANYING LIN*
*Institute of Semiconductors, Chinese Academy of Sciences, P.O.Box:912, Beijing, 100083,
CHINA   Email:yjhan@red.semi.ac.cn
** Beijing Synchrotron Radiation Laboratory, Institute of High Energy Physics, Chinese
Academy of Sciences, 100039, Beijing, CHINA

ABSTRACT

Experimentally observed X-ray reflectivity curves show bi-crystal(twin) characteristics.
The study revealed that there was defect segregation at the twin boundary. Stress was
relaxed at the edge of the boundary. Relaxation of the stress resulted in formation of twin
and other defects. As a result of formation of such defects, a defect-free and stress-free zone
or low defect density and small stress zone is created around the defects. So a twin model
was proposed to explain the experimental results. Stress(mainly thermal stress), chemical
stoichiometry deviation and impurities nonhomogeneous distributions are the key factors
that cause twins in LEC InP crystal growth. Twins on (111) face in LEC InP crystal were
studied. Experimental evidence of above mentioned twin model and suggestions on how to
get twin-free LEC InP single crystals will be discussed.

INTRODUCTION

Twins, which often cause serious problems in single crystal growth, are planar defects
commonly observed in as-grown InP. These defects reduce the yield of usable single-
crystalline material grown by different techniques. Among the III-V compounds InP has a
low stacking fault energy, which together with the high ionicity of bonding increases the
probability of twin formation. Several reasons were proposed for twin formation during
crystal growth, including nonstoichiometry, impurities and solid particles at the
solid/liquid interface, the temperature gradient at the solid/liquid interface, the cone angle
of the crystal, the quality of raw materials, B_2O_3 encapsulant, mechanical shock, sticking
effect of crucibles and rapid diverging crystal surfaces and so on [1-5]. Twinning occurs on
{111} facets, usually during shoulder growth. Most of these problems have been solved
practically, as a result of continuous efforts. However, there are remaining questions
concerning formation and atomic structure of twins in InP crystals with the sphalerite
structure.
X-ray topography is a well established technique to characterize growth- or process-
induced defects. With the advent of synchrotron-radiation sources, X-ray topography has
been used widely in material characterization.

EXPERIMENT

Several different as-grown (100)-oriented and (111)-oriented iron-doped sulfur-doped
and nominally undoped InP single crystals grown by LEC were investigated by high
energy synchrotron radiation X-ray white beam topography and by X-ray monochromatic reflected diffraction at Beijing Synchrotron Radiation Facility(BSAF). For

77

chemical etching, samples were cut to plates containing a twin at the center, so that the top and bottom faces of the plates were {111} plates parallel to the twin plate. The top and bottom faces of the plates were lapped then, mechanically polished with Al₂O₃ powders, chemically polished with Br₂-CH₃OH.

RESULT

InP has the zinc blende(cubic ZnS) structure in which In(111) and P(111) planes stacked alternatively following the fcc stacking sequence. LEC crystals were partly polycrystalline,

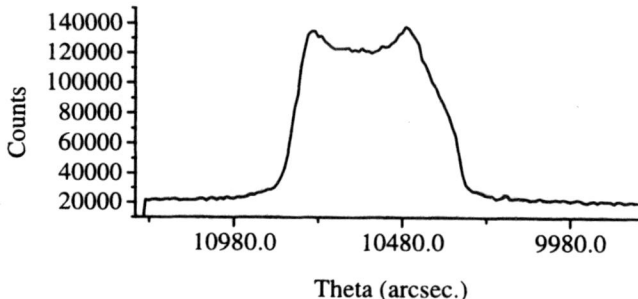

Fig. 1 X-ray diffraction peak recorded in reflection with monochromatic radiation of partly polycrystalline and twinned InP.

and twins or twin lamellae were mostly observed in the polycrystal region. Twinning on {111} facets was investigated by x-ray diffraction topography or by monochromatic x-ray diffractometry. A small orientation difference exists across the twin interfaces, and this is seen in X-ray topographs as well as in rocking curves. The 003 InP diffraction pattern shown in Fig. 1 was recorded with λ=0.106 nm at θ=15.717° and show the presence of twins. The separation between peaks was less than 300 arcseconds, and the large number of counts between the maxima indicate the range of orientations present. Figure 2 and Fig.3 reveal that the main defects in these samples are twin and grain boundaries. Fig.4 and Fig.5 reveal that there was defect segregation at the twin boundary causing twin-free and boundary defect-free region. This twin and boundary depletion area was caused by a stress relaxation that extends a few micrometers away from the boundary. In fact, the formation of twin in InP was strongly influenced by stoichiometry fluctuations during crystal growth, a non-homogenous impurity distribution, a rapidly diverging solid/liquid interface[6,7]. Impurity segregation in the melt can cause compositional non-homogeneities leading to stoichiometry inhomogeneities.

The LEC crystal puller used in InP single crystal growth has many adjustable parameters which greatly affect the formation of twins. These parameters include pull rate, crucible rotation, crystal rotation, position of the melt surface, shape of interface at solid/liquid, pressure and type of ambient gas, heater and magnetic field strength. Previous

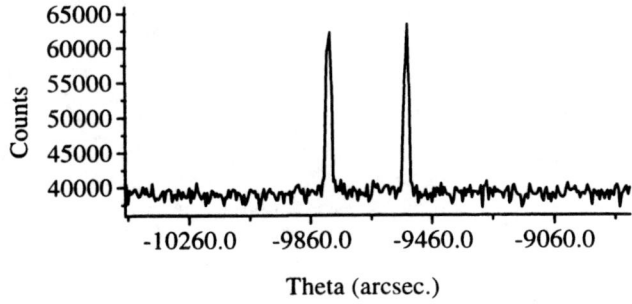

Fig.2 X-ray monochromatic reflected
diffraction in twinned InP.

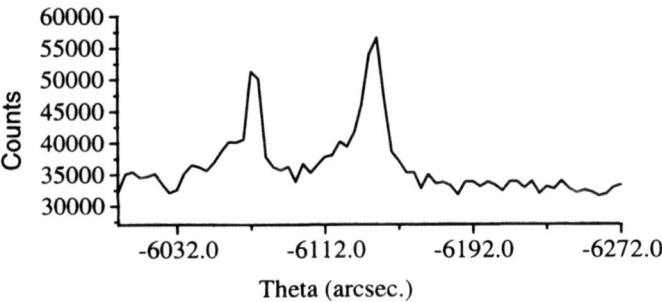

Fig.3 X-ray monochromatic reflected
diffraction in twinned InP.

Fig.4 X-ray reflection topography of twinned InP.

studies of crystallinity breakdown in LEC GaAs have attempted to correlate the probability of single-crystal growth with various growth parameters. Thomas[8] reported the incidence of twinning to be reduced by growing with gradual cone angles, by using dry B_2O_3, and by avoiding rapid diameter changes. These empirical correlations have resulted

Fig.5 X -ray reflection topography of twinned InP.

in improved single crystal growth yield, but have not identified the mechanisms which cause crystal structure breakdown. Good diameter control is critical to avoiding gallium droplet penetration and the resultant grain boundary formation in GaAs[7]. To achieve reproducible growths of twin-free crystals a growth procedure was adopted, which included methods to control the stoichiometry and impurity distributions, as well as to reduce residual stress in the crystal.

CONCLUTION

Twins and grain boundaries were studied by X-ray topography or by X-ray mono-chromatic reflected diffraction. Formation mechanisms of twins in InP were ascribed to stress, non-stoichiometry during crystal growth and non-homogenous impurity distributions.

REFERENCES

1. G. W. Iseler, J. Crystal Growth 54, p.16(1981).
2. W. A. Bonner, Mater. Res. Bull. 15, p.63(1980).
3. W. A. Bonner, J. Crystal Growth 54, p.21(1981).
4. K. J. Bachmann, E. Buehler, J. L. Shay and A. R. Strnad, J. Electron. Mater. 4, p.389(1975).
5. S. Shinoyama, C. Uemura, A. Yamamoto and S. Tohno, J. Electron. Mater. 10, p.941(1981).
6. Masatomo Shibata, Yukio Sasaki, Tomoki Inada and Shoji Kuma, J. Crystal Growth 102, p.557(1990).
7. J. P. Tower, R. Tobin, P. J. Pearah and R. M. Ware, J. Crystal Growth 114, p.557(1990).
8. R. N. Thomas, H. M. Hobgood, G. W. Eldridge, D. L. Barrett, T. T. Braggins, L. B. Ta and S. K. Wang, in: Semiconductors and Semimetals, Vol. 20, Eds. R. K. Willardson and A. C. Beer(Academic Press, New York, 1984), p.1.

REAL TIME *IN SITU* X-RAY TOPOGRAPHIC OBSERVATION OF DEFORMATION OF SINGLE CRYSTALS AND THIN FILMS

Z. B. Zhao*, J. Hershberger*, A. Chiaramonti*, Z. U. Rek** and J. C. Bilello*
*Center for Nanomaterials Science, Department of Materials Science & Engineering, University of Michigan, Ann Arbor, MI 48109-2136.
**Stanford Synchrotron Radiation Laboratory, Stanford University, CA 94309

ABSTRACT

An experimental apparatus, which is capable of performing real time in situ X-ray topographic observation of deformation process via synchrotron white beam topography, has been developed. This device enables both tensile data (load-displacement) and topographic images to be recorded simultaneously. It has been utilized to study the deformation behaviors of crystals of Mo and W. These specimens have been subject to mechanical cycling with increasing load, and their deformation processes have been observed in real time and *in situ* via x-ray topography. This leads to the observation of several phenomena, which would have been difficult to reveal by other experimental techniques. They include stress concentration, microyielding, reversible variation of contrasts and stress relaxation. In addition, the deformation behaviors of small angle grain boundaries have also been examined. Furthermore, the specimens can be heated through a heating device attached to the tensile stage, which allows high temperature topography to be performed in real time. The technique has been applied to the Ta films on Si (100) substrates. With increasing temperature, the topographic observations have revealed that the Ta films yield, fracture and then proceed to delaminate from their substrates.

INTRODUCTION

X-ray diffraction topography is an imaging technique that allows various defects such as dislocations, stacking faults and other types of boundaries to be visualized [1-3]. X-ray topography is not only nondestructive, but also enables large sample areas to be probed [4]. With the well-collimated and high intensity synchrotron radiation, x-ray topography can also be used to perform *in-situ* observations, in real time, of many dynamic processes of materials such as motions of defects, crack initiation and propagation, crystal growth and phase transformations [5]. For example, x-ray topography has been utilized to study the deformation behavior of the silicon crystals deformed by a high temperature deformation stage [6,7], and by a high temperature indentation technique [8]. Earlier studies of deformed metallic crystals using x-ray topography include the imaging of strain filed near the bicrystal boundaries [9], and the observation of dislocation arrangements [10]. In this work, *in-situ* real time synchrotron X-ray topographic observations have been made of the deformation processes of various single crystals and bicrystals using a specially designed system. The unique feature of the apparatus is the simultaneous acquisition of both tensile data and topographic images, which allows us to make correlation between them if there is any.

We also used a high temperature X-ray topography method to have performed real time observations of cracking and delamination of thin films. The magnetron sputtered Ta films deposited on Si single crystals substrates were selected for investigating cracking and delamination phenomena. Earlier work indicated that significant stress buildup occurs in Ta films during thermal processing [11,12]. Thus, Ta films seemed to be a suitable system for high temperature X-ray topographic observations.

EXPERIMENTS

Fig. 1 shows the setup of the apparatus on beamline 2-2 at Stanford Synchrotron Radiation Laboratory in a transmission Laue mode. Basically, the system consists of two subsystems based on their respective functions: the mini-tensile device, which deforms the material and records displacements and the X-ray imaging system, which observes the deformation process. The detailed description of the apparatus has been reported elsewhere [13,14]. The specimens for the

81

Fig.1: The setup on BL 2-2 at SSRL

Fig.2(A): Mechanical cycling of a Mo single crystal.

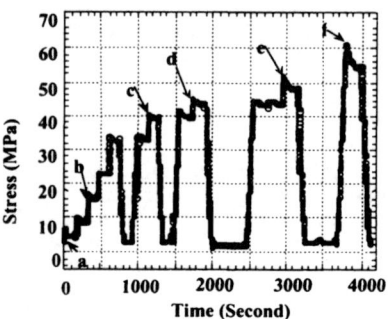

Fig.2(B) and (C): Stress relaxation at different stress levels (B) and variations of topographic images as a function of mechanical loading shown in Fig.2(A).

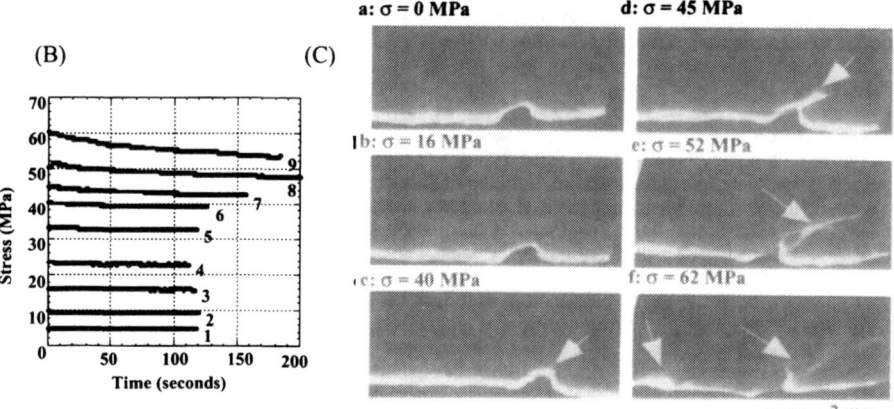

tensile study include the crystals of refractory metals Mo and W. Their final dimensions are roughly 5 mm x 10 mm x 150 μm after a series of preparation steps [13,14].

As a new feature, a specimen-heating device[15] has been added for performing high temperature X-ray topographic experiments. The specimens used here were 400 μm Si(100) wafers coated with 200 nm Ta films. These films were deposited by DC magnetron sputtering in a system with base pressure of 2×10^{-6} Torr. The pre-cut Si wafers, mounted onto a rotating plate (ϕ=12" and RPM=20), were placed 10 cm from the 4" magnetron sputter guns. Prior to deposition, the Ta target was pre-sputtered for 5 minutes to remove the surface oxides and thus reduce the impurities in the Ta films. These films were deposited at Ar pressure 3 mTorr and power 460 W.

During the experiments, the surfaces of specimens were always kept to be normal to the incident beam. A slit rectangular aperture was used only to illuminate the specimen area of interest. A 26μm thick Nb filter was put in front of the specimen to limit the spectral range of the white radiation. A beam stopper and a Al filter were used to prevent the detector from oversaturating, hence optimizing the image quality.

RESULTS AND DISCUSSION

Fig. 2 shows results from a Mo single crystal with a pre-cut notch. As indicated in Fig.2(B), stress relaxation is negligibly small when the nominal stress is below 40 MPa. Correspondingly, the topographic images (a, b and c in Fig.2(C)) exhibit little variation in such a stress range. As the nominal stress further increases, the stress concentration initiates near the notch. The strong strain contrast starts to occur locally when the nominal stress reaches 45 MPa (d in Fig.2(C)). At higher stresses, the region near the notch appears to become more deformed, and deformation is observed to extend to other areas as well. These structural variations are well correlated to the data in Fig.2(B), which indicate the remarkable stress relaxation at higher stresses. Probably more defects, such as dislocations, have been generated at higher stresses, as indicated by the topographs. The motions of these defects allow the stress to be relaxed more rapidly at higher stresses.

Fig.3: The mechanical cycling (A), stress relaxation (B) and topographs (C) of a W single crystal.

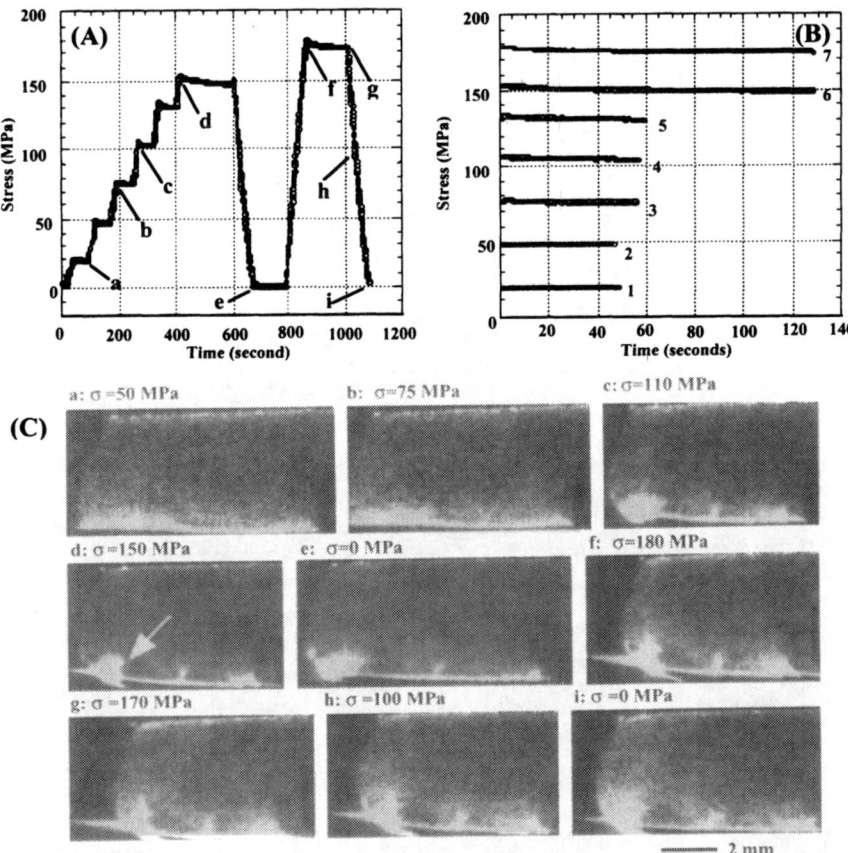

Fig. 3 shows the results of a W single crystal under the mechanical cycling. Although its stress relaxation is not as obvious as Mo, the similar effect has been observed. While the stress relaxation is negligibly small at low stresses, the associated topographs (Fig.3(a) and (b)) display little variation. Upon the increase of stress, the stress concentration appears near the bottom corner of the specimen and microyielding commences at σ~110 MPa. The average microyielding stress (taking other W single crystals into account), is ~15-30% of the macroyielding value. At higher stresses (σ>150 MPa), a sharp spike with bright contrast appears near the area of stress concentration. While this spike-shaped contrast becomes larger in size as stress increases, it disappears almost completely upon the release of load. In other words, the variation of the spike-shaped contrast is nearly reversible. Such a phenomenon is also observed in a W bicrystal, which undergoes the mechanical cycling shown in Fig.4(A). A small angle grain boundary, separating two nearly perfect grains, is clearly resolved in the topographic images presented in Fig.4(B). Under increasing tensile stress, plastic deformation commences in grain 1, and one can observe that the fine slip bands uniformly distribute over the entire sample area. This suggests that dislocation motion proceeded quite homogeneously in the whole volume of this crystal (grain 1). The dislocation slip bands in grain 1 seem to terminate at the small angle grain boundary. The second crystal (grain 2) is not affected by its highly deformed adjacent grain and shows no evidence of deformation. This observation suggests that this particular small angle grain boundary served as a dislocation barrier. For this W bicrystal, one also observes the occurrence of a spike-shaped contrast and its nearly reversible variation, similar to the case of W single crystal. The contrast can be probably interpreted as from the "focusing" effect of the localized region that has a bending curvature. However, whether this observed phenomenon results from simple elastic bending or from other dislocation mechanisms awaits further studies.

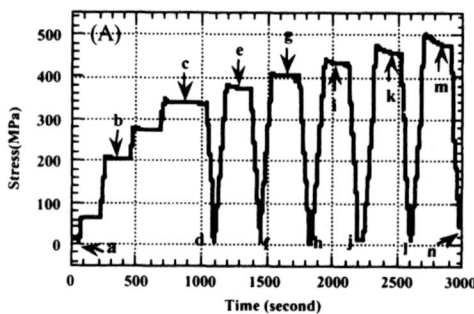

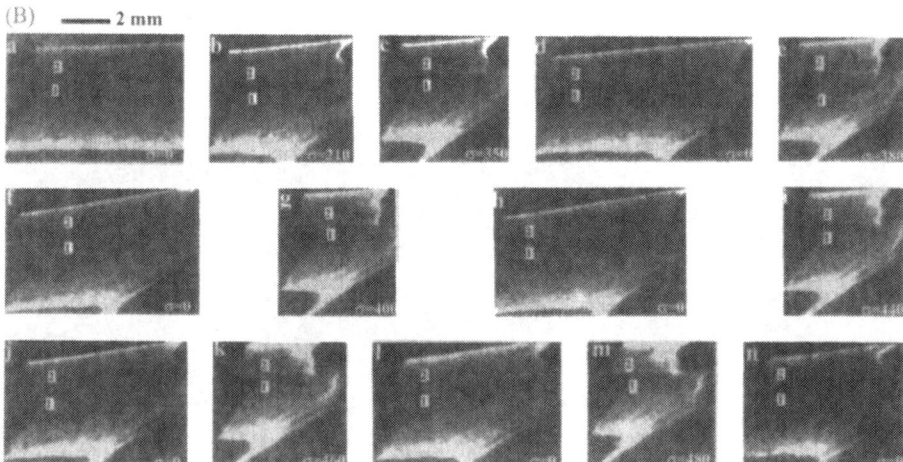

Fig.4: Mechanical cycling of a W bicrystal (A) and the variations of associated topographs (B).

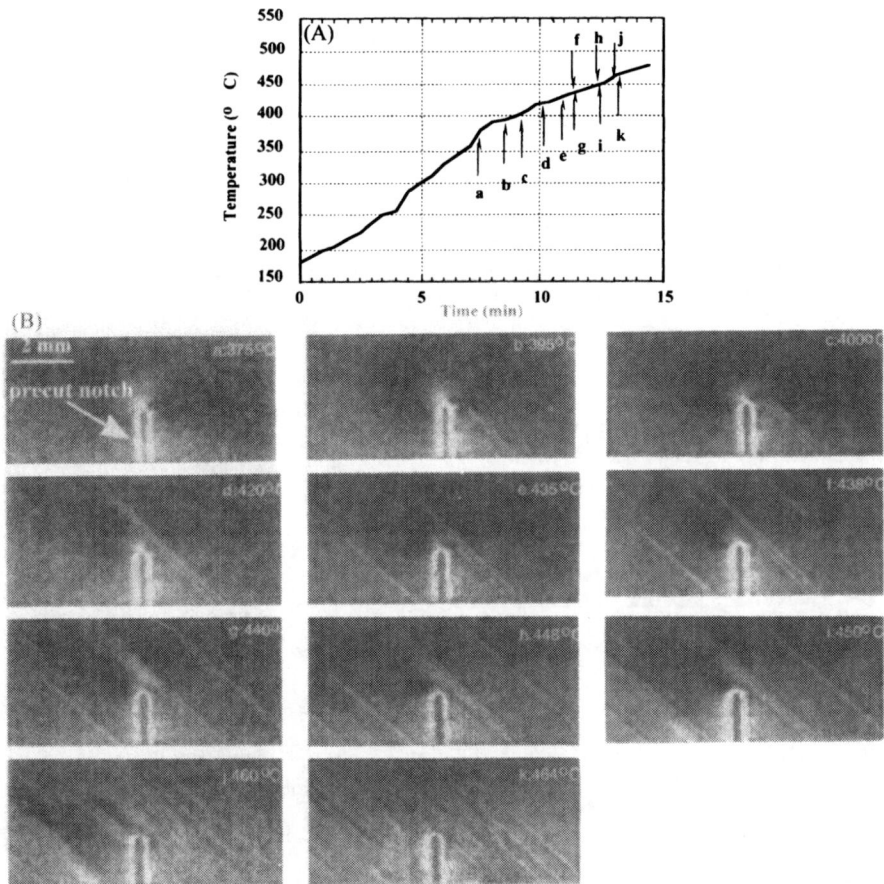

Fig.5: (A) shows the temperature of a Ta/Si specimen as a function of time, and (B) presents a series of representative topographs of with increasing temperature.

Fig.5 shows the results of high temperature topography on a Ta/Si specimen. The real time variations of the topographic images with increasing temperature are captured by a x-ray CCD detector and recorded. A series of representative examples for a Ta/Si specimen are presented in Fig.5. Fig.5 (a) is the topographic image obtained at room temperature. Below 375 °C, the image shows little variation as the specimen is heated. At about 375 °C, a couple of straight lines, which seem to initiate near the edges of the specimen, start to appear. These straight lines, forming a 45° angle with specimen edges, increase their lengths rapidly by propagating towards the interior of the specimen and then channel through the entire image width (see Fig.5 (b) and (c)). With a further increase in temperature, more lines with similar appearance form and propagate with increasing frequency. Starting from 450°C (from Fig.5(i)-5(k)), the formation of these striations, becomes expansive. The features revealed by the recorded x-ray topographic images can be summarized in the following: (a).Nearly straight striations form at 45° angles with the edges of the specimens, (b). Most of these lines initiate near the edges of specimen and then propagate towards the interior, (c). These striations form a very regular pattern. These features strongly suggest that the line contrasts arise from shear bands in the film. The driving force for shearing must be related to the residual growth stress and thermal stress in the film. Since Ta films have a larger coefficient of thermal expansion than Si, a component of the compressive thermal stress could build up with increasing temperature, and then superimpose with the internal growth stress (also compressive under the given deposition conditions). The thermal stress increases (nearly) linearly with temperature. Here we propose the following mechanism for the observed phenomenon. In the interior of the specimen, the stress in the Ta film is biaxially balanced, thus there is no component of shear stress. Near the specimen edges, however, the stress in the Ta film is nearly uniaxial. This gives rise to shear stresses; its maximal component should be in the direction that forms 45 ^{o} with the specimen edges. Thus when the stress reaches a critical level, shear of the Ta film initiates near the edge of the specimen and the shear bands propagate towards its interior. The film cracks along these shear bands and then proceeds to delaminate from the substrate.

ACKNOWLEDGMENT

This work is sponsored by DURIP-ARO-DAAH04-96-1-0260. The experiments were performed on beamline 2-2 at SSRL. The authors also wish to thank Mr. Mark Wallace for his assistance in designing the heating device.

REFERENCES

1. J. C. Bilello, H. A. Schmitz and D. Dew-Hughes, J. Appl. Phys., 65, P.2282 (1989).
2. D.K. Tanner, X-ray Diffraction Topography, (Pergamon, Oxford, 1976).
3. H. A. Schmitz, J. C. Bilello and Z. U. Rek, Materials Science and Engr., 81, P.283 (1986)
4. J. C. Bilello, Mat. Res. Soc. Symp. Proc., 82, P.197 (1987).
5. B. Lengeler, Mikrochim. Acta [Wien], 1, P.455 (1987).
6. A. George and G. Michot, J. Appl. Cryst., 15, P.412(1982).
7. G. Michot, M. L. L. de Oliveira and A. Goerge, Materials Science and Engr., A176, P.99(1994).
8. M. Yoshioka and K. Kawamura, Jpn. J. Appl. Phys., 34, p. L1553 (1995).
9. M. Polcarova and J. Bradler, J. Appl. Cryst., 21, P. 169(1988).
10. W. Wasserbach, Phys. Stat. Sol., (a), 151, P.61(1995).
11. C. Cabral, Jr; L. A. Clevenger-LA and R. G. Schad, J. Vac. Sci. & Technol. B, 12, P.2818 (1994).
12. K. Kondo, M. Nakabayashi, K. Kawakami and T. Chijimatsu, J. Vac. Sci. Technol. A, 11, P.3067(1993).
13. Z. B. Zhao, J. Hershberger, Z. U. Rek and J. C. Bilello, SSRL 1996 Activity report, A 266, 1996.
14. Z. B. Zhao, J. Hershberger, Z. U. Rek and J. C. Bilello, Mat. Res. Soc. Symp: In Situ Diagnostics and Materials Processing, 1997, in press
15. Z. B. Zhao, J. Hershberger, A. Chiaramonti, Z. U. Rek and J. C. Bilello, SSRL 1997 Activity report, in press.

Part II

Diffraction and Scattering

STRAIN AND SHAPE IN SELF-ASSEMBLED QUANTUM DOTS STUDIED BY X-RAY GRAZING INCIDENCE DIFFRACTION

I. KEGEL*, T.H. METZGER*, J. PEISL*, P. FRATZL**, A. LORKE*, J.P. KOTTHAUS*, J. M. GARCIA***, P. M. PETROFF***

* Sektion Physik, Ludwig-Maximilians-Universität München, D-80539 München, Germany
** Erich Schmid Institute of Solid State Physics, Austrian Academy of Sciences, University of Leoben, A-8700 Leoben, Austria
*** Materials Department, University of California, Santa Barbara, CA 93106, USA

ABSTRACT

We have developed a method to determine the relationship between strain and lateral size of coherent self-organized quantum dots. In our approach, X-ray grazing incidence diffraction is used to collect information on strain and shape effects in the vicinity of a prominent surface reflection. We demonstrate that for highly strained nano-scale islands it is possible to separate strain-induced and form factor-induced scattering without comparing different reflections. Experimental data from InAs on GaAs(100) quantum dots is discussed with respect to this model. Reciprocal space mapping around the (220) surface reflection shows a linear relationship between relaxation from the substrate lattice parameter and the outer perimeter of the dot. In addition, the functional form of the gradient of relaxation is found to be non-monotonous and rapidly increasing towards the tip of the dot.

INTRODUCTION

Dislocation-free nano-crystallites formed in heteroepitaxy of highly mismatched material systems have received widespread attention as potential next-generation semiconductor devices [1]. The experimental determination of the spatial configuration of these three-dimensional islands grown in the coherent Stranski-Krastanow-mode is a key element for the understanding of the growth process. Improvements of the zero-dimensional electronic properties of such „quantum dots" are expected to be triggered by advances in the mastering of the complicated self-organized growth. The recent controversy about the correct assessment of thermodynamic and kinetic effects [2] shows the necessity for experimental techniques to determine structural properties of quantum dots.

X-ray grazing incidence diffraction (GID) has been shown to be ideally suited for the structural characterization of thin layers [3]. Buried interfaces and strain distributions in heteroepitaxial layers have been successfully investigated using this technique. Depth and surface sensitivity of GID allows for the detection of near-surface structures and defects which cannot be resolved in a large angle setup due to the dominant presence of substrate scattering. This advantage particularly applies to quantum dots. Furthermore, GID-experiments yield the change in *in-plane* lattice parameter which quantifies the relaxation of the dot material with respect to the substrate lattice parameter.

In this paper we propose a novel interpretation of scattering data which directly extracts from the measured intensities the relationship between the height-dependent *lateral dimension* and the *relaxation* within axially symmetric dots. As an interesting quantity we obtain the vertical gradient of the relaxation describing the structure of the dots in growth direction.

EXPERIMENT

InAs on GaAs(100) quantum dots have been grown at 530° C substrate temperature and As partial pressure of 1×10^{-5} Torr. An AFM-analysis yields dot heights of 100 ± 10 Å, dot diameters of 340 ± 60 Å and a Poisson-like lateral distribution with a mean distance of 500 Å.

The X-ray measurements were performed at the BW2 beamline at HASYLAB/DESY with a wavelength of 2.07 Å. A schematic view of the two-crystal grazing incidence setup is given in Fig. 1. To assure zero vertical momentum transfer for comparison with simulated intensities, a position sensitive detector was used to select exit angles up to the critical angle of total external reflection. Scans are labeled according to the axes shown in Fig. 2.

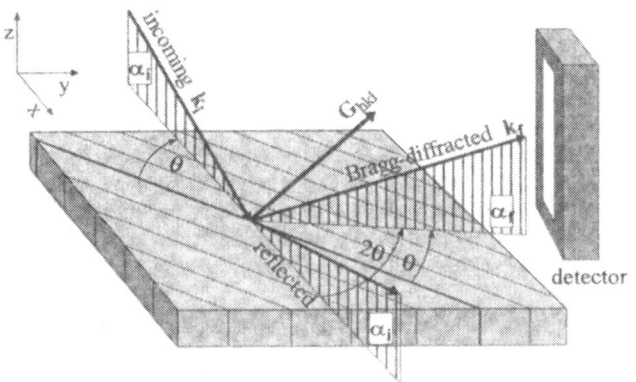

Fig. 1: Setup for Grazing Incidence diffraction

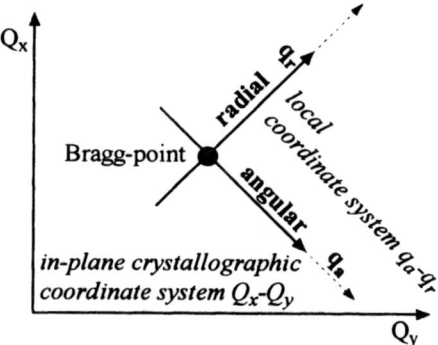

Fig. 2: Axes and scan directions in reciprocal space

MODEL DESCRIPTION

To arrive at an analytical expression for the scattering intensities from quantum dots, several simplifying approximations have to be made. The full-featured strain distribution in a single dot is reduced to the dependence of the lattice parameter a on the height z above the substrate. For axially symmetric dots, the outer shape can be described by a height-dependent radius $R(z)$ (Fig. 3). The structure of the dot is thus expressed by two scalar functions of height, which are assumed to be monotonous. This implies that there is a one-to-one relationship between radius R and reciprocal lattice parameter $g = 2\pi / a$.

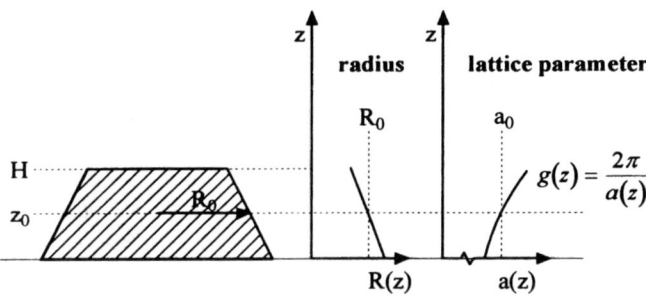

Fig. 3: Structure of the dot described by two scalar functions of height above the substrate

Hence, the structure factor can be written as a coherent integration along z over form factors of disks with the corresponding radii centered at the appropriate lattice parameters given by $g(z)$.

$$F(q_a, q_r) = \int_0^H \pi R^2(z) \frac{2 J_1\left(R(z) \sqrt{[q_r - g(z)]^2 + q_a^2}\right)}{R(z) \sqrt{[q_r - g(z)]^2 + q_a^2}} dz \tag{1}$$

The further evaluation of the integral requires two approximations regarding the change of lattice parameter with height and the mutual dependence of radius and lattice parameter. Within this model, for a correct description of the angular dependence of the structure factor at a radial position $q_r^{(0)} = g(z_0)$ (see Fig. 3) the conditions

$$\frac{g''(z_0)}{[g'(z_0)]^2} \frac{2\pi}{R_0} \ll 1 \quad \text{and} \quad \left.\frac{dR}{dg}\right|_{g=q_r^{(0)}} \frac{2\pi}{R_0^2} \ll 1 \tag{2}$$

have to be fulfilled (primes indicate differentiation with respect to z).
The final result of the calculation

$$I(q_a, q_r) = \underbrace{I_0 \left(\frac{4\pi R(q_r)}{g'(q_r)}\right)^2}_{I_{max}(q_r)} \left(\frac{\sin R(q_r) q_a}{R(q_r) q_a}\right)^2 \tag{3}$$

predicts that the functional form of the angular dependence of the scattering intensity is only determined by the radius at the corresponding relaxation, whereas the radial scan $I_{max}(q_r)$ at $q_a = 0$ is determined by the gradient of the reciprocal lattice parameter, which is equivalent to the gradient of relaxation from the substrate lattice parameter. Both radius and gradient of relaxation are determined as functions of q_r, i.e. of relaxation itself.

RESULTS AND DISCUSSION

Since InAs has a bigger lattice parameter than GaAs (mismatch 7%), the relaxed parts of the dots scatter at negative q_r. The reciprocal space mapping (Fig. 4) shows a monotonous decay of intensity from the position at the GaAs peak to the position of fully relaxed InAs. In the angular direction an oscillatory behavior is observed whose period is getting larger for greater relaxations. Since the more relaxed parts of the dots are supposed to be found at the tip of the dot, this means that its lateral dimension of is decreasing with height above the substrate, as expected for any type of coherent island.

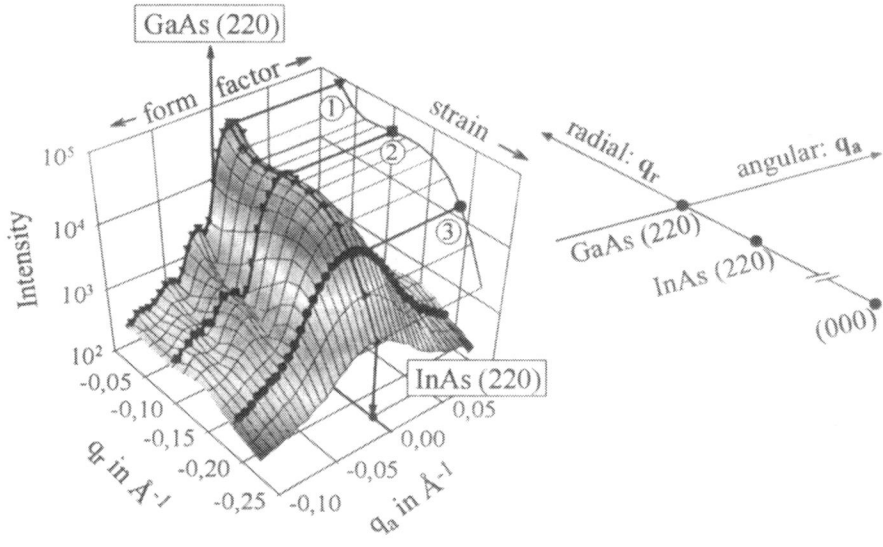

Fig. 4: GID measurements between the 220 reflections of GaAs and InAs. For reasons of clarity, the very intense GaAs peak as been omitted

As a typical fit of the scattering intensity using Eq. 3, the highlighted angular trace No. 1 in Fig. 4 is shown in Fig. 5 together with the fit for a single disk. The central part is described well by both approaches, whereas the side wings are not compatible with disk-scattering. The remaining discrepancies between our model and the measured intensities are attributed to higher order effects neglected in our derivation. A log normal size distribution for the lateral dot dimensions was included in the fitting routine which yielded an optimal FWHM of about 25 % for all relaxations. The results for radius and gradient of relaxation are shown in Fig. 6.

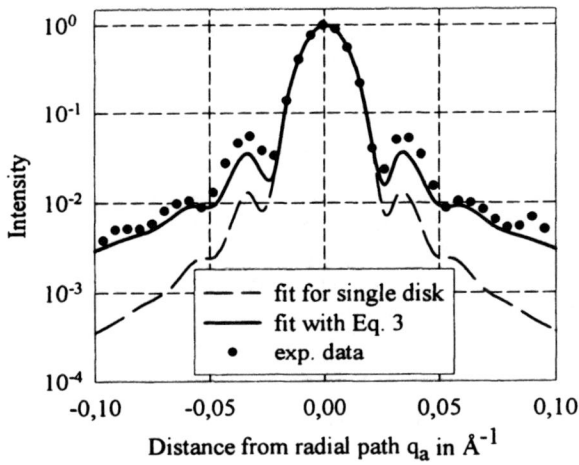

Fig. 5: Typical fit of angular scattering distribution

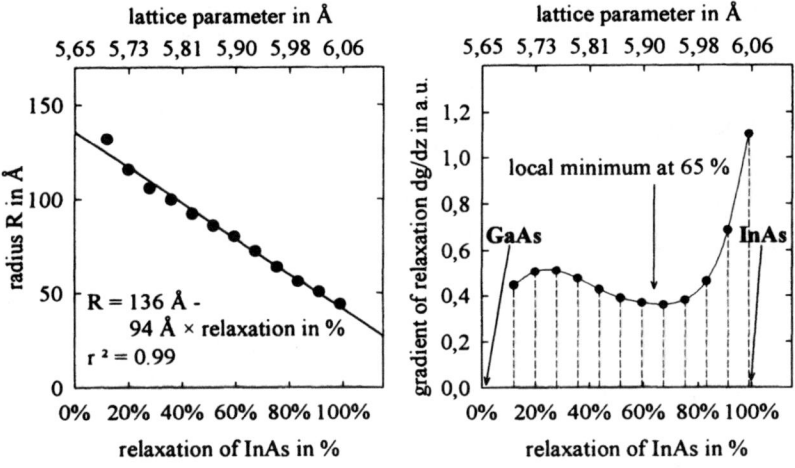

Fig. 6: Results for radius and gradient of relaxation within the dots

The relationship between radius and relaxation is found to be linear with the only significant deviation at the lowest relaxation, corresponding to the region where the dot connects to the substrate. Here the strain distribution is known to be more complicated with non-diagonal components of the strain tensor playing a major role. The gradient of relaxation is determined in arbitrary units. Its functional form is non-monotonous with a local minimum at about 65% of the full relaxation of InAs and displays a steep increase towards the tip of the dot.

CONCLUSION

We have demonstrated a method to determine the relationship between strain and shape within self-assembled coherent quantum dots. Radius and gradient of relaxation are obtained as functions of relaxation from the substrate. For InAs on GaAs(100) islands the radius is linearly dependent on the local relaxation. Our results are important for band structure calculations as well as for the verification of growth models.

ACKNOWLEDGEMENTS

We would like to acknowledge the partial support by the VW Stiftung under grant number I/72 512 and the help in traveling expenses by HASYLAB. We appreciate very much the technical support from the HASYLAB staff at beamline BW2. The LMU-UCSB collaboration is supported by the BMBF and a Max-Planck Forschungspreis.

REFERENCES

1. S. Fafard, K. Hinzer, S. Raymond, M. Dion, J. McCaffrey and Y. Feng, S. Charbonneau, Science **274**, 1350 (1996)
2. H.T. Dobbs and D.D. Vvedensky, A. Zangwill, J. Johansson, N. Carlsson, W. Seifert, Phys. Rev. Lett. **79**, 897 (1997)
3. U. Pietsch, H. Metzger, S. Rugel, B. Jennichen, and I. K. Robinson, J. Appl. Phys. **74**, 2381 (1993)

X-RAY DIFFUSE SCATTERING INVESTIGATION OF DEFECTS IN ION IMPLANTED AND ANNEALED SILICON

C.H. Chang[1,3], U. Beck[2], T.H. Metzger[2] and J.R. Patel[1,3]

[1] SSRL/SLAC, Stanford University, Stanford, CA 94309
[2] Sektion Physik der Universität München, 80539 München, Germany
[3] ALS, Lawrence Berkeley National Laboratory, Berkeley, CA 94720

ABSTRACT

To characterize the point defects and point defect clusters introduced by ion implantation and annealing, we have used grazing incidence x-rays to measure the diffuse scattering in the tails of Bragg peaks (Huang Scattering). An analysis of the diffuse scattered intensity will allow us to characterize the nature of point defects or defect clusters introduced by ion implantation. We have also observed unexpected satellite peaks in the diffuse scattered tails. Possible causes for the occurrence of the peaks will be discussed.

INTRODUCTION

The nature of defects introduced by ion implantation and their subsequent interaction during high temperature processing with the implanted ion species is of considerable technological and recent scientific interest [1-4]. If ion implantation is to continue being used for future device generations [5], it is imperative that we understand and attempt to control the phenomenon of Transient Enhanced Diffusion (TED). TED is observed upon annealing ion-implanted silicon, we notice that the diffusion of the implanted ion species is enhanced by several orders of magnitude compared to the bulk diffusivity. In addition, the phenomenon is transient and of limited duration depending on time and temperature. The phenomenon is illustrated in Fig. 1, from the work of Michel et al [3] which shows the concentration profiles of boron implanted silicon versus depth as a function of annealing time at 800°C. Note the saturation of the diffusion profile at long annealing times.

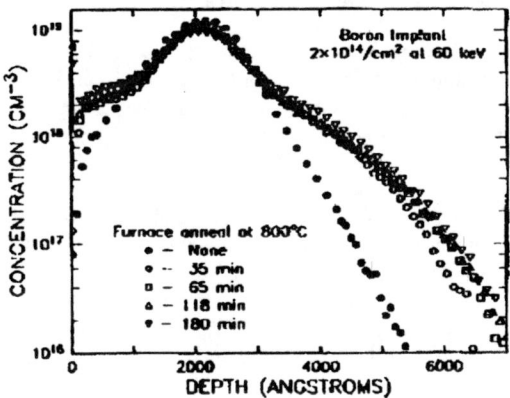

Figure 1. Boron profiles for several annealing times at 800°C [3].

At SSRL we have initiated a program in collaboration with our industrial colleagues at Intel Corporation and scientific colleagues at the University of München, Stanford University, the University of Illinois and the Lawrence Livermore National Laboratory to try to understand in a systematic way the defects introduced by ion implantation and their subsequent interaction with implanted ions. In particular we concentrate on identifying and characterizing point defects and point defect clusters too small to be observed by transmission electron microscopy. A powerful method for detecting point defects or defect clusters is to analyze the diffuse scattering in the tails of Bragg peaks or the weak scattering between Bragg reflections [6-8]. In metals the use of these methods allows unambiguous characterization of point defects and clusters such as dumbbell pairs [6].

EXPERIMENTAL TECHNIQUE

Since the implanted layers extend about 2000Å from the crystal surface, we used grazing incidence geometry to restrict penetration of the incident x-rays and to suppress bulk scattering. In grazing incidence diffraction, the plane of diffraction runs parallel or nearly parallel to the crystal surface. Both radial and angular scans were taken around the (400) and (220) Bragg peaks using a position sensitive detector to map out a region in reciprocal space around the Bragg tails.

In order to produce samples with a preponderance of one defect type we have adopted the following procedure for sample preparation. Floating zone dislocation-free silicon crystals were cleaned and implanted with a high boron dose at 32 keV. The wafers were then Rapid Thermal Annealed (RTA) at 1070°C in order to anneal the implant damage and to locate the boron predominantly in substitutional sites. The wafers were then implanted with a lower dose of silicon at 80 keV and annealed at 750°C for various times. We expect that defects introduced by the silicon implant will be captured during annealing by the substitutional boron already present and thus produce a defect population largely consisting of boron, B-Si interstitial pairs or small clusters.

RESULTS

Diffuse Scattering Around Bragg Peaks in Grazing Incidence

The contours of diffuse scattering intensity around (400) and (220) Bragg peaks are shown in Fig. 2 for the crystal containing a high B concentration (6E15 cm^{-2}) annealed at 1070°C for 10 sec after subsequent silicon implant (4E14 cm^{-2}) and annealing at 750°C for 15 min. Besides the expected diffuse scattering around the (220) reflection, we notice a rather unusual feature. There are lobes of intensity displaced from the Bragg peak in the [220] direction. For the (400) reflection the peaks are not as clear, however, we can clearly see four lobes of intensity extending in the [-220] and the [220] directions. The intensity lobes are more clearly seen for crystals with an initial concentration of B (6E14 cm^{-2}) that is an order of magnitude less than the previous case shown in Fig. 2. After annealing at 1070°C for 10 sec and subsequent Si implantation (2E14 cm^{-2}) anneal at 750°C for 15 min, the intensity contours for the (400) and (220) reflections are shown in Fig. 3. For the (220) reflection the lobes in the [220] direction are clearly visible. Around the (400) reflection four lobes of intensity are observed in the two {220} directions.

Intensity Profiles in {220} Directions

In Fig. 4 we plot the intensity profile in the [220] direction. Besides the tail of the (220) Bragg peak satellite peaks on either side of the main peak are clearly visible. From the spacing **q** of the satellite peaks we obtain a period distance of 74Å for the high B sample shown in Fig. 4(a). A similar profile for the low B sample shows satellite peaks somewhat lower in intensity and a periodicity of 85Å, Fig. 4(b).

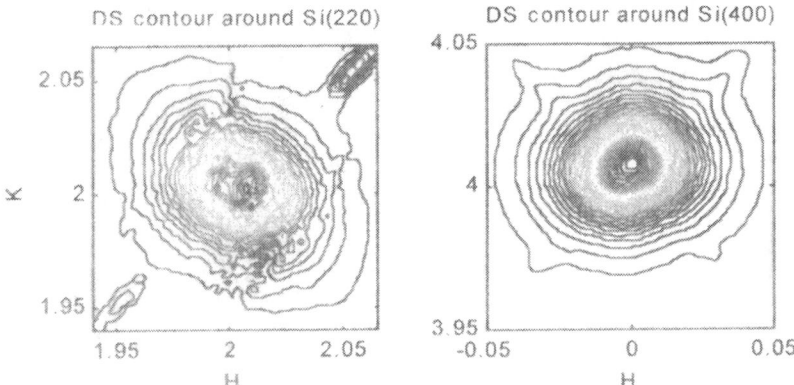

Figure 2. Contour of diffuse scattering intensity from silicon containing a high B concentration (6E15 cm^{-2}) annealed at 1070°C 10 sec after subsequent silicon implant (4E14 cm^{-2}) and annealed at 750°C 15 min.

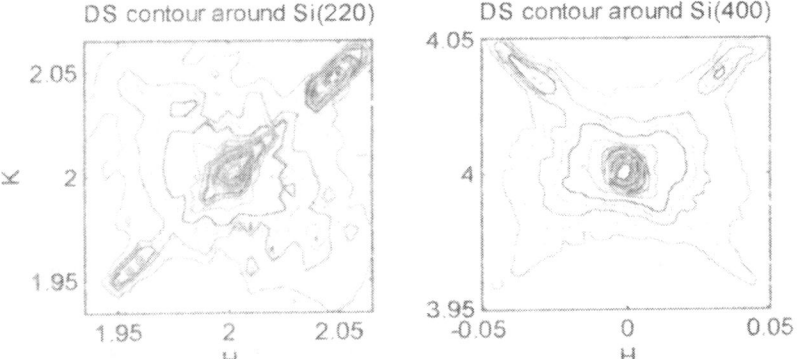

Figure 3. Contour of diffuse scattering intensity from silicon containing a low B concentration (6E14 cm^{-2}) annealed 1070°C 10 sec after subsequent silicon implant (2E14 cm^{-2}) and annealed at 750°C 15 min.

The results from the two (400) and (220) measurements are internally consistent since the (400) reflection shows as would be expected pairs of satellite peaks in the [-220] and [220] directions. This is especially clear for the low B case and less so for the high B (400) results.

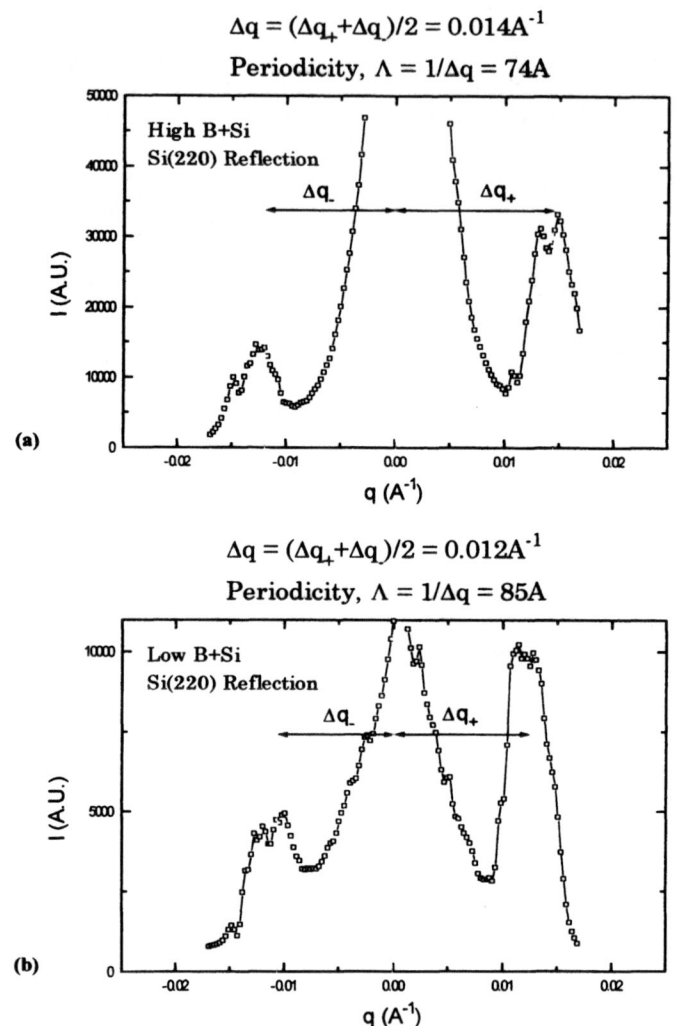

$$\Delta q = (\Delta q_+ + \Delta q_-)/2 = 0.014 A^{-1}$$

Periodicity, $\Lambda = 1/\Delta q = 74A$

$$\Delta q = (\Delta q_+ + \Delta q_-)/2 = 0.012 A^{-1}$$

Periodicity, $\Lambda = 1/\Delta q = 85A$

Figure 4. Intensity profile along (220) direction. (a) High boron concentration; (b) Low boron concentration.

CONCLUSIONS

While the results clearly indicate a periodicity in the <220> directions, it is difficult at this stage to propose a model for the origin of these satellites. At present we can only speculate that the data are consistent with a lateral modulation parallel to the crystal surface on the scale of 75-85Å. Since the implanted layer is only 2000Å from the surface, a surface-mediated strain induced periodicity may be the cause for the observed satellites. We intend to look at reflections other than those investigated in this report to shed further light on this rather unexpected result.

The original aim of this work was to characterize the defects introduced by ion implantation by analyzing the diffuse intensity in the Bragg tails. To aid in the defect characterization we have used molecular dynamics simulation methods to predict the defect types and impurity profiles. The predicted profiles will be compared to experimentally obtained SIMS (Secondary Ion Mass Spectroscopy) profiles and refined to fit the experimental data. Preliminary work indicates that the defects are mainly point defects as well as B3I and B4I2 clusters, where (I) refers to interstitial silicon [9]. Atomic displacements around such clusters have been calculated using first principles methods by our collaborators at the Lawrence Livermore National Laboratory [10,11]. All of this information will be used to calculate directly the diffuse scattering under Bragg reflections and compared to the results we have in contour form in Figs. 2 and 3. Experimental verification of the simulation results is important since it will provide a sounder basis for the simulations and provide additional confidence for including such data in process codes that are widely utilized in the semiconductor industry.

ACKNOWLEDGEMENTS

Our colleagues in this rather extensive project are listed below. At München the support and participation of J. Peisl is gratefully acknowledged. We are grateful for a travel grant from NATO that made this cooperative effort possible. Our close collaborators in the myriad problems raised in these studies are our Stanford colleagues P.B. Griffin and J. Plummer, at Lawrence Livermore T. Diaz de la Rubia and M. Caturla, at Illinois B. Averback and K. Nordlund, and at Intel M. Giles and B. Doyle. We acknowledge gratefully many conversations and guidance from H-J. Gossmann, G.H. Gilmer and D. Eaglesham at Lucent, Bell Labs. The diffuse scattering experiments were carried out at ESRF, Grenoble and SSRL, Stanford. We gratefully acknowledge the expert assistance of J. Arthur, S. Brennan and G. Grubel. Finally one of us (JRP) would like to thank J. Carruthers of Intel for suggesting the TED problem and his encouragement and support for this work.

REFERENCES

1. P.M. Fahey, P.B. Griffin and J.D. Plummer, Rev. Mod. Physics **61**, 289 (1989)
2. H-J. Gossmann, C.S. Rafferty, H.S. Luftmann, F.C. Unterwald, T. Boone and J.M. Poate, Appl. Phys. Lett. **63**, 639 (1993)
3. A.E. Michel, W. Rausch, P.A. Ronsheim and R.H. Kastl, Appl. Phys. Lett. **50**, 417 (1989)
4. L. Pelaz, G.H. Gilmer, M. Jaraiz, H-J. Gossmann, C.S. Rafferty, D.J. Eaglesham, and J.M. Poate, Mat. Res. Soc. Symp. Vol. **469**, 341 (1997)
5. M.D. Giles, S. Yue, H.W. Kennel, and P.A. Packan, Mat. Res. Soc. Symp. Proc. Vol. **469**, 253, (1997)
6. P.Ehrhart, J. Nucl. Mat. **216**, 174 (1990)
7. S. Grotenhans, G. Wallner, E. Burkel, H. Metzger, J. Peisl and H. Wagner, Phys. Rev. **B39**, 8450 (1989)
8. K. Nordlund, P. Patrtyka, and R.S. Averbach, Mat. Res. Soc. Symp. Proc. **469**, 199 (1997)
9. M-J. Caturla (Personal communication)
10. J. Zhu, Mat. Res. Soc. Proc. Vol. **469**, 151 (1997)
11. M-J. Caturla, T. Diaz de la Rubia, J. Zhu and M. Johnson, Mat. Res. Soc. Symp. Vol. **469**, 335 (1997)

GROWTH AND STRUCTURE OF NANOMETRIC IRON OXIDE FILMS

E. GUIOT*, S. GOTA*, M. HENRIOT*, M. GAUTIER-SOYER*, S. LEFEBVRE**
*DSM / DRECAM / SRSIM - CEA SACLAY - 91191 - Gif sur Yvette Cedex - France,
**LURE, Batiment 209 D, Centre Universitaire Paris Sud - 91405 - Orsay cedex - France.

ABSTRACT

Nanometric films of iron oxides (Fe_3O_4, α and γ Fe_2O_3) of high crystalline order and purity are epitaxially grown on α-Al_2O_3(0001) by atomic oxygen assisted MBE. A complete characterization of the films structure has been performed by in situ LEED and RHEED, and ex situ GIXRD using synchrotron radiation. The films grown at room temperature and post annealed at 400°C and 700°C ($p_{O2}=10^{-6}$ Torr) are respectively metastable γ-Fe_2O_3 (111) and α-Fe_2O_3 (0001). For a substrate temperature of 450°C during growth, Fe_3O_4 (111) is directly obtained. GIXRD shows an in-plane expansion of the films, which decreases with thickness (0.8 and 0.2% for film thickness of 20 and 80 Å, respectively).

INTRODUCTION

Transition metal oxide thin films present unique magnetic properties, with exciting potential applications. Only few oxides present both full spin polarization at the Fermi level (« half metallic ferromagnets ») and magnetic ordering temperatures above 300 K. Among them, magnetite Fe_3O_4 (T_C = 854 K) is a good candidate for the development of new spin polarized current injection devices. In the case of nanometric films, surfaces and interfaces play an essential role in the film properties. So, there is great interest in the development of film growth procedures permitting to obtain films of controllable stoichiometry and high crystalline order and purity. The first ordered iron oxide films were obtained by Somorjai, Weiss and co-workers[1]. The method consists in depositing first a monolayer of Fe on platinum, followed by an oxidation at high temperature under controlled p_{O2}. Other groups used an alternative method consisting in the deposit of metallic Fe from a Knudsen cell in the presence of oxidizing agents more reactive than molecular oxygen, as NO_2[2,3] or O^+ ions produced by an electron cyclotron resonance (ECR) plasma source[4,5].

In this paper, we present the synthesis and characterization of high quality Fe_3O_4, α and γ-Fe_2O_3 thin films epitaxially grown on an α-Al_2O_3(0001) substrate. The films were prepared by molecular beam epitaxy (MBE) assisted by an atomic oxygen source. A complete characterization of the films structure was performed by combining several techniques : in situ low energy electron diffraction (LEED) and reflection high-energy electron diffraction (RHEED), and ex situ grazing incidence x-ray diffraction (GIXRD) using synchrotron radiation.

CRYSTAL STRUCTURES

Both α-Al_2O_3 and α-Fe_2O_3 exhibit a rhombohedral corundum crystal structure with Fe^{3+} located in distorted octahedra. The crystal structure of γ-Fe_2O_3 is quite complicated and closely related to that of the inverse spinel Fe_3O_4. In this latter compound, there are 8 Fe^{3+} ions per unit cell in tetrahedral sites and equal amounts of 8 Fe^{3+} and 8 Fe^{2+} ions in octahedral sites. Maghemite γ-Fe_2O_3 can be described as a « non stoichiometric » magnetite $Fe_{3-\delta}O_4$ with a vacancy parameter δ = 1/3 in which ferrous ions are converted to ferric ones obtaining the highest oxidized spinel possible. In the four oxides, surface oxygen sub-lattices have the same symmetry but slightly different cell parameters. The expected epitaxial orientation relationship between hematite and alumina is α-$Fe_2O_3(0001)[10\bar{1}0]//\alpha$-$Al_2O_3(0001)[10\bar{1}0]$, with a lattice mismatch of 5.4%. The expected epitaxial orientation relationship between maghemite or magnetite and alumina is γ-Fe_2O_3 or $Fe_3O_4(111)$ $[0\bar{1}1]//\alpha$-$Al_2O_3(0001)$ $[10\bar{1}0]$ with a lattice mismatch of $\approx$ 8%. In view of symmetry considerations (i. e., neglecting for the moment the difference in cell parameters) we can consider that the surface unit cells of $Fe_3O_4(111)$ and γ-$Fe_2O_3(111)$ correspond to a $(2/\sqrt{3}x2/\sqrt{3})R30°$ superstructure with respect to the corundum (0001) one, labeled (1 x 1). Figure 1 shows a schematic superposition of the related reciprocal lattices. This figure will be useful for discussing the diffraction results.

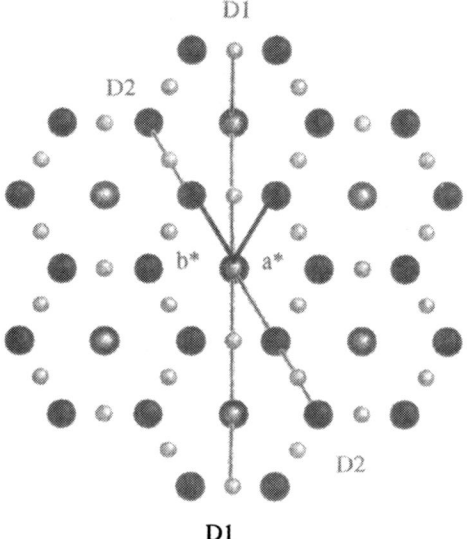

Fig.1. Superposition of the two-dimensional reciprocal lattices of α-$Fe_2O_3(0001)$ (large black balls) and γ-$Fe_2O_3(111)$ or $Fe_3O_4(111)$ (small grey balls). When the RHEED electron beam is along the $[1\bar{1}00]$ (resp.$[10\bar{1}0]$) direction, we probe the reciprocal lattice rods along the D1 (resp. D2) direction.

EXPERIMENTAL DETAILS

Samples growth and in-situ characterization

Alumina single crystals oriented (0001) are used as substrates. After the cleaning procedure described in ref.[6], followed by an in situ annealing at 600°C ($p_{O2}=10^{-6}$ Torr), the surface is free of any contaminant, as checked by Auger spectroscopy. The MBE chamber is fitted with an iron Knudsen-type cell and an oxygen plasma source (OPS) with both iron and oxygen fluxes directed towards the substrate. The OPS consists basically in a U shaped quartz tube with two Pt electrodes. The oxygen plasma is created by a DC glow discharge between them. The oxygen flux is introduced in the MBE chamber through a 150 µm diameter effusion hole. The active species for oxidation is mostly atomic oxygen, produced in the plasma via electronic dissociation[7]. The iron evaporation rate is 1 Å/min. RHEED measurements can be performed during film growth. The base pressure in the MBE chamber is 5.10^{-11} Torr, and never rise above 5.10^{-8} Torr during deposition. During the growth, the substrate can be either kept at room temperature (RT) or heated up to 600°C. In an interconnected chamber, films can be post-annealed under controlled p_{O2} and analyzed by LEED.

In this paper we consider four films prepared as follows :
(S1) growth at RT followed by annealing at 400°C ($p_{O2}=10^{-6}$ Torr), 80 Å thickness.
(S2) growth at RT followed by annealing at 700°C ($p_{O2}=10^{-6}$ Torr), 80 Å thickness.
(S3) growth at 450°C, 80 Å thickness.
(S4) growth at 450 °C, 20 Å thickness.

Ex situ grazing incidence x-ray diffraction

The measurements were carried out at the D23 beamline of the LURE-DCI storage ring in Orsay, France. The $\lambda = 1.7712$ Å radiation was selected from a bending magnet source using a Si(111) double monochromator. The incident angle $\phi = 0.6°$ was chosen close to the critical angle for total reflection ϕ_c. The width and height of the beam were fixed by slits. In our experimental geometry, the sample was horizontal, and the detector could move only in this plane (grazing detection). The diffracted beams were selected using a Soller collimator (opening 1.43 mrad) and detected in a NaI solid detector. Since the angle of incidence is close to ϕ_c, the penetration depth is only a few 100 Å to a few 1000 Å and the diffracted intensity is relatively high. As in conventional diffraction, θ-2θ scans are performed, with the sample rotating with half the angular speed of the detector. In epitaxial films, diffraction takes place only at discrete values of φ, the azimuthal angle of the film around its normal, and one has to search the diffracted beams by varying φ with a fixed angle 2θ of the detector. From the Bragg angle we obtain the lattice parameter and strain [8].

EXPERIMENTAL RESULTS

Figure 2 shows the LEED patterns obtained on S2 (a) and S3 (b) samples. The LEED pattern of the corresponding alumina substrate is shown as inset. The LEED patterns of S1 and S4, not shown, looked similar to (b).

The RHEED patterns were recorded with the incident beam along the azimuthal $[10\bar{1}0]$ and $[1\bar{1}00]$ directions referred to the direct space of alumina, probing the (D2) and (D1) lines of the two-dimensional reciprocal space sketched in figure 1. Figure 3 (b) and (e) show the RHEED patterns of the alumina substrate along these two directions. They look like transmission patterns, likely due to the presence of steps. In contrast, the RHEED patterns of the films show sharp streaks, characteristic of flat surfaces. The upper RHEED patterns (figure 3 (a) and (d)) were obtained on S2 sample, and the lower ones (figure 3 (c) and (f)) on S3 sample. The RHEED patterns of S1 and S4 samples, not shown, looked similar to those of S3.

As concerns the GIXRD experiments, we have searched all the diffraction peaks corresponding to planes perpendicular to the surface of the three possible iron oxide phases. The in-plane cell parameters deduced from diffraction peaks of the four studied samples are gathered in table 1. The upper part of figure 4 (a) (b) and (c) shows the θ-2θ scans corresponding to the S1 and S3 samples. The lower part of figure 4 (d) and (e) show the θ-2θ scans corresponding to the S2 one.

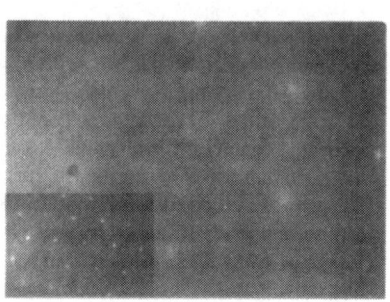

Fig.2. LEED patterns corresponding to sample S2 (a) and sample S3 (b). The primary energies are 87eV and 140eV respectively. The LEED patterns of the 0001-oriented alumina substrates shown as inset are taken at the same energy.

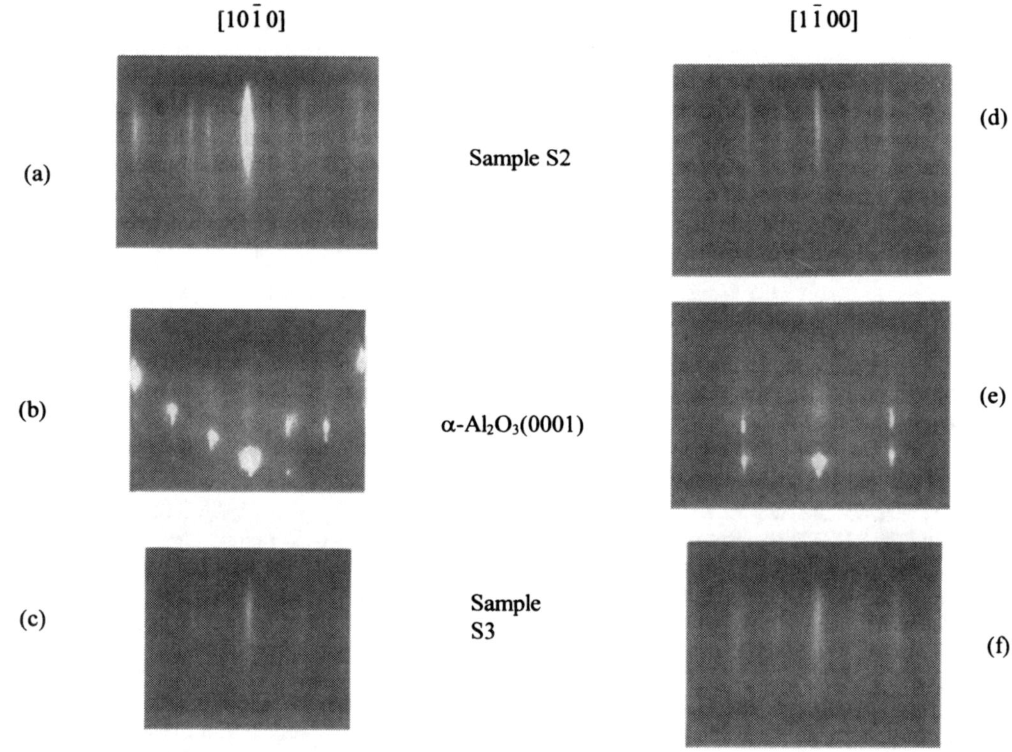

$[10\bar{1}0]$ $[1\bar{1}00]$

(a)   Sample S2   (d)

(b) α-Al$_2$O$_3$(0001)   (e)

(c)   Sample S3   (f)

Fig.3. RHEED zero-order patterns at Ep=30 keV along the $[10\bar{1}0]$ direction (a,b,c) and the $[1\bar{1}00]$ direction (d,e,f).

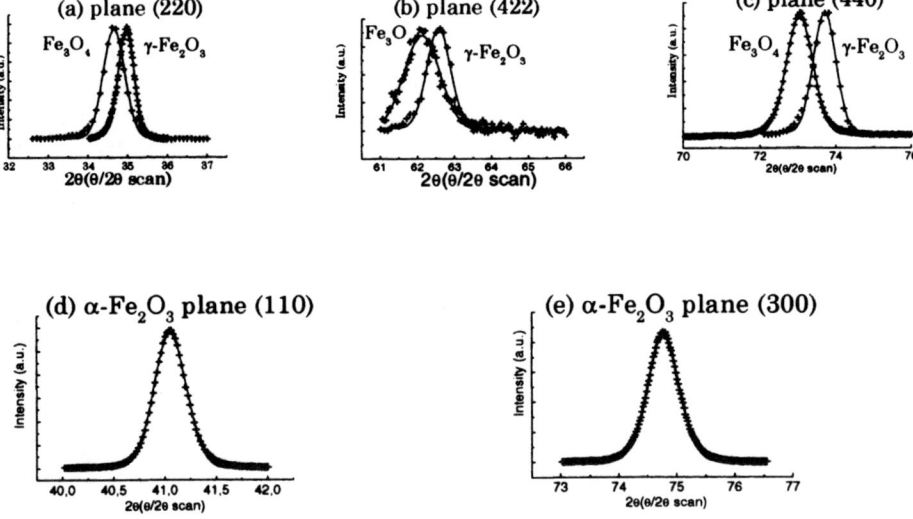

Fig.4. GIXRD θ/2θ scans corresponding to samples S1, S3 (a,b,c) and S2 (d,e).

	Thickness	Plane	d_{hkl}^{bulk}(Å)	d_{hkl}^{exp}(Å)	Δ(*)
α-Al₂O₃(0001)	substrate	(110)	2.379	2.379	0%
γ-Fe₂O₃(111)	80Å	(220)	2.9512	2.9553	0.1%
Sample S1		(440)	1.4755	1.4777	0.2%
		(422)	1.7040	1.7070	0.2%
α-Fe₂O₃(0001)	80Å	(110)	2.5188	2.5258	0.3%
Sample S2		(300)	1.4543	1.4586	0.3%
Fe₃O₄(111)	80Å	(220)	2.9677	2.9736	0.2%
Sample S3		(440)	1.4838	1.4881	0.3%
		(422)	1.7133	1.7166	0.2%
Fe₃O₄(111)	20Å	(220)	2.9677	2.9885	0.7%
Sample S4		(440)	1.4838	1.4972	0.9%
		(422)	1.7133	1.7270	0.8%

(*)Δ=(d_{hkl}^{exp} - d_{hkl}^{bulk})/ d_{hkl}^{bulk}

Table 1 : Summary of the in-plane cell parameters deduced from diffraction peaks positions for the different epitaxial thin films.

DISCUSSION

From GIXRD measurements the film grown at RT and post-annealed at 400°C ($p_{O2}=10^{-6}$ Torr) (S1) film can be identified as γ-Fe$_2$O$_3$ (111). Indeed, GIXRD measurements allow to discriminate between the γ-Fe$_2$O$_3$ (111) and Fe$_3$O$_4$ (111) phases. Their cell parameters are so close that these two phases cannot be distinguished by LEED and RHEED patterns. In contrast their common (220), (422) and (440) diffraction peaks present enough different θ-2θ positions for being resolved, as shows the upper part of figure 4 (a), (b) and (c). Maghemite γ-Fe$_2$O$_3$ is a metastable phase in the bulk phase diagram and has been successfully stabilized in a thin film configuration by a non equilibrium MBE growth technique.

From GIXRD measurements the film deposited at RT and post-annealed at 700°C ($p_{O2}=10^{-6}$ Torr) (S2) can be identified as α-Fe$_2$O$_3$ (0001). The specific (110) and (300) diffraction peaks have been measured (see figures 4(d) and (e)). Referring to the sketch of figure 1, its LEED pattern 3(a) is also characteristic of a α-Fe$_2$O$_3$ (0001) surface. Around the main spots, we note the presence of satellite structures, which are under investigation right now. Similar satellite structures have been already observed at the surface of bulk iron oxide single crystals by combined LEED/STM and interpreted as a complex surface reconstruction[9]. In the RHEED patterns 3(a) and 3(d), we observe in addition to the streaks belonging to α-Fe$_2$O$_3$ (0001), streaks which can be attributed to the presence of either to γ-Fe$_2$O$_3$ (111) or Fe$_3$O$_4$ (111) (see figure 1) at the extreme surface. From GIXRD (results not shown) this minority phase was characterized as γ-Fe$_2$O$_3$ (111). It is known that the $\gamma \rightarrow \alpha$ phase transition occurs at 325 °C in the bulk[10]. Comparing the structures of samples S1 and S2, we can infer that the $\gamma \rightarrow \alpha$ transition temperature is higher for the thin films.

The films grown at a substrate temperature of 450°C (S3 and S4) can be identified as pure Fe$_3$O$_4$ (111) from GIXRD measurements (see figures 4 (a), (b) and (c)). The corresponding RHEED and LEED patterns of figures 2(b) and 3 (c) (f) are in agreement with GIXRD results.

From the width of the θ-2θ diffraction peaks, a size of 200-300 Å for ordered domains can be deduced.

We see in table 1 that the 80 Å-thick films all show an in-plane expansion of about 0.2-0.3%, compared to the bulk value. The 20 Å-thick film is more expanded (0.8%). This is an unexpected result, as a contraction would be predicted when comparing the lattice parameters of Al$_2$O$_3$ (0001) and the three iron oxides. This in-plane expansion decreases when the film thickness increases, suggesting that the most distorted part of the film is the interface region, as already observed by Fujii et al[11].

CONCLUSION

Nanometric films of iron oxides (Fe$_3$O$_4$, α and γ Fe$_2$O$_3$,) of high crystalline order and purity are epitaxially grown on an α-Al$_2$O$_3$(0001) substrate by atomic oxygen assisted MBE. GIXRD using synchrotron radiation coupled with RHEED and LEED permits characterization of micro-structural properties of thin films such as crystallographic orientation, ordered domain size, and strains. The films grown at room temperature and post annealed at 400°C and 700°C ($p_{O2}=10^{-6}$ Torr) are respectively metastable γ-Fe$_2$O$_3$ (111) and α-Fe$_2$O$_3$ (0001). For a substrate temperature of 450°C during growth, Fe$_3$O$_4$ (111) is directly obtained. GIXRD shows an in-plane expansion of the films, which decreases with thickness and which is likely due to a distorted interface region.

ACKNOWLEDGMENTS

The authors would like to thank Drs. M. Touzeau and D. Pagnon from « Laboratoire de Physique des Gaz et des Plasmas », Orsay, France for the help in the development and adjustment of the OPS. We are very grateful to Drs. M. Goldmann and M. Bessière from « Laboratoire pour l'Utilisation du Rayonnemment Electromagnétique » (LURE), Orsay, France for assistance in the setting of the GIXRD experiments and very helpful discussions.
T. Hibma (University of Groningen, The Netherlands) and T. Fujii (Okayama University, Japan) are acknowledged for stimulating discussions.

REFERENCES

[1] W. Weiss, A. Barbieri, M. A. Van Hove and G. A. Somorjai, Phys. Rev. Lett 71 (1993), 1848 ;.

[2] F. C. Voogt, T. Hibma, P. Smulders and L. Niesen, J. Crystal Growth 174 (1997), 440.

[3] T. Fujii, D. Alders, F. C. Voogt, T. Hibma, B. T. Thole and G. A. Sawatzky, Surf. Sci. 366 (1996), 579.

[4] D. M. Lind, S. D. Berry, G. Chern, H. Mathias and L. R. Testardi, Phys. Rev. B 45 (1992), 1838.

[5] Y. Gao and S. A. Chambers, J. Cryst. Growth 174 (1997), 446.

[6] R. F. C. Farrow, G. R. Harp, R. F. Marks, T. A. Rabedeau, M. F. Toney, D. Weller and S. S. P. Parkin, J. Cryst. Growth 133 (1993), 47.

[7] P. Luzeau, X. Z. Xu, M. Laguës, N. Hess, M. Nanot, F. Queyroux, M. Touzeau and D. Pagnon, J. Vac. Sci. Technol. A 8 (6) (1990), 3938.

[8] A. Segmüller, J. Sci. Technol. A9 (1991) 2477..

[9] N. G. Condon, F. M. Leibsle, A. R. Lennie, P. W. Murray, D. J. Vauhan and G. Thornton, Phys. Rev. Lett. 75 (1995), 1961.

[10] Y. Gao, Y.J. Kim, S. Thevuthasan, S. A. Chambers and P. Lubitz, J. Appl. Phys. 81 (1997) 3253.

[11] T.Fujii, M. Takano, R. Kakano, Y. Isozumi, Y. Bando, J. of Magnetism and Magnetic Materials 135 (1994) 231.

DEGREE OF CRYSTALLINITY AND STRAIN IN B_4C AND SiC THIN FILMS AS A FUNCTION OF PROCESSING CONDITIONS

[1]J. HERSHBERGER, [2]Z. U. REK, [3]F. KUSTAS, [1]S. M. YALISOVE, AND [1]J. C. BILELLO
[1]Center for Nanomaterials Science, University of Michigan, Department of Materials Science & Engineering, Ann Arbor, MI 48105. jhersh@engin.umich.edu
[2]Stanford Synchrotron Radiation Laboratory, Stanford, CA 94309
[3]Technology Assessment and Transfer, Annapolis, MD 21401. Now at Lockheed-Martin, Denver CO

ABSTRACT

Amorphous and crystalline content in sputtered B_4C and SiC thin films has been analyzed by synchrotron grazing incidence x-ray scattering (GIXS). GIXS provided quantitative information on the average structure while TEM was used to find inhomogeneities such as small volume fraction phases. GIXS results were compared to simulations to determine average particle size or bond length for crystalline or amorphous phases respectively. In this work, we compared results from films deposited with, and without, an RF bias applied to the substrate during deposition. Results indicated that SiC can be described as strained polycrystalline material with particle size of approximately 13 Å for biased samples and 9Å for unbiased samples. Boron carbide deposited without bias was completely crystalline with a particle size of approximately 30 Å, while the data suggested that B_4C deposited with bias is amorphous. The scattering from the biased materials was Fourier transformed to yield radial distribution functions (RDF). This provided nearest neighbor distances, and it was demonstrated that the technique can be used to determine full three-dimensional strain tensors in amorphous thin films.

INTRODUCTION

A low-power bias applied to the substrate during thin film growth has previously been shown to affect the surface morphology [1, 2] and chemistry [3]. Application of an RF bias reduces the density of electrons near the substrate, giving rise to a DC sheath which accelerates ions towards the substrate. It is also known that ion bombardment (whether due to substrate biasing or to low working gas pressure) affects the residual stress, by modifying the microstructure (removing tension-inducing voids) [4] or by inducing compressive stresses through the peening mechanism [5]. Magnetron sputtered and CVD films of both B_4C and SiC have previously been found to be amorphous when the deposition rate is high and the substrate temperature is low [6, 7]. This study focuses on the effect of bias on the atomic structure, i.e. particle size and amorphous content, using both X-rays and TEM for quantitative particle sizes and for local inhomogeneities. Fourier transform X-ray analysis techniques used for bulk amorphous materials have also been successfully applied to GIXS data from B_4C and SiC thin films.

EXPERIMENT

Thin films of B_4C and SiC were grown on quartered 100mm wafers of (100) oriented Si by DC and RF magnetron sputtering respectively. Depositions were performed at the Wear Sciences Group of Technology Assessment and Transfer Inc. as described elsewhere [2]. Base pressure in the chamber was better than $5*10^{-6}$ Torr for all depositions and 2.5 mTorr Ar was used as the sputtering gas. The substrates were placed on a rotating carousel within the chamber and the 9 cm by 20 cm targets faced inwards. Prior to sputtering, a high power RF bias was applied to all substrates to improve adhesion, and this bias was also used at low power during growth of some of the films. The thicknesses of the films grown were 2520±50Å and 2380±200Å for biased and unbiased B_4C respectively, and 800±20Å and 1260±100Å for biased and unbiased SiC respectively.

TEM cross sections were prepared by gluing the film surfaces together and then grinding the sandwich to form a wedge using the tripod polishing technique. The samples were mounted

Mat. Res. Soc. Symp. Proc. Vol. 524 © 1998 Materials Research Society

on Mo rings and ion milled for 2 to 30 minutes at an 8° incident angle with an Ar beam of 0.3 mA at 5.5 kV in a Gatan dual ion mill. The experiments were performed on a JEOL 2000 microscope operating at 200kV.

Asymmetric and symmetric mode GIXS experiments were carried out on beamline 7-2 of Stanford Synchrotron Radiation Lab, an 8-pole focused wiggler beamline, as described elsewhere [8]. A double crystal Si(111) monochromator was used to select the wavelength 1.2395Å (10 keV). It was necessary to surround the samples with a He environment purged of air four times before each experiment in order to reduce air scattering, which was otherwise large compared to the scattering power of the samples. A vacuum chuck secured a part of the sample 2 cm from the illuminated area to reduce warping. The azimuthal angle phi (ϕ) of the sample was chosen to minimize the contribution of the substrate to the signal. Bare Si substrates were scanned separately and that intensity was subtracted from that of the film.

RESULTS AND DISCUSSION

Figure 1 shows bright field cross section TEM micrographs of biased (a) and unbiased (b) B_4C and biased (c) and unbiased (d) SiC. In figure 1a, a region of defected Si was visible below the film; this may have been due to the high stress in the film and the mechanical action of polishing. An inset in each figure (a-d) with enhanced contrast showed the presence of alternating light and dark layers visible within the films. This represented either varying elastic scattering power diverting intensity out of the transmitted beam, or, more likely, varying absorption due to density or chemistry. X-ray reflectivity results from the biased samples (to be published elsewhere) revealed evidence of this periodicity on a small spatial scale as well as the oscillations due to the film thickness. This indicated a periodic variation of electron density, and therefore of density or chemistry. It is likely that the layering resulted from the fact that the deposition was intermittent, with material deposited at a high rate as the substrate passed under the target but with the bias operating continuously when applied. The fact that this layering appears even in the unbiased B_4C sample (1b) implies that it is not strictly a result of the influence of the bias on surface mobility and impurities, although it does not rule out mobility and impurities as possible causes.

GIXS data and simulations for estimation of particle size are presented in figure 2 for biased (a) and unbiased (b) B_4C and biased (c) and unbiased (d) SiC. The simulations were not intended to provide an exact match to the data. They were performed to give a qualitative estimate of the particle size from the peak widths. The simulations were based on ICDD (formerly JCPDS) listings for $B_{13}C_2$ and (cubic) SiC, with gaussian peak widths determined by the Scherrer relation and peak locations shifted by strain [9]. The simulation peak widths matched the data most closely for a 15Å particle size in the case of biased B_4C (a) and 13Å for biased SiC (c). Unbiased SiC most closely matched the simulation for a 9Å particle size; a 30Å particle size was optimal in the case of unbiased B_4C (b).

Since the unit cells of SiC and B_4C are 4.36 Å and 12.2 Å (c axis) respectively, the simulations were outside the regime where they would provide a detailed representation of the data. The unbiased B_4C was best described as crystalline. The biased B_4C was amorphous,

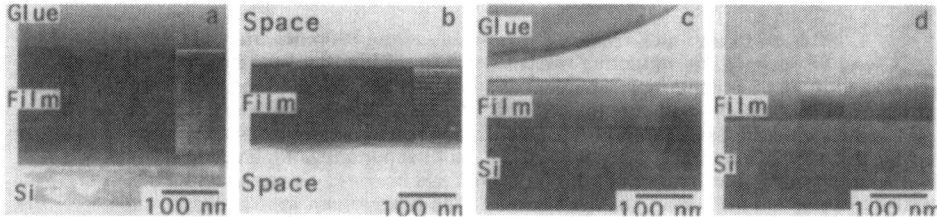

Figure 1. Bright field TEM cross section micrographs of biased (a) and unbiased (b) magnetron sputtered B_4C and biased (c) and unbiased (d) SiC are shown. The inset in each micrograph had enhanced contrast to show light and dark layering within the film thickness.

since it could not have had more than one repeating unit cell, although it possessed short-range order on a length scale of ~4 Å as evidenced by the intensity at ~18°2θ. In this case, the simulations based on ICDD listings of bulk materials could not provide a good match to the data, indicating that the structure is not that of the bulk. Biased SiC was describable as either amorphous or crystalline, with as many as three repeated unit cells in the average structure, but the unbiased SiC would have only two and so was better described as amorphous. The scattered X-ray intensities from SiC displayed the classical behavior of amorphous materials, oscillating around the curve of atomic scattering factor squared plus Compton scattering. The strain incorporated into the simulations to match peak locations indicated a lattice expansion of approximately 4% relative to the ICDD values in all cases except that of the unbiased B_4C where no strain was used. This value was well beyond what would be expected from strain in a perfect crystal and had the opposite sign from the stress, which was compressive in all cases as measured by the substrate curvature technique Double Crystal Diffraction Topography (DCDT) [10]. This suggested a highly defected structure, possibly with a large concentration of interstitial impurities. Preferred crystallographic orientation would affect the elastic compliances of the materials and therefore the strains but is unlikely in these cases considering their nanocrystalline or amorphous structure.

Figure 3 shows RDF results for biased B_4C (a) and SiC (b). The data reduction process has been described fully elsewhere [10]. In brief, the data from the film was isolated by subtracting the separately measured substrate response, and, where necessary, an analytical curve

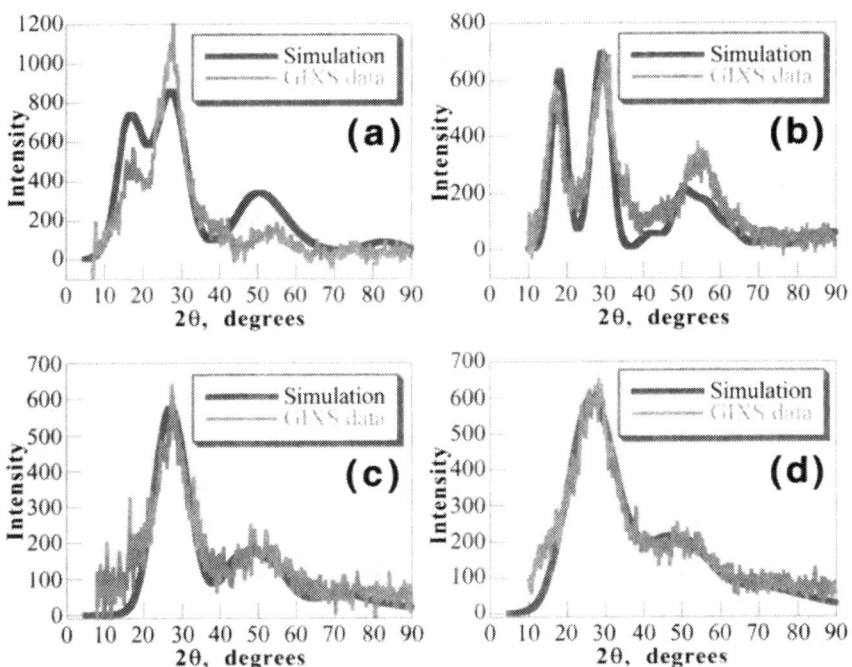

Figure 2. GIXS data and simulations based on ICDD listings of bulk structures are shown for biased (a) and unbiased (b) B_4C and biased (c) and unbiased (d) SiC with λ = 1.2395Å. The simulations took into account broadening due to particle size and peak shift due to strain. The particle sizes used in the simulations were 15Å, 30Å, 13Å, and 9Å respectively. With the exception of the unbiased B_4C case, the simulations used a lattice expansion of 4% relative to the ICDD standard to match the peak locations. The unbiased B_4C simulation used no strain.

describing the scattering from a mixture of air and He. The resulting data set was corrected and normalized to yield a reduced intensity i(k) and a Fourier transform was performed. The result gives the correlation function, in which the nearest neighbor peaks were measured, and the radial distribution function (RDF), which gave coordination numbers. The first and second nearest neighbor distances determined for biased B_4C were 1.668±.002Å and 3.057±.003Å respectively. The nearest neighbor distances for biased SiC were 1.767±.001Å and 3.140±.004Å. The error values given for nearest neighbor distances represent peak fit confidences. Other sources of error will be considered in later publications. In the bulk crystalline ICDD standard materials, the first nearest neighbor distances were 1.814Å for B_4C and 1.890Å for SiC. Comparison of these values indicated a substantial contraction (~6-8%) of the first nearest neighbor shells of the biased films relative to the bulk materials for both B_4C and SiC. This was consistent with the compressive sign of the stress measured by DCDT, but the magnitude of the contraction observed may also indicate a change in coordination. The fact that the simulations in figure 2 required lattice expansion relative to the ICDD listing to match the data can be explained by a large concentration of interstitial defects.

Figure 4 shows the geometry (a) and data (b) for determination of full three-dimensional strain tensors using first nearest neighbor distances. Figure 4a schematically shows six independent scattering vectors defined by three azimuthal angles each in both asymmetric and symmetric mode GIXS. Figure 4b presents six data sets collected from a biased B_4C sample using this technique, with each set after the first offset upwards by 10 electron units of intensity. Symmetric mode is represented in gray in (a) and as symbols in (b). These six data sets will be Fourier transformed to give d spacings (first nearest neighbor distances) in six directions. These will be geometrically transformed to form a d tensor. The values of d along the principal directions of the wafer, d_{xx}, d_{yy}, and d_{zz}, will be used with a measured stress to determine the stress free interatomic spacing d_0:

$$\sigma_{xx} = E\left[\frac{d_{xx} - v(d_{yy} + d_{zz})}{d_0} - (1 - 2v)\right] \quad (1).$$

The value σ_{xx} for one *in-plane* stress has been measured by DCDT [10], and bulk values adjusted for density will be used for the elastic constants. In the plane stress case ($\sigma_{zz} = 0$) the form of the equations eliminates E from calculation of d_0. Once d_0 is known, the d tensor will be used to form the strain tensor.

CONCLUSIONS

Application of an RF bias to the substrates during growth of magnetron sputtered B_4C

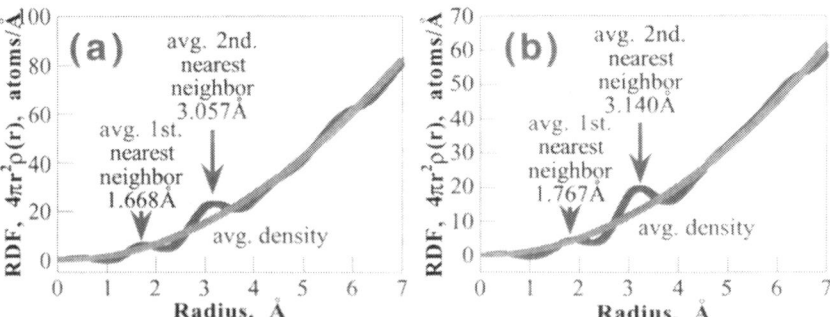

Figure 3. Plots a and b give RDF results from GIXS experiments on biased B_4C and SiC respectively. The average first and second nearest neighbor distances measured from the correlation function were 1.668±.002Å and 3.057±.003Å respectively for B_4C, and 1.767±.001Å and 3.140±.004Å for SiC. The error values represent peak fit confidence.

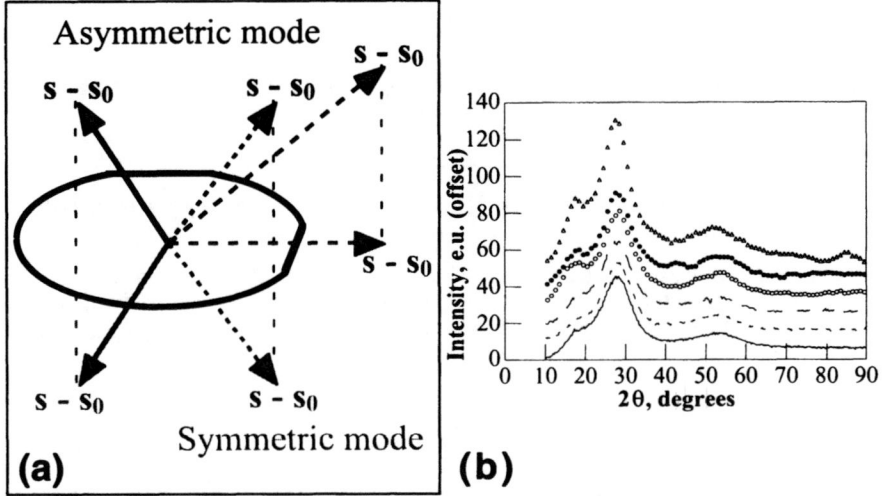

Figure 4. The schematic in figure 4 (a) shows the geometry of six independent GIXS scattering vectors (s - s₀) and their directions relative to the sample, pictured as a Si wafer with flats. These scattering vectors use three azimuthal angles each for asymmetric and symmetric mode GIXS. Figure 4 (b) shows six corresponding data sets collected from a sample of biased B_4C using λ = 1.2395Å. In both figure 4 (a) and (b), asymmetric mode is shown in black lines. Symmetric mode is shown in gray in fig. 4 (a) and as symbols in fig. 4(b).

resulted in a structure best described as amorphous, while deposition without bias yielded crystalline material with a 30Å particle size. GIXS data from magnetron sputtered SiC best matched simulations with particle sizes of 13Å and 9Å in the biased and unbiased cases respectively, indicating that their structures were amorphous or borderline crystalline. Amorphous phase average nearest neighbor distances have been determined from Fourier transforms of GIXS data. This analysis will be performed for six independent scattering vectors and combined with separately measured stress data to determine full three-dimensional strain tensors in amorphous thin films of B_4C and SiC.

ACKNOWLEDGEMENTS

The authors wish to thank Bart Johnson and Sean Brennan of SSRL for assistance on beamline 7-2. This work was done (partially) at SSRL, which is operated by the Department of Energy, Office of Basic Energy Sciences.

REFERENCES

1. E. Chason, T. M. Mayer, B. K. Kellerman, D. T. McIlroy, and A. J. Howard, Phys. Rev. Lett. **72** (19), 3040 (1994).

2. J. Hershberger, T. Ying, F. Kustas, L. Fehrenbacher, S. M. Yalisove, J. C. Bilello, Surf. Coat. Tech. **86-87**, 237 (1996).

3. D. M. Mattox, Thin Solid Films **53**, 81-96 (1978).

4. J. A. Thornton, J. Vac. Sci. Technol. A **4** (6), 3059 (1986).

5. H. Windischmann, Crit. Rev. Sol. State and Mat. Sci. **17** (6), 547 (1992).

6. M. A. McKernan, Surf. Coat. Tech. **49**, 411 (1991).

7. Y. Hirohata, Y. Nemoto, T. Hino, and T. Yamashina, Thin Solid Films **214**, 150 (1992).

8. J. Hershberger, Z. U. Rek, S. M. Yalisove, J. C. Bilello, "Structure of an Amorphous Thin Buried Layer", accepted for publication in <u>SSRL 1997 Activity Report </u>Proposal 2346.

9. L. H. Schwartz, J. B. Cohen, <u>Diffraction from Materials</u>, 2nd. ed., Springer-Verlag Berlin, Heidelberg, 1987, p. 377.

10. J. Hershberger, F. Kustas, Z. U. Rek, S. M. Yalisove, and J. C. Bilello, "Structure Determination of B_4C And SiC Thin Films Via Synchrotron High-Resolution Diffraction", accepted for publication in <u>Thin Films-Stresses and Mechanical Properties, MRS Proc. Fall 1997.</u>

PROBING STRESS STATE AND PHASE CONTENT IN ULTRA-THIN TA FILMS

J.F. WHITACRE[1], Z.U. REK[2], S.M. YALISOVE[1], AND J.C. BILELLO[1]

[1]Center for Nanomaterials Science, Department of Materials Science and Engineering, University of Michigan, Ann Arbor MI 48109

[2]Stanford Synchrotron Radiation Laboratory, Stanford University, Stanford, CA 94309

ABSTRACT

Ta films 25Å to 200Å in thickness were sputter-deposited using different sputter gas (Ar) pressures and cathode power settings. The average in-plane stresses were determined using double crystal diffraction topography (DCDT). X-ray analysis (using the grazing incidence x-ray scattering (GIXS) geometry) was performed using a synchrotron light source. To study microstructure and phase content, transmission electron microscopy (TEM) and transmission electron diffraction (TED) were used. Well resolved x-ray patterns were collected for all of the films. The DCDT stress data was found to be consistent with stress effects evident in the GIXS data. In general, residual stress state was not strongly dependent upon Ar pressure. The strongest evidence of amorphous content was found in both x-ray and TED data taken from 25Å thick films deposited using 2mTorr Ar pressure and 460 W cathode power. These results show that it is possible to create and study ultra-thin Ta films which posses a range of residual stresses and phase compositions.

INTRODUCTION

The ability to study and control residual stress state and phase content in very thin metallic films is important; these layers are used in many applications, including tribological coatings, microelectronic circuits, multilayer films, and x-ray optics [1,2,3]. Previously, very thin Ta films were found to display extremely high compressive stresses (up to ~-3 GPa) [4,5]. Grazing incidence x-ray scattering (GIXS) analysis showed that Ta films 25Å to 50 Å thick were partially amorphous in structure [5]. The factors controlling the formation of this type of microstructure could include impurity incorporation, adatom kinetics (and energetics), deposition rate and geometry, and non-equilibrium phase formation due to large residual stress. Described below is a study aimed at understanding how two deposition parameters, sputter gas pressure and cathode power, affect the phase content and stress state in very thin Ta films.

PROCEDURE

Ta films were sputter-deposited (planar magnetron) onto silicon (100) test-grade wafers. Growth was conducted at room temperature in a custom-designed Denton™ vacuum chamber (for a detailed description of the sputter system, see reference [6]). The substrates rested on a rotating platen, allowing multiple films to be grown simultaneously. Four sets of four films (~25Å, ~50Å, ~100Å, and ~200 Å thick) were grown under the following conditions: (1) 2 mTorr, 100 W, (2) 10 mTorr 100 W, (3) 2 mTorr, 460 W, (4) 10 mTorr, 460 W. X-ray reflectivity data was used to calibrate film thickness for each set of deposition parameters. The average residual stress in all films was calculated from lattice curvature measurements of the substrates acquired via double crystal diffraction topography (DCDT) [7]. Two or three films at each thickness for each deposition parameter were analyzed.

Standard lab x-ray sources are not powerful enough to directly study such thin films. The Stanford Synchrotron Radiation Laboratory (SSRL) was used instead. All experiments used the eight-pole focused wiggler station on beamline 7-2 (BL 7-2) (3 Gev and 100 mA at fill). A silicon(111) monochrometer was used to select the wavelength of the x-ray beam, 0.124 nm (10 keV), in focused mode. The vertical and horizontal divergence of the beam were 0.2 mrad and 3 mrad, respectively. 1 by 1 mm slits were used to define the incident beam and 1 mrad soller slits limited diffracted beam divergence. Diffracted x-ray intensity was measured using a (liquid nitrogen cooled) solid-state Ge detector tuned to 10 keV. Samples were manipulated using an automated Huber four-circle diffractometer. Diffractometer motion and data collection were coordinated using a dedicated unix workstation.

The incident beam dose was measured by a scintillation detector placed upstream (off of a beam-splitter) from the incident beam slits. All measurements were conducted using a set dose for each data point, eliminating any possible experimental error due to beam fluctuation.

Since the GIXS geometry was used, a very low incident angle created a relatively large volume of diffracting material. 2θ was scanned from 10 to 90 degrees while the incident beam was held at an incident angle of ~0.4° (just above total external reflection). Scans were conducted in both symmetric (diffraction vector scanned in the plane of the film) and asymmetric (diffraction vector scanned normal to the film). The (single crystal) silicon substrates were oriented at a particular in-plane angle with respect to the incident beam for all scans. This orientation was selected such that there were minimal diffraction peaks from the substrate in the 2θ range of interest. A scan of a bare silicon (100) wafer was used for background (surface and air) scattering subtraction. To minimize noise from atmospheric scattering, some scans were collected in a He environment.

TEM and TED data was collected using a JEOL 2000 TEM (accelerating voltage = 200 kV). Plan-view TEM samples were created by cutting the substrates into 3mm discs, mechanically thinning them from the back, and chemically etching them to perforation with a HNO_3:HF:CH_3COOH mixture (3:5:3 ratio). The films of interest and a very thin (100Å to 500Å) Si layer remained in electron transparent areas of the sample. Bright field, dark field, and diffraction patterns were collected.

RESULTS, DISCUSSION

Figures 1a and 1b show the results of DCDT stress analysis. For films grown at a cathode power of 460 W, the thinnest films (~25Å) are under less residual stress than thicker films grown under the same conditions. For films grown using cathode power = 100W, the stress is greatest for the 25Å thick films and decreases for thicker films.

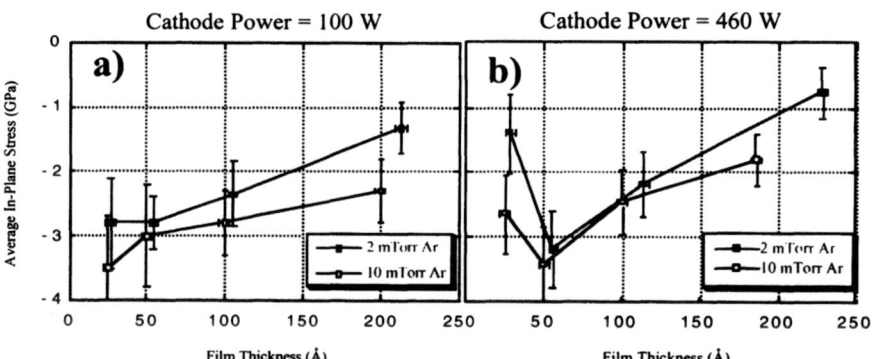

Figure 1: Double crystal diffraction topography (DCDT) analysis for films of various thicknesses grown using either 2 or 10 mTorr of Ar. (a) Cathode power = 100W. (b) Cathode power = 460 W.

In general, there is little difference in stress state with respect to Ar pressure. This is much in contrast to previous studies, which have shown a strong dependence of residual stress on sputter gas pressure during DC magnetron sputtering of Ta [3,4], These studies, however, were of films which were grown to at least 100nm in thickness before analysis.

Figures 2a and 2b show symmetric GIXS scans of 25Å, 50Å, and 100Å thick films deposited under two different sets of deposition conditions. In figure 2a, the cathode power was 100 W and the Ar sputter gas pressure was 10 mTorr, while figure 2b shows the same analysis for films grown using 460 W at 2 mTorr Ar pressure.

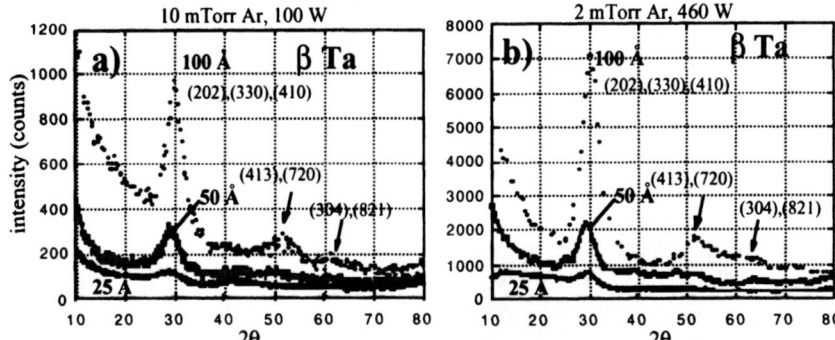

Figure 2: Asymmetric GIXS data collected from Ta films 25 Å, 50 Å and 100 Å thick deposited using: (a) 10 mTorr Ar, 100 W, and (b) 2 mTorr Ar, 460 W.

These two parameter sets were chosen because they created the most contrasting stress states in their respective 25 Å thick films (according to DCDT analysis).

The 2θ positions of the diffraction peaks in both cases are consistent with those expected from β Ta, which has a tetragonal unit cell. The first large maxima at 2θ ~ 30° is a superposition of the (410), (330), and (202) peaks, while the maxima at 2θ ~ 55° is from a superposition of the (413) and (720) peaks.

There is a significant shift in peak location with respect to thickness in the films deposited using 100 W cathode power and 10 mTorr Ar pressure (figure 1). In the thinner films, the main peak shifts to lower 2θ, indicating that the d-spacings out of the plane of the substrate (as measured by the asymmetric GIXS diffraction vector) are expanding. This implies that there is a compressive in-plane stress, as observed using DCDT analysis. The broad peaks for the 50 Å and 25 Å films consistent with films that are nanocrystalline or amorphous in nature, though could also be caused by the existence of microstrains. For similar films grown under 460 W cathode power and 2 mTorr Ar pressure (figure 2b), there is also shift of peak location with respect to film thickness. These peaks, however have shifted less for the 50 and 25 Å films, as expected from DCDT analysis.

Figures 3a and 3b show a comparison between symmetric and asymmetric GIXS data recorded (in He ambient) from 25Å Ta films grown under the same parameters as shown in figure 2.

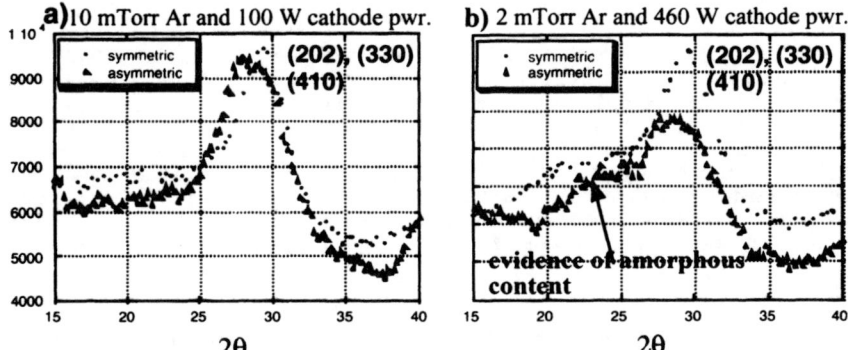

Figure 3: Detail of symmetric and asymmetric GIXS data collected for 25Å thick films deposited using: (a) 10 mTorr, 100 W, and (b) 2 mTorr, 460 W.

The relative peak shifts are consistent with a highly compressive in-plane stress: in the symmetric scans, peaks are shifted to higher 2θ (smaller in-plane d spacing), and to lower 2θ in the asymmetric scans, as previously described. There is a significant difference between these two scans in that the film deposited under 2mTorr Ar and 460 W shows a well defined, broad maxima at $2\theta \sim 23°$. This diffracted intensity is not expected by any crystalline Ta phase and is not a result of substrate or background scattering. This broad peak is consistent with the existence of amorphous material in the film.

Figures 4a and 4b show the TEM/TED analysis of the same two films.

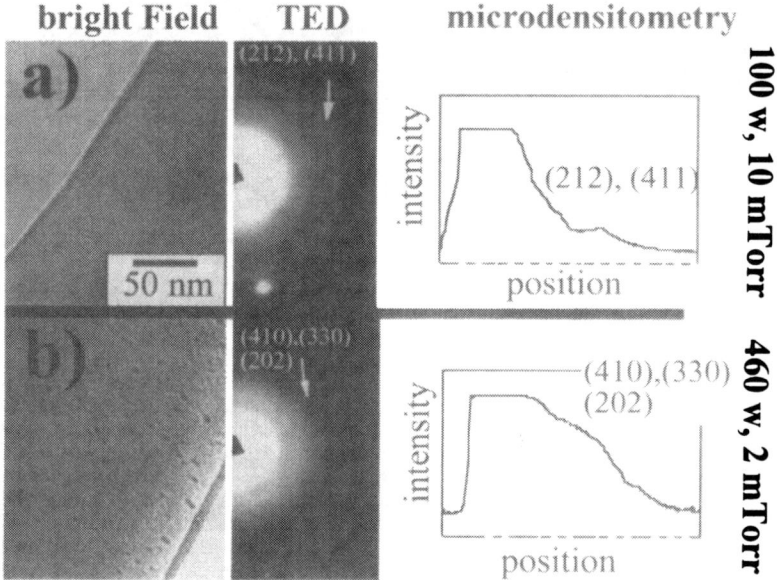

Figure 4: TEM analysis of 25 Å thick Ta films. Shown are bright field images, transmission electron diffraction (TED) and microdensitometry images for films grown using (a) 100 W, 10 mTorr, and (b) 460 W, 2 mTorr. There is more amorphous scattering in the 460 W, 2 mTorr film.

Different microstructures are observed. The film grown under 10 mTorr and 100 W has small (< 3 nm) equaxial grains, whereas the film grown using 2 mTorr and 460 W has grains which vary in size. The electron diffraction patterns and corresponding microdensitometry plots show a significant difference in crystalline content between the two films. There is a well defined diffraction ring corresponding to the (212) and/or (411) d-spacings for the film grown in 10 mTorr 100W. The diffraction pattern for other film (2 mTorr, 460W) has no well-defined ring, which is consistent with diffraction observed from amorphous materials. The TED results agree with the GIXS results: there is strong evidence that the ~25Å film grown under 2 mTorr and 460 W has a higher amorphous content than the 25Å film grown using 10 mTorr and 100 W.

CONCLUSIONS

The effects of sputter cathode power and sputter gas pressure on the formation of very thin Ta films has been studied. DCDT stress analysis and GIXS were successful probes in examining films as thin as 25Å. Sputter gas pressure and residual stress state were not strongly linked for these films. Phase content was found to vary. The films deposited using 2mTorr Ar and 460 W (residual stress = ~ -1.5 GPa) displayed strong evidence of amorphous scattering using GIXS.

Those films grown using 10 mTorr and 100 W (residual stress = ~-3.5 GPa) did not display the same degree of amorphous scattering. These results were verified using TEM/TED analysis. Those films under higher stress do not posses higher amorphous content, as might be expected. Further studies will focus on quantitative analysis of the GIXS data. Models which describe the sputter deposition process will be considered and related to measured stress state and crystalline content values.

ACKNOWLEDGMENTS

Thanks to J. Hershberger for significant aid in data collection at SSRL.

This work was supported under ARO contract number DAAH 04-95-1-0120.

Work done (partially) at SSRL, which is operated by the Department of Energy, Office of Basic Energy Sciences.

REFERENCES

1. K. Holloway, P.M. Fryer, Appl. Phys. Lett. **57** (17) (1990) p 1736.
2. D. Dietsch, Th. Holz, S. Hopfe, H. Mai, R. Scholz, B. Schoneich, H Wendrock. Fresenius J. Anal. Chem. (1995) 353, p 383.
3. L.A. Clevenger, A. Mutscheller, J.M.E. Harpter, C. Carbral, Jr., and K. Barmak, J. Appl. Phys. **72** (10) (1992) p 4918.
4. H. Windischmann, Critical Review in Solid State and Materials Science, **17**, (1992) p 547.
5. L. K. Parfitt, Stress and Impurities in Ultra-thin Ta films. Unpublished work (Ph.D. thesis)
6. O.P. Karpenko, J.C. Bilello, S.M. Yalisove, J. Appl. Phys. **76** (8) (1994) p 4610.
7. J. Tao, L.H. Lee, and J.C. Bilello, J. Elect. Mat. **20** no 10 (1990) p 819.

CRYSTALLOGRAPHIC ANALYSIS OF CVD FILMS
BY USING X-RAY POLYCHROMATIC RADIATION

B. LAVELLE*, L. BRISSONNEAU, E. BAGGOT AND C. VAHLAS

Laboratoire Matériaux et Interfaces, INPT-CNRS, ENS Chimie de Toulouse,
118, route de Narbonne, 31077 Toulouse cedex 4, France, cvahlas@ensct.fr
* Centre d'Elaboration de Matériaux et d'Etudes Structurales, CNRS,
29, rue J. Marvig, 31055 Toulouse cedex 4, France, lavelle@cemes.fr

ABSTRACT

The Energy Dispersive X-ray Diffractometry (EDXD) technique was tested for in-situ crystallographic characterization of nickel films processed by chemical vapor deposition (CVD). The diffracted beam at low Bragg angle was analyzed in energy by a solid state detector. A nickel reference sample was used to face the problems of EDXD background signal and uncertainty of sample location. The relative accuracy on lattice parameters measurements is $1.5.10^{-3}$, to be compared to $0.5.10^{-3}$ for classical (monochromatic) X-Ray diffraction. Texture measurements yields results in agreement with those obtained from recent texture goniometer. Finally, an estimation of the thickness was obtained from the intensity of nickel fluorescence peak. In view of the obtained results, EDXD appears to be a promising technique for in-situ studies. Although less powerful compared to the synchrotron facility, it is more flexible and can be applied at lower cost.

INTRODUCTION

Chemical vapor deposited (CVD) films are typically studied ex-situ by monochromatic X-ray diffraction, often in grazing incidence. When the evaluation of the characteristics of the films during their growth is necessary, in-situ measurements can also be performed using a synchrotron radiation. Energy Dispersive X-ray Diffractometry (EDXD) was also proposed in early '80 for stress determination [1], despite its poor accuracy at that time. Further attention is actually paid to this technique, due to recent improvements of solid state detector part [2]. Recent reflectivity measurements with a similar technique show that information about few nanometers thick interfacial layers can be collected in-situ[3]. As the diffracted beam is energy dispersive analyzed, several crystallographic planes can be observed at the same time for a fixed Bragg angle. Moreover, by tilting the detector, information on the texture and thickness of the deposit can be collected. In the present work, the application of EDXD using the polychromatic radiation issued from a classical X-ray tube, is investigated as an alternative to synchrotron radiation based techniques for the in-situ studies of CVD films. Results obtained by this new technique are compared with results obtained by monochromatic XRD and scanning electron microscopy (SEM) concerning a textured pure nickel sheet and two CVD nickel coatings.

EXPERIMENT

Experimental device

In order to investigate the feasibility of in-situ crystallographic analysis of CVD films by EDXD, a texture goniometer was equipped, as shown on Fig.1a, with crossed Soller slits and a germanium solid state detector. This setup was originally built for residual stress measurements [4]. The sample was set in the same relative position, versus incident and diffracted beams, as it would be in a CVD reactor (Fig.1b). The Ψ rotation on the goniometer simulates the motion of the detector on a circle which center lies on the transmitted beam and radius is a function of the Bragg angle Θ (Fig.1b). Using synchrotron radiation, the detector location would be the same. With monochromatic X-ray, the detector would move in the dashed area of figure 1b), which limits are related to the range of the investigated Bragg angles.

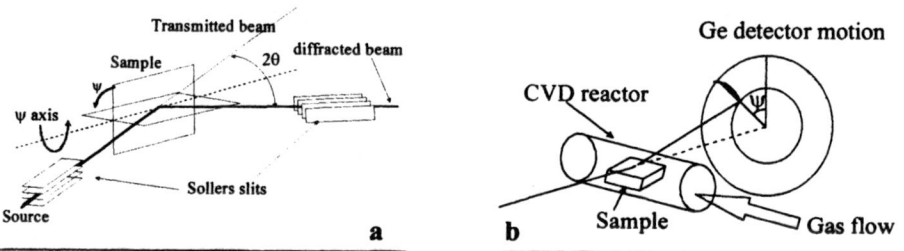

Figure 1 : a) Scheme of the experimental device ; b) configuration for in-situ measurements

The X-ray source was a classic Ag tube, supplied by 50 KV and 20 mA. The divergence value of the Soller slits is 0.3°. This is a compromise between the intensity and the width of the energy peak in the diffracted beam spectrum. The Bragg angle θ was chosen in such a manner that the diffraction peaks do not interfere with the characteristic radiations of the X-ray tube, with the fluorescence peaks of the sample, and with their respective escape peaks depending on the used detector. The relatively high energy radiations, which are necessary for the beam to pass through the silica walls of the reactor in an in-situ application, lead to a small value for θ angle. This characteristic is an advantage for the coupling of the EDXD with a cold wall CVD reactor in view of the space occupation of the heater device of the latter.

The inter-reticular distances of the deposits were measured by Θ-Θ experiments on a Seifert XRD 3000 TT. Texture was studied on a Seifert XRD 3000 PTS texture goniometer. The sample thickness was measured on LEO-435 SEM micrograph.

Materials studied

We studied two thin films of nickel processed by MOCVD from nickelocene (Ni(η_5-C_5H_5)$_2$) decomposition in He/H$_2$ atmosphere. We used glass substrates on a graphite susceptor heated at 175°C by HF induction, under partial (100 Torr) or atmospheric pressure. The film processed at low pressure is contaminated with 10 at.% of carbon, whereas the carbon contamination of the other film is lower than 3at.% [5]. A pure, thick (30 μm), highly textured and stress-free nickel sheet was used as reference sample. For nickel characterization, a θ angle near 10° is convenient. Five different ψ angles: 0, 28, 42, 54.5 and 70° were used.

Procedure for spectrum analysis

The spectrum (Fig.2) was collected on a multi-channel analyzer of 2048 channels. By contrast with a synchrotron radiation, the incident beam intensity varies with energy. In particular, two dominant peaks can be detected, corresponding to the silver fluorescence (22.1054 KeV for K$_\alpha$ and 24.9424 KeV for K$_\beta$). The detected beam is composed of :
i) the diffusion of the incident beam by the layer and the substrate,
ii) fluorescence peaks of the atoms of the deposit only, as the atoms of the substrate are too light,

iii) the diffraction of the nickel layer only, as the substrate is amorphous.

For each measurement, the uncertainty on energy due to the detector is about 5%. In order to increase the accuracy for the calculation of the inter-reticular distance, a mathematical model has to be chosen to describe the peaks after background correction. It was shown in a previous work, that a relative accuracy of about 10^{-3} can be obtained [4] by a gaussian fitting. After reducing the electronic noise by a seven point smoothing, the background appears to be related to the ψ angle value. It was also considered that the higher contribution to the background intensity comes from the diffusion from the nickel layer. It was not possible to estimate the nickel diffusion contribution from aqueous solution of nickel salts or metallic amorphous nickel sheet. Thus, the above mentioned reference sample was used to find an empirical background model. On this sample, all the energy peaks are known. In order to measure background intensity versus the ψ values five energies were selected far enough from the expected peaks. Under 24 KeV, the background was weak. Above 24 keV, its intensity I, as a function of energy E and the angle ψ, is well described by the following relation :

$$I = \sqrt{(a_0 + a_1 E + a_2 E^2)\psi + b_0 + b_1 E} \qquad (1)$$

where a_0, a_1, a_2, b_0, and b_1 are experimental constants

After background subtraction and gaussian fitting of each peak, a precise energy calibration of each spectrum for the different ψ was performed, using silver and nickel K_α and K_β peaks, which are not affected by geometrical considerations. The actual Θ angle value for the experiment at ψ = 0° was calculated to be 9.834° ($\pm 4.10^{-3}$) from energies measured for diffraction peaks of (111), (200), (220), (311), and (222) planes, and using lattice parameters from JCPDS databank. The location of the sample with respect to the geometrical center of the goniometer induces a variation on Θ when ψ varies. This variation can be expressed in degrees by the following relation [6] :

$$\delta\theta = \frac{180 * \cos\theta}{\pi * R} \cdot \left[z * \left(\frac{1}{\cos\psi} - 1 \right) - e * \tan\psi \right] \qquad (2)$$

where e and z are respectively the distances between the rotation axis ψ and the plane of diffraction, and between the rotation axis ψ and the sample surface (Fig; 3), and R is the goniometer radius. e and z have been evaluated from the variation of $\delta\Theta$ versus the ψ values on the reference sample to be respectively 0.21mm and 0.12mm. Thus for the used Θ value, a relative accuracy of 10^{-3} on inter-reticular distances can be obtained only for an accuracy on e and z values better than 10^{-2} mm.

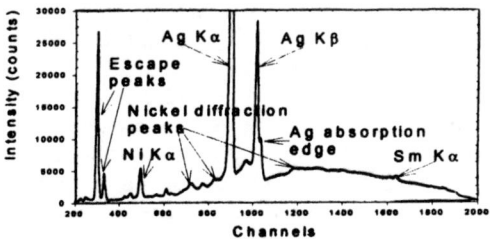

Figure 2 : Diffracted beam spectrum at $\Theta \simeq 10°$

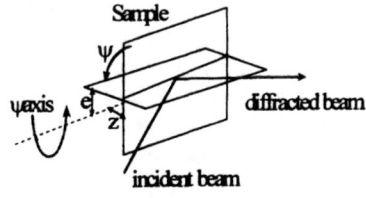

Figure 3 : Definition of e and z parameters for sample location

RESULTS AND DISCUSSION

Inter-reticular distances

Equation (1) was used to select the energies from which the background intensity could be measured. Thus, it means that a linear relationship exists between the ψ value and the square of the intensity I. Some evident fluorescence energies were identified among the energies that exhibit high values for the correlation coefficient of the above mentioned relationship. So a second criterion was added, following which the selected energies are positioned on a local minimum. The intensity of the background presents a discontinuity at the silver absorption edge (25.5 KeV). Above this edge, the background intensity can be fitted by a second order polynomial. For lower energies, two third order polynomials with one common point around 13 KeV were used. The obtained spectrum, background simulation and treated spectrum are shown on Fig.4 for the low pressure CVD sample. After background subtraction, the peaks were fitted by gaussian curves.

The inter-reticular distances (d) for (111) and (200) planes were calculated with Θ, e and z values previously measured. They are compared in Fig5. with those obtained by a classical Θ-Θ monochromatic characterization. By using this latter technique, a relative increase of $2.5.10^{-3}$ with a relative accuracy of $0.5.10^{-3}$ was evidenced for the lattice parameters [7]. In EDXD experiments, the relative accuracy for the best defined energy peak ((111) planes) is measured as $1.5.10^{-3}$. Thus, EDXD appears to be a suitable technique to appreciate "in-situ" the lattice parameters variations.

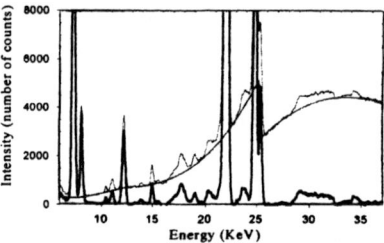

Figure 4 : Experimental spectrum (gray), background simulation (dark gray) and treated spectrum (dark black)

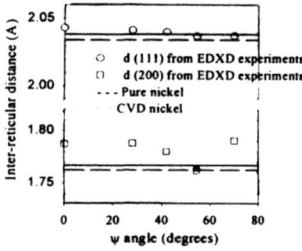

Figure 5 : Variation of calculated (d) as a function of the Ψ angle, compared with monochromatic Θ-Θ and JCDPS databank

Texture analysis

Intensities of EDXD peaks in transmission mode are related to the sample texture [4]. On the reference nickel sample, classical X-ray texture measurements were performed, with Copper $K\alpha$ radiation, on (111) and (200) planes. A strong [200] fiber texture was determined. Mean cuts along the radius of measured pole figures are shown on fig 6a. On the EDXD spectra, the intensities of energy peaks related to (200), (220), (311), and (222) planes were measured, for the five Ψ values. For each peak, the background was evaluated by a straight line in the peak vicinity; and for each spectrum, the intensity of the $K\alpha$ nickel fluorescence peak was measured in the same way. The ratio between peak intensity of each plane and fluorescence peak intensity versus the Ψ values is also plotted in Fig6a,. The correlation between these results and the classically determined texture demonstrates the feasibility of texture analysis in reflection by EDXD.

For the film processed at atmospheric pressure, the same figure (fig. 6b) can be obtained from both techniques. As the intensity of all planes increases in the same way versus Ψ, this sample appears thin and non-textured.

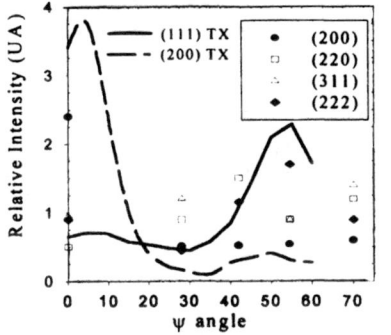 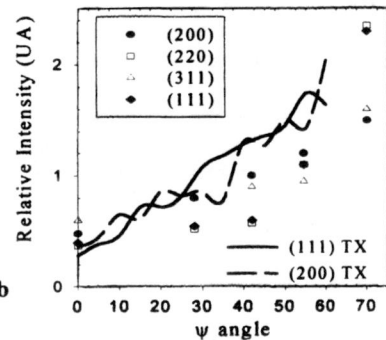

Figure 6 : Comparison of mean cut of pole figures from classical texture measurements (TX, lines) and relative intensities from EDXD measurements versus ψ angle (points) for :
    a) reference sample, b) CVD deposit.

Thickness Measurements

Fluorescence peaks can also yield information on the composition of the sample. However, due to the lower energy limit of the detector, this holds only for the heavy elements. In that case, the thickness of the film can also be evaluated. After calibration of the apparatus by using the nickel reference sample, the classical equation used in fluorescence analysis [8] was applied. At Θ near 10° and $\psi = 0$°, this technique can yield thickness of nickel films up to 2μm. The accuracy of this measure depends on the film roughness and its exact composition. The thickness of the film processed at atmospheric pressure is 0.6 μm to be compared with the mean value of 0.6μm obtained for ten measures on the SEM micrograph shown Fig.7. Consequently EDXD could be used for the estimation of the film growth rate during the process.

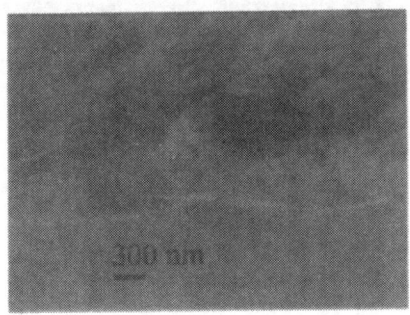

Figure 7 : SEM micrograph of a nickel film processed at atmospheric pressure. A precise value can't be attributed to the thickness (0.45 to 0.65 μm) because of the nodular morphology

125

CONCLUSION

The EDXD technique was found to be promising for in-situ crystallographic studies of films processed by cold wall CVD. It can provide in a single experiment information on lattice, parameters, texture and mean thickness, if a one hour time of exposure can be accepted. Compared with monochromatic X-ray diffraction, the motion of the detector is clearly simpler. For the lattice parameters measurements, the EDXD accuracy is lower than the classical X-ray diffraction one, but sufficient enough to measure a relative variation more than 2.10^{-3}, if the sample location is known with an error less than 10^{-2} mm. The sample location can be measured with the needed accuracy by using a reference sample.

For texture analysis, the results agree with those obtained from a classical specific X-ray equipment. Finally, the sample thickness can be appreciated as in fluorescence studies.

Compared to the synchrotron facility, EDXD is clearly less accurate and can't be used for stress determination. But its higher flexibility and its far lower cost are advantages for pre-studies saving synchrotron time.

ACKNOLEDGMENTS

Bruno Lavelle thanks the Seifert society and Alexandre-Labo for technical support.

BIBLIOGRAPHY

[1]   C.J. Bechtoldt, R.C. Placious, W.J. Boettinger and M. Kuriyama, *Advances in X-Ray Analysis* **25** (1982) 329-337.

[2]   H. Ruppersberg and I. Detemple, *Materials Science and Engineering* **A161** (1993) 41-44.

[3]   E. Chason and M. Chason, *Journal of Vacuum Science and Technology A* **12** (1994) 1565-1568.

[4]   B. Lavelle and J. Jaud, Transmission X-ray stress determination using a polychromatic radiation on aluminum alloys and stainless steel, presented at ICRS5, Linköping1997.

[5]   L. Brissonneau, A. Reynes and C. Vahlas, Morphology and carbon content of nickel films deposited from nickelocene, presented at Chemical Vapor deposition XIV and EUROCVD-11, Paris,97-25, 1997.

[6]   F. Convert and B. Miege, *Journal of Applied Crystallography* **25** (1992) 384-390.

[7]   L. Brissonneau, C. Vahlas and Reynes, Processing of pure nickel MOCVD films, presented at Symposium I, MRS Spring Meeting, San-Francisco1998.

[8]   J.-P. Eberhart, in Analyse structurale et chimique des matériaux, 2nd (Dunod, Paris, 1997), p.319-340.

IN SITU MONITORING OF THE ELECTROCHEMICAL ABSORPTION OF DEUTERIUM INTO PALLADIUM BY X-RAY DIFFRACTION USING SYNCHROTRON-WIGGLER RADIATION

D.D. DOMINGUEZ, P.L. HAGANS, E.F. SKELTON, S.B. QADRI AND D.J. NAGEL
U.S. Naval Research Laboratory, Washington, DC

ABSTRACT

With low energy x-rays, such as those from a Cu x-ray tube, only the outer few microns of a metallic sample can be probed. This low penetrating power prohibits structural studies from being carried out on the interior of an electrode in an electrochemical cell because of absorption by the cell material, electrodes and the electrolyte. The work described in this paper circumvents this problem by utilizing high energy, high brightness x-rays produced on the superconducting wiggler beam line, X-17C, at the National Synchrotron Light Source (NSLS) at Brookhaven National Laboratory. The penetrating power of the higher energy x-rays allowed Pd diffraction spectra to be obtained *in-situ* on a 1 mm diameter Pd wire cathode during electrolysis of heavy water. Moreover, the beam (28 x 28 μm in cross-section) allowed diffraction spectra to be acquired as a function of distance across the sample. Spectra were recorded in 50 μm steps from the edge of the Pd wire to its core. This was done at 2 minute intervals as a function of electrolysis time. The α-β phase transition induced in the Pd while deuterium was electrochemically absorbed was observed by monitoring the Pd-(422) diffraction peaks. Results allowed the diffusion rate and the diffusivity of deuterium atoms in the Pd wire to be determined. Other features of the structural changes associated with the absorption of deuterium into Pd are reported.

INTRODUCTION

As hydrogen or deuterium are incorporated into the palladium metal lattice at room temperature, the material passes through three different crystallographic regions. For PdH_x or PdD_x, these are the pure α-phase for $x<0.03$, the α,β-mixed phase region (known as the miscibility gap) for $0.03<x<0.07$, and the pure β-phase at $x>0.7$. Pure Pd and all the PdH_x and PdD_x phases are face-centered cubic. The work described in this paper was initiated to obtain *in situ* crystallographic information during the electrochemically induced absorption of deuterium into a Pd cathode and, thereby, ascertain information about the mechanism of deuterium diffusion and formation of the different phases. Our approach takes advantage of the highly intense radiation provided by a synchrotron equipped with a superconducting wiggler magnet and a beam that defines a microspot 1.8×10^{-13} m^3 in volume on the Pd electrode. The data presented represents the first time that Pd lattice expansion has been measured in an electrochemical cell as a function of electrolysis time and distance across the electrode.

EXPERIMENT

Electrochemistry. The absorption of deuterium into a 1 mm diameter Pd wire cathode was carried out electrochemically in a 1 x 1 x 4.5 cm polyethylene cuvette that served as the cell. The wire cathode (3 cm long) was rigidly mounted and centered in the cell with Teflon blocks located at the top and bottom of the cuvette. The cathode was surrounded by a 0.254 mm thick graphite foil anode. The anode and cell materials were selected because of their transparencies to x-rays. Both the Pd cathode and the foil anode held platinum lead wires that extended through the Teflon cell top. The Pt lead wires and three additional Pt wires that were attached to the cathode for resistance measurements were isolated with heat shrinkable Teflon tubing inside the cell. A schematic of the cell and its contents (electrodes, wires and centering blocks) is shown in Figure 1.

The cell was filled with enough electrolyte to cover the entire length of the cathode. The electrolyte was 0.1 M LiOD in D_2O made by dissolving Li metal in D_2O. Heavy water electrolysis occurred when a charging current was applied to the cell and deuterium, formed during the electrolysis, was electrochemically deposited onto the surface of the Pd wire. Initially, the current density on the cathode was 1 mA/cm^2; it was gradually increased to 300 mA/cm^2 over the course of the experiment (36 hours). Electrolysis gases (D_2 and O_2) were allowed to exit the cell through a hole in the Teflon block at the top of the electrodes. Electrolyte in the cell was replenished periodically with D_2O. The frequency of additions varied depending on the applied current, but the electrolyte level was always kept above the electrodes. Throughout the experiment the change in the axial electrical resistance of the Pd cathode was continuously monitored *in situ* via a standard four-point probe resistance measurement technique. Resistance data were used to estimate the bulk concentration, x, of deuterium in the Pd wire from published R/R_0 vs. D/Pd curves [1]. These data are reported elsewhere [2].

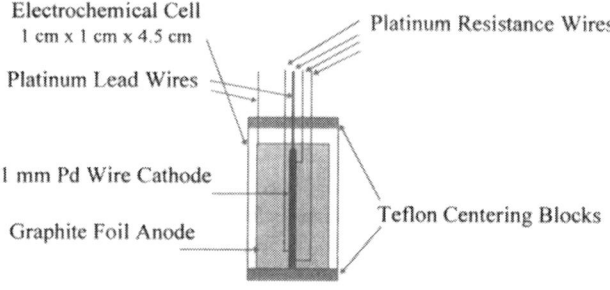

Electrochemical Cell
1 cm x 1 cm x 4.5 cm

Platinum Resistance Wires

Platinum Lead Wires

1 mm Pd Wire Cathode

Graphite Foil Anode

Teflon Centering Blocks

<u>Overall Electrochemical Reactions</u>

Pd Cathode: $2D_2O + 2e^- \longrightarrow D_2 + 2OD^-$

Graphite Anode: $2D_2O \longrightarrow O_2 + 4D^+ + 4e^-$

Figure 1. Schematic diagram of the electrochemical cell showing the Pd cathode, the graphite foil anode, Teflon centering blocks, platinum lead wires and platinum resistance wires. Also shown are the electrochemical reactions that take place at the two electrodes.

X-Ray Diffraction Measurements. Energy dispersive x-ray diffraction (XRD) experiments were carried out on the superconducting wiggler-beamline, X17C, at the National Synchrotron Light Source (NSLS), Brookhaven National Laboratory. X-ray photons available on X17C have energies in excess of 80 keV. Energies of this magnitude were necessary to determine structural changes in a metal lattice *in situ* during an electrochemical experiment and to obtain diffraction data from the interior of a wire electrode.

Our XRD experimental configuration is illustrated in Figure 2. The electrochemical cell was rigidly mounted on a translational stage which allowed movement in the three orthogonal directions (x,y,z). The incident x-ray beam was collimated to 28 x 28 μm in cross-section by polished tungsten blocks. The scattered radiation was measured with an intrinsic Ge energy-sensitive detector at a fixed scattering angle (2θ). The 2θ angle of the diffractometer was held at 13.000° throughout the experiment; it was calibrated to within ±0.001° using a gold foil standard. The energy scale of the detector electronics and multichannel analyzer was calibrated using a set of known fluorescence peaks excited with an Am^{241} radioactive source. Diffraction data were collected from a spot on the Pd wire defined by the intersection of the collimated x-ray beam and the beam path defined by apertures in front of the Ge detector. The volume of the microspot was 1.8 x 10^{-13} m^3. During the experiment, the scattering geometry remained fixed and the cell (and hence the Pd wire) was systematically translated in 50 μm steps normal to the beam. At each position across the wire, the diffraction pattern from the volume element was recorded at 2 minute intervals as a function of electrolysis time. Lattice volume changes were determined from the unit cell parameters. The unit cell parameters were determined from the energies of the α- and β- Pd-D (422) peaks as a function of deuterium absorption from the edge to the center of the Pd wire. Results are reported as volume change $\Delta V/V_{Pd}$ where $V_{Pd}=a^3$ where a is the Pd lattice (unit cell) parameter and $\Delta V=V_{PdD}-V_{Pd}$ where V_{PdD} is the volume of a particular PdD phase.

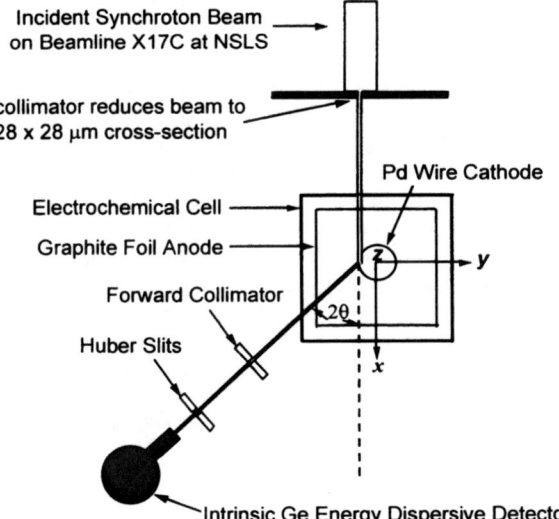

Figure 2. Schematic diagram of the XRD experimental configuration on X17C at the NSLS, showing the collimated beam, the electrochemical cell, and the energy dispersive detector at a fixed 2θ scattering angle.

RESULTS

Figure 3 shows the energy dispersive x-ray diffraction pattern obtained from a spot near the edge of the Pd wire cathode before D_2O electrolysis began. A unit cell parameter of a_{Pd}=3.8874 ± 0.0006 Å was determined from the 10 diffraction peaks shown in the figure. This is slightly less (0.05%) than the accepted value of a_{Pd} (3.8898 Å) [3]. In contrast, Figure 4 shows the x-ray diffraction spectrum recorded from the center of the Pd wire cathode shortly after the start of electrolysis. In this region of the wire, the (422) diffraction peaks of the α- and β-phase Pd-D were the lowest energy (65-70 keV) peaks observed; peaks of lower energy were lost due to absorption by the electrolyte and the Pd. Since the (422) diffraction peak was found in the spectrum of pure Pd (Figure 3) and in both Pd-D phases (Figure 4) this peak was used to monitor the crystallographic changes that occurred during the absorption of deuterium into the Pd lattice.

A sequence of XRD spectra was recorded in 50 μm steps across the Pd wire as a function of increasing electrolysis time. The sequence shows the decay of the α-phase and the growth of the β-phase with time. In the α,β-mixed phase region these data were used to determine diffusion rates for deuterium atoms through the wire. This was accomplished by using a least-squares fitting routine to determine the linear curves representing the time dependence of the normalized integrated intensities of the (422) peaks. The slopes of the lines gave an average diffusion rate of 57 ± 8 nm/s at a current density of 1 mA/cm^2. Assuming this rate, the total time for the formation of the pure β-phase in the 500 μm diameter wire was 2.9 hrs. This is approximately five times longer than what was calculated by Fick's second law of diffusion for a cylindrical electrode geometry [4]. The integrated intensities of the diffraction peaks from each phase were also used to compute the diffusivity, D, of the deuterium atom in Pd lattice according to Fick's second law [5]. From our data, D was determined to be 2-3 x 10^{-11} m^2/s [5] which is close to the literature values determined from different types of diffusion experiments [6,7]. Details of our calculations are reported elsewhere [2].

A plot of lattice volume changes that occur as a function of electrolysis time for α- and β- Pd-D phase formation at the edge and center of the Pd wire electrode is shown in Figure 5. Two important observations are made from this figure. First, during the initial 150 minutes of electrolysis, the α- and β-phases coexist at the surface and at the core of the wire; after 200 minutes only the pure β-phase is observed in these regions. Second, β-phase lattice expansion decreases at the edge of the wire with increasing electrolysis time while it increases in the center. Transition from the α,β-mixed phase region to the pure β-phase occurs at the time when the volume change is equal in both regions. In addition, the time of the transition to pure β-phase Pd-D coincides with the disappearance of the α-phase in the center of the wire. While the later is not surprising, the former observation implies that the transition from the α,β-mixed phase region to the pure β-phase occurs simultaneously in both the edge and the center of the wire.

CONCLUSIONS

In summary, the results of our *in situ* XRD measurements on a Pd cathode during heavy water electrolysis represent the first time that the Pd lattice expansion has been measured in an electrochemical cell as a function of electrolysis time and distance across the cathode. The data presented provide information about Pd lattice expansion, the diffusion rate and the diffusivity of the deuterium atom through the Pd lattice.

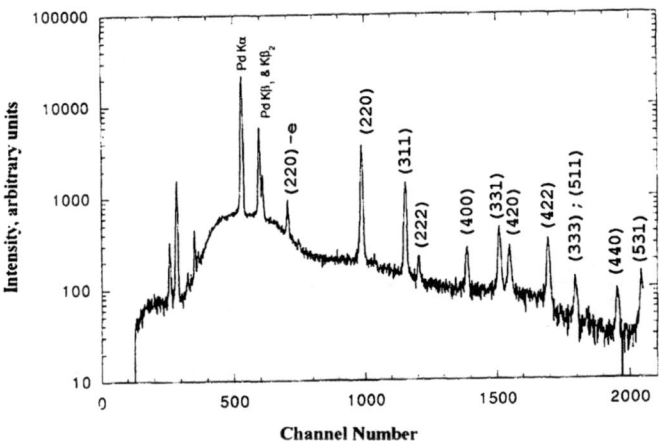

Figure 3. The energy dispersive x-ray diffraction spectrum recorded from a spot near the edge of the 1 mm diameter Pd wire cathode before D₂O electrolysis began. (Data collection time was 2.00 min.)

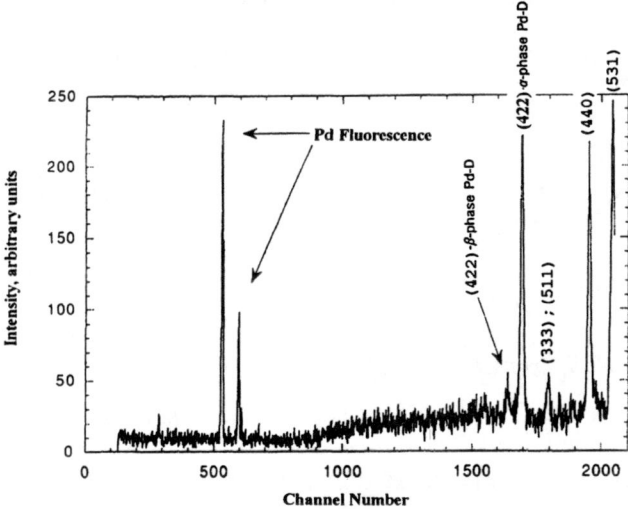

Figure 4. The energy dispersive x-ray diffraction spectrum recorded from a spot near the center of the 1 mm diameter Pd wire cathode 35 minutes after the start of electrolysis with 1 mA/cm² current density on the cathode. (Data collection time was 2.00 min.)

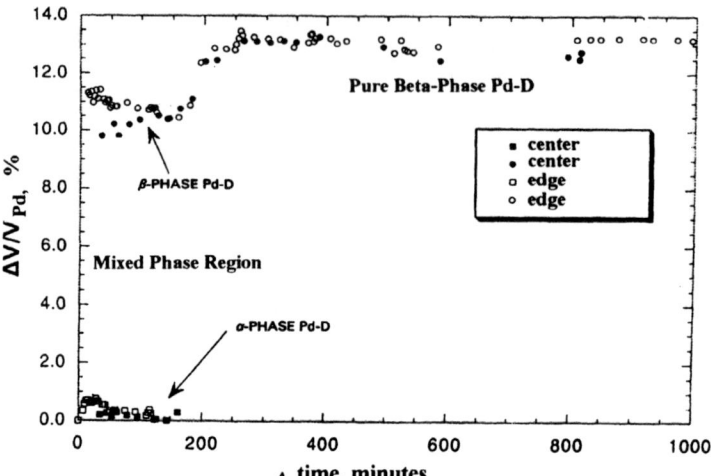

Figure 5. Percent volume change measured as a function of the total elapsed electrolysis time for α- and β-phase Pd-D$_x$ formation at the edge and center of a 1 mm diameter Pd wire cathode.

ACKNOWLEDGMENTS

        The authors gratefully acknowledge A.C. Ehrlich of NRL for assistance with the resistance measurements and J.Z. Hu of The Carnegie Institution of Washington for assistance with the XRD measurements. We also acknowledge Brookhaven National Laboratory for use of the NSLS facilities. This research was financially supported by the Office of Naval Research.

REFERENCES

1. M.C.H. McKubre, S. Crouch-Baker, R.C. Rocha-Filho, S.I. Smedley, and F.L. Tanzella, J. Electroanal. Chem. **368**, 55-66 (1994).

2. E.F. Skelton, P.L. Hagans, S.B. Qadri, D.D. Dominguez, A.C. Ehrlich, and J.Z. Hu, submitted to Phys. Rev.-B.

3. R.W.G. Wyckoff, <u>Crystal Structures</u>, Vol. 1, (John Wiley & Sons, New York, 1960), p.10.

4. E.A. Moelwyn-Hughes, Physical Chemistry (Pergamon Press, 1968).

5. H.S. Carslaw and J.C. Jaeger, <u>Conduction of Heat in Solids</u>, (Oxford: The Claredon Press, 1959), p.188.

6. G.L. Powell and J.R. Kirkpatrick, Phys. Rev.-B **43** (9), 6968-6976 (1991).

7. T.B. Flanagan and W.A. Oates, Annu. Rev. Mater. Sci. **21**, 269-304 (1991).

HIGH PRESSURE SYNCHROTRON DIFFRACTION OF $KCa_2Nb_3O_{10}$, A LAYERED PEROVSKITE COMPOUND

K. A. STEINER, W. T. PETUSKEY
Department of Chemistry and Biochemistry, Arizona State University, Tempe, AZ, 85287

ABSTRACT

High pressure synchrotron x-ray diffraction experiments were conducted on $KCa_2Nb_3O_{10}$, to determine lattice constants as a function of pressure. A diamond anvil cell was used to produce pressures up to 66 GPa. A phase transition occurred at 13.5 GPa. From the lattice constants, linear compressibilities of 8.55×10^{-4} GPa^{-1} in the a direction, -9.40×10^{-4} GPa^{-1} in the b direction, and 142×10^{-4} GPa^{-1} in the c direction, and a bulk modulus of 68.5 GPa were found for the lower pressure orthorhombic phase.

INTRODUCTION

Layered perovskites have crystal structures consisting of cation layers in the basal plane every 2 - 7 perovskite octahedra [1]. $KCa_2Nb_3O_{10}$ has a perovskite slab three $\{Nb\}O_6$ octahedra thick separated by layers of potassium in trigonal prismatic coordination [2]. Due to this layered nature, the physical properties are highly anisotropic, as shown by linear compressibilities found in this study.

Layered perovskites are proposed as an interface material in ceramic fiber-ceramic matrix composites. The attractive qualities of layered oxides are ease of fracture along basal planes and stability in air. $KCa_2Nb_3O_{10}$, has been shown to exhibit considerable fracture anisotropy [3]. We use linear compressibility as a function of crystallographic direction as another way to quantify anisotropy.

From the diffraction data, it appears that a phase change occurs when pressure is increased above 6.83 GPa. *In-situ* diamond anvil cell Raman spectroscopy was performed to confirm the presence of a phase transition and the pressure at which this transition takes place.

EXPERIMENT

$KCa_2Nb_3O_{10}$ was synthesized by combining stoichiometric Nb_2O_5 and $CaCO_3$, with 25% excess K_2CO_3, and heated at 1250 °C for 72 hours. The powder was washed with water after synthesis to remove the excess alkali. The powder was determined to be phase pure by x-ray diffraction.

$KCa_2Nb_3O_{10}$ powder was loaded into a diamond anvil cell with gold flakes as a pressure calibrant. No pressure medium was used in this experiment.

X-ray diffraction was carried out at beamline X17C at the National Synchrotron Light Source at Brookhaven National Laboratory. The first scan was taken at ambient pressure with a 2θ value of 12.5°. The pressure was increased and a scan was taken at a 2θ of 12.5°. Another scan was taken at this pressure with a 2θ value of 10°. All subsequent scans were taken at $2\theta = 10°$.

When diamond anvil cells are loaded without a pressure medium, the pressure is nonhydrostatic, which results in a pressure gradient across the cell. However, gold diffraction peaks were visible in all of the $KCa_2Nb_3O_{10}$ scans, so there should be no error in calculated pressure due to a pressure gradient. Pressure was calculated from the gold unit cell volume and the Birch-Murnaghan equation of state [4].

The incident radiation at NSLS has an intensity profile which is not constant; the radiation with low energy has a higher intensity than the radiation with high energy [5]. In addition, Compton scattering from diamond affects the intensity of the incident radiation. Therefore, the intensities of diffracted peaks cannot be used in a refinement. We chose to do a LeBail refinement, rather than a standard Rietveld refinement. LeBail refinement differs from Rietveld refinement in that peak intensities are not used to calculate atom positions, only lattice parameters are refined.

Before a refinement can be completed, the 2θ angle was precisely calibrated. Slight errors in 2θ can affect lattice parameter determination. To calibrate the 2θ angle, scans of gold were taken

133

at ambient pressure. The ambient pressure lattice parameter and space group of gold are well known [6]. The calibration scans, one at 12.5° and one at 10°, were read in to GSAS, a refinement program [7]. The 2θ angle was changed until the calculated pattern fit the experimental data. The true angles are 12.545° and 10.015°.

Fluorescence peaks from the steel gasket and niobium in the sample dominated the spectrum below 20 keV. True diffraction peaks existed in this energy range when the 2θ angle was 12.5°, but these peaks were shifted out of this range when 2θ was set at 10°. Because of the fluorescence peaks, the refinement neglected any peaks below 20 keV. From the angle calibration patterns of gold at ambient pressure, it was determined that the energy calibration was inaccurate above 65 keV. For this reason, the refinement excluded energies above 65 keV.

There are scans at four different pressures prior to the phase transition. GSAS was used to do a full pattern LeBail refinement of the low pressure phase. Lattice parameters for $KCa_2Nb_3O_{10}$ and gold were determined. These gold cell parameters were used to calculate the pressure in the diamond cell. From the cell parameters of $KCa_2Nb_3O_{10}$, linear compressibilities and a bulk modulus were calculated.

To confirm the diffraction results, *in-situ* diamond anvil cell Raman spectroscopy was conducted at Arizona State University. The Raman spectrometer uses a 488 nm line of a coherent 90-5 Ar⁺ laser. The optics are mounted around an Olympus BH-2 petrographic microscope. The spectrometer is an Instrument S. A. S300 triple spectrometer with a Princeton Instruments liquid nitrogen cooled CCD detector PI-1100. 180° backscattering geometry was used. Depolarized light was used.

$KCa_2Nb_3O_{10}$ powder was loaded non-hydrostatically in a diamond anvil cell with ruby chips. The pressure was determined by fluorescence of these chips. Raman spectra were taken every 2-3 GPa from 7.6 to 25.2 GPa. After the highest pressure was achieved, a decompression study was performed, and spectra were taken every 2-3 GPa in pressure until the pressure was released.

RESULTS

The lattice parameters and cell volume of the orthorhombic phase of $KCa_2Nb_3O_{10}$ are given in Table I, along with respective pressures.

Table I. $KCa_2Nb_3O_{10}$ lattice parameters with pressure

Pressure (GPa)	a (Å)	b (Å)	c (Å)	V (Å³)
0	3.8717(09)	3.8521(08)	29.4950(6)	439.894(28)
1.51	3.8694(3)	3.8577(3)	29.1646(25)	435.338(36)
2.67	3.8590(4)	3.8479(5)	29.0250(34)	430.995(46)
4.30	3.8805(3)	3.8328(4)	27.9286(45)	415.390(62)
6.83	3.8421(5)	3.8863(8)	26.7539(79)	399.468(69)

The linear compressibilities of $KCa_2Nb_3O_{10}$ are highly anisotropic, as expected from the crystal structure. The a and b directions (in the perovskite slab) are nearly incompressible, while there is considerable compression in the c parameter.

Figure 1 shows percent expansion along the three crystallographic directions. From this figure, the anisotropy in compressibilities is evident. The a and b directions are nearly incompressible, relative to the c direction, as suggested from the structure. The compression in the a and b directions is dominated by the stiff perovskite slabs and is not sensitive to the potassium ion layer, so the properties in this plane should be similar to an isotropic perovskite. However, the c direction includes the weakly bound potassium ion layer. The high compressibility in this direction includes a large contribution from the alkali.

Linear compressibilities and a bulk modulus were calculated from lattice parameters. The bulk modulus of $KCa_2Nb_3O_{10}$ is 68.5 (± 5.5) GPa. The compressibility in the a direction is 8.55 x 10^{-4} (± .71 x 10^{-4}) GPa⁻¹. The compressibility in the b direction is -9.40 x 10^{-4} (± .78 x 10^{-4})

GPa^{-1}. The compressibility in the c direction is much greater than in the basal directions, at 142 x 10^{-4} (± 10 x 10^{-4}) GPa^{-1}. The slightly negative compressibility in the b direction may be an artifact of experimental scatter. Suffice it to note that compression is very small in the a and b directions.

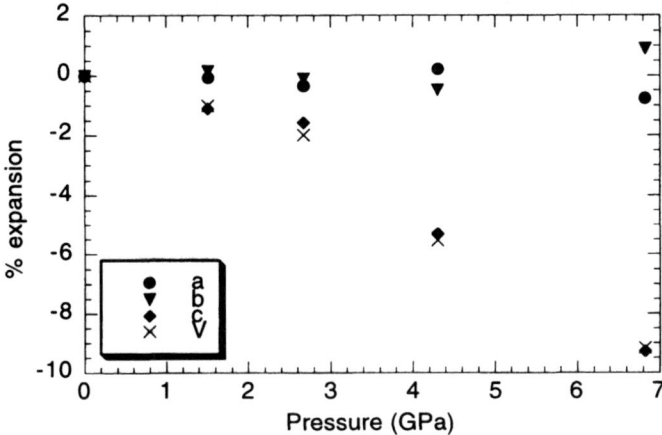

Figure 1. Compressibilities of KCa$_2$Nb$_3$O$_{10}$

The bulk modulus of KCa$_2$Nb$_3$O$_{10}$ compares well with the bulk moduli of other layered materials. The bulk modulus of brucite, a layered mineral, has been reported in the range of 54.3 GPa [8] to 39.6 GPa [9]. The linear compressibilities of brucite are 42 x 10^{-4} GPa^{-1} in the a direction and 183 x 10^{-4} GPa^{-1} in the c direction [9]. The bulk modulus of graphite is 35.8 GPa, with linear compressibilities of 9.62 x 10^{-4} GPa^{-1} in the bonding plane and 270 x 10^{-4} GPa^{-1} as an interplane (c-axis) compressibility [10]. The compressibilities of KCa$_2$Nb$_3$O$_{10}$ are similar to graphite, showing the effect of the layering.

A dramatic difference in the diffraction pattern was noted between 6.83 GPa and 13.5 GPa. This change in the diffraction pattern is indicative of a phase change. Figure 2 shows these diffraction patterns.

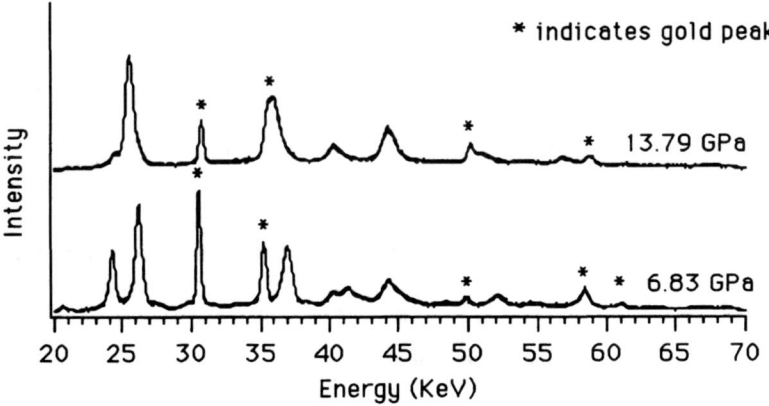

Figure 2. Synchrotron diffraction patterns of KCa$_2$Nb$_3$O$_{10}$

In-situ diamond anvil cell Raman spectroscopy was performed to further study this phase change. The first pressure studied was 7.6 GPa. This Raman spectrum was identical in form to published results obtained at one atmosphere [11]. A change in the Raman spectrum was observed between 10.6 and 13.5 GPa (Figure 3). This change in Raman spectrum occurs in the same pressure range as seen by x-ray diffraction, and affirms that a phase change is occurring. Upon decompression, the high pressure form is retained down to 2.7 GPa (Figure 4). This indicates a large hysteresis in which the transformation occurs. The spectrum taken after all pressure was released is very similar, but not identical, to the spectrum of noncompressed powder. This indicates that the $KCa_2Nb_3O_{10}$ returns to nearly the same structure as the ambient pressure form in the beginning. However, no detailed crystallographic studies have yet been carried out to determine the slight difference.

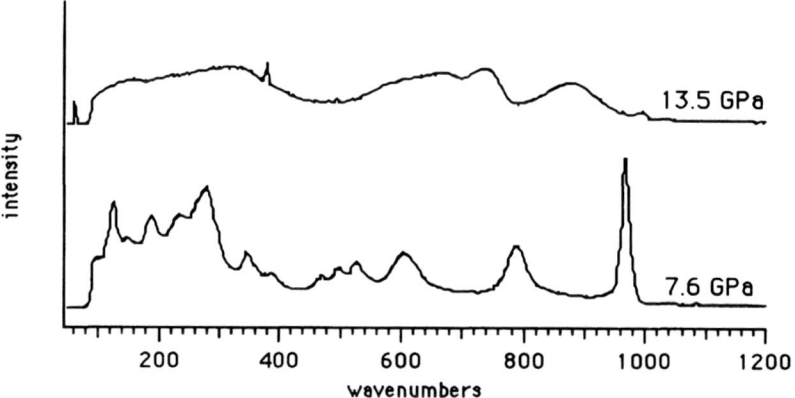

Figure 3. Raman spectrum of $KCa_2Nb_3O_{10}$ upon compression

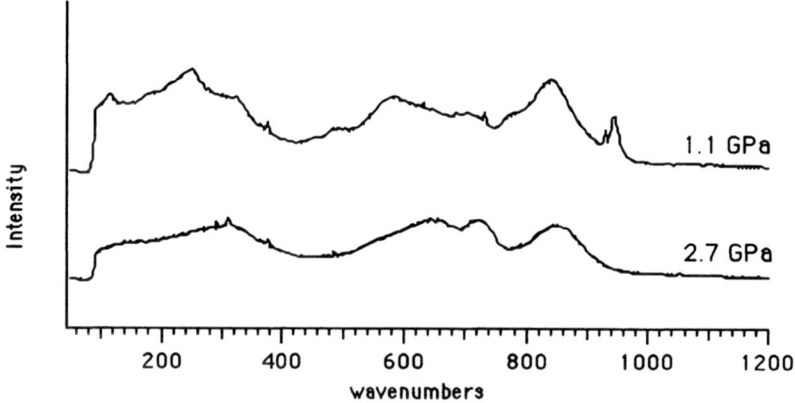

Figure 4. Raman spectrum of $KCa_2Nb_3O_{10}$ upon decompression

CONCLUSIONS

The linear compressibilities of $KCa_2Nb_3O_{10}$ show a large degree of anisotropy. The a and b parameters were nearly incompressible, with an average compressibility of $-.473 \times 10^{-4}$ GPa^{-1}, while the c parameter shows a tremendous compressibility over 6.83 GPa. This is expected from the highly anisotropic structure of $KCa_2Nb_3O_{10}$.

A phase transition occurs between 6.83 and 13.5 GPa as is evident from the reduced number of x-ray diffraction peaks. This reduction in peaks is indicative of a change to higher symmetry, such as a tetragonal cell. This transition was confirmed by high pressure Raman spectroscopy in a diamond cell apparatus.

ACKNOWLEDGMENTS

The authors would like to thank Cynthia Polsky for her assistance with the Raman spectroscopy and Kurt Leinenweber for his assistance with GSAS. Funding was provided by the Air Force Office of Scientific Research grant F49620-95-1-0155. Beamlines X17B and X17C at the National Synchrotron Light Source at Brookhaven National Laboratory are managed jointly by the Geophysical Laboratory in Washington, D. C. and the Materials Science and Research Center at Arizona State University. The latter is funded by the National Science Foundation under grant DMR 9632635.

REFERENCES

1. A. Jacobson in <u>Chemical Physics of Intercalation II</u>, edited by P. Bernier, et al., Plenum Press, New York, 1993, p. 117-139.

2. M. Dion, M. Ganne and M. Tournoux, Mat. Res. Bull. **16**, p. 1429 (1981).

3. S. Sambasivan (private communication).

4. D. L. Heinz and R. Jeanloz, J. Appl. Phys. **55**, 885 (1984).

5. L. W. Finger in <u>Modern Powder Diffraction</u>, edited by D. L. Bish and J. E. Post, Mineralogical Society of America, Washington, D. C., 1989, p. 309-331.

6. Swanson and Tatge, Natl. Bur. Stand., Circ. 539, I 33 (1953).

7. A. C. Larson and R. B. Von Dreele. GSAS-Generalized Crystal Structure Analysis System, Los Alamos National Laboratory Report No. LA-UR-86-748 (1987).

8. Y. Fei, and H. K. Mao, J. Geophys. Res. **98**, p. 875 (1993).

9. X. Xia, D. J. Weidner, and H. Zhao, American Mineralogist, **83**, p. 68 (1998).

10. Y. X. Zhao and I. L. Spain, Physical Review **B40**, p. 993 (1989).

11. Jih-Mirn Jehng and Israel E. Wachs, Chem. Mater., **3**, p. 100 (1991).

HIGH TEMPERATURE X-RAY DIFFRACTION IN TRANSMISSION UNDER CONTROLLED ENVIRONMENT

L. MARGULIES *, M.J. KRAMER *, J.J. WILLIAMS *, E.M. DETERS *, R.W. McCALLUM *, D.R. HAEFFNER **, J.C. LANG **, S. KYCIA ***, A.I. GOLDMAN *
*Ames Laboratory, Iowa State University, Ames, IA 50011, hanuman@iastate.edu
**APS, Argonne National Laboratory, Argonne, Il 60439
***CHESS, Cornell University, Ithaca, NY 14853

ABSTRACT

A compact tube furnace has been developed for high temperature X-ray diffraction studies using high energy synchrotron radiation. The furnace design has a low absorption path in transmission yet allows for a high degree of control of the sample atmosphere and a minimal temperature gradient across the sample. The design allows for a maximum temperature of 1500°C with a variety of atmospheres including inert, reducing, and oxidizing. Preliminary results obtained at the SRI-CAT 1-ID undulator line (60keV) at the APS facility and the A2 24 pole wiggler line (45keV) at CHESS on the $Ti_5Si_3Z_5$ (Z = C, N, O) system will be presented to demonstrate the feasibility of this approach.

INTRODUCTION

High temperature structural refinement of polycrystalline materials has received little attention in synchrotron x-ray diffraction studies due to a number of experimental difficulties. Furnace design is a major consideration. Stable temperature control and minimizing thermal gradients across the probed sample volume are critical. Similarly, if meaningful and consistent data is to be collected the sample atmosphere must be well controlled, especially in the case of samples which are highly oxidizing at elevated temperatures. The high brightness of the third generation synchrotron sources allows high energy x-rays (in excess of 150 keV) to be used while still maintaining adequate flux on the sample[1]. The use of such high energy x-rays has a number of advantages. Bulk sampling of large volumes can be achieved. This is particularly important in composite samples or in studying peritectic reactions where liquid and solid phases segregate macroscopically. Higher values in Q space can be scanned at these energies, allowing local atomic structure to be probed in amorphous or quasi-periodic materials[2,3]. Such high Q scans also allow greater reliability in fitting thermal parameters (especially problematic in highly anisotropic materials) which is often a major limitation in achieving meaningful structural refinements at high temperatures. This paper presents initial work done at the SRI-CAT 1-ID beamline at the Advanced Photon Source (APS) and the A2 beamline at CHESS on the $Ti_5Si_3Z_5$ (Z = C, N, O) system to address issues in developing a reliable method for high temperature structural modeling.

EXPERIMENT

Furnace design

A schematic of the furnace design is shown in Figure 1. The furnace has been constructed to be easily portable and to fit a standard Huber four circle diffractometer. The furnace heating elements consist of Pt-Rh wire wound on an alumina tube with an opening for

139

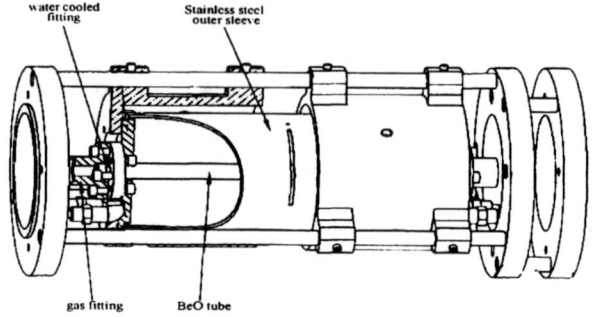

Figure 1: A schematic of the furnace design

the incident beam and a slit to pass diffracted beams at angles of up to 90° 2θ. An inner BeO tube acts as the sample chamber allowing the use of a variety of atmospheres including inert, reducing, and oxidizing atmospheres. The x-ray absorption due to the BeO tube is minimal at the energies of interest (>45keV). The furnace has been designed as a scaled down version of standard tube furnaces which are routinely used in many research laboratories. In this way we hope to mimic actual processing environments as closely as possible. Depending upon the particular sample and furnace atmosphere, a variety of sample holders can be used including MgO, Al_2O_3, and BN. The use of high energy x-rays offers greater flexibility in choosing sample holders while still maintaining good sample to holder signal ratios. This is especially important when studying highly reactive partial melts. The furnace has been tested to temperatures up to 1450°C. Temperature calibration was done using an NIST traceable Pt/Pt10Rh thermocouple as well as by observing solid state phase transformations in $BaCO_3$ and $SrCO_3$. Thermal gradients across the sample area were measured at less than 1°C/mm at 1000°C, which is acceptable for the typical sampling lengths of 1-2 mm.

Beam optics

The beam optics are shown in Figure 2. A Si (111) double crystal monochromator was tuned to the Tm K absorption edge (59.39 KeV). A NaI scintillator detector was scanned vertically in 2θ to collect diffraction data using Debye Scherrer geometry. Due to the large size

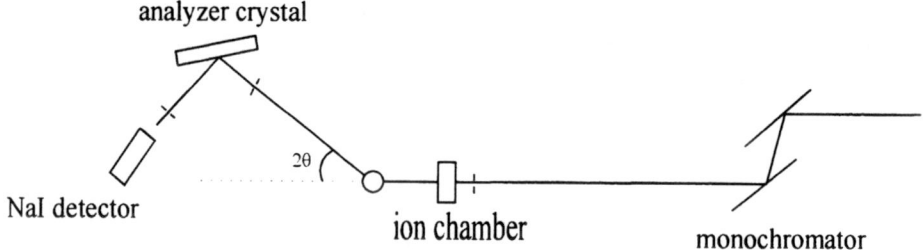

Figure 2: A schematic of the instrumentation and diffraction geometry

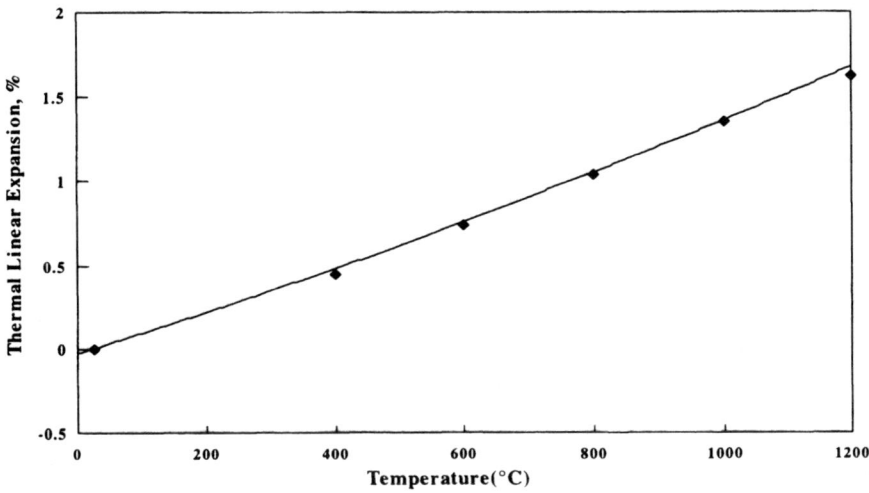

Figure 3: A comparison of thermal expansion data collected with this apparatus on MgO with reported values (solid line).

of many of the samples (up to 3 mm in diameter) it was determined that an analyzer crystal was necessary. The analyzer crystal acts as an effective angular slit in front of the detector yielding superior resolution and reduced background, as well as removing any sample displacement effects[4]. Using a Ge (111) analyzer crystal with this detector geometry at APS(60keV), the (200) reflection of MgO gave a full width at half maximum of .004° 2θ. The same reflection at CHESS(45keV) using a Si(111) analyzer crystal gave a full width of .007° 2θ. A step size of .001° and counting time of .5 seconds led to typical scanning times of 30-50 minutes. The beam energy and zero 2θ were fit using well characterized standards (Si, Al$_2$O$_3$). Reflections from the BeO tube and crucible materials also acted as internal standards to monitor instabilities in the beam energy. The thermal expansion of MgO was measured in order to check for any systematic errors in either the temperature or 2θ values. The results in Figure 3 show good agreement between our measured values on MgO and the linear thermal expansion values reported in the literature[5].

RESULTS

As a first attempt at collecting real data with this apparatus, we chose to examine the system Ti$_5$Si$_3$Z$_{.5}$ (Z = C, N, O). Ti$_5$Si$_3$ is considered a potential high temperature structural material due to its low density and high melting point. The P6$_3$/mcm structure of Ti$_5$Si$_3$ is particularly interesting due to its ability to accommodate a wide range of interstitial atoms[6]. This interstitial doping can dramatically affect the bonding of the structure leading to changes in thermal expansion and high temperature oxidation resistance[7]. It is important to understand how the dopant atoms affect the high temperature structure in order to get a better grasp on this

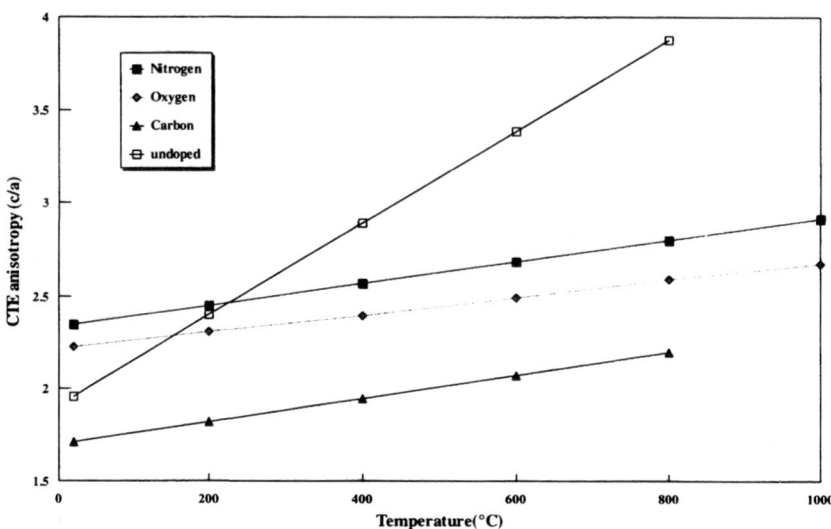

Figure 4: The anisotropy in the coefficient of thermal expansion shows significant dependence on interstitial doping.

effect. Specifically, the coefficient of thermal expansion (CTE) is of importance in consideration of potential high temperature structural uses. In particular, an understanding of changes in the anisotropy of thermal expansion with temperature and doping is critical due to problems with fracturing during thermal cycling of these materials. Anisotropy in the thermal expansion can lead to residual stresses between and within grains during cooling of hot consolidated material causing transgranular or intergranular microcracking[8]. Decreasing the grain size of the consolidated material below a critical grain size minimizes microcracking[9]. A knowledge of the anisotropy in the coefficients of thermal expansion is necessary in order to model this critical grain size. Figure 4 shows thermal expansion data measured up to 1000°C on powdered samples in a He atmosphere. It is clear that the interstitial dopants have a dramatic effect on the thermal expansion anisotropy. In all cases the CTE anisotropy was reduced by addition of the interstitial dopants, with C having the most dramatic effect. It has previously been shown that C doping increases the oxidation resistance of Ti_5Si_3 at high temperatures. Both these effects are most likely related to changes in the metal-metal bonding within the $P6_3/mcm$ structure. A number of problems prevented the accurate measurement of structure factors needed to properly model changes in bond lengths. It is likely that complete powder averaging was not achieved and that scans were not taken to sufficiently high angles to assure reliable fitting of thermal parameters at high temperatures.

The combination of the low angular divergence of the undulator beam and the high angular resolution achievable by the use of the analyzer crystal revealed structure in the diffraction patterns that is not easily seen in Cu radiation scans using a conventional source. Figure 5 shows the (102) reflection of undoped Ti_5Si_3 at both 8 keV (Cu K_α) and 59.4 keV scaled in Q space. It is clear that the details we see in the synchrotron data are obscured by the

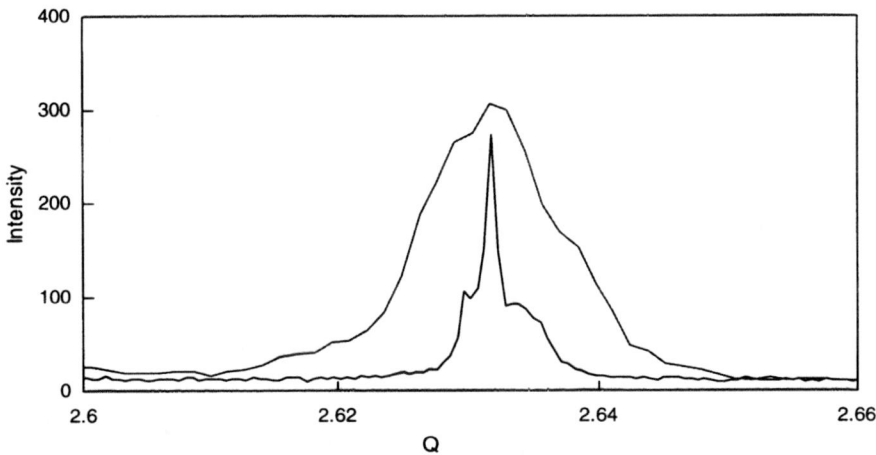

Figure 5: The (102) reflection of Ti_5Si_3 recorded with Cu K_α radiation and 59.4 KeV synchrotron radiation. The increased resolution and lack of the $K_{\alpha 1}/K_{\alpha 2}$ doublet reveals structure that is masked with conventional sources.

$K_{\alpha 1}$, $K_{\alpha 2}$ doublet of the Cu radiation. This effect was most pronounced for the undoped material. It has been suggested that the undoped Ti_5Si_3 does not exist as a single phase. If this is the case,the structure observed may indicate the coexistence of a number of distinct doping levels in the nominally "undoped" material.

CONCLUSIONS

Although we were able to fit reasonable lattice parameters to the data collected, high temperature structural fits of atomic positions could not be achieved using Reitveld refinement. After attempting structural refinement it became clear that data was not collected far enough out in Q space to allow a good fit of the thermal parameters. It is clear that in the future much larger data sets will need to be taken in order to adequately fit atomic positions. We are also currently working to modify the apparatus to allow sample spinning in order to alleviate any possible problems due to lack of powder averaging. Despite this, we have demonstrated the ability to collect data of sufficient quality to determine lattice parameters and phase content in controlled atmospheres at temperatures of up to 1200°C in scan times on the order of 30-50 minutes. In examining fast reactions in which phase identification is the prime objective, position sensitive detectors will be required to speed up data acquisition with an acceptable loss in resolution. In many cases the resolution achieved in this study is greater than necessary. The use of analyzer crystals with larger mosaic structures would allow larger step sizes and therefore faster scans when this is appropriate. The potential for obtaining novel structural and chemical information on materials at high temperature in controlled environments using high energy synchrotron radiation has only begun to be realized. Continued work needs to be done to recognize the full power of such an approach.

ACKNOWLEDGMENTS

The work at Ames Laboratory was supported by the U.S. Dept. of Energy through Iowa State University under contract No. W-7405-ENG-82. The work at APS was supported by the U.S. Dept. of Energy, BES-Materials Sciences, under contract No. W-31-109-ENG-38. The work conducted at the Cornell High Energy Synchrotron Source (CHESS) was supported by the National Science Foundation, under Award No. DMR-9311772.

REFERENCES

1. S.D. Shastri, R.J. Dejus, and D.R. Haeffner, J. Synchro. Rad., Vol 5, pp. 67-71 (1998).

2. Y.S. Badyal, M.L. Saboungi, D.L. Price, D.R. Haeffner, and S.D. Shastri, Europhys. Lett. **39**, pp. 19-24 (1997).

3. T. Egami, S.J.L. Billinge, S. Kycia, W. Dmowski, and A.S. Eberhardt, Synchrotron Radiation Instrumentation: Tenth US National Conference, edited by E. Fontes, 1997, pp. 209-213.

4. D.E. Cox, J.B. Hastings, W. Thomlinson, and C.T. Prewitt, Nucl. Instrum. Methods **208**, pp. 573-578 (1983).

5. Thermophysical Properties of Matter, Vol. 13: Thermal Expansion, Non-Metallic Solids, edited by Y.S. Touloukian (Plenum Press, New York, 1977), pp. 288-289.

6. E. Garcia and J.D. Corbett, Inorgan. Chem. **27**, p. 2353 (1988).

7. A.J. Thom, M.K. Meyer, Y. Kim, and M. Akinc in Processing and Fabrication of Advanced Materials for High-Temperature Applications-III, edited by T.S. Srivatsan and V.A. Ravi (Min. Met. Mater. Soc. Symp. Proc., 1994) pp. 413-438.

8. A.G. Evans, Acta. Metall. **26**, pp. 1845-1853 (1978).

9. R.W. Rice and R.C Pohanka, J. Am. Ceram. Soc. **62**, pp. 559-563 (1979).

INTERFACIAL EFFECTS IN MULTILAYERS

TROY W. BARBEE, JR.
Lawrence Livermore National Laboratory
Chemistry and Materials Science Department
Livermore, CA   94550, USA

ABSTRACT

   Interfacial structure and the atomic interactions between atoms at interfaces in multilayers or nano-laminates have significant impact on the physical properties of these materials. A technique for the experimental evaluation of interfacial structure and interfacial structure effects is presented and compared to experiment. In this paper the impact of interfacial structure on the performance of x-ray, soft x-ray and extreme ultra-violet multilayer optic structures is emphasized. The paper is concluded with summary of these results and an assessment of their implications relative to multilayer development and the study of buried interfaces in solids in general.

INTRODUCTION

   There are many physical characterization approaches[1] which evaluate a limited set of structural elements in multilayers: they study a single interface[2,3,4]; they study a single layer of material; they study a very small sample of a multilayer[5,6]. On a broader basis, the interference phenomena on which the performance of x-ray optic multilayers is based integrates over the full area/volume[7,8,9] of the multilayer illuminated. In order to gain understanding of the impact of imperfections on multilayer performance it is necessary to develop an experimental approach that provides detailed information about the effects of interfaces in the multilayer obtained when the multilayer is being applied in a manner directly related to application. Additionally, it is also of interest to determine the breadth of application of any such experimental approach to the general study of interfaces in solids.

   The primary goal in this research was to develop an experimental methodology to quantitatively characterize both the physical and electronic characteristics of interfaces in multilayer structures The approach was to fabricate multilayers from three elements so that one monolayer or less thick "marker layers" were selectively deposited on a given set interfaces in the multilayer. These "marker layers" could then interrogated by scattering and fluorescence techniques for their distribution, for their atomic arrangements relative to the thicker layers and for their electronic state at the interfaces as affected by the thicker layer materials. WC/C multilayers with one monolayer (2.33 Å) of tantalum at the WC on C and the C on WC interfaces were fabricated and studied. Ta was selected as the marker layer material as its L_3 absorption edge is at 9879 eV, more than 300 eV less than the W L_3 edge at 10200 eV. Reflectivities at 9850 eV, 9879 eV and 9950 eV were measured: Ta layers standing wave fluorescence on the multilayer Bragg peak at these energies and fluorescence EXAFS of the Ta layers were also obtained. These results are modeled and the implications of the results for x-ray optic structures and the study of buried interfaces in solids discussed.

EXPERIMENTAL DESIGN AND SAMPLE FABRICATION

It is well known that at the Bragg peak a x-ray standing wave[10] is established in a periodic structure such as a multilayer. At fixed x-ray wavelengths this standing wave sweeps a distance of half the multilayer period through the multilayer as the Bragg peak is scanned in angle space. The nodes and anti-nodes of this standing wave field sweep over the interfaces and, in principle, provide a probe for sampling the interface structure. Samples in this study were designed to determine the effects of interface imperfection on mutlilayer performance using this characteristic of the diffraction process. The objective with these samples was to determine the impact of imperfection position on the two interfaces characteristic of multilayers on performance. A second objective was to extend the experiments demonstrate the capability to study atomic arrangements and electronic structure at interfaces.

In the following a set of samples with a monolayer thick marker layer deposited at specified interfaces that may be fluoresced by the standing wave field as it sweeps through the mutlilayer are described and experimental results obtained at the Stanford Synchrotron Radiation Laboratory BL - 10 reported. Two (No's 110 & 111) special multilayers designed to use the standing wave to sample interface structure were fabricated and characterized. The multilayers were synthesized using standard magnetron sputter deposition techniques from carbon (C), tungsten carbide (WC) and tantalum (Ta) and had 60 periods of 32.28 Å (110) and 32.78 Å (111). The structures synthesized are:

No. 110 - (111) Silicon Substrate/C/**Ta**/WC/C/**Ta**/WC/C/**Ta**/......./**Ta**/WC/C/Ambient
No. 111 - (111) Silicon Substrate/C/WC/**Ta**/C/WC/**Ta**/C/WC/**Ta**/....../WC/**Ta**/C/Ambient.

The design layer thicknesses deposited are t_C = 15.9 Å, t_{WC} = 15.6 Å and t_{Ta} = 2.33 Å. It is expected that the Ta will react with 1.25 Å of the carbon to form TaC resulting in the observed multilayer periods. The selection of these materials was made on the basis of three materials properties. First, WC is the highest carbon containing compound in the W/C binary system so that compositionally abrupt and potentially smooth interfaces are expected for the WC/C system. Second, it was expected that the Ta would react with the C to form TaC which is isostructural with WC. Third, and most important, the L_3 absorption edges of Ta and W are 9881 eV and 10,207 eV respectively. This difference in the absorption edge energies of 326 eV made possible at the Stanford Synchrotron Radiation Laboratory's BL - 10 scanning in angle of incidence of the first order Bragg peaks of these samples at photon energies below, at and above the L_3 edge of Ta at 9881 eV. In these experiments reflectivity was measured at energies of 9850, 9879 and 9950 eV. Total standing wave fluorescence was also measured by placing a detector above and parallel to the multilayer surface and monitoring the fluorescent intensity as a function of angle of incidence at the same energies. The tantalum L lines will only be excited at energies at or above approximately 9881 eV so that scans as described above will enable assessment of the effects absorbers/imperfections at specific interfaces in a x-ray optic multilayer.

EXPERIMENTAL RESULTS AND MODELING

X-ray reflectivity and standing wave experiments on multilayer samples were performed on BL 10-2 at the Stanford Synchrotron Radiation Laboratory at extreme grazing angles of incidence (0 to 3 deg.) by scanning angle at fixed photon energy or by scanning photon energy at fixed angle of incidence. There are strongly absorbing white lines on the $L_{3,2}$ edges of Ta and W which dramatically enhance their effect on the reflectivity at those energies. The Ta L_3 white line energy of 9879 eV used in the remainder of this paper was determined by performing an energy scan (EXAFS) at fixed angle of incidence over the energy range 9780 eV to 9980 eV and detecting the total fluorescent signal as shown in Figure 1. Multilayer reflectivities for sample No's 110 and 111 were measured at incident x-ray energies of 9850 eV, 9879 eV and 9950 eV. These were selected as the Ta absorption increases in the order 9850 - 9950 - 9879 eV. The effect of Ta absorption could thus be determined at three mass absorption values resulting in a more reliable data set.

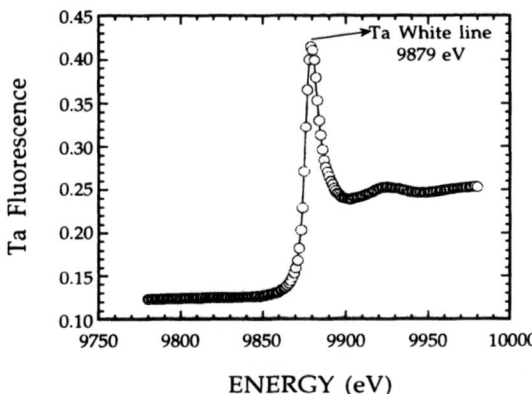

Figure 1. EXAFS scan of Sample 111 showing the Ta L_3 edge white line position at 9879 eV.

Experimental and calculated reflectivities at 9850 eV, 9879 eV and 9950 eV for Sample 110 are compared in Figure 2. These results show that the effect of the interfacial Ta monolayer in this multilayer is negligible. Very different results are found for Sample 111 where the Ta monolayer is placed at the other interface in the multilayer as shown by the experimental and calculated reflectivities presented in Figure 3. These properties are explained by calculations of the standing wave fields in these multilayer as shown in Figure 4. The Ta layers lie at the nodes of the standing wave field in Sample 110 and at the antinodes in Sample 111. This results in a strong absorption at the x-ray energies 8979 eV and 9850 eV for Sample 111. The important conclusion to be drawn here is that, for x-ray optics, control of the structure of one interface set in a multilayer is critical to performance. In this case, it is the WC on C interface.

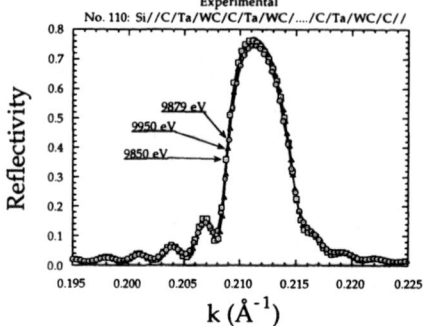
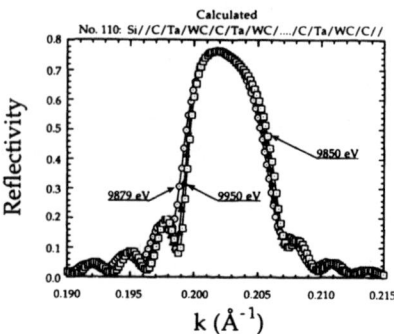

Figure 2   Experimental and Calculated reflectivities at 9850 eV, 9879 eV and 9950 eV for Sample 110 are compared. The Ta interfacial monolayer has essentially no effect on the reflectivity when at the WC on C interfaces.

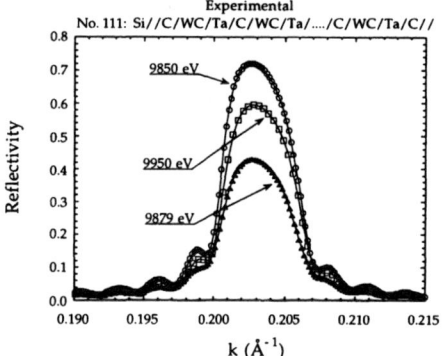
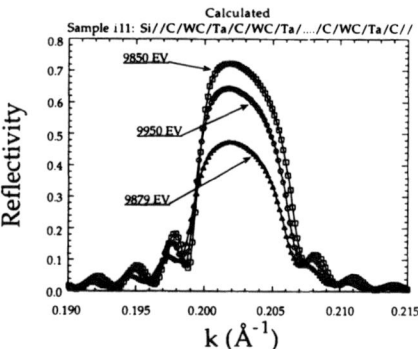

Figure 3. Experimental and calculated reflectivities for Sample 111 showing the impact on the reflectivity due to the Ta monolayer at the C on WC interfaces. This Ta position is at the antinode of the standing wave in the multliayer where an absorber or an imperfection would have the largest effect on the reflectivity.

Since there is absorption in the Ta layer at 8979 eV which significantly impacts reflectivity there must also be Ta fluorescence which is determined by the distribution of Ta at the interfaces. The Ta fluorescence was detected using a 2 cm^2 solid state ion chamber placed above and parallel to the sample surfaces. In Figure 5 the standing wave fluorescence generated by 8850 eV and 8979 eV x-rays is shown as a function of the angle of incidence on the samples. At 8850 eV the fluorescence from the two are essentially identical. At 8979 eV the signal from Sample 111 is

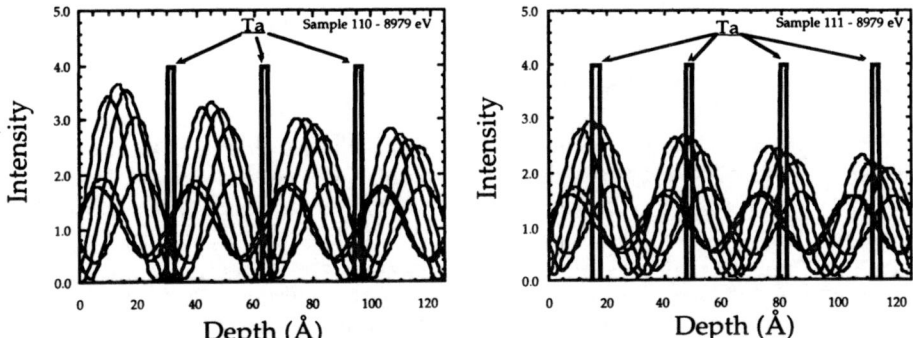

Figure 4. Calculated standing wave fields (SWF) in Samples 110 and 111 at the Bragg peak are shown. The Ta layers in Sample 110 fall at the nodes of the SWF thus minimizing their impact on the reflectivity. In contrast, the Ta layers in Sample 111 fall at the antinodes of the SWF maximizing their impact on reflectivity.

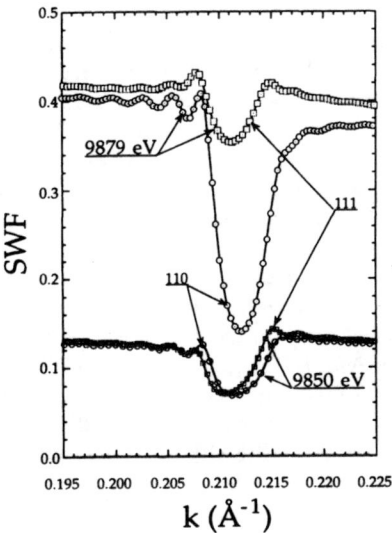

Figure 5. Experimental standing wave fluorescence intensities for No's 110 and 111 at x-ray energies of 9850 eV and 9879 ev are shown as a function of k (= $4\pi \sin\theta/\lambda$). The SWF at 9850 eV is essentially the same for 9850 eV x-rays. At 9879 eV No. 110 shows strong primary extinction on the Bragg peak as expected for an efficient structure. No. 111 exhibits a strongly attenuated primary extinction that indicates there is a strong loss mechanism operating at 8979 eV - absorption in the Ta layer.

clearly larger and less affected by primary extinction as expected from the reflectivity data. Modeling of this SWF data enables experimental measurement of the distribution of the Ta normal to the interfaces in the multilayer. Additionally, the magnitude of the SWF signals in these experiments indicates that very small interfacial concentrations can be detected and used to study interface atomic arrangements and electronic structure. This approach can be extended to fluorescence studies in the soft x-ray domain at energies less than 3 keV.

DISCUSSION AND CONCLUSIONS

The primary conclusion from these experiments is that the perfection of a specific set of interfaces in XR, SXR and EUV multilayers is crucial to their performance. In the work presented here an absorber, Ta, was applied as a representatve imperfection at multilayer interfaces. Absorption in the Ta layers acted as the "imperfection" enabling the role of imperfection position in the multilayer to be demonstrated. Just as interesting is the potential that extrapolation of these observations to the study of buried interfaces holds. In this case the approach would be to decrease the concentration of the "marker element" to minimize its effects on the nature of the interfaces in the multilayer.

Calculations for multilayers with 2.3Å (1 monolayer[ml]), 1Å (0.43 ml), 0.23Å (0.1 ml), 0.1Å (0.023 ml) and 0.023Å (0.01 ml) of Ta at the interfaces were performed and showed that the effect of 0.01 monolayers is to decrease the reflectivity by approximately 0.4%. The effect of 1 monolayer of Ta is to decrease the reflectivity by about 30%. The decrease in Ta absorption at 0.01 monolayers is therefore about a factor of 75. As stated earlier the detector used had an area of 2 cm^2. The detector area can be increased by a factor of 50 to 100 cm^2 so that the fluorescent signal from the 0.01 monolayer Ta structure will be greater than 25% of that from the 1 monolayer Ta structure. This demonstrates that the "interfacial marker atom" approach can yield detailed atomic arrangement information at interfaces, chemical reaction data at interfaces at the atomic level and potentially, local electronic band structure characterization at interfaces. This exploratory modeling only applies to x-ray energies above 2 keV. Calculations of the reflectivity of a WC/C multilayer at 1 keV with 1 and 0.1 monolayer of copper at the interfaces were also performed. These calculations predicted a 6% absolute reduction in reflectivity at 1 monolayer Cu and 0.8% at 0.1 monlayer. This again indicate that the "interfacial marker element" technique will be effective in the soft x-ray regime (0.4 to 2.0 keV).

The calculations directed to the exploration of the breadth of application of the "interfacial marker layer/interfacial marker element" technique have demonstrated the feasibility of the application of these approaches to the experimental study of the atomic level nature of atomic arrangements at interfaces, of chemical interactions at interfaces and potentially of electronic band structure effects at interfaces. Electronic band structure effects will be accessible for the transition elements through study white lines at their $L_{3,2}$ edges as the strength of these edge features is directly related to the d band occupancy[11] in the 4th (SXR), 5th (SXR and XR) and 6th (XR) rows of the Periodic Chart. These opportunities are a direct consequence of the development of multilayer synthesis technology over the past two decades. This technology has made possible the conceptulization of monolayer scale control of composition in macroscopic samples. The world wide availability of intense scanning monochromatic x-ray and soft x-ray sources at synchrotrons such as the Stanford Syncrotron Radiation Laboratory is also clearly crucial as such experiments would not be feasible without such facilities.

150

Therefore, the "interfacial marker layer/interfacial marker element" technique has the clear potential to yield atomic level experimental data on the nature of interfaces in solids. Such results will be valuable to theorists in that they will provide a basis for existing and to be developed formalisms. I note that at the lowest interfacial marker element concentrations (< 0.01 monolayer) this "experimental marker element" approach is a heterogeneous analogy to the homogeneous embedded atom[12] theoretical methodology and, in principle, should provide a sound testing ground for that methodology.

ACKNOWLEDGMENTS

This work was performed under the auspices of the U.S. Department of Energy by the Lawrence Livermore National Laboratory under Contract No.W-7405-ENG-48. UC-National Laboratory Participating Research Team facilities (BL 10-2) at the Stanford Synchrotron Radiation Laboratory (SSRL) were applied in this study. SSRL operates under the support of the U. S. Department of Energy.

REFERENCES

1. P. Dhez and C. Weisbuch, Eds., *Physics, Fabrication and Applications of Multilayered Materials*, Plenum Press, New York (1988).

2. Ph. Houdy, "Kinetic Ellipsometry Applied to Soft X-ray Multilayer Growth", Revue Phys. Appl. **23**, 1653-1659, (1988).

3. M. Yamamoto and T. Namioka, "In Situ Ellipsometric Study of the Optical Properties of Ultra Thin Films", Appl. Opt. **31**, 1612-1621, (1992).

4. J. Slaughter, P. Kearney and C. Falco, "Characterization of Pd-B, Ag-B and Si-B Interfaces", Proc. SPIE **1547**, 71-79, (1991).

5. W. M. Stobbs, "Techniques for Characterizing Artificial Layer Structures Using the Fresnel Method", in Multilayers: Synthesis, Properties and Non-Electronic Applications, T. W. Barbee, Jr., F. Spaepen and L. Greer, Eds., MRS Sym. Proc. **103**, 121-131,(1988).

6. W. C. Shih and w. M. Stobbs, "Measurement of Roughness of W/Si Multilayers by the Fresnel Method", Ultramicroscopy **32**, 219-239, (1990).

7. E. Spiller, *Soft X-ray Optics*, SPIE Optical Engineering Press (1994).

8. T. W. Barbee, Jr., "Multilayers for X-ray Optics", Opt. Eng. **25**, 898-915, (1986).

9. E. Spiller, "Characterization of Multilayer Coatings by X-ray Reflection", Revue Phys. Appl. **23**, 1687-1700, (1988).

10. T. W. Barbee, Jr. and W. K. Warburton, "Evanescent and Standing Wave Fluorescence from a Multilayer Structure", Materials Letters 3, 17-25, (1984).

11. D. H. Pearson, C. C. Ahn and B. Fultz, Phys. Rev. B47, 8471-8478 (1993).

12. M. S. Daw, S. M. Foiles and M. I. Baskes, Mat'ls. Sci. Rpts 9, 251-310 (1993)

COMPARISON OF A MOSAIC-CRYSTAL SPECTROMETER TO A HIGH-PERFORMANCE SOLID-STATE DETECTOR FOR X-RAY MICROFLUORESCENCE ANALYSIS

J.-S. CHUNG*, S. ISA*, C. J. SPARKS*, G. E. ICE*, S. McHUGO** and A. THOMPSON**
*Oak Ridge National Laboratory, P.O. Box 2008, Oak Ridge TN 37830
**Lawrence Berkeley National Laboratory, 1 Cyclotron Rd., Berkeley CA 94720

ABSTRACT

The minimum-detectable-limit of a compact double-focusing graphite mosaic-crystal spectrometer is compared to the minimum-detectable-limit from a high-performance Ge solid-state detector. The solid angle and efficiency of the solid-state detector is much greater than for the crystal spectrometer. However, the better signal-to-noise of the spectrometer and its insensitivity to matrix fluorescence and scattering can give it a better minimum-detectable-limit for trace element analysis. The relative advantages of the two detectors are illustrated for some simple test samples. The performance of the crystal spectrometer compared to the solid-state detector increases as the flux in the x-ray probe increases. This makes crystal spectrometers especially interesting for use with new high intensity 3rd generation synchrotron microprobes. An estimate is made of the source and sample conditions favored for each detector.

INTRODUCTION

X-ray fluorescence analysis with small electron, proton or x-ray beams is widely used for the identification of elemental distributions.[1-3] The figure-of-merit for fluorescence analysis is the minimum-detectable-limit: literally the minimum concentration which can generate a significant signal above background. Minimum detectable limit (MDL) depends on the intensity and nature of the excitation probe, the efficiency of the detector, the data collection time, the detector bandpass, and the sample composition or "matrix". We adopt the definition of minimum detectable limit given by Sparks[4]. The MDL can be determined from signal and background of a known standard,

$$C_{MDL} = 3.29 C_Z \frac{\sqrt{N_b}}{N_s}. \tag{1}$$

Here C_{MDL} is the minimum detectable limit, C_Z is the mass fraction in a measured standard, N_b is the background counts beneath the standard fluorescence signal and N_s is the net counts at the fluorescence energy.

X-ray excitation typically creates a much smaller background than charged particle probes (Ref 4) and monochromatic x-ray sources have the best signal-to-noise. Until recently however, monochromatic x-ray beams with μm^2 cross sections have been very weak. The advent of intense 3rd generation synchrotron sources however, has now stimulated renewed interest in x-ray excited microfluorescence analysis.[5,6]

To efficiently use these powerful new sources it is important to match the detector to the source. Fluorescence spectra are typically measured with a solid-state detector[7] although curved and mosaic crystal analyzers[8,9] are also used. In addition to the natural background generated in the sample, a solid-state detector contributes background due to insufficient charge collection. A solid-state detector also has (in general) poorer energy resolution than a crystal spectrometer and its count-rate limit is set by the major elements in the sample matrix rather than by the trace elements. However a solid-state detector has a large advantage in terms of total solid angle and efficiency compared to a crystal spectrometer. A solid-state detector also simultaneously measures all elements in the sample matrix with Z>18..

Here we compare the MDL which can be achieved with a high-performance Ge solid state detectors to the MDL achievable with a double-focusing mosaic-graphite crystal spectrometer. The aim of this paper is to define the experimental conditions which favor the two detectors.

EXPERIMENTAL SETUP

Source and geometry

Measurements were made on beamline 10.3.1 at the Advanced Light Source (ALS). This beamline utilizes graded elliptical Kirkpatrick-Baez (KB) multilayers to focus the beam. The multilayer optics reflect an ~6% bandpass from the incident white synchrotron radiation beam. Selectable mirrors allow several bandpass options. For the present measurements the multilayer mirrors were an ~12.5 keV pair which focused ~10^9 photons into ~1x1.5 μm^2. For white beam measurements the upstream slits of the monochromator tank were used to collimate the white beam to ~50x250 μm^2. The intensity of Fe fluorescence excitation was ~200x greater with the collimated white beam than with the focused beam. Hence the flux density in the focused spot was ~50x greater for the focused beam even though it had only a 4% bandpass.

The detector was placed perpendicular to the incident x-ray beam in the plane of the ring. The sample was placed in the bisecting geometry. This orientation minimized elastic and Compton scattering. Measurements were made with both focused monochromatic x-rays (4% bandpass) and with collimated white beams.

Samples

Two standards were used to measure the MDL over a range of experimental conditions. A sample with 0.4 Wt. %Fe in an Al matrix was prepared to study the case of a trace impurity in a low Z matrix. Al and. Fe powder were mixed, then pressed into a pellet. A powder pellet formed from ultra-pure Al (99.999%) was also made as a background standard. For the case of a trace impurity in a high Z matrix, a sample of Fe in W was prepared. Fe_2O_3 powder and WO_3 powders were mixed together in the mass fractions which made 1 Wt. % of Fe in total mass. A pure WO_3 pellet was made for a background standard.

Detectors

Ge Solid State Detector. The solid-state (energy dispersive) detector was a single-element high-purity Germanium detector from EG&G ORTEC (Model No. IGLET-11145S). This detector incorporates an ultra-thin detector entrance window which allows for good collection efficiency for low Z elements and high peak-to-background ratios. The preamplifier electronics are based on a pulsed reset circuit which allows for high count rate. For Fe 55 radiation with an Mn K_α fluorescence energy of 5.9 keV, the IGLET detector has a resolution of 137 eV with a 10 μs amplifier shaping time and a 110 mm^2 active area. With an active area of 110 mm^2 the solid angle which can be accepted by this detector is ~$110/d^2$(mm).

Solid-state detectors have a trade-off between count rate and energy resolution. With a 10 μs shaping constant, the effective detector deadtime is ~30 μs. For a non-paralyzable detector this results in an ~10% deadtime at 3,000 cps. The amplifier shaping constant can be reduced to improve the count rate, but only at the expense of energy resolution. For example, at 1 μs shaping constant a 10% deadtime will not occur until 30,000 cps. With a 1 μs amplifier shaping constant however the energy resolution degrades by a factor of ~2 at 5.9 keV. From Eq. 1 it is clear that higher count rate is favored over good energy resolution for an isolated fluorescence line. For example, if background scales linearly with energy resolution, a 1 μs shaping constant compared to a 10 μs shaping constant, will allow a factor of 10 high count rate with only a factor of 2 higher background. This should yield an ~2.2 times smaller MDL assuming the solid-angle or probe intensity can be scaled to maintain the count-rate at ~10% deadtime.

In actual trace-element experiments however the background can often increase much faster than linearly with energy resolution. This is especially true if the spectra near the characteristic line of interest contains an intense peak. Furthermore, the total deadtime of a solid state detector depends on the integrated deadtime over the whole spectrum and not on the count rate at the fluorescence line of interest. This means that the count-rate of a trace-element measurement with a solid-state detector is limited by fluorescence from *all* of the elements in the sample matrix rather than from the trace elements of interest.

One approach to the problem of matrix fluorescence is to use a graphite prefilter to a solid state detector. With large (~10^4 μm^2) beams and long (~10^3 s) counting times this approach has been successfully used to study heavy metals in polymers and toxic trace metals with 10-100 PPB sensitivity.[10,11] Another interesting approach uses a Kumakov lens to collimate the x-ray beam before energy analysis with a mosaic crystal.[12] Best signal-to-noise and lowest MDL (for non count-rate limited measurements) however is realized with a focusing wavelength-dispersive-spectrometer (WDS) of which our mosaic crystal spectrometer is one example.[9,12]

Mosaic-Crystal Spectrometer. The mosaic crystal (wavelength dispersive) spectrometer is a compact version of a mosaic-crystal spectrometer described in a previous paper.[9] A schematic of the spectrometer is shown in Fig. 1. The basic elements of the spectrometer are a monochromator housing, a curved graphite monochromator crystal, and a compact linear position sensitive detector. The curved graphite crystal had a sagittal (out of the diffracting plane) curvature of ~29 mm, and a mosaic spread of ~0.8 degrees. X-rays are parafocused onto the detector in a 1:1 magnification geometry. The condition for optimum sagittal focusing is given by,[9] $F_1 = R_s/\sin(\theta_B)$. Here R_s is the sagittal radius, F1 is the sample to crystal distance and θ_B is the Bragg angle . The useful solid angle of the detector is determined by the effective width and length of the mosaic crystal(~10x35 mm^2), the distance to the sample, and the Bragg angle, $\sin(\theta_B) = 1.8478/E(kev)$. The effective solid angle of the mosaic crystal is reduced because of the ~40% reflectivity of the mosaic crystal. The effective solid angle, S_E , of wavelength-dispersive spectrometer is,

$$S_E = 0.4 \, x \, 350(mm^2) \frac{1.8478}{E(keV)} \left\{ \frac{1.8478(mm)}{29(mm)E(keV)} \right\} 2 = \frac{1.05}{E(keV)^3} \quad . \tag{2}$$

At 8 keV, the solid angle of the mosaic crystal detector is ~2×10^{-3} whereas the solid angle of the Ge detector at d=50 mm from the sample is ~4×10^{-2} (20x higher).

The linear detector is a specially designed ORDELA model 1010X Xe linear proportional counter. This detector has a 1 cm long by 0.8 cm wide active area with a spatial resolution of ~40 μm. The theoretical energy resolution of the wavelength-dispersive spectrometer (WDS) as a function of x-ray energy and x-ray probe size is calculated based on reference 9 (Fig. 2). The measured energy resolution was ~46 eV which is about 40% worse than anticipated.

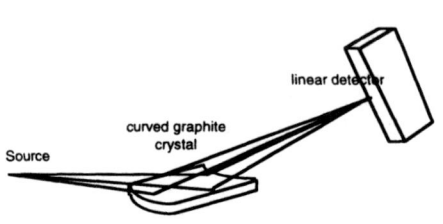

Fig. 1. Schematic of mosaic crystal spectrometer showing the essential components.

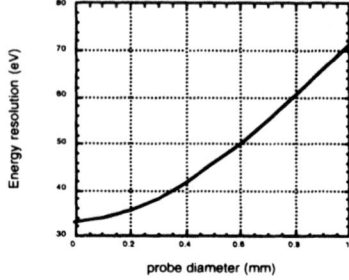

Fig. 2. Theoretical energy resolution at 6.4 keV as a function of source size.

EXPERIMENTAL MEASUREMENTS

The mosaic analyzer crystal was mounted 10.2 cm from the sample to optimize the focus and signal-to-noise. The Ge detector was moved as close as possible to the sample without exceeding 10% deadtime. An amplifier shaping constant of 10μs was used. Data were collected and compared for the two detectors with both white (collimated) and monochromatic (focused) beams. With the Ge detector the lowest MDL was measured with the focused monochromatic beam; monochrmatic beams have lower intrinsic background and the detector was already count rate limited. With the WDS detector best MDL was achieved with the white beam (see description below). All spectra, background counts and fluorescence signals were compared in the region of +/- 3 σ. Here σ's were determined by least square fittings to Gaussian curves.

Crystal Spectrometer

Fig. 3 illustrates the performance of the WDS. The bottom of Fig. 3 superimposes the measured spectra with Fe impurity on the spectra measured off ultra pure Al. As can be seen, the crystal spectrometer background analysis is complicated by the significant Lorentzian component of the Kα line and by the non-uniform energy response of the spectrometer; the crystal spectrometer background exhibits a broad Gaussian shape arising from the increasing efficiency of the crystal spectrometer near the nominal center of its range. A 99.999% pure Al powder sample was used to measure the background. The background is dominated by a broad Gaussian which corresponds to elastic scattering within the 0.5 degree mosaic spread of the graphite crystal. As shown in Fig. 3 this background is much broader than the peak associated with Fe K alpha radiation.

A further complication for this particular setup was the unfortunate overlap of x-ray elastic scattering from the graphite (00●4) reflection at ~12.8 keV onto the detector. Because the energy bandpass of the multilayer mirrors was almost exactly twice the energy of the fluorescence line of interest, the monochromatic beam had an unusually high λ/2 background. About 80% of this background was suppressed by gating the linear proportional counter sum pulse. However, the major signal-to-noise improvement expected for monochromatic beam as compared to white beam was not realized with this setup due to the $\lambda/2$ contamination.

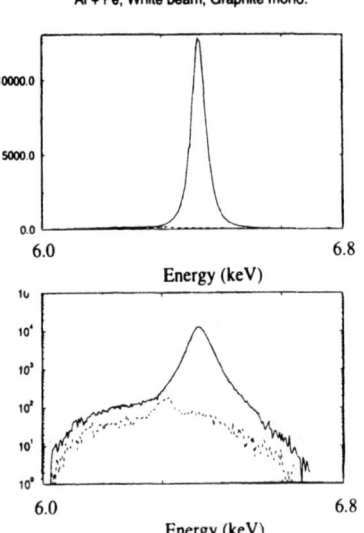

Fig. 3 Fe Kα line as measured by the graphite spectrometer. The top plot shows the Fe Kα line on a linear scale. The bottom line superimposes the background from ultra pure Al on the measured spectra from the standard.

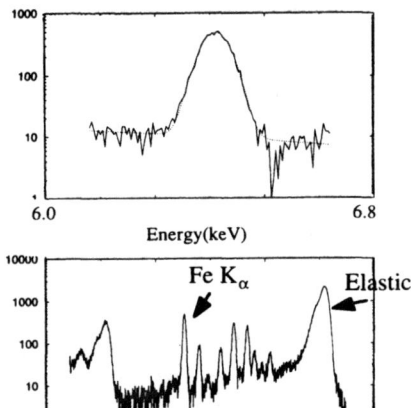

Fig. 4 Bottom figure shows the overall spectra measured by the Ge detector whilethe top figure shows an enlargement of the Fe Kα region.

The mosaic-crystal spectrometer spectra off the Fe_2O_3 in a WO_3 matrix is similar to the spectra observed with an Al matrix. However, the heavy Z matrix has lower signal-to-noise and henceahigherminimum-detectable-limit.

Ge Detector

The spectra from the Fe-Al sample, measured with the Ge detector, is shown in Figure 4. This spectra was collected with a focused monochromatic beam with ~200 times less fluorescing power than the collimated white beam. Trace element count rates are limited primarily however by deadtime restrictions imposed by the elastic scattering peak in the FeAl sample. In the Fe_2O_3-WO_3 sample, W fluorescence restricted the acceptable solid-angle of the detector. In addition, for the Fe_2O_3 sample, the Fe Kα line lies on top of a sloping background from the W lines.

The count rates, signal-to-background and minimum detectable limits for the three samples are summarized in Table 1. As can be seen, because the count rate in the SSD is limited by the matrix signal, the WDS can achieve better MDL even with a white beam. Much better signal-to-noise is possible if a monochromatic x-ray beam ~10% above the absorption edge of the trace elements is used. The MDL for the solid-state detector however will remain virtually fixed with increased signal because it has already reached its deadtime limit. Improved MDL for the SSD will require multiplexing the detector. We note that ~400 solid state detectors are needed to match the MDL of the single mosaic crystal spectrometer.

With the experimental setup used in this example, the MDL can be estimated as a function of beam intensity. We ratio the flux to $I_0 \equiv$ flux in the focused monochromatic beam on 10.3.1 (~10^9 12.5 keV x-rays). As can be seen the solid-state detector outperforms the crystal spectrometer at low flux due to its much higher collection efficiency. For high intensity x-ray beams however the deadtime limitations of the solid-state detector restrict the MDL which it can achieve. The estimated flux at the APS is ~ 4 orders of magnitude more than on beamline 10.3.1.

The solid-state detector does particularly well compared to the crystal-spectrometer on systems with low Z matrices and for weak incident beams. It is also at a big advantage if more than one element must be monitored simultaneously.

Table 1. Summary of the performance of the two detectors. Counting time was 1000 seconds.

	Mosaic-Crystal Spectrometer (collimated white beam)			Solid-state detector (focused monochromatic beam)		
	Counts	Signal-to-noise	MDL (ppm)	Counts	Signal-to-noise	MDL (ppm)
Al powder	208,000	75:1	2.3	3750	14:1	40
$Fe_2O_3+WO_3$	564,600	28:1	5.8	14200	2.6:1	121

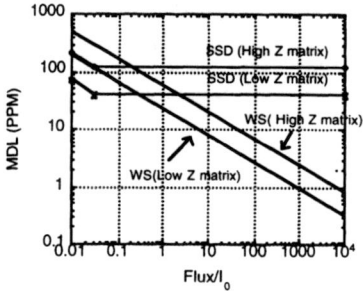

Fig. 5 MDL/s for an x-ray microprobe as a function of flux. I_0 is the nominal intensity of the focused x-ray beam on ALS beamline 10.3.1 (~10^9 photons/s). Note the MDL in table 1 is for 1000 s.

CONCLUSIONS

For some simple test samples, a compact mosaic crystal spectrometer has been shown to yield lower minimum-detectable-limits than an advanced solid-state detector. The relative advantages of the two detectors depends on the sample being measured and on the intensity of the source. For the most advanced x-ray microprobe beams, mosaic crystal spectrometers are favored because they are not paralyzed by the intense fluorescence of major matrix components of the sample, and because they have better signal-to-noise.

ACKNOWLEDGMENTS

Research sponsored by the Laboratory Directed Research and Development Program of the Oak Ridge National Laboratory and by the Office of Basic Energy Sciences, Division of Materials Sciences, U.S. Department of Energy under contract DE-AC05-96OR22464 with Lockheed Martin Energy Research Corporation. Research performed in part at beamline 10.3.1 at the ALS which is sponsored by Basic Energy Sciences U.S. Department of Energy.

REFERENCES

1. J.R. Chen, E.C. T. Chao, J.A. Minkin, J. M. Back, K.W. Jones, M.L. Rivers and S.R. Sutton, *Nucl. Instr. and Meth.* **B49** 533 (1990).

2. D.R. Beaman and L.F. Solosky, *Anal. Chem.* **44** 1598-1610 (1972).

3. G.J.F. Legge and A. Saint, *Nucl. Inst. and Meth* **B49** 418 (1990).

4. C.J. Sparks, "X-ray Fluorescence Microprobe for Chemical Analysis", in Synchrotron Radiation Research, edited by H. Winick and S. Doniach, 459 (Plenum Press 1980); L. A. Currie, *Anal. Chem.* **40** 587 (1968).

5. P. Chevallier and P. Dhez, Hard X-ray Microbeam: Production and Application .*in* <u>Accelerator based atomic physics techniques and applications</u> (S.M. Shafroth and J. Austin eds.) pp. 309-348 AIP Press, New York.

6. G.E. Ice, *X-ray Spectrometry* **26** 315-326 (1997).

7. A.D. Smith, G.E. Derbyshire, R.C. Farrow, A. Sery, T.W. Raudorf and M. Martini, *Rev. Sci. Instrum.* **66** 2333 (1995).

8. F. Folkmann and F. Frederiksen, Nucl. Inst. and Meth. **B49** 126 (1990).

9. G. E. Ice and C.J. Sparks, *Nucl. Inst. and Meth.* **A291** 110-116 (1990).

10. C.J. Sparks, L.A. Harris and O.B. Cavin, "Development of High Sensitivity X-ray Fluorescence for Analysis of Trace Toxic Elements", ORNL-NSF-EATC-1 (Progress Report)1972.

12. V. Baryshev, Y. Kolmogorov, G. Kulipanov, and A. Skrinsky, *Nucl. Inst. and Meth* **A246** 739 (1986).

13. J. P. Kirkland, V.E. Kovantsev, C.M. Dozier, J.V. Gilfrich, W.M. Gibson, Q.F. Xiao, K. Umezawa, *Rev. Sci. Inst.* **66** 1410-1412 (1995).

ANISOTROPY OF NH₄AP CRYSTAL X-RAY SUSCEPTIBILITY FOR BRAGG REFLECTION NEAR CK ABSORPTION EDGE

A.V. OKOTRUB*, G.S. BELIKOVA**, T.N. TURSKAYA**, L.N. MAZALOV*.
*Institute of Inorganic Chemistry SB RAS, Novosibirsk, Russia, spectrum@che.nsk.su
**Institute of Crystallography RAS, Moscow, Russia

ABSTRACT

Bragg reflection (001) of high-quality crystal of ammonium acid phthalate (2d = 52.28 Å) is investigated in the region of CK absorption edge. First order reflection for this crystal is forbidden. Due to anisotropic additions, it appears only near CK edge. Two-crystal optical scheme was used for the measurement of the crystal reflectivity. As shown, the change of crystal azimuth in X-ray spectrometer leads to ability to control the resolution. Rotation effects of X-ray beam polarization on 90° angle is considered for the forbidden reflection.

INTRODUCTION

The possibility of appearance of forbidden reflections caused by anisotropy of chemical bonding was discussed in [1,2]. The symmetry considerations showed that specific forbidden reflections appear due to the anisotropic additions. As the value of these additions is usually very small, the intensity of the reflections is not high also. The measurements of these reflection types were carried out, e.g., on V_3Si and potassium bromate [3,4].

Study of the reflection efficiency for crystals of alkali metals acid phthalate showed the reflection anomalies near the OK absorption edge [5]. The maximum at the energy of 532 eV was caused by the peculiarities of resonance scattering of photons on oxygen atoms of CO_2-groups of phthalic acid.

A weak forbidden reflection with the energy of 270-285 eV was found in investigation of the X-ray characteristics of ammonium acid phthalate crystal (NH₄AP) [6]. The authors explained the occurrence of this reflection by the crystal imperfection, namely, by the presence of dislocations that resulted in the breakage of the reflection prohibition in odd orders. Our study of the possibility of using the NH₄AP crystal in ultrasoft X-ray area found out a strong forbidden reflection which was used to register the carbon emission spectra [7,8]. Measurement of the X-ray bremsstrahlung radiation of copper anode revealed that high efficiency of reflection in the area of CK absorption edge was observed when the optical axis of spectrometer coincided with the X direction of the crystal (Fig.1). When the optical axis of spectrometer coincided with the Z axis of the crystal the intensity of OKα line did not change, while the reflection in the first order disappeared.

X-ray susceptibility is proportional to the third power of the radiation wave length (λ). The forbidden reflection of NH₄AP crystal is observed near the CK absorption edge with wave lengths of λ~44 Å. This leads to the significant intensity of reflection compared to the ones observed in [3,4] and can be of the same magnitude or even excess the intensity of the permitted reflections. As a result, the high quality NH₄AP crystals can be used to obtain the high resolution spectra of carbon compounds.

In this work we present the results of the forbidden reflection measurements for the first order of (001) reflection of ammonium acid phthalate crystal.

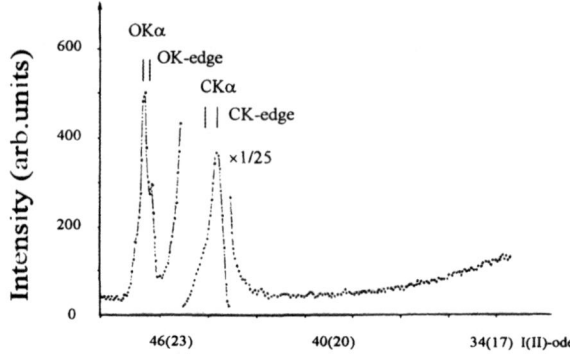

Fig.1 Intensity of bremsstrahlung radiation of Cu anode, measured by using NH₄AP crystal

OKα
OK-edge
CKα
CK-edge
×1/25

46(23)   40(20)   34(17) I(II)-oder λ(Å)

THEORETICAL BASIS

The NH₄AP crystals were grown by a technique of temperature decrease in the Institute of Crystallography of RAS. The growth technique, and the peculiarities of crystal structure and their mechanical properties are given in [9]. Unlike the KAP and RbAP crystals which belong to D^3_2-$P2_12_12$ space group with crystal lattice parameter $c \approx 13$ Å, the NH₄AP crystal belongs to D^{15}_{2h} - Pcab space group with parameter $c \approx 26.14$ Å [10, 11]. Bragg reflection for odd orders, including the first order of reflection, is forbidden for the crystals of this group. This prohibition is caused by the screw-axis rule as there are two twofold screw-axes in a unit cell of NH₄AP crystal which are perpendicular to the cleavage plane (Fig. 2). The structural dislocations, thermal fluctuations, chemical bonding anisotropy can weaken this type of X-ray reflection prohibition [1].

To determine intensity and polarization properties of individual reflections, the structure amplitude is presented as a Fourier component of X-ray susceptibility tensor:

$$F= -v\chi(s)/r_0\lambda^2 \equiv -v/r_0\lambda^2 \int\chi(r)exp(2\pi isr)dr \qquad (1)$$

where v is the unit cell volume, $r_0 = e^2/mc^2$, λ is the X-ray radiation wave length.

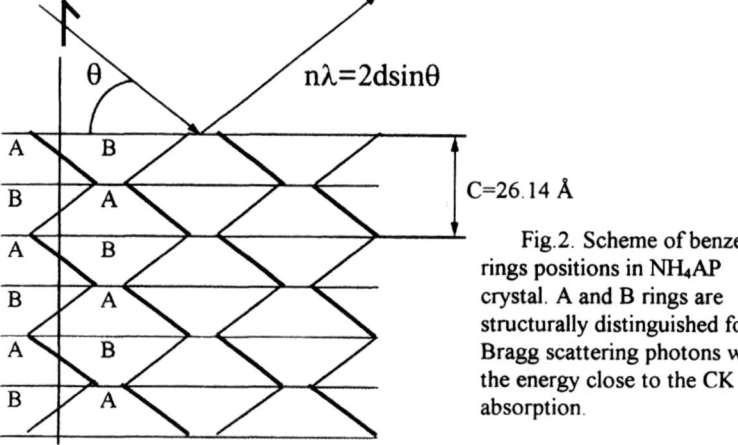

$$n\lambda=2dsin\theta$$

C=26.14 Å

Fig.2. Scheme of benzene rings positions in NH₄AP crystal. A and B rings are structurally distinguished for the Bragg scattering photons with the energy close to the CK edge absorption.

162

Structure amplitude may be divided into isotropic and anisotropic parts:

$$F^s = F^s I + \Delta F^s \qquad Sp(\Delta F^s) = 0 \qquad\qquad (2)$$

F^S is a standard structure amplitude, I is a unit matrix.

In case of a crystal with the screw-axis the structure amplitude of (001) reflection is presented as following:

$$F^{001} = -v/r_0\lambda^2 \int\chi(r)\exp(2\pi i l z)dxdydz \qquad\qquad (3)$$

Taking into account the invariance of X-ray susceptibility tensor in the case of forbidden reflection in the crystals with twofold axes, it was obtained in [2]:

$$F_{xx} = F_{yy} = F_{zz} = 0; \ F_{xy} = F_{yx} = 0; \ F_{xz} = F_{zx} = F_1; \ F_{yz} = F_{zy} = F_2 \qquad (4)$$

Intensity of such forbidden reflection depends on polarization of X-ray radiation falling on the crystal (α) and reflecting from it (β):

$$I\alpha\beta = |A|^2|\beta^*F^s\alpha|^2, \qquad\qquad (5)$$

where A is the constant determined by the crystal structure and the conditions of the experiment. Under such approach the reflection intensity is [2]:

$$I = |A|^2\cos^2\theta_b[|F_1|^2\sin^2\varphi_s + |F_2|^2\cos^2\varphi_s - Re(F_1F_2^*)\sin2\varphi] \qquad (6)$$

F_1 and F_2 are the components of structural factor tensor that are caused by the anisotropic additions in the latter.

In the ammonium acid phthalate crystals, occurrence of forbidden reflection in the area before CK absorption edge shows that the anisotropy of X-ray susceptibility arises due to the anisotropy of chemical bonding in a crystal, namely, due to the structure and location of benzene rings. The angle of inclination of benzene molecules to the cleavage plane is ~64°, which is very close to the angle of Bragg reflection ($61.9° > \theta > 56.7°$) for photons with energies of 270-285 eV. Thus, X-ray radiation has an optimal opportunity for resonance scattering on π-system of benzene rings of one layer, and is practically absent in the next one (Fig. 2).

EXPERIMENTS

X-Ray Anisotropy Measurement

According to (6), the rise of the forbidden reflection caused by anisotropic additions should lead to azimuthal dependence of reflection efficiency. Measurement of the crystal reflection efficiency dependence in the azimuthal plane was carried out for NH$_4$AP crystals of size $10\times10\times0.2$ mm^3. Plane crystal was fixed in a holder providing its rotation in the azimuthal plane. The measurements were carried out on a laboratory X-ray spectrometer «Stearat». X-ray beam was collimated by two slits. The spectrometer resolution was ~ 1 eV in the area of CKα line. X-ray tube operation parameters were 1.2 kV, 0.25 A. Reflection intensity dependence on crystal orientation near CK absorption edge ($\theta_B = 58°$) is shown in Fig. 3.1. When the optical axis is parallel to Z-axis of a crystal, the reflection intensity is very low. Such azimuthal dependence is typical for «forbidden» reflections [12]. Study of the dependence allows to determine the value of F^H tensor components (F_1 and F_2 parameters) and their relative phase. The dependence may be presented as a direction diagram of reflection efficiency for NH$_4$AP crystal (Fig. 3.2).

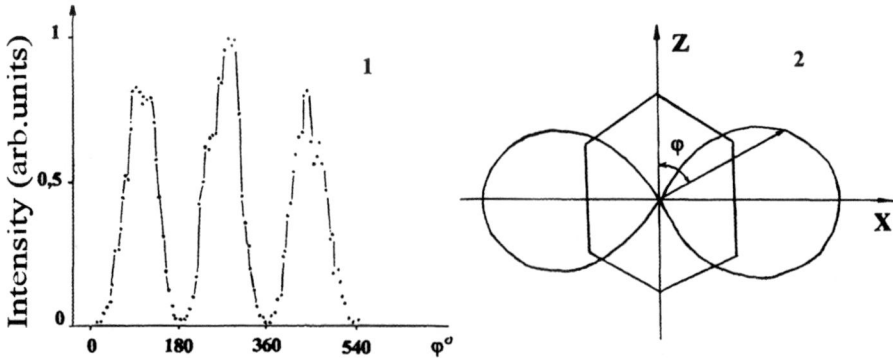

Fig. 3. Reflected intensity dependence of azimuthal angle (3.1) and diagram of (001) Bragg reflectivity of NH_4AP crystal in the projection on cleavage plane.

<u>Efficiency of Reflectivity Measurements</u>

Study of reflection efficiency for NH_4AP crystal near CK absorption edge was done using the scheme of two-crystal spectrometer. Spectrum of bremsstrahlung radiation in an ultrasoft X-ray area is a smooth continuous function. X-ray radiation of a copper anode reflects from the first analyzing crystal, goes through the two-slit collimating system, falls on the second crystal, and finally, is reflected and registered by the gas proportional counter. The reflection spectra obtained using two NH_4AP crystals are presented in Fig. 4.1. The intensity of Bragg reflection is seen to decrease quite fast when moving off the absorption edge. Comparison of the reflection curves normalized to intensity maximum is shown in Fig. 3.2. One can note different widths of the reflection curves. Resolution of NH_4AP crystal in the range of 280 eV is about 0.1 eV. The additional broadening of maxima (3) and (4) is caused by resolution of spectrometer optical scheme. These lines are ~ 1 eV in width. The broadening of the reflection curves (1) and (2) up to 2-3 eV is an evidence for 10-20 times broadening of the proper reflection curve of NH_4AP crystal in the energy range of 285-289 eV.

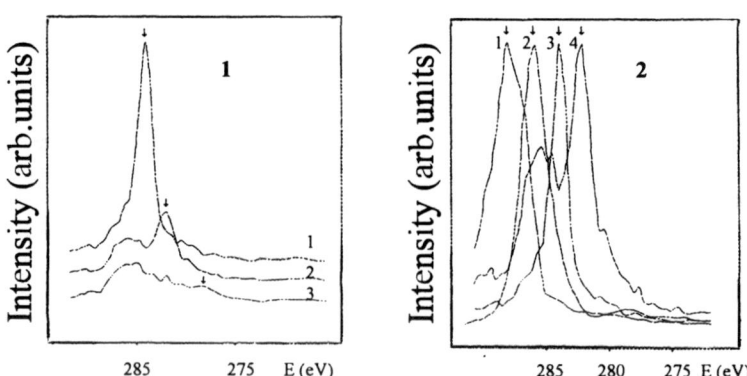

Fig. 4. X-ray reflection spectra using the two-crystal scheme. Photon energies corresponding to Bragg reflection of the first crystal are indicated by the arrows. Reflection spectra for three positions of the first crystal NH_4AP correspond to the energy range of 278-284 eV (4.1). Four reflection spectra near the CK edge (281-287 eV) normalized to the maximum-(4.2).

The measurements show that the high intensity of NH₄AP crystals reflection near CK absorption edge is caused by the anisotropic additions. As a result, the depth of penetration of the radiation into a crystal decreases, the reflection curve broadens, and the intensity of the reflected X-ray radiation increases (Fig. 4).

Measurements of Polarization Rotation

The X-ray radiation reflection that takes place due to the anisotropic additions has unique polarization properties [2]: a σ-polarized incident beam can produce a π- polarized diffracted one and *vica versa*, a π- polarized beam when reflected can give σ-polarized radiation. To verify this assumption, a device containing three crystals was constructed. Crystal **1** of mono-ester octadecyl hydrogen maleate (OHM) is used as a monochromator with the wave length of 44.2 Å. As Bragg reflection angle of this radiation is close to 45°, the radiation reflected by this crystal is σ-polarized. This radiation, collimated by the two slits, falls on the crystal **2**. The X- axis of the crystal is parallel to the optical axis of the spectrometer. The reflected X-ray beam falls on the crystal **3** (OHM) used as a polarization analyzer. This crystal and the gas counter may be oriented as (‖) or (⊥), i.e. the radiation reflected from crystal **3** may be in the plane of the optical scheme or in the perpendicular one. In case of π-polarized radiation falling onto this crystal, its reflection efficiency must be extremely low. 90° change in polarization of the falling beam results in the maximum efficiency of reflection. Rotation of crystal **3** and the radiation detector about the axis of X-ray beam reflected from **2**, allows to determine the polarization of radiation falling onto the crystal **3**. The measurements were carried out for two cases with monocrystals of NH₄AP and OHM used as crystal **2**. The X-ray tube operating regime was: $U = 4$ kV, $I = 0.5$ A. The data obtained is shown in Table 1.

Table 1. The X-ray intensities measured with three crystal scheme.

Crystal 2	Crystal 3 (⊥)	Crystal 3 (‖)
OHM	2 ± 37	78 ± 37
NH₄AP	107 ± 33	33 ± 31

As one can see, with high degree of reliability the polarization of X-ray beam reflected from NH₄AP and OHM crystals is opposite. X-ray radiation reflected from a NH₄AP crystal is π-polarized. The complexity of the experiment causes relatively high statistic error of the experiment due to low efficiency of the three-crystal scheme considered above and limited time of the experiment — about 30 minutes. Running experiment for longer time results in crystal thermal destruction, since the latter is placed directly in the X-ray tube.

CONCLUSION

We have shown that the reflection efficiency for the certain types of organic crystals is mainly determined by anisotropic contribution. These crystals have to possess a structure characterized by specifically oriented conjugated chemical bonds. NH₄AP crystal was shown to be one of the perspective materials for X-ray optic, as it presents a possibility to control its' resolution efficiency through rotation of the reflection plane. The experimental data obtained confirm the theoretical assumption that the polarization of X-ray beam changes into the opposite when the considered forbidden reflection takes place.

REFERENCES

1. V.E.Dmitrienko, Acta Cryst. **A40**, pp.89-95 (1984).

2. V.E.Dmitrienko, Acta. Cryst. **A39**, pp.29-35 (1983).

3. B.Borie, Acta Cryst. **A37**, pp.238-241 (1980).

4. D.H.Templeton, L.K.Templeton, Acta. Cryst. **A41**, pp.133-142 (1985).

5. R.L.Blake, in <u>Advances X-Ray Analysis</u>, edited B.L.Henke, J.B.Newkirk, G.R.Mallett, **13**, pp.352-360 (1970)

6. D.M.Barrus, R.L.Blake, A.J.Burek *et al.*, Phys.Rev.B **22**, pp.4022-4037 (1980).

7. V.D.Yumatov, A.V.Okotrub, L.N.Mazalov et al, Zh. Strukt.Khimii **26**, N4, pp.59-64 (1985)

8. A.V.Okotrub, V.D.Yumatov, L.N.Mazalov et al, Zh. Strukt.Khimii **29**, N2, pp.167-170 (1988)

9. G.S.Belikova, T.N.Turskaya, L.E.Kraeva, Patent USSR, № 1515787 (1989)

10.Y.Okaya, R.Pepinsky, Acta. Cryst. **10**, pp.324-328 (1957).

11.R.A.Smith, Acta Cryst. B **31**, pp.2508-2509 (1975).

12.D.H.Templeton, L.K.Templeton, Acta. Cryst. **A43**, pp.573-574 (1987).

Part III

Photoemission and Microphotoemission;
Fluorescence and Microfluorescence

SPECTROSCOPIC STUDIES OF LOW DIELECTRIC CONSTANT FLUORINATED AMORPHOUS CARBON FILMS FOR ULSI INTEGRATED CIRCUITS

Yanjun Ma and Hongning Yang
Sharp Microelectronics Tech., 5700 NW Pacific Rim Blvd, Camas, WA 98607

J. Guo, C. Sathe, A. Agui, and J. Nordgren
Physics Department, Uppsala University, Uppsala, Sweden

ABSTRACT

Performance of future generations of integrated circuits will be limited by the RC delay caused by on-chip interconnections. Overcoming this limitation requires the deployment of new high conductivity metals such as copper and low dielectric constant intermetal dielectrics (IMD). Fluorinated amorphous carbon (a-CFx) is a promising candidate for replacing SiO_2 as the IMD. In this paper we investigated the structure and electronic properties of a-CFx thin films using high-resolution x-ray absorption, emission, and photoelectron spectroscopy. The composition and local bonding information were obtained and correlated with deposition conditions. The data suggest that the structure of the a-CFx is mostly of carbon rings and CF_2 chains cross-linked with C atoms. The effects of growth temperature on the structure and the thermal stability of the film are discussed.

INTRODUCTION

With shrinking device feature size and increasing circuit complexity, the semiconductor industry is facing serious challenge from many technology fronts. Among them is the need for many new materials in the integrated circuits manufacturing process. According to the 1997 National Technology Roadmap for Semiconductors,[1] new gate dielectrics and gate electrodes, low dielectric constant insulators and high conductivity metals for interconnect, and high dielectric constant materials for DRAM capacitors need to be developed in the next few years. Indeed currently there are already intense industry-wide efforts in searching for new low dielectric constant (low-κ, with κ<2.5) materials to replace silicon dioxide (κ ~ 4) as the interlayer dielectrics in multilevel interconnect schemes. Together with the use of copper metallization, the interconnect RC delay can be substantially reduced in ULSI circuits. There is great demand for the characterization of these new materials using all available techniques. In this paper, we report the studies of low dielectric constant films using x-ray techniques. We will demonstrate that the application of several spectroscopic techniques in characterizing low dielectric constant fluorinated amorphous carbon films gives a fairly complete information on the structure of the amorphous film.

There are a number of requirements for the new low-κ materials, such as low dielectric constant, high thermal stability (400°C or higher) in order to survive later processing, electrically insulating, high mechanical strength, low moisture absorption and low solubility in water, and good adhesion to neighboring layers. These stringent requirements have reduced the candidates to porous silica materials and a few carbon-based polymers. Fluorinated amorphous carbon (a-CFx) [2-3] and parylene [4-5] polymer thin films showed significant promise. Since these materials are new in IC applications, a complete understanding of the properties of these films is essential before they can be reliably used in future ULSI circuits. This is especially true for the amorphous CFx films where a range of composition can be found just by varying the substrate temperature during growth. These films are amorphous, and their structural and electronic properties are not well known. The deposition conditions need to be optimized to yield the best quality film. Here we present high-resolution x-ray spectroscopy studies of a-CFx and parylene thin films. We will show that C 1s x-ray photoelectron spectra and C K-edge near-edge-x-ray-absorption-fine-structure (NEXAFS) provide detailed information on the local bonding environment of carbon atoms. While unoccupied states were provided by the absorption spectra, we used soft x-ray emission spectra to probe the occupied states to get a more complete picture of the electronic structure of these films and correlated them with their electrical properties.

Fluorinated amorphous carbon films attracted attention in ULSI technology only recently.[2-3] A relative dielectric constant as low as 2.2 and good thermal stability were reported. These films are usually prepared by a plasma enhanced chemical vapor deposition (PECVD) technique where many variables, such as substrate temperature and bias, plasma power, feeding gas composition and flow rate, could affect the film growth. For example, substrate temperature during deposition plays a significant role in determining the thermal stability of a-CFx films. We will relate the structural information obtained from spectroscopic studies to the thermal stability of the a-CFx film grown at different temperatures.

Parylene (poly-p-xylylene) and its derivatives have dielectric constants ranging from 2.25 to 3.1 and thermal stability of up to 450°C.[4,5] Parylene N (Pa-N) is the un-substituted normal type parylene and has dielectric constant of about 2.8 and thermal stability of about 400°C. Even though these properties are not as good as those of fluorinated parylene films such as Pa-F,[5] Pa-N is attractive, because it contains no fluorine atoms. Fluorine atoms may be undesirable, because they may be very corrosive to the neighboring layers during subsequent processing steps.

EXPERIMENT

Thin films (100-400 nm) of a-CFx were prepared by PECVD using a parallel plate reactor with a RF power of 175 Watts on the top plate. The feed gas was a mixture of octafluorocyclobutane (C_4F_8) and methane(CH_4), with a flow ratio of 6.7. This flow ratio is a result of compromise between low dielectric constant and thermal stability. Films were grown on Si wafers with (100) orientation. Substrate temperatures of room temperature, 180, 250, and 350°C were used in this study. The film prepared at 350°C has a dielectric constant of 2.35 and thermal stability of better than 400°C. The room temperature film has a lower dielectric constant (<2.1) but also a much lower thermal stability (<250°C). The PECVD film growth rate is typically about 100nm/min, depending upon the substrate temperature. At room temperature the growth rate is much faster than at elevated temperatures. Pa-N film was vacuum deposited on to Si wafers held at room temperature using dimer precursor. It showed a thermal stability of better than 380°C and a dielectric constant of about 2.8.

C1s absorption and K emission experiments were performed at Beamline 7.0 of the Advanced Light Source (ALS) at Lawrence Berkeley National Laboratory. The beamline comprises of a 99-pole, 5 cm period undulator and a spherical-grating monochromator [6]. F K-edge absorption and emission experiments were performed at beamline W3 of Hamburg Synchrotronstrahlungslabor (HASYLAB).[7] The beamline uses a SX-700 monochromator. NEXAFS spectra were obtained by measuring the total electron yield from the sample. The resolution of the monochromator was set to 0.15 eV at the C K-edge. The NEXAFS spectra were normalized to the incident photon current using a clean gold mesh to correct for intensity fluctuation of the photon beam. The XES spectra were recorded using a high-resolution grazing-incidence x-ray fluorescence spectrometer [8]. During the XES measurement, the overall resolution was about 0.5 eV and 0.65 eV at the C K, and F K-edge, respectively.

RESULTS AND DISCUSSION

The a-CFx films were characterized by x-ray photoelectron spectroscopy (XPS) (performed by Evans East, Plainsboro, NJ) using a monochromatic Al x-ray source. The overall energy resolution of the measurement is about 1 eV. Because of the sample charging, an electron flood gun was used during the XPS measurement. The C1s spectra for a-CFx films grown at room temperature and 250°C are shown in Fig. 1. Also indicated in the figure is the tentative identification of the spectral features.

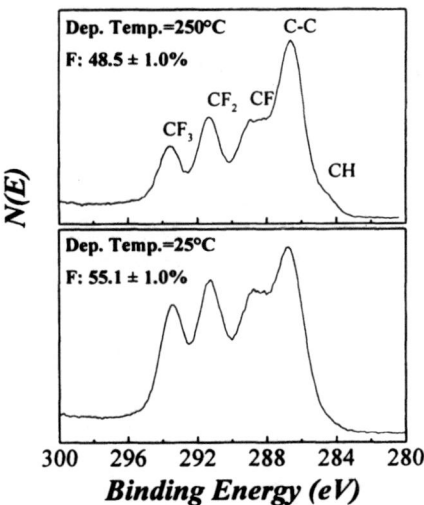

Figure 1. XPS spectra of a-CFx films prepared at room temperature and 250°C. Because of the charging effect, the two spectra are shifted with respect to each other, only the relative position of the peaks is relevant. The spectra were analyzed by curve fitting, and the results are listed in Table 1.

Table 1. Results of C1s XPS analysis for a-CFx films prepared at RT and 250°C.

Peak Assignment	Energy (eV)	27°C carbon at%	27°C fraction carbon	250°C carbon at%	250°C fraction carbon
C*-H	284.8	0.5	1%	1.8	3.5%
C*-C	286.8	13.8	31%	20.9	40%
C*-HF	288.4	4.6	10%	6.2	12%
C*-F	289.3	7.6	17%	7.3	14%
C*-F2	291.4	9.9	22%	9.7	19%
C*-F3	293.7	8.1	18%	6.2	12%
Total Carbon (at%)		44.5	100%	52.1	100%
F Measured		55.1		48.5	

The peaks in the spectra are well separated and can be identified to be due to carbon atoms in C-F, C-F$_2$, C-F$_2$, and C-C bonding configuration. Fitting the spectra using Gaussian line profiles, the relative abundance of different configurations can be obtained. Curve fitting also resulted in two additional peaks. The peak at 284.8 eV can be attributed to C-H configuration. On the other hand, the origin of the peak at 288.4 eV is somewhat uncertain. Since its energy is between those of C-C and C-F. We tentatively assigned it to carbon atoms with one F and one or more H neighbors. The fitting results are listed in Table. 1, where we listed peak assignment/position, relative weight (fraction of carbon atoms), as well as the overall atomic percent of carbon. The composition of these films was obtained by comparing the C1s XPS intensity with that of a Teflon (CF$_2$ chains) standard.

From Table 1, first we notice that the composition of the two films is different. The a-CFx film grown at room temperature has x>1.0 while the film grown at 250°C has x<1.0. Breaking down according to the carbon configurations, there are fewer carbon atoms in the CF$_2$ configuration and more carbon atoms in the non-fluorinated configurations at the elevated temperature. The drop in the relative population of CF$_3$ group is especially large. This may partially explain the poor thermal stability of the room temperature grown film. CF$_3$ forms a terminal connection to a CF network such as a polymer chain, more CF$_3$ presence in the structure means a larger surface to bulk ratio, indicating a smaller structure, e.g., shorter polymer chain, therefore the lower the thermal stability of the film.

Figure 2 shows the C K-edge absorption spectra of a Pa-N and three a-CFx films grown at RT, 180, and 350°C. These near edge absorption features directly measure the unoccupied states of carbon atoms. Remarkably detailed local bonding information of these new films can be obtained from comparing with existing literature on simple fluorocarbon molecules.[9] Specifically, the peaks at 285 and 287 eV are due to the electronic transition from C1s to a π^* orbital and is a measure of the presence of unsaturated C-C bonds, e.g., C=C double bond.[9] For example, this peak shows up prominently in the C K-edge spectra of graphite where carbon atoms has sp^2 hybridization and out of plane π bonds, but not in that of diamond with only sp^3

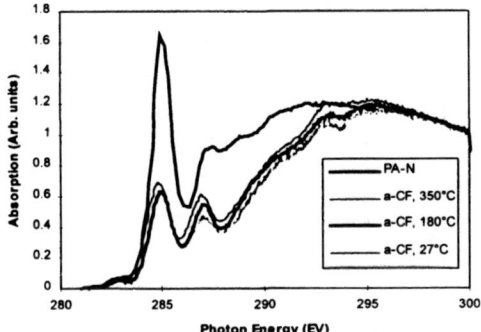

Figure 2, C K-edge absorption spectra of Pa-N and a-CFx films prepared at substrate temperature of room temperature, 180°C, and 350°C. These spectra were normalized at 297 eV.

hybridization.[10,11] In addition, the position of the π^* resonance depends on the other neighbors of the excited carbon atom C^*. For carbons with only H or C neighbors, e.g., $C=C^*H$ or $C=C^*C$, the resonance is at around 285 eV. For C with one F neighbor, the resonance shifted to around 287 eV. The resonance shifts to around 290 eV for carbon with two F neighbors.[9] The absence of any peak at around 290 eV in Figure 2 indicates that there is negligible amount of $C=CF2$ groups in the a-CFx structure. Since XPS results (Table 1) suggest that up to 20% of the carbon atoms has two F neighbors, they must predominantly bond in $C-C^*F2-C$ configuration, i.e., only with single bond (sp^3 hybridization). This is an important observation and, as we will discuss later, it will have implications on the structure of the a-CFx films. Of course, there cannot be double bonds for carbon atoms with three F neighbors.

The π^* peak is very intense in the Pa-N film, reflecting the fact that this film is primarily phenyl rings connected by CH_2 groups.[4] Six carbon atoms, belonging to the phenyl ring, out of the total eight carbon atoms have double bonds, i.e. 75% of the carbon atoms in Pa-N have double bonds. The intensity of the 285 eV π^* resonance of a-CFx films, which is associated with C with no F neighbors, is only about 40% that of the Pa-N. We can roughly estimate that about 30% of these carbon atoms have double bonds (in sp^2 hybridization). From Table 1, we can see that typically about 30-40% of the carbon atoms in a-CFx have no F neighbors. This means that over 70% of these carbon atoms are in sp^2 hybridization with double bonds. We should caution that because of issues such as normalization, bond orientation, surface sensitivity of the measurement technique, and the complexity of the bonding environment in a-CFx, the above estimate is highly qualitative. Still, from this qualitative argument we can conclude that the majority of non-fluorinated carbon atoms have double bonds with sp^2 hybridization.

From the XPS result we see that the percentage of non-fluorinated carbon atoms increased from about 30% to about 40% for the film grown at 250°C, compared to the film grown at room temperature. However, as shown in Fig. 2, the intensity of the 285 eV resonance is about the same for all three samples. This suggests that raising the growth temperature has the effect of producing more non-fluorinated atoms in sp^3 hybridization.

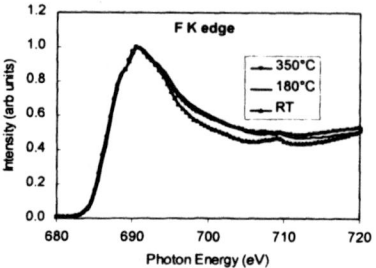

Figure 3, F K-edge NEXAFS spectra for a-CFx films grown at RT, 180°C, and 350°C.

In contrast, the intensity of the π* resonance at 287 eV is lower for the films with higher growth temperature. Since this peak is associated with a C=C*FX (X=C or H) configuration, the reduction in intensity suggests that this kind of environment is not favorable during high temperature deposition. On the other hand, we know from Table 1 that the intensity of the CF peak only decreased marginally, i.e., the number of carbon atoms with one fluorine neighbor has not decreased noticeably. Again we found that raising the growth temperature will produced more carbon atoms with sp^3 hybridization, this time among singly fluorinated carbon atoms by converting sp^2 to sp^3 hybridization.

Contrary to the C K-edge NEXAFS result, F K-edge spectra offer relatively little information. In Figure 3 we plotted the F K-edge absorption spectra for the three a-CFx films. This is attributable to the simpler electronic structure of the fluorine atom.

In Figure 4 C K-emission spectra of a-CFx film grown at 350°C and Pa-N are shown. These spectra were taken with monochromatic photon excitation at excitation energies listed in the figure. The most prominent peaks found at 275.4, 284.9, 286.9, in these figures are due to elastic scattering. Complementary to the absorption spectra, occupied electronic states were probed in the emission spectra. With different excitation energy, occupied states associated with different carbon configuration were probed. For example, with excitation tuned at 284.9 eV, only non-fluorinated carbon atoms with double bonds were excited. The resulting emission spectra in Fig. 4 (a) and (b) show the occupied states of those atoms. On the other hand, at the excitation energy of 286.9 eV, carbon atoms in the =C*FX (X=C, H) configuration were excited, and the corresponding emission spectrum in Fig. 4(a) shows the occupied states of those atoms. Finally, at high energy such as 295 eV, essentially all the carbon atoms were excited, and we obtain a composite picture of the unoccupied states.

An additional piece of information can be extracted from Fig. 4(b). The excitation energy of 275.4eV is below the C K absorption threshold. Thus the resulting spectra can be interpreted as an energy loss spectra. Appreciable intensity appears at about 6 eV below the elastic peak which corresponds to the excitation of an electron across the band gap, i.e. we can estimate that the energy bandgap of Pa-N is about 6 eV.

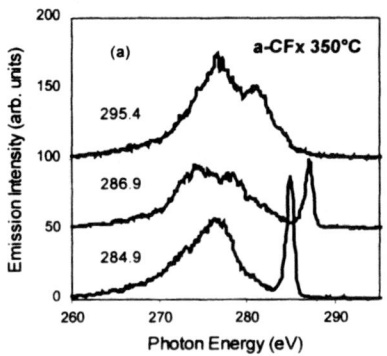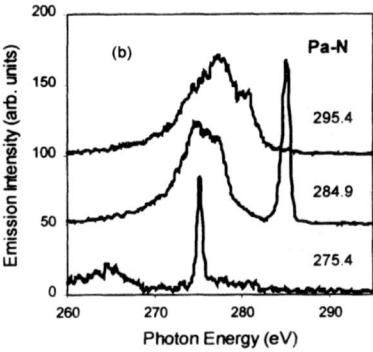

Figure 4. C K-emission spectra for a-CFx (a) and Pa-N (b) films measured at different excitation energy.

In Figure 5, we compared the K-emission spectra of a-CFx, Pa-N with that of amorphous carbon and graphite. From comparing with a-C and from Figure 4(a), we can attribute the emission intensity centered around 275 eV to C-C interactions. More importantly, in the region of 282-285 eV, there is little intensity in the spectra of Pa-N and amorphous carbon. We conclude that the intensity in the spectra of a-CFx is mostly due to F induced states and to the amorphous nature of the films. Since these states are at the top of the valence bands, this has important implications in the electrical properties of these a-CFx films, it narrows the band gap. Because the absorption spectra of a-CFx and Pa-N films have similar a threshold, we can argue that the bandgap of a-CFx is about 2 eV smaller than that of the Pa-N. Again because the band edge is not well defined for amorphous films, this conclusion is very qualitative. Qualitatively, this conclusion is supported by the fact that a-CFx is not as good an insulator as Pa-N. While the Pa-N films is practically insulating, a 2000 Å thick a-CFx films showed a leakage current of 10^{-8} A/cm2 at an electrical field of 0.5 MV/cm.[12]

From the XPS and NEXAFS results, one can obtain a fairly good picture of the structure a-CFx. In general, CF compounds are characterized by either chain or ring structures. Unsaturated bonds are typically associated with the ring structure. Since the majority of the non-fluorinated carbon atoms are found with unsaturated bonds, they are mostly associated with ring structures. We also found that CF_2 groups are mostly present with saturated bonds in the C-CF_2-C environment. This is typical of chain structures similar to, e.g., TEFLON, which is composed of CF_2 chains. Note also that CF_2 groups are found only among about 20% of the carbon atoms, while non-fluorinated carbon atoms are found in about 30-40% of the carbons. The structure that emerges from this picture is non- fluorinated carbon and CF groups in mostly ring structures linked by CF_2 groups and terminated by CF_3 groups.

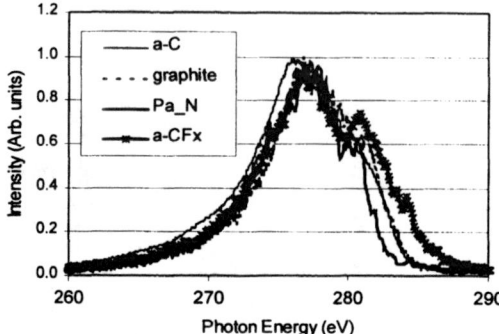

Figure 5. C K-emission spectra, at excitation energy of 295 eV, of amorphous carbon, graphite, Pa-N, and fluorinated amorphous carbon prepared at 350°C.

The thermal stability of a-CFx films has been found to be strongly dependent on the substrate temperature during deposition. For example, the room temperature deposited films loses about 50% of their weight after annealing at 400°C, while the 350°C deposited films remain essentially unchanged. From XPS and NEXAFS results, we can see at least three effect of the higher growth temperature: (1) there are fewer carbon atoms in the CF_3 configuration; (2) more non-fluorinated carbon atoms. They are mostly found with single bonds in sp^3 hybridization; (3) even though there are similar amount of carbon atoms with one fluorine neighbor, more are found with only single bonds in the higher temperature films. All three effects point to a larger/longer and more cross-linked ring/chain structures in the higher temperature films.

Because of the complexity of the PECVD process, one can only qualitatively understand the reason for the observed temperature effect. Decomposition of C_4F_8 in the plasma resulted in various radicals: C, CF, CF_2, which are good building blocks for long chain/larger rings; CF_3, which as a terminal block blocks the growth of the structure; F, and F_2 tend to etch away already deposited atoms. Addition of methane creates H and H_2 species, which tend to bond with F to form HF, thereby reducing the etching of the films. Growth of the film is a result of competition between the attachment of CF species and the etching by the F species. These are strong functions of substrate temperature, plasma power, reactor design, and presence of other gases such as the CH_4. Here we demonstrated the effect of substrate temperature. By properly adjusting the plasma power and the gas mixture ratio, we have recently improved the thermal stability of the a-CFx film even further.[12]

CONCLUSIONS

In conclusion, we performed spectroscopy studies on the fluorinated amorphous carbon films. By combine XPS and NEXAFS results, important structural information is obtained. The structure we derived is that of C-C, CH, and CF ring and chain structures linked by CF_2 and non-fluorinated carbon groups. Increasing the deposition temperature has a strong effect on the structure of the film: the amount of CF_3 configuration is reduced; amount of C-F with sp^3 hybridization is increased; and non-fluorinated carbon with sp^3 hybridization is also increased. The overall effect is a more cross-linked structure with improved thermal stability. The band gap of the a-CFx film is smaller than that of the parylene film, resulting in higher electrical leakage.

ACKNOWLEDGEMENTS

YM and HNY would like to thank Dr. S.T. Hsu for encouragement and support. The synchrotron part of the work was supported by the Swedish Natural Science Research Council (NFR), the Göran Gustavsson Foundation for Research in Natural Sciences and Medicine. The experimental work at ALS, Lawrence Berkeley National Laboratory was supported by the Director, Office of Energy Research, Office of Basic Energy Sciences, Materials Sciences Division of the U. S. Department of Energy, under Contract No. DE-AC03-76SF00098.

REFERENCES

[1]. The National Technology Roadmap for Semiconductors, Semiconductor Industry Association, San Jose, (1997).

[2]   K. Endo and T. Tatsumi, J. Appl. Phys. **78**, p. 1370 (1995);   H. Kudo, R.Shinohara, and Y. Yamada, Mat. Res. Soc. Proc. **381**, p. 105 (1995).

[3] H. Kudo, R. Shinohara, S. Takeishi, N. Awaji, and Y. Yamada: Jpn. J. Appl. Phys. **35**, 1583 (1996); A. Grill, V. Patel, K.L. Saenger, C. Jahnes, S.A. Cohen, A.G. Schrott, D.C. Edelstein, and J.R. Paraszcak, Mat. Res. Soc. Proc. **443**, p. 155 (1996); T.W. Mountsier and D. Kumar, Mat. Res. Soc. Proc. **443**, p. 41 (1996).

[4] T.M. Lu, J.F. McDonald, S. Dabral, G.R. Yang, L. You, and P. Bai, Mat. Res. Soc. Proc. **181**, p55 (1990); Y. Lu, Ph.D. Thesis, Rensselaer Polytechnic Inst. (1993).

[5] M.A. Plano, D. Kumar, T.J. Cleary, Mat. Res. Soc. Symp. Proc. **476**, 213 (1997).

[6] T. Warwick, P. Heimann, D. Mossessian, W. McKinney and H. Padmore, Rev. Sci. Instr. **66**, 2037 (1995).

[7] T. Möller, Synchrotron Radiation New 6, 16 (1996).

[8] J. Nordgren, G. Bray, S. Cramm, R. Nyholm, J. E. Rubensson and N. Wassdahl, Rev. Sci. Instr. 60, 1690 (1989).

[9] J. Stohr, NEXAFS Spectroscopy, (Springer Series in Surface Sciences, No 25, Springer, NY, 1992), p109; R. McLaren, et al, Phys. Rev. A 36, 1683 (1987).

[10] Y. Ma, N. Wassdahl, P. Skytt, J. Guo, J. Nordgren, P.D. Johnson, J.E. Rubensson, T. Boske, W. Eberhardt, and S. Kevan, Phys. Rev. Lett. 69, 2598 (1992).

[11] Y. Ma, P. Skytt, N. Wassdahl, P. Glans, D.C Mancini, J. Guo, and J. Nordgren, Phys. Rev. Lett. 71, 3725 (1993).

[12] H.N. Yang, D. Tweet, Y. Ma, T. Nguyen, D.R. Evans, S.T. Hsu, 1998 MRS Proceedings, to be published.

ANGLE-RESOLVED PHOTOEMISSION STUDY OF THE ELECTRONIC STRUCTURES OF AuAl₂ AND PtGa₂

L.-S. HSU *, J. D. DENLINGER **, J. W. ALLEN **
*Department of Physics, National Chang-Hua University of Education, Chung-Hua, Taiwan, ROC
**Randall Laboratory, University of Michigan, Ann Arbor, Michigan 48109-1120, USA

ABSTRACT

Synchrotron-radiation-excited angle-resolved photoemission spectra of AuAl₂ and PtGa₂ are presented. Experimental dispersion relations from normal emission spectra are compared to semi-relativistic augmented-plane-wave band-structure calculations. For PtGa₂, the Pt 5d bands show good agreement within a few tenths of an eV, while for AuAl₂, the experimental Au 5d band width is ≈0.5 eV greater than theory. In addition, polar-angle spectra and Fermi-edge intensity mapping allow the band dispersions of weak s-p bands to be revealed, and a hole pocket centered on the L-point is observed.

INTRODUCTION

AuAl₂ and PtGa₂ are intermetallic compounds of both technological and scientific importance. The former has potential applications as a selective solar absorber [1], and the latter was proposed to be used as a thermodynamically stable conducting contact to GaAs [2]. They both crystallize in the fluorite structure, and superconduct at low temperatures [3]. They are also prototype materials for studying the Au and Pt 5d bands in intermetallic compounds, since group-III metals contribute only s-p states to the valence band. The purple and gold colors of AuAl₂ and PtGa₂, respectively, also strongly motivated researchers to study the electronic structures of these two compounds. Many electronic and physical properties of these two compounds have been measured, and a majority of these works was recently reviewed by Hsu [3]. However, no angle-resolved photoemission study of these two materials has been reported so far. This paper will focus on just such a study.

EXPERIMENTS

Procedures for preparation of the (100) and (111) faces of AuAl₂ and the (111) face of PtGa₂ were the same as those reported in a previous paper [4]. Experiments were performed at the undulator beamline 7.0 at the Advanced Light Source using a system customized for automated x-ray photoelectron diffraction [5]. Photon energies in the range of 80 to 220 eV were selected with a 150 lines/mm spherical grating monochromator, and a Physical Electronics 137-mm hemispherical analyzer was used at 3 eV pass energy with total instrumental resolution of less than 80 meV. The angular resolution was less than ±1°, corresponding to a momentum resolution of ≈0.2 Å⁻¹ at 125 eV kinetic energy. The analysis chamber had a base pressure of 2×10^{-10} Torr and was equipped with low-energy electron diffraction (LEED) and a dual-anode Mg/Al K_α x-ray source for non-synchrotron photoemission and photoelectron diffraction (XPD) measurements. Sample surfaces were cleaned by repeated cycles of sputtering with Ar ions and annealing to 500°C until no traces of O or C contamination were detected. Annealing also restores the surface stoichiometry from an Al- or Ga-deficient surface after sputtering due

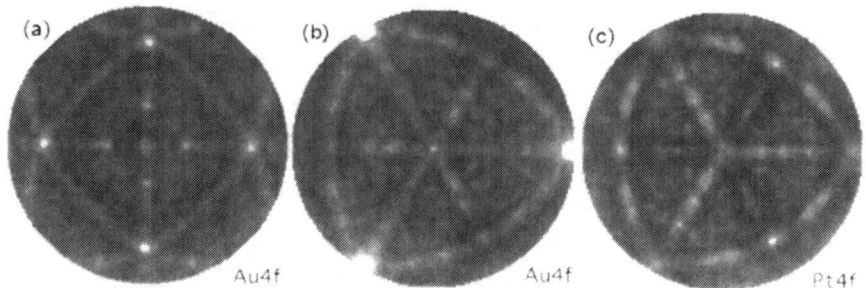

Fig. 1. X-ray photoelectron diffraction patterns from (a) AuAl$_2$(100), (b) AuAl$_2$(111), and (c) PtGa$_2$(111) for Au (Pt) 4f emission. The maximimum polar angle is 75° for each emission pattern.

to differential sputtering cross sections. Long-range surface order was confirmed with LEED, and the sample orientation of normal emission and azimuth was determined by high-kinetic Au (Pt) 4f energy photoelectron diffraction polar and azimuth scans. Stereographic two-dimensional angular intensity XPD patterns taken with Mg K$_\alpha$ excitation for AuAl$_2$(100), AuAl$_2$(111), and PtGa$_2$(111) are shown in Fig. 1. The strongest intensity peaks correspond to inter-atomic axes in the fluorite structure highlighted by forward-scattering enhancement of high-kinetic energy ($\approx$1 keV) electrons by near-neighbor atoms [6].

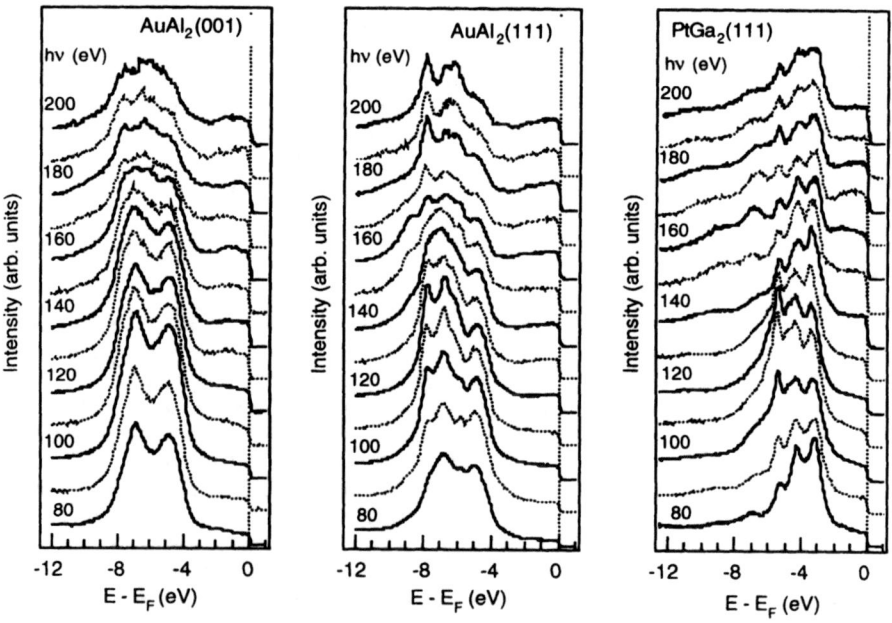

Fig. 2. Normal emission angle-resolved photoemission spectra from (a) the (001) face of AuAl$_2$, (b) the (111) face of AuAl$_2$, and (c) the (111) face of PtGa$_2$.

RESULTS AND DISCUSSION

Figure 2 shows a set of normal emission valence photoemission spectra from the (100) and (111) faces of AuAl$_2$ and from the (111) face of PtGa$_2$. The photon energy (hv) was varied so as to probe the entire fourth Brillouin zone (BZ) along the [100] direction, and the entire fourth and half of the fifth BZ along the [111] direction. The valence spectra are comprised of narrowly dispersing 5d bands (between 4-8 eV binding energy for AuAl$_2$ and between 3-6 eV for PtGa$_2$) in between rapidly dispersing s-p bands at higher and lower energies. The main splitting of the 5d-bands (most prominent in AuAl$_2$(100) spectra) is due to the spin-orbit interaction. For PtGa$_2$, the three-peak structure of the Pt 5d bands with binding energies of 3.40, 4.35, and 5.45 eV at Γ (hv$\approx$110 and 200 eV) is in close agreement with previous low-energy angle-integrated spectra [7]. The crystal-field and spin-orbit parameters of PtGa$_2$ were determined to be 0.90 and 0.56 eV, respectively, from the spacing of these three bands (Γ_8^+, Γ_7, Γ_8^-) in the same manner as was done previously [4].

By using a direct-transition model and assuming a free-electron conduction-band structure [7], the E vs. k curves can be plotted. The values of the inner potential used for AuAl$_2$ and PtGa$_2$ were estimated to be 12.1 and 10.8 eV, respectively. These values were taken to be the differences between the muffin-tin zeros of energy in the augmented-plane-wave calculations [8] and the vacuum level as determined from the work functions of these two compounds. In Figures 3 and 4, experimental bands of AuAl$_2$ and PtGa$_2$ derived from the normal emission data are compared with bands obtained from semirelativistic band-structure calculations for AuGa$_2$ and PtGa$_2$ [8]. The agreement between experiment and theory is poor for AuAl$_2$, but qualitatively good for PtGa$_2$.

The experimental d-band width for AuAl$_2$ is observed to be approximately 0.5 eV larger on the higher energy side than the theoretical calculation for AuGa$_2$. Since AuGa$_2$ has only a 1.3%

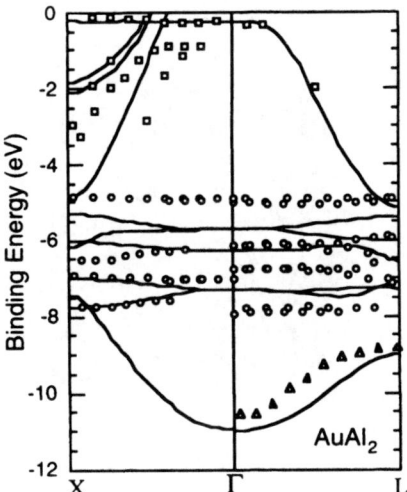

Fig. 3. Band structure of AuAl$_2$ along X-Γ-L. Open symbols (solid lines) are experimental (theoretical) dispersion curves.

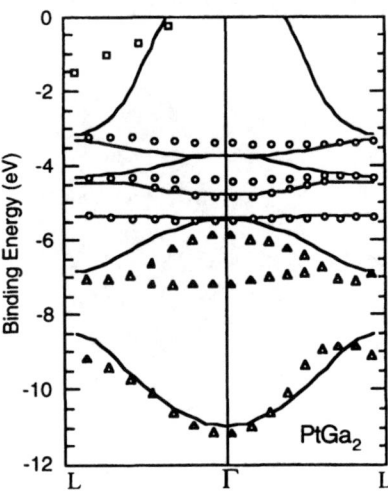

Fig. 4. Band structure of PtGa$_2$ along Γ-L: Open symbols (solid lines) are experimental (theoretical) dispersion curves.

larger lattice constant (a=6.076 Å) than AuAl$_2$ (a=5.998 Å), the band structures are expected to be quite similar. However, the 0.06 Å smaller Au-Au distance (a/√2) for AuAl$_2$ would result in greater d-band overlap, shallower d-band binding energy and a larger bandwidth, and, hence, could be be responsible for the quantitative disagreement with the AuGa$_2$ calculation. Comparison of non-relativistic band structure calculations for AuAl$_2$ and AuGa$_2$ [9,10], though, do not exhibit differences as great as 0.5 eV. The non-relativistic calculations for AuAl$_2$ [9-11] have a more severe disagreement (greater than 1 eV smaller d-band width) with the experiment mainly due to lack of spin-orbit interactions which split and widen the d-bands. For PtGa$_2$, the theory quantitative agrees much better (within a few tenths of and eV) with the experimental Pt 5d bands as observed in Fig. 4. Similar to AuAl$_2$, the calculated band width is smaller than experiment and the discrepancy is at the top of the d-band.

The dispersions of the s- and p- bands are not very discernable in the normal emission spectra of Fig. 2, in part, due to the weak s-p cross sections at energies greater than 50 eV and to the decreasing k-space resolution with increasing photon energy at a fixed angular acceptance. Only for PtGa$_2$, are the deeper s-bands easily observed and the dispersions quantifiable. In addition to normal emission spectra, automated rotations of the sample allowed detailed angle-dependent spectra and constant-binding energy intensity maps to be performed. In this manner, the structure of the s-p bands in the plateau region between the d-bands and the Fermi-level were explored. Two examples for PtGa$_2$ are presented here.

First we illustrate the use of Fermi-edge (E$_F$) intensity mapping in Fig. 5. The polar and azimuth angles of the sample were scanned with a 0.4 eV detector window centered at E$_F$ to obtain more than half of the image in Fig. 5 (20° maximum polar angle and 240° azimuth range). The final image results from 3-fold symmetrization and contrast enhancement. Intensity maxima (bright) in the angle map originate primarily from bands dispersing across the Fermi-level. Three bright circular rings are observed to dominate the intensities in Fig. 5. To understand these rings, we plot that the angular data set, which represents a curved surface in k-space, as a projection

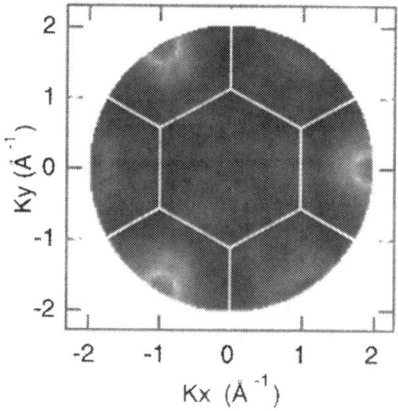

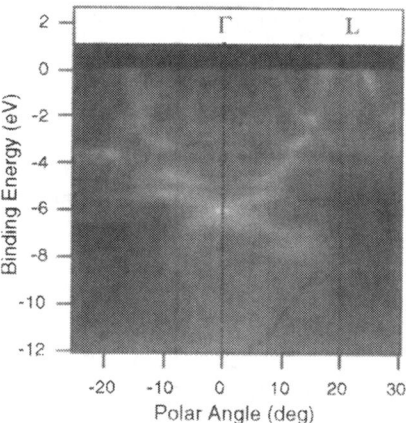

Fig. 5. Fermi-edge intensity map for PtGa$_2$(111) at 105 eV projected onto the kx-ky plane. The image center is near G and the bright rings are centered at the L-point in the second BZ.

Fig. 6. Band dispersions for PtGa$_2$(111) at 105 eV along a polar angle corresponding to a horizontal line in Fig. 5. The bright ring at L is revealed to be a hole pocket.

onto the k-parallel plane. The photon energy (105 eV) was selected to be near Γ at normal emission (center) and cross through the L-point in the second BZ at the edge along the [112] azimuth. By correlating the orientation of the Fermi-edge intensity pattern to the photoelectron diffraction in Fig. 1, we learn that the bright rings are exactly centered on L.

To further illuminate the nature of these rings more detailed angle and energy maps centered on one ring were performed. In addition, spectra were measured for a detailed polar angle cut (1° increments) through the L-point ring. The spectra were merged to form the image in Fig. 6 which reveal the dispersions of s-p bands as well as the d-bands. Division by an average of all the spectra was performed in order to put the d-bands and the s-bands on the same intensity scale. The polar-spectra image reveal that the bright rings centered at the L-point originate from bands converging towards L and hence can be described as hole pockets. A polar-photon E_F intensity map, not shown, shows that the E_F ring does not close in the k_z direction and hence is not a closed ellipsoidal Fermi surface.

CONCLUSIONS

Synchrotron-based angle-resolved photoemission spectra of $AuAl_2$ and $PtGa_2$ are presented and compared to semi-relativistic band structure calculations. The experimental and theoretical d-bands for $PtGa_2$ show generally qualitative agreement, but quantitative disagreement of the order of a few tenths of an eV. A larger disagreement for $AuAl_2$ band width exists between the photoemission data and a band structure calculation for $AuGa_2$. In addition, Fermi-edge intensity maps and detailed angle-dependent spectra are shown to reveal Fermi-surface features (a hole pocket at the L-point is given as an example) and s-p band dispersions, which will allow additional checks of the accuracy of the theoretical band calculations.

ACKNOWLEDGMENTS

We thank R. J. Baughman and R. S. Williams for supplying the samples, and G-H. Gweon for simulating discussions. This work was supported by the National Science Council, Taiwan, R.O.C., the U.S. Department of Energy under Contract No. DE-FG-02-90ER45416, and the U.S. National Science Foundation under Contract No. DMR-94-23741.

REFERENCES

1. R. E. Hahn and B. O. Seraphin, Phys. Thin Films 10, 1 (1978).
2. Y. K. Kim, D. A. Baugh, D. K. Shuh, R. S. Williams, L. P. Sadwick, and K. L. Wang, J. Mater. Res. 5, 2139 (1990).
3. L.-S. Hsu, Mod. Phys. Lett. B 8, 1297 (1994), and references therein.
4. L.-S. Hsu and K.-L. Tsang, Phys. Rev. B 56, 6615 (1997).
5. J. D. Denlinger et al., Rev. Sci. Instrum. 66, 1342 (1995).
6. W. F. Egelhoff, Jr, Crit. Rev. Solid State Mater. Sci. 16, 213 (1990).
7. J.G. Nelson, W.J.Gignac, S.Kim, J.R. Lince, and R.S. Williams, Phys. Rev. B 31, 3469 (1985).
8. K. J. Kim, B.N. Harmon, L.-Y. Chen, and D. W. Lynch, Phys. Rev. B 42, 8813 (1990).
9. A. C. Switendick and A. Narath, Phys. Rev. Lett. 22, 1423 (1969).
10. S. Kim, J. G. Nelson, and R. S. Williams, Phys. Rev. B 31, 3460 (1985).
11. A. Gupta and R. Sen Gupta, Phys. Stat. Sol. B168, 455 (1991).

PRELIMINARY RESULTS FROM A NEW SPIN SPECTROMETER

J.G. Tobin[a], P.J. Bedrossian[a], T.R. Cummins[b], G.D. Waddill[b], S. Mishra[c], P. Larson[d], R. Negri[d], M. Miller[d], E. Peterson[d], P. Boyd[e]and R. Gunion[f]

a.Lawrence Livermore National Laboratory, Livermore, CA 94550; b.University of Missouri-Rolla, Department of Physics, Rolla, MO 65401-0249; c.Virginia Commonwealth University, Dept of Physics, Richmond, VA 22384-2001; d.Physical Electronics, Inc., Eden Prairie, MN; e.Boyd Technologies, Livermore, CA 94550; f.ESG Associates, Pleasanton, CA 94550

ABSTRACT

The first preliminary results from a novel spectrometer for elementally-specific measurements of magnetic surfaces and ultrathin films are presented here. The key measurements are based upon spin-resolving and photon-dichroic photoelectron spectroscopy. True spin-resolution is achieved by the use of a Mini-Mott detection scheme. The photon-dichroic measurements include the variant magnetic x-ray linear dichroism (MXLD). Both a multi-channel, energy dispersive collection scheme as well as the spin-detecting Mini-Mott apparatus are used in data collection. The "Spin Spectrometer" is based at the Spectromicroscopy Facility (Beamline 7) at the Advanced Light Source.

DISCUSSION

In the last several years, it has become clear that multi-element materials and devices are crucial to the continued miniaturization of magnetic information storage. Examples of these include colossal magneto-resistive[1,2] (CMR), giant magneto-resistive[3–5] (GMR), and spin-valve[6–8] materials. To sort out the underlying physics and promote the optimization of these structures for applications in devices, probes that combine elemental specificity and magnetic sensitivity are essential. In the case of ultrathin films and surfaces, one very powerful probe is core-level photoelectron spectroscopy[9] (PES). Core-level PES is intrinsically elementally specific. By utilizing spin-resolving detection[10,11] and/or photon-polarization dichroism[12–19], one is able to achieve a direct magnetic sensitivity in the photoemission measurement. Furthermore, it is possible to extend the PES measurements to include photoelectron diffraction[20] (PED) as well, thus, generating a sensitivity to local atomic ordering (nanoscale geometric structure). However, this appealing combination of characteristics does not come without a price. Small cross sections and low instrumental efficiencies (e.g., the Figure of merit of a Mini-Mott spin detector[21,22] can be as low as 10^{-4}) militate that spectrometer design and beamline integration must be optimized to achieve workable data counting rates. Here, we make the first preliminary report of a novel spectrometer, located at Beamline 7 of the Spectromicroscopy Facility of the Advanced Light Source, which is dedicated to spin-resolving and photon-dichroic photoelectron spectroscopic measurements of magnetic ultrathin films and surfaces.

The design requirements for the analyzer and spectrometer include the following criteria. First, it is necessary that the total resolution and effective throughput of the spectrometer be sufficient to permit the minimization (and preferably elimination) of instrumental broadening from the observed core level peaks while still retaining sufficient counting rates to collect spectral data sets in a reasonable amount of time, preferably on the order of an hour or less. The spin analyzer must include a capability to resolve the vertical component of spin, perpendicular to the horizontal interaction plane typical of magnetic x-ray linear dichroism experiments[19]. It must also be capable of resolving spin along the electron emission direction, especially at normal emission. This is to facilitate the study of "perpendicularly" magnetized samples[15,17,20]. The spectrometer must also be capable of providing an ultrahigh vacuum environment in which magnetic surfaces and ultrathin films can be prepared and characterized, prior to magnetic analysis. (Some of these capabilities can be seen in Figure 1.) Finally, it is crucial to utilize the appropriate source of photons. In our case, the source is Beamline 7 of the Spectromicroscopy Facility at the Advanced Light Source[23]. Here, the high brightness radiation produced by the U5.0 undulator is essential for optical matching with the electron analyzer described below. The tunability and high resolution provided by the spherical grating monochromator is also essential. While U5.0 undulator is a linearly polarized source, circular polarization is provided at lower energies (hv = 95 eV) by a soft x-ray phase retarder[24]. Next, the analyzer itself and the initial test results will be discussed.

The configuration of the spectrometer and analyzer are shown in Figure 1. The high angular and energy resolution with high throughput is achieved via the use of an 11-inch mean diameter hemispherical analyzer[25] supplied by Physical Electronics. Included in this package is an electron collections lens stack with an adjustable aperture, permitting selection of various angular and sample spot sizes.

The novel aspect of our PHI analyzer is that the multi-channel detector has a hole in the center[26], permitting the direct passage of energy analyzed electrons into the electron optics without resorting to an electron switch yard. The presence of the hole does cause some problems when the multi-channel (non-spin) detection is being used: an increase in dark and background counts. Dark counts are defined as non-zero electron counting that occurs when the multi-channel detection is "on" but no excitation is striking the sample. Background counts are the counts underlying the elastic photoelectron peaks, e.g., a core-level, when actual collection is underway. Regardless, under many conditions, these problems are inconsequential. An example of our data collection is shown is Figure 2.

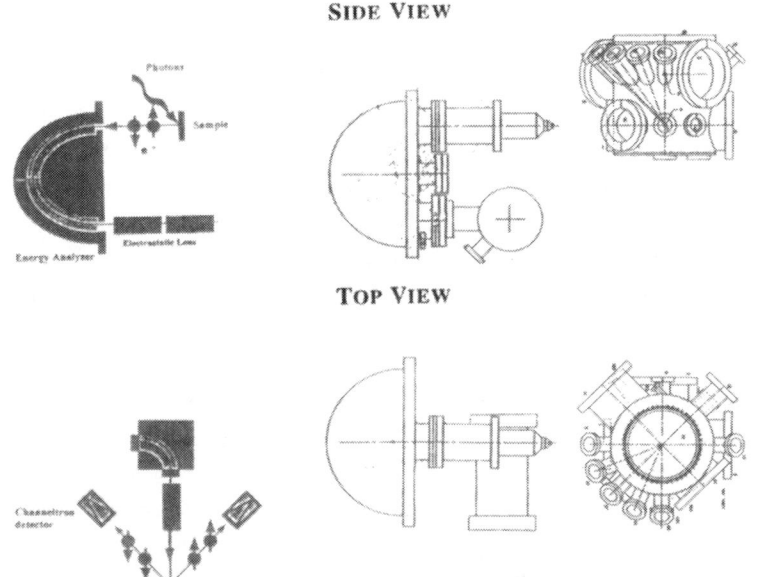

TOP VIEW

FIGURE 1. Top and side views of the energy analyzer and vacuum vessel. Ports include access for LEED, magnetizing coils, evaporation, and sample manipulation. The analyzer has a mean radius of 5.5 inches. The analyzer and vacuum vessel are scaled identically. The analyzer has an acceptance lens system on the top and the lens stacks, 90° sector and Mini Mott on the bottom.

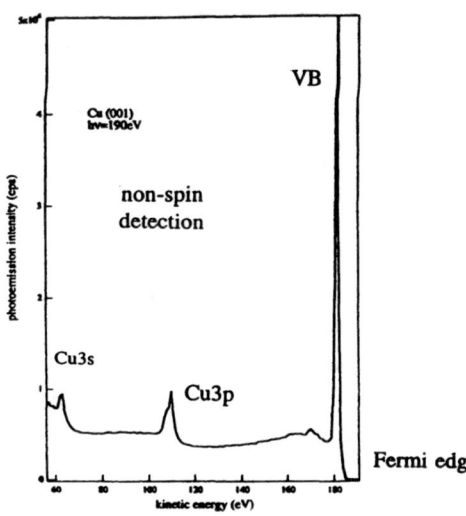

FIGURE 2 Here, the measurement was made using Cu(001) as the sample, with the photon energy being 190 eV and data collection along the sample normal, i.e., normal emission. The total energy resolution (both analyzer and monochromator) was 0.1 eV and the electron angular resolutions ±2°. The counting rate in the valence bands (VB) was 5 million counts/sec and the Cu3p was 1.0 million counts/sec. (B^F = 75, 77 eV.) The Cu3s is at B^F = 122 eV.

One measure of the dark and background counting rates is the number of counts per second observed above the Fermi edge ($B^F < 0$). As can be seen in Figure 2, they are essentially zero relative to the Cu valence bands (VB) and Cu3p intensities. It should be noted that linearity in the detector electronics tends to degrade at counting rates over 1×10^6 counts/sec. Usually, experiments will be done with either even higher resolution or less dense samples. In terms of core levels, an energy resolution of 0.1 eV is almost ideal, generally rendering instrumental energy broadening inconsequential relative to lifetime and peak asymmetry broadening.

Spin resolution is achieved by directing the electrons through the optics and into the Mini-Mott detector. In this case, the high voltages on the channel plates are turned off and the channel plates and anode assembly become part of the first lens stack, directing the electrons into the 90° spherical sector. The 90° sector is run at a relatively high pass energy: energy resolution is provided solely by the hemisphere and the photon monochromator. Because the multi-channel detection is at the exit plane of the hemisphere, the imaging of the hemisphere entrance slit[27,28] onto the multi-channel analyzer is unperturbed and high resolution non-spin counting is achieved. Furthermore, since the spin resolving detection does not require precise imaging, the burden of high resolution spatial imaging is lifted from the 90° sectors. This allows the 90° sectors to be run at high pass energies, optimizing throughput. The 90° sector serves one salient purpose: it allows the simultaneous resolution of both the vertical spin (z) and the spin along the electron emission direction(x). As described above, these two components are of particular importance in our experiments. After the 90° sector, the electrons travel through another lens stack, into the Mini-Mott detector[21,22]. In the Mini-Mott, the electrons are accelerated to 24 kV, with four channeltrons positioned horizontally and vertically used for electron counting. While suffering from a relatively low figure of merit[21,22], the Mini-Mott has two key advantages to our design. First, it provides essentially turn-key operation with the requirement of no special preparation. Second, it has been shown that the electron optical matching of a spin detector to a large hemisphere can be optimized by the use of a Mini-Mott[28].

To test the spin resolving capabilities of our spectrometer, we performed an MXLD study of a non-magnetic material, Cu(001), following the example of Kisker et al in Ref 29. In a nonmagnetic system, there will be no MXLD effect in photoemission unless the individual spin components can be resolved. We have seen dichroism in both the valence bands and in the Cu 3p core level spectra. The Cu 3p core level spectra in the Z channels are shown in Fig. 3. As previously observed by Kisker et al [Ref 29], the spin dichroism effect is only observed in the z channels, perpendicular to the interaction plane containing the Poynting vector and linear polarization vector of the x-rays, as well as the electron emission vector along the sample normal and the [110] plane of the Cu(001). Of course, the absence of the dichroic effects in the interaction plane, along the x direction, is consistent with the requirement of a transverse chirality for MXLD photoemission. As a test for instrumental asymmetry, we performed similar measurements upon the Cu 3s level. Here, no effect should be observed in either the x or z channels, and in fact none was observed. It is important to point out that the data in Figure 3 is essentially "raw data, " with no smoothing or manipulation beyond the subtraction of a linear background from each spectrum and assuming an approximately equal peak size for each member of the pair.

FIGURE 3.

These Cu 3p core level spectra were collected in the Z spin channels using a Cu(001) sample. The magnetization is along the Z axis in the dichroism experiment configuration. The upper panel shows the "raw" data with only a linear background subtracted from each spectrum. The lower panel shows the difference between the two spectra. The mono slits were 140 μ x 140 μ. The photon energy was 160 eV, the analyzer pass energy was 23.5 eV and the total energy resolution was about 0.5 eV. Angular resolution was ± 2°. Counting rates in the peak were about 5000 CPS or better. Note that the spin-orbit splitting is easily observable.

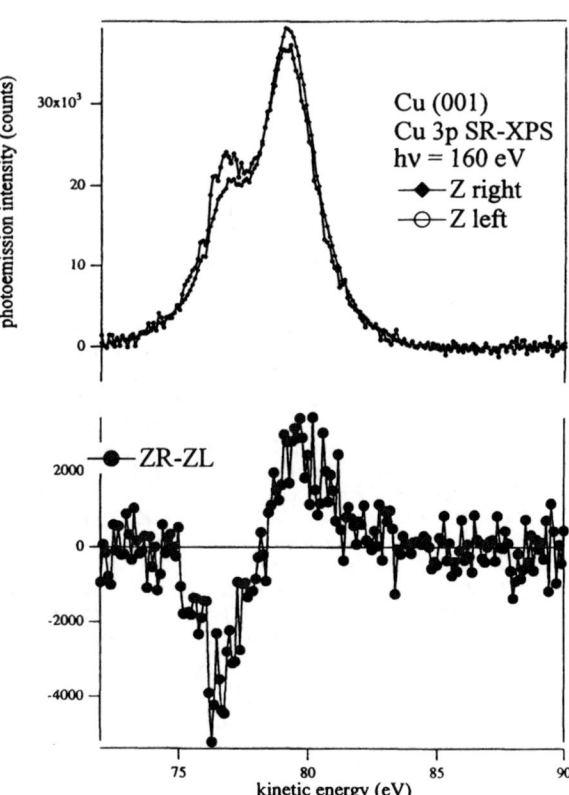

We have performed preliminary measurements upon Cu(001) using our new Spin Spectrometer at the Spectromicroscopy Facility at the ALS. Both the multichannel (non-spin) and MiniMott (4 spin channels) have been demostrated as functional. Further work is in progress.

This work was performed under the auspices of the U.S. Department of Energy by the Lawrence Livermore National Laboratory under contract number W-7405-ENG-48. The Spectromicroscopy Facility and the Advanced Light Source were constructed and are operated with the support of the Office of Basic Energy Sciences in the Department of Energy. Karen Sitzberger provided clerical support for this work. We would also like to thank D.P.Pappas of NIST in Boulder, CO for his help with this project.

REFERENCES

1. G.H. Jonker and J.H. Van Santen, Physica 16, 337 (1950) and G.H. Jonker, *ibid.* 22, 707 (1956); J. Volger, Physica 20, 49 (1954); C.W. Searle and S.T. Wang, Can. J. Phys. 47, 2703 (1969); *ibid.* 48, 2023 (1970).
2. R. von Helmholt,et al, Phys. Rev. Lett. 71, 2331 (1993); S. Jin, et al, Science 264, 413 (1994); M.A. Subramanian,, et al, Science 273, 81 (1996).
3. R. Pool, Science 261, 984 (20-AUG-93); T.L. Hylton, et al, Science, 261, 1021 (20-AUG-93).
4. S.S.P. Parkin, Phys. Rev. Lett. 71, 1641 (1993); A.C. Ehrlich, Phys. Rev. Lett. 71, 2300 (1993).
5. V. Grolier, et al, Phys. Rev. Lett. 71, 3023 (1993).
6. B.A. Gurney, et al, Phys. Rev. Lett. 71, 4023 (1993).
7. B. Dieny, et al, J. Appl. Phys. 69, 4774 (1991).
8. B. Dieny, et al, Phys. Rev. B 43, 1297 (1991).
9. L.M. Falicov, et al, J. Mat. Res. 5, 1299 (1990).
10. C. Carbone and E. Kisker, Solid State Commun. 65, 1107 (1988); F.U. Hillebrecht, R. Jungblut, and E. Kisker, Phys. Rev. Lett. 65, 2450 (1990).
11. C. Carbone, T. Kachel, R. Rochow, and W. Gudat, Z. Phys. B 79, 325 (1990).
12. C.H. Roth, et al Phys. Rev. Lett. 70, 3479 (1993); Solid State Commun. 86, 647 (1993); F.U. Hillebrecht, et al, Phys. Rev. Lett. 75, 2883 (1995); F.U. Hillebrecht, et al, Phys. Rev. B. 53, 12182 (1996).
13. G. Rossi, et al Solid State Commun. 90, 557 (1994); F. Sirotti and G. Rossi, Phys. Rev. B 49, 15682 (1994).
14. X. Le Cann, C. Boeglin, B. Carriere, and K. Hricovini, Phys. Rev. B 54, 373 (1996).
15. W. Kuch, et al, Phys. Rev. B 51, 609 (1995); C.M. Schneider, et al, Phys. Rev. B 45, 5041 (1992).
16. L. Baumgarten, et al, Phys. Rev. Lett. 65, 492 (1990).
17. G.D. Waddill, et al Phys. Rev. B 46, 552 (1992); E. Tamura, et al, Phys. Rev. Lett. 73, 1533 (1994).
18. K. Starke, et al, Phys. Rev. 50, 1317 (1994); K. Starke, et al, Phys. Rev. B 48, 1329 (1993).
19. J.G. Tobin, et al, J. Appl. Phys. 79, 5626 (1996); J. Vac. Sci. Tech. B 14, 3171 (1996).
20. G.D. Waddill, et al, Phys. Rev. B 50, 6774 (1994); J.G. Tobin, et al, Surf. Rev. Lett. 3, 1429 (1996).
21. J. Unguris, et al Rev. Sci. Instrum. 57, 1314 (1986), and references therein.
22. D.P. Pappas, J. Vac. Sci. Tech. B 14, 3203 (1996), and references therein.
23. P. Heimann, et al, Rev. Sci. Instrum. 66, Feb (1995); T. Warwick, et al, Rev. Sci. Instrum. 66, Feb (1995).
24. J.B. Kortright, et al., Appl. Phys. Lett. 60, 2963 (1992); J.B. Kortright, and J.H. Underwood, Nucl. Instrum. and Methods A291, 272 (1990).
25. Physical Electronics, OMNI Package.
26. Physical Electronics, X-Ray Emission Spectrometer.
27. C.E. Kuyalt and J.A. Simpson, Rev. Sci. Instrum., 38, 103 (1967); E.M. Purcell, Phys. Rev. 54, 818 (1938).
28. D.-J. Huang, et al, Rev. Sci. Instrum. 64, 3474 (1993).
29. Ch. Roth, F.U. Hillebrecht, W.G. Park, H.B. Rose, and E. Kisker, Phys. Rev. Lett. 73, 1963 (1994).

THE SPIN POLARIZED BAND STRUCTURE OF STRAINED THIN FILMS OF GADOLINIUM

C. WALDFRIED*, E. VESCOVO**, P. A. DOWBEN*
*Department of Physics & Astronomy and the Center for Materials Research and Analysis, University of Nebraska-Lincoln, Lincoln, NE 68588-0111.
**National Synchrotron Light Source, Brookhaven National Laboratory, Upton, N.Y. 11973.

ABSTRACT

The magnetic properties of strained thin films of gadolinium are characterized by a wave vector and thickness dependence of the exchange splitting. The spin-resolved band structure has been mapped by spin polarized photoemission, and provides considerable insight into the relationship between magnetism of local moment systems, and band structure.

INTRODUCTION

For more than 30 years theorists [1] have predicted that the magnetic coupling and exchange splitting of elemental local moment magnetic systems is wave vector dependent and strongly affected by the band structure. We have now the first direct experimental evidence of wave vector dependent exchange splitting in an elemental local moment system. The spin-polarized band structure of strained thin films of gadolinium is found to exhibit a compelling wave vector and thickness dependence of the magnetic exchange splitting [2].

Repeatedly there has been a tendency to assume that both, magnetic coupling and the correlation energy U, are wave vector independent. The possibility that the magnetic coupling or the correlation energy is anything other than a scalar is often ignored. Some recent studies [3] even suggest a wave vector independent exchange splitting of gadolinium. There is no a priori reason for this [1].

Magnetic anisotropy along distinct crystallographic directions is a consequence of spin-spin interactions (or so called dipole-dipole interactions) and spin-orbit interactions. The spin-orbit interactions make the spin sensitive to the crystal lattice and are generally the dominant effect [4,5]. Gadolinium is a ferromagnet where coupling to the crystal lattice is traditionally expected through spin-spin coupling (crystal field effects dominate) while spin-orbit coupling is expected to be weak because of the half filled 4f shell. The strong dipole coupling will also manifest in an anisotropy relative to specific crystal directions (magneto-crystalline anisotropy), and in

191

conjunction with some spin-orbit contributions can result in a wave vector dependence of the magnetic coupling. This also implies that the magnetic coupling of an elemental local moment system is expected to not only depend upon lattice spacing and bond angles but, as we will discuss in this paper, may also be strongly affected by the valence electron localization and the spin-polarized band structure. It is important to understand the significance of wave vector dependent magnetic behavior.

EXPERIMENT

Key to our understanding of the relationship between electronic structure and rare earth magnetism may be in the investigation of surfaces other than the basal plane (0001) [6] and in strained lattices. Since the electronic structure is expected to be different for different surfaces, so should be the magnetic behavior. Strained thin films of gadolinium with an increased lattice constant of approximately 4% as compared to Gd(0001) have been obtained by growing Gd on the corrugated surface of Mo(112) [2]. Ultra-thin (3 ML < d < 10 ML) and thin (d > 10 ML) films

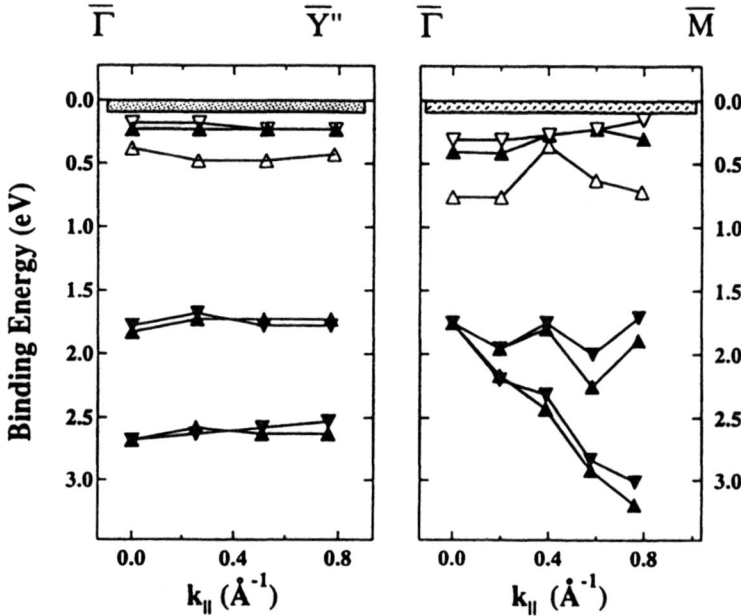

Fig. 1: Spin resolved band structure of strained Gd(10$\underline{1}$2) (left) and Gd(0001) (right) at approximately 145 K. The filled symbols indicate bands with bulk character while the open symbols mark the binding energy positions of states with surface character. The spin resolved band structure indicates majority (Δ) and minority (∇) bands. The hatched region near the Fermi level indicates the limited resolution.

of Gd order in well defined rectangular and hexagonal surface unit cells, that resemble strained Gd(10$\underline{1}$2) and strained Gd(0001), respectively. Spin-polarized photoemission experiments were carried out at the new U5UA undulator beamline of the National Synchrotron Light Source (NSLS) at the Brookhaven National Laboratory in Upton, NY. The details of the experimental setups are described elsewhere [7,8]. The surface and bulk character of the bands has been determined from chemisorption studies and photon energy dependence, while the symmetry of the bands has been ascertained from the light polarization dependence.

RESULTS

The electronic band structure represents a key factor for the description of the magnetic behavior. Based on indirect measurements it has been postulated [9-12] that the exchange splitting of a local moment magnetic system is not the same for all points in the Brillouin zone, but has a strong wave vector dependence. This has been suggested by theory for Gd [13] and is clearly the case for thin films of strained Gd(0001) (Fig. 1b). It can be seen that the Gd $5d_{xz,yz}$ or $5d_{x^2-y^2}$ bulk bands at 1.8 eV below E_F at $\overline{\Gamma}$ disperse and split along $\overline{\Gamma\Sigma M}$ of the hexagonal surface Brillouin zone, while at the same time the exchange splitting gradually increases.

The wave vector dependent magnetic behavior of the Gd bulk bands becomes even more clear when plotting two indicators of magnetism - exchange splitting and spin asymmetry - for

Fig. 2: The exchange splitting (top) and spin asymmetry m i n u s background (bottom) of the bulk spin subbands as a function of wave vector. Data is shown for a 4 ML thick film (open symbols) and a 40 ML thick film (filled symbols) of strained Gd. The data points have been extracted from spin-polarized photoemission spectra.

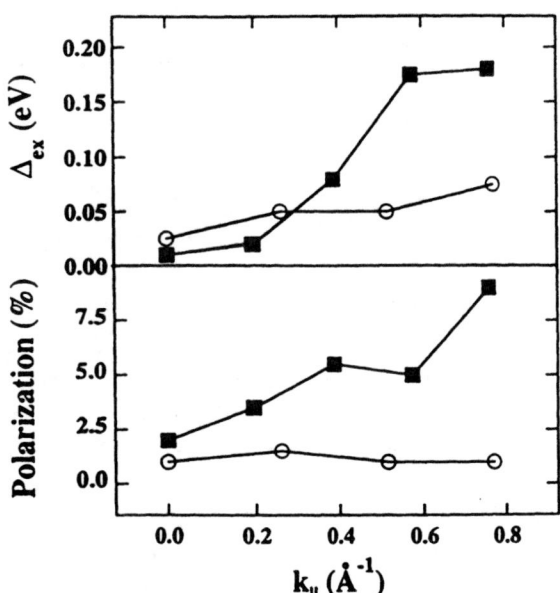

193

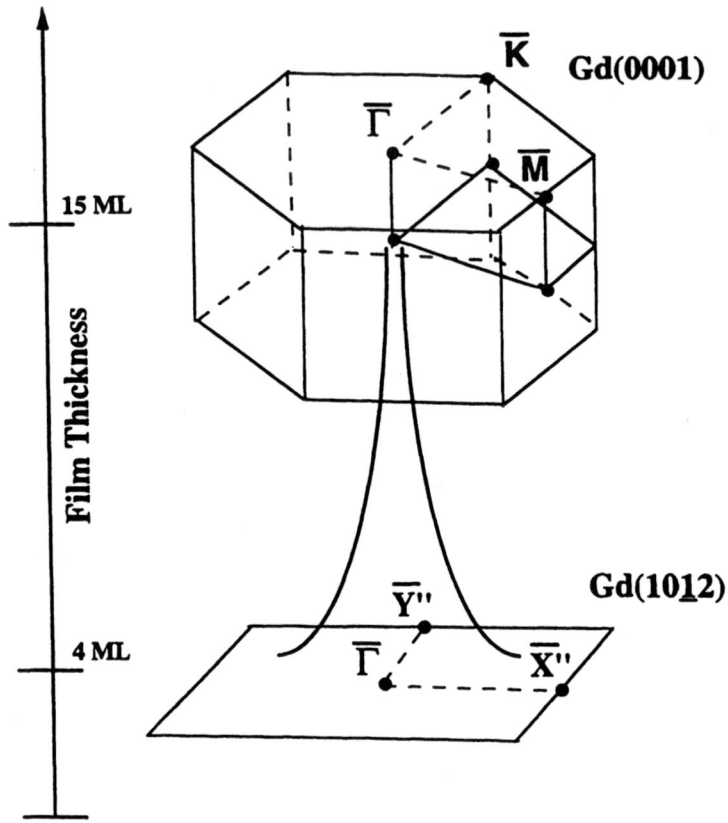

Fig. 3: Schematic of the Brillouin zone volume in which the bulk bands are paramagnetic, as function of film thickness. The Brillouin zone changes from a rectangular shape that resembles strained Gd(10$\bar{1}$2) (bottom) to the hexagonal structure of strained Gd(0001) (top).

different points of the surface Brillouin zone (Fig. 2). While the magnitude of the measured exchange splitting is an indirect indication of the Gd 5d moment, the polarization (above background) provides insight into the extent of the magnetic activity (Stoner-like and rigid band ferromagnetism). The Δ_{ex} vs. k plot (Fig. 2) confirms the increase in exchange splitting of the Gd $5d_{xz,yz}$ or $5d_{x^2-y^2}$ bulk bands with increasing wave vector from less than 0.05 eV at $\bar{\Gamma}$ to 0.2 eV at $\bar{M}$. The spin asymmetry behaves similarly, with minimal asymmetry at $\bar{\Gamma}$ and large polarization at $\bar{M}$. Both indicators suggest a change of the bulk bands from paramagnetic-like behavior at the Brillouin zone center $\bar{\Gamma}$ to strong ferromagnetism at the zone edge $\bar{M}$.

The spin-polarized band structure and the wave vector dependent exchange splitting are dominated by increasing electron localization for the thinner strained gadolinium films. In Fig. 1a we plot the spin-polarized band structure of an ultra-thin strained Gd film (d ≈ 4 ML) from $\bar{\Gamma}$ to $\bar{Y}'$ of the corresponding rectangular surface Brillouin zone (see Fig. 3). The bands show diminished dispersion. The exchange splitting of the bulk bands is negligible nearly throughout the Brillouin zone, with some indication for exchange splitting in the vicinity of the Brillouin zone edge (Fig. 2). The minimal exchange splitting, accompanied by a spin polarization that is very close to that of the background (Fig. 2), also indicates paramagnetic bulk band behavior over nearly the entire Brillouin zone.

CONCLUSIONS

Our thickness dependent spin-polarized band structure of strained thin films of gadolinium indicates that with decreasing film thickness the bulk bands become more and more localized and occupy increasingly more paramagnetic volume in the bulk Brillouin zone. In other words, the bulk bands take an increasingly "passive" role in the thinner films, while in contrast the surface remains magnetically "active" This is indicated by the significant exchange splitting (Fig. 1a) and large spin asymmetries of the surface in the ultra-thin films. The wave vector and thickness dependent distribution of para- and ferromagnetic bulk band behavior in the Brillouin zone of strained Gd is schematically shown in Fig. 3.

ACKNOWLEDGMENTS

This work was supported by NSF through grant # DMR-92-21655 and DMR-94-96131. The experiments were carried out at the National Synchrotron Light Source which is funded by the DOE. The authors like to thank Dulip Welipitiya, and Tara McAvoy for their assistance in the experiments undertaken as part of this work.

REFERENCES:

1.   R. E. Watson, and A. J. Freeman, Phys. Rev. Lett. 14, 695 (1965); R. E. Watson, and A. J. Freeman, Phys. Rev. 152, 566 (1966); R. E. Watson, A. J. Freeman, and S. Koide, Phys. Rev. 186, 625 (1969); R. E. Watson, and A. J. Freeman, Phys. Rev. 178, 725 (1969); R. E. Watson, and A. J. Freeman, J. Appl. Phys. 39, 1100 (1968).

2.   C. Waldfried, D. Welipitiya, T. McAvoy, E. Vescovo, and P. A. Dowben, submitted to Phys. Rev. Lett.

3.   Bongsoo Kim, A. B. Andrews, J. L. Erskine, Kwang Joo Kim, and B. N. Harmon, Phys. Rev. Lett. 68 (1992) 1931.

4.   B. D. Cullity, *Introduction to Magnetic Materials* (Addison-Wesley Series in Metallurgy and Materials, Reading, Mass., 1972).

5. S. V. Vonsovskii, *Magnetism* (John Wiley and Sons, NewYork, 1974).

6. S. D. Barrett, Surf. Sci. Reports **14** (1992) 271

7. E. Vescovo, et. al., Activity Report 1996, Nat. Synch. Light Source, A-25 (1997).

8. P. D. Johnson et. al., Rev. Sci. Instrum. **63**, 1902 (1992); J. Unguris, D. T. Pierce, and R. J. Calotta, Rev. Sci. Instrum. **57**, 1314 (1986).

9. Dongqi Li, Jiandi Zhang, P. A. Dowben, and M. Onellion, Phys. Rev. B **45** (1992) 7272.

10. Dongqi Li, Jiandi Zhang, P. A. Dowben, Rong-Tzong Wu, and M. Onellion, J. Phys. Condens. Matter **4** (1992) 3929.

11. M. Donath, Appl. Phys. A **49**, 351 (1989).

12. V. Korenman, R. E. Prange, Phys. Rev. Lett. **44**, 1291 (1980); V. Korenman, *Metallic Magnetism*, ed. by H. Capellmann (Springer, Berlín 1987)

13. W. Nolting, T. Dambeck and G. Borstel, Z. Phys. B**94** (1994) 409; W. Nolting, G. Borstel, T. Dambeck, T. Fauster and A. Vega, J. Magn. Magn. Mat. 140-144 (1995) 55.

EVIDENCE FOR THE PHOTOEMISSION NATURE OF Gd 4f RESONANT PHOTOEMISSION

S. R. MISHRA[1], T. R. CUMMINS[2], W. J. GAMMON[1], G. D. WADDILL[2], G. VAN DER LAAN[3], K. W. GOODMAN[4], AND J. G. TOBIN[4]

[1]Department of Physics, Virginia Commonwealth University, Richmond, VA 23284
[2]Department of Physics, University of Missouri-Rolla, Rolla, MO 65401
[3]Daresbury Laboratory, Warrington WA4 4AD, United Kingdom
[4]Lawrence Livermore National Laboratory, Livermore, CA 93550

ABSTRACT

The constructive interference between direct and indirect channels above the absorption threshold of a core level leads to a massive increase in the emission cross section leading to a phenomenon called "resonant photoemission". Using novel magnetic linear dichroism in angular distribution photoelectron spectroscopy experiment we have tried to understand the nature of the resonant photoemission process in Gd metal. The presence of dichroism in Gd 4f photoemission intensity at a photon energy corresponding to resonant photoemission clearly demonstrates the photoemission-like nature of the resonant photoemission process.

INTRODUCTION

The photoemission (PE) of 4f electrons from rare-earth metals and their compounds is strongly enhanced when the photon has enough energy to excite a 4d electron to an unoccupied 4f level, leading to a process called resonant photoemission (REPES). In the generic picture, the resonant photoemission is interpreted as due to a process where a 4d electron in the initial state is first excited to the unoccupied 4f level, forming a tightly coupled bound intermediate state, 4d core hole plus 4f electron, and then decay via autoionization into the final state, thus producing a state identical to that obtained by a direct photoemission process for the ejected electron[1]. The transition rate is greatly enhanced if the decay of the excited state is a super-Coster-Kronig (sCK) process. At photon energy corresponding to resonant photoabsorption the direct and resonant photoemission processes interfere constructively resulting in a dramatic enhancement of the photoelectron emission intensity, see figure 1. Quantum mechanically, the sequence of events leading to resonance photoemission requires treating the transition between initial and final states as a single process. Here we demonstrate that magnetic linear dichroism in angle resolved photoemission can probe the true nature of the resonant photoemission.

The availability of synchrotron x-rays has added a new tool for the study of magnetism, attracting considerable interest both from experimental and theoretical point of view. The magnetic properties of surface, subsurface, and interfaces have been probed using linearly[2] polarized light in photoemission and circularly[3,4] polarized light in both photoemission[5-9] and photoabsoption[10] modes. The interaction of circularly or linearly polarized light with the ferromagnetic atom excite a core electron thus leaving behind a core hole. The spin-orbit coupling and the exchange interaction of the core hole with the polarized valence electron lead to what is called the magnetic dichroism. The shape and intensity of photoemission or photoabsorption spectra excited with circularly or linearly polarized light, depend on the relative orientation of

Mat. Res. Soc. Symp. Proc. Vol. 524 ©1998 Materials Research Society

photon polarization and sample magnetization. The dependence of photoemission line shapes, and thus the dichroism, on the direction of magnetization when using linearly polarized light leads to what is called the magnetic linear dichroism in the angular dependence (MLDAD)[11]. As this effect arises due to the interference between the emission channels l+1 and l-1, it specifically depends upon the emission direction and vanish under the angle integration.

MLDAD, which probes the core-levels of atom, can provide wealth of information on electronic and magnetic nature of atoms as the shape of dichroic spectra depends on the core spin-orbit interaction and the exchange interaction. MLDAD experiments have been performed on transition metal[3,12-14] and rare earth metals[15] to understand magnetic properties of these metallic system. Herein we report the use of MLDAD effect in the 4f core-level photoemission from ferromagnetic Gd metal to investigate the nature of the resonantly emitted photoelectrons. The resonance effect at 3p threshold has been observed in transition metal due to 3p-3d transition[16]. Here the resonant enhancement is less prominent because 3d states of transition metals are less localized than 4f state of rare-earth metals, leading to a smaller matrix elements between the initial and the intermediate state. Thus, the attraction to study rare earth Gd metal

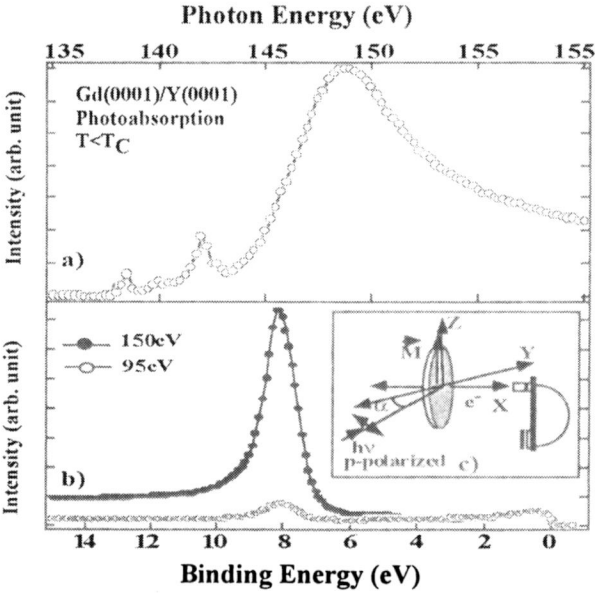

Figure 1(a). An x-ray absorption spectrum of Gd(0001)/Y(0001) recorded in a total electron yield mode. The pre-peak structure occurs around photon energies of 135-143 eV and the giant resonance is at photon energy of 148 eV. (b) A comparison of Gd 4f photoemission spectra in non-resonant vs. 4d-4f resonant photoemission. The spectrum shown with open circles was taken at off resonance with photon energy of hv = 95 eV, and the spectrum shown with filled circles was taken near the 4d-4f resonant maximum at photon energy of hv = 150 eV. (c) comes from its well localized nature of the 4f states which allows resonant Synchrotron radiation PES experimental geometry. The synchrotron x-rays are incident at an angle $\alpha=30^\circ$ relative to the YY axis. The photoelectrons are collected along the sample normal[17].

comes from its well localized nature of the 4f states which allows resonant photoemssion being treated as an atomic process [18].

EXPERIMENT

The photoemission experiments were performed with linearly polarized soft x-ray at the undulator Beamline-7 at the Advanced Light Source Facility in Berkeley at Lawrence Berkeley National Laboratory[19]. Epitaxial Gd (0001) metal films approximately 100 Å thick were prepared by e-beam evaporation onto a Y(0001) substrate at room temperature. An approximately 100 Å film was grown because the Curie point is expected to be significantly reduced from the bulk value of 293 K for thin Gd films with thickness less than 15 monolayer (ML)[20]. Subsequent annealing for 45 sec at 710 K resulted in well-ordered Gd films. The quality of the substrate as well as of the film was checked via x-ray photoelectrons excited using Mg Kα and Al Kα radiation, respectively. The crystallinity of the film was monitored via low energy electron diffraction (LEED) and x ray photoelectron diffraction (XPD)[21]. The Gd films were magnetized remanently in plane along [11-20] by the pulses of 100 kOes from a nearby solenoid. From previous studies of Gd(0001)/W(0001), the magnetization of the film is expected to lie in the plane in a single domain state[22]. All photoemission measurements were made in remanence at approximately 250 K. The angle resolved photoemission spectra were collected in

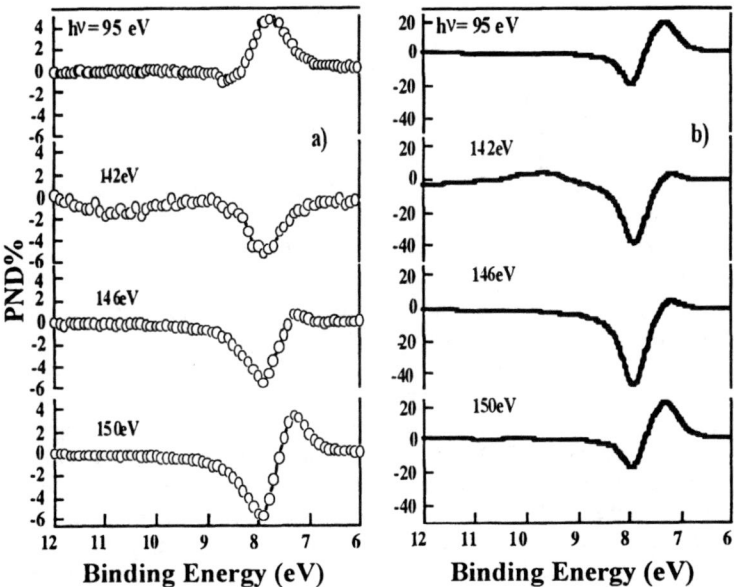

Figure 2(a). Experimental MLDAD in Gd 4d-4f REPES as the photon energy is tuned through the photoabsorption region, see figure 1(b). (b). Theoretical MLDAD spectra calculated in intermediate coupling scheme using Cowan's relativistic Hartee-Fock code.

the chiral geometry, see figure 1(c), using a Perkin Elmer hemispherical energy analyzer with +/- 2° acceptance angle and with the energy resolution of 250 meV (obtained from Fermi level cut off). The photoelectron intensity is normalized to the photon flux to account for instabilities in the synchrotron ring current. MLDAD spectras are recorded by reversing the magnetization, by rotating the sample by 180°, perpendicular to the data collection plane[21].

RESULTS AND DISCUSSION

The total electron yield spectrum from metallic Gd is displayed in figure 1(a). A group of weak narrow peaks appear near the 4d absorption edge, and a broad strong absorption feature appears at higher energy, around 150 eV, far beyond the 4d absorption edge. The strong intermediate coupling arising from the exchange and coulomb interaction between 4d hole and 4f electrons result in intermediate splitting of $4d^9 4f^8$ configuration. These interactions are very large due to the similar radii of the 4d and 4f wave functions. Features in the 4d-4f absorption curve arise from the transition from the ground level of $4d^{10} 4f^7$ configuration to the numerous intermediate levels of $4d^9 4f^8$ configuration. The broad maximum, the so called giant absorption, arises from the sCK decay of the intermediate states from $4d^9 4f^8$ configuration into $4d^9 4f^7$ ϵl[23]. These type of giant resonance absorption has been observed before in partially filled 5f, 4f, and 3d shell elements[1].

Figure 1(b) shows a set of angle resolved EDC's of Gd taken in normal emission at photon energies corresponding to on and off resonance of 4d-4f absorption threshold. The REPES effects are distinguished by comparing photoemission intensity of spectra taken on (150 eV) and off (95 eV) resonance, as determined from the absorption spectrum. It is evident that the strong increase in the Gd 4f peak photoemission intensity around 8 eV is correlated with the onset of the Gd 4d-4f giant-resonance absorption. The REPES is caused by the constructive interference between direct photoemission channel and indirect photoemission channel[24]. In the direct 4f photoemission process, an electron in the half filled f^7 ground state absorbs a photon and is ejected into the vacuum leaving behind the f^6 ion core. The resonant mechanism is viewed to consist of a two step process, $4d^{10} 4f^7 + h\nu \rightarrow 4d^9 4f^8 \rightarrow 4d^{10} 4f^6 + \epsilon l$, where ϵl represents the electron in the continuum state. In the first step, as the photon energy is tuned through the 4d absorption edge, a d shell electron gains enough energy to be absorbed into one of the empty spin minority 4f states just above the vacuum level. In the second step, the $4d^9 4f^8$ intermediate state decays via autoionization mechanism such that 4d core-hole is refilled by a 4f shell electron and another electron is simultaneously ejected into the vacuum. However, the latter event taking the system to the final state cannot be treated as individual, time-order energy-conserving event. Quantum mechanics treats the transition between initial and final state as a single step process.

Figure 2(a) compares MLDAD in Gd 4f photoemission at photon energies between 95 and 150 eV, both off and on photoabsorption resonance region and includes both experimental and theoretical spectrum. The MLDAD effect is represented as peak normalized difference (PND%), obtained by subtracting background from both spectras, and then normalizing the difference in intensity at each binding energy with the sum of peak maximum intensity of the two spectra, i.e. PND% = $((I^+ - I^-)/(I^+ + I^-)_{max}) * 100$. The MLDAD simulations, figure 2(b), are the result of theoretical spectra calculated in intermediate coupling scheme using Cowan's relativistic Hartee-Fock code[25]. Radiative transitions were taken into account to first order and sCK transitions to infinite order[26]. Line broadening of the photoelectron state and experimental resolution were included by a convolution with a Lorentzian and Gaussian, respectively. The values of the parameters can be found elsewhere [27]. For the 4f emission the interference effects between the different photoemission final states and between direct and resonant channel were fully included.

CONCLUSION

The MLDAD effects in Gd 4f resonant photoemission is solely a photoemission, not an absorption driven process. This is because the chirality which gives rise to magnetic sensitivity is due to the vectorial configuration in MLDAD as opposed to the intrinsic chirality of circularly polarized x rays in magnetic x ray circular dichroism (MXCD) techniques. In absorption, where there is an averaging over all emission angles, the vectorial chirality is lost. Thus, MLDAD is the perfect measurement to distinguish between photoemission and absorption processes. Angle-resolved photoemission in a magnetic system should show an MLDAD effect: x ray absorption and thus absorption driven processes such as Auger emission will not show MLDAD effect. Our results clearly demonstrate that Gd 4f resonant photoemission is truly photoemission in nature.

ACKNOWLEDGMENT

This work was performed under the auspices of the U.S Department of Energy by Lawrence Livermore National Laboratory under contract no. W-7405-Eng-48. Experiments were carried out at the Spectromicroscopy Facility (Beamline 7.0) at the Advanced Light Source Facility, built and supported by the U.S. Department of Energy. Stimulating discussion with Dr. D. Pappas is greatly acknowledge.

REFERENCES

1. Extensive references by J. Allen in Synchrotron Radiation Research, ed. R.Z. Bachrach (Plenum Press, New York, 1992), Vol. I., p. 253., Giant Resonances in Atoms, Molecules, and Solids, eds. J. P. Connerade, J.-M. Esteva, and R. C. Karnatak, NATO ASI series B (Plenum Press, New York, 1987).
2. L. Baumgarten, C. M. Schneider, H. Petersen, F. Schafers, and J. Krischner, Phys. Rev. Lett. 65, 492 (1990); C. M. Schneider, D. Venus, and J. Krischner, Phys. Rev. B 45, 5041 (1992).
3. Ch. Roth, H. B. Rose, F. U. Hillebrecht, and E. Kisker, Phys. Rev. Lett. 70, 3479 (1993).
4. F. Sirotti and G. Rossi, Phys. Rev. B 49, 15 682 (1994); G. Rossi, F. Sirotti, N. A. Cherepkov, F. C. Farnouz, and G. Panaccione, Solid State Commun. 90, 557 (1994).
5. C. M. Schneider, M. S. Hammond, P. Schuster, A. Cebollada, R. Miranda, and J. Krischner, Phys. Rev. B 44, 12 066 (1991).
6. G. D. Waddill, J. G. Tobin, and D. P. Pappas, Phys. Rev B 46, 552 (1992).
7. B. T. Thole and G. van der Laan, Phys. Rev. B 44, 12 424 (1991).
8. K. Starke, E. Navas, L. Baumgarten, and G. Kaindl, Phy. Rev. B 48, 1329 (1993).
9. K. Starke, E. Navas, E. Arenholz, Z. Hu, L. Baumgarten, G. van der Laan, C. T. Chen, and G. Kaindl, Phy. Rev. B 55, 2672 (1997).
10. C. T. Chen, F. Sette, Y. Ma, and S. Modesti, Phys. Rev. B 42, 726 (1990).

11. Ch. Roth and F. U. Hillebrecht, Phys. Rev. Lett. **70**, 3479 (1993).

12. F. U. Hillebrecht, H. B. Rose, Ch. Roth, and E. Kisker, J. Magn. Magn. Materials **146**, 49 (1995).

13. J. G. Tobin, K. W. Goodman, G. J. Mankey, R. F. Wills, J. Denlinger, E. Rotenberg, and A. Warwick, J. Appl. Phys. **79**, 5626 (1996).

14. D. P. Pappas, G. D. Waddill, and J. G. Tobin, J. Appl. Phys. **73**, 5936 (1993).

15. T. Kinoshita, H. B. Rose, C. Roth, F. U. Hillebrecht, and E. Kisker, J. Electron Spec. Related Phen. **78,** 333 (1996).

16. C. Cuillot, Y. Ballu, J. Paigue, J. Ecante, K. P. Jain, P. Thiry, R. Pinchauz, Y. Petroff, and L. M. Falikov, Phys. Rev. Lett. **39**, 1632 (1977), L. C. Davis and L. A. Feldkamp, Phys. Rev. **A 17**, 2012 (1978).

17. After the M.S. Thesis of W. J. Gammon, Virginia Commonwealth University, 1996

18. S.-J.Oh and S. Donaich, Phy. Rev. **B 26**, 1859 (1982).

19. J. D. Denlinger, E. Rotenberg, T. Warwick, G. Visser, N. Nordgen, J. -H. Guo, P. Skytt, S. D. Kevan, K. S. McMCutcheon, D. Shuh, J. Bucher, N. Edelstein, J. G. Tobin, B. P. Tonner, Rev. Sci. Instrum. **66**, 1342 (1995).

20. M. Farle, K. Babershke, U. Stetter, A. Aspelmeier, and F. Gerhardter, Phys. Rev. **B 47**, 11 571 (1993).

21. S.R. Mishra, T. R. Cummins, G. D. Waddill, K. W. Goodman, J. G. Tobin, W. J. Gammon, T. Sherwood, and D. P. Pappas, J. Vac. Sci. Techn. **16**, xxxx (1998).

22. A. Berger, A. W. Pang, and H. Hopster, Phys. Rev. **B 52**, 1078 (1995).

23. J. L. Dehmer, A. F. Starace, U. Fano, J. Sugar, and J. W. Cooper, Phys. Rev. Lett. **26**, 1521 (1971); J. Sugar, Phys. Rev. **B 5**, 1785 (1972), A. F. Starace, Phys. Rev. **B 5**, 1773 (1972); L. C. Davis and L. A. Feldkamp, Phys. Rev. **A 17**, 2012 (1978).

24. F. Gerken, J. Barth, and C. Kunz, Phys. Rev. Lett. **47**, 993 (1981).

25. R. D. Cowan, The Theory of Atomic Structure and Spectra (University of California Press, Berkeley, 1981).

26. G. van der Laan, M. Surman, M. A. Hoyland, C. J. F. Flipse, B. T. Thole, Y. Seino, H. Ogasawara, and A. Kotani, Phys. Rev. **B 46**, 9336 (1992).

27. G. van der Laan, E. Arenholz, E. Navas, Z. Hu, E. Mentz, A. Bauer, and G. Kaindl, Phys. Rev. **B 56**, 324 (1997)

ESCA MICROSCOPY ON ELETTRA: CHEMICAL CHARACTERIZATION OF SURFACES AND INTERFACES WITH SUB-MICRON SPATIAL RESOLUTION

M. KISKINOVA, L. CASALIS, L. GREGORATTI, S. GUNTHER, M. MARSI
Sincrotrone Trieste, Area Science Park, Basovizza, 34012-Trieste, Italy

ABSTRACT

The chemical sensitivity, spatial resolution and data acquisition rates achieved with the scanning photoelectron microscope built at ELETTRA, a third-generation synchrotron facility in Italy has made it an indispensable tool for surface analysis of laterally heterogeneous materials. The information on the composition and evolution of coexisting phases, mass transport and other surface processes occurring on a microscopic scale can be obtained by combining elemental and chemical mapping with core and valence level photoelectron spectromicroscopy with spatial resolution better than 0.15 μm and energy resolution better than 0.5 eV. Selected recent results will be presented to illustrate the capabilities and some of the possible applications of synchrotron radiation spectromicroscopy for characterization of complex spatially heterogeneous materials and studies of dynamic processes related to the interface chemical structure.

INTRODUCTION

The scientific and technological impact of core and valence level spectromicroscopy comes from the fact that it provides all the power of electron spectroscopy for chemical analysis (ESCA) with spatial resolution less than one micrometer. Recently, thanks to the ultrahigh brightness of the third generation synchrotron sources, the synchrotron radiation ESCA microscopy has achieved spatial resolution comparable and even better than scanning Auger microscopes (SAM). This, combined with superior chemical sensitivity, has opened the opportunity for material science to perform microcharacterization on a spatial scale comparable to that of the processes and the phases occurring on morphologically and chemically complex surfaces, such as catalysts, electronic and magnetic devices etc.

The selected results presented in the following sections aim at illustrating several classes of experiments that have become possible thanks to the high performance level, as far as spatial resolution and photoemitted (PE) signal, achieved by the synchrotron radiation scanning photoemission microscope (SR-SPEM) built on ELETTRA. The experiments were performed using model systems and were part of the commissioning and testing of this rapidly developing technique. The emphasis is to demonstrate the variety of useful information that can be obtained by examining the chemical and electronic properties of micron-sized features on semiconductor interfaces, metal surfaces and supported catalysts, thereby identifying local surface processes related to chemical interactions or mass transport.

THE INSTRUMENT

The scanning photoemission microscope (SPEM) has operated at ELETTRA since the fall of 1995 [1]. The tuneable coherent x-ray radiation is provided by an 4.5 m, 56 mm period undulator. The requirements for the source, to preserve the central brightness and cut the incoherent part of the photon flux, are met by a modified version of a spherical grating monochromator (SGM) with fixed entrance and exit slits designed to work in the 250-1000 eV photon range. Basic components of the microscope are the photon focusing system for demagnification of the photon beam to a submicrometer size, specimen positioning and scanning systems and a hemispherical capacitor electron analyser mounted at 70 degrees with respect to the incident beam and sample normal. A combination of a Fresnel zone plate lens (ZP) with a central stop and an order sorting aperture is used as a photon optics system. The main advantage of the chosen ZP focusing element is that it can cover a wide photon energy range (200-1000 eV) and the spatial resolution is independent of the kinetic energy of the emitted photoelectrons [2]. The scanning photoemission microscope is UHV connected with sub-chambers where low energy electron diffraction (LEED), Auger electron spectroscopy (AES), mass-spectrometry and basic facilities for sample cleaning (Ar ion sputtering, annealing) and preparation (gas-in let system, metal evaporators, etc) are available. More details for the beamline set up and experimental station can be found in refs. 1 and 3.

The microscope can operate in two modes: imaging and spectroscopy. In the imaging mode the sample surface is mapped by simultaneously collecting photoelectrons of a selected kinetic energy while scanning the sample with respect to the focused photon beam. Thus each pixel of the two-dimensional micrograph corresponds to the monitored photoelectron signal, which in turn translates into elemental or chemical contrast. For each map the spatial variation of the contrast reflects the variation of the photoelectron yield from the selected elemental electronic level and can be used as a measure of the local elemental concentration. When one element is present in several chemical states, determined by interactions with other constituents, the binding energies of the core and valence electrons shift. These shifts can be used for producing maps of the chemical states of each element. The chemical contrast of complementary maps probing different chemical states of the elements provides the lateral distribution of the compounds on the sample surface. It should be pointed out that the grey level of the contrast can be used directly as a measure for the elemental concentration only in the cases of polished flat surfaces. When the sample surface contains more pronounced topographical features, a shadow on one side and contrast enhancement on the other side of these features is a common artifact also observed in SAM and secondary electron microscopy (SEM) [4]. Another undesired effect on the photoelectron yield can be caused by diffraction, back-scattering and deflection of the outgoing photoelectrons. These effects should be considered in the specific cases when the sub-surface substrate structure and composition of the probed areas vary substantially and when there is a local electrostatic field due to sample charging. Image processing and analysis require spectral processing methods namely, background subtraction applied for each pixel, removal of the topographical contrast, noise reduction etc, described in more details in [3, 5]. The second operation mode is photoelectron spectroscopy from a microspot, i.e. a detailed spectroscopic examination of features selected from the two-dimensional map by measuring energy distribution curves (EDC). The collected core and valence level spectra can be deconvoluted when they contain more than one component in order to evaluate the lateral changes in the chemical status and electronic structure of the probed samples.

RESULTS

Spatial variation in the reactivity of metal surfaces due to structural differences

The correlation between surface structure and reactivity has long been recognized and probed by extensive studies on single crystal metal and alloy surfaces with different orientations. However single crystal surfaces are far from the real world. Until now, detailed investigations on the local variation of the chemical composition of polycrystalline surfaces, when treated in different active substances, was almost impossible because of the lack of an adequate probing technique which combines high chemical surface sensitivity with high spatial resolution. Up to now the reactivity variations related to structural and/or compositional changes were extensively studied by means of photoemission electron microscopy (PEEM), imaging the local work function changes which determine the contrast level of the maps [6, 7]. These studies can evidence lateral variations in the reactivity only assuming that there is direct correlation between the changes of the work function and the local composition of the surface. This limits the application of PEEM because competing processes like adsorption, absorption and compound formation can induce reverse work function changes. In such cases PEEM images do not even provide qualitative chemical information.

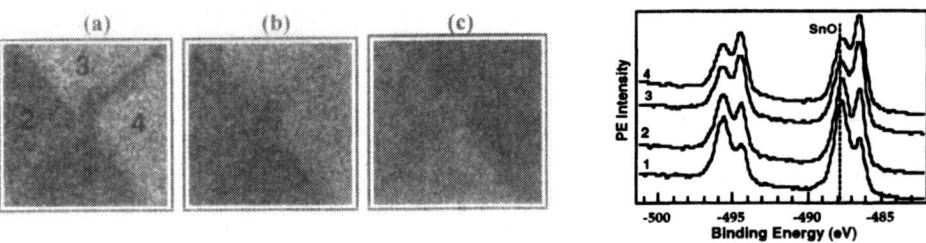

Fig. 1 (a) Sn^0 $3d_{5/2}$ map (rough data) containing topographic features; (b) and (c) Corrected Sn^0 and Sn^{2+} $3d_{5/2}$ maps showing pure chemical contrast.

Fig. 2 Sn 3d spectra taken from the four distinctive grains numbered 1 to 4 in Fig. 1(a). The dashed line indicates the position of the chemically shifted Sn^{2+} state. Photon energy: 592 eV.

The first SPEM investigations of the structural impact on the local surface reactivity were performed with polycrystalline metal surfaces (Sn, Pb, Zn, Cd) probing the correlation between the grain structure and the local oxidation rate [8-10]. Figure 1 shows two-dimensional micrographs which represent chemical maps corresponding to the intensity variations of the metallic (a and b) and oxide (c) components of the Sn $3d_{5/2}$ core levels, respectively. The clean Sn sample was exposed to ~ 1000 L oxygen at room temperature. With the energy resolution of the microscope used in these experiments (~0.5 eV), an accurate chemical mapping with photoelectron kinetic energy tuned to the metallic or oxide component of the Sn $3d_{5/2}$ core level was possible, since the change in the Sn chemical state on oxidation results in a relatively large chemical shift of the Sn 3d levels, by ~ +1.5 eV eV for SnO (Sn^{2+}) and +2.2 eV for SnO_2 (Sn^{4+}). The geometry of the microscope with an analyser mounted at a very grazing angle with respect to the probed surface (20 degrees) has an enhanced sensitivity to surface topography so that boundaries between the adjacent grains are visible in most cases. The maps in Fig. 1 are taken on an area centred close to the corner of four grains: the darker wide band along the grain boundary between the grains numbered 3 and 4 in Fig. 1(a) has a pure topographic origin. The image processing using the correction procedure for removing topographic artifacts described in ref. [5] resulted in the (b) and (c) images where the reverse contrast is purely due to the difference in the Sn oxidation state. The results shown in Fig. 1 also confirm that grain boundaries do not always divide two differently oriented crystallographic planes; two out of the four grains (3 and 4) show the same grey scale, supposing the same crystallographic planes. This is in excellent agreement with the photoelectron Sn 3d spectra in Fig. 2, which clearly illustrate that the relative weight of the oxide component varies in accordance with the grey scale level in the Sn^{2+} chemical map (Fig. 1(c)). The chemical shift of the Sn 3d core levels indicates that under the formation conditions applied in the experiment, namely exposure of the Sn sample to 1000 L O_2 at room temperature, the oxidation is limited to formation of a SnO oxide phase. After deconvolution of the Sn 3d spectra, the relative intensity of the oxide and metallic components can be used for evaluating the thickness of the formed oxide layer, considering the escape depth of the Sn 3d photoelectrons ($\lambda \sin\theta = 2.4$ Å for this experimental set-up). For these mild oxidation conditions our estimations revealed that even for the most heavily oxidized grains the thickness of the top oxide layer does not exceed ~ 4 Å, i.e. 2 ML. Normalizing against the highest grey level of the contrast in the Sn^{2+} map shown in Fig. 1(c) we evaluated that the relative oxide concentration changes from 1 for grain 1 to 0.67 for the least oxidized grains 3 and 4. This means that the reaction rate changes up to $\sim 30\%$ between the three different planes displayed in the figure.

Morphologically complex semiconductor interfaces

Spatial variations of the electronic interface parameters in both metal/semiconductor and semiconductor/semiconductor systems related to lateral inhomogeneities in the interface composition and/or morphology can exert a dramatic effect on the transport properties. Often the initial stages of formation of metal/semiconductor interfaces prepared using reactive epitaxy (RE) or solid phase epitaxy (SPE) involve nucleation and growth of multiple phases of varying composition [11]. A common phenomenon in such interfaces is the growth of 3D islands, the sizes of which vary from a few nm to a few microns. Depending on the chemical affinity between the metal and the semiconductor constituents, the 3D islands can be simply metal agglomerates or a compound formed as a result of chemical interaction between the metal and the substrate. These 3D islands, although containing most of the deposited metal, often cover only a very small fraction of the surface and can also be unevenly distributed on the surface. Thus their contribution is negligible when the characterization of the interfaces was performed with laterally averaged surface sensitive techniques and compositional information about the 3D islands, which have microscopic dimensions is lacking [12, 13]. Besides the academic interest, recent diffusion and electromigration results have shown that the 3D islands can act as a source (or trap) of diffusing atoms playing an important role in the formation dynamics at the interfaces [14, 15]. In electronic devices which contain large numbers of finite laterally confined wiring and contacts, mass transport and other interactions occurring on a very local scale can degrade the performance of the devices, a phenomenon long ago recognized in interfacial engineering.

This motivated our studies of several morphologically complex semiconductor interfaces which also served as a test for the capability of SPEM to analyse, on a microscopic scale, the compositional and structural inhomogeneity of their surface and near-surface regions [16-18]. We identified, with submicrometer spatial resolution stoichiometrically different 2D and 3D phases and followed their evolution under varying formation conditions (temperature, coverage or presence of compositional discontinuities). Here some of our most recent results on the Ni/Si, Ni/SiO$_x$/Si and Ge/Si interfaces will be presented.

Ni/Si and Ni/Si-oxide/Si systems

The Ni/Si interface is a typical example of an interface where transition metal (TM) silicides are formed. These silicides combine metallic conductivity, relative chemical inertness and high melting temperatures and are widely used as interconnections and ohmic contacts in integrated silicon devices. Considering the processes involved in production of silicon-based devices, an important property of the TM silicides is their behaviour in oxidizing ambient and high temperature environments. In most of the cases SiO_2 films are reported to be formed on top of the silicide layers without significant changes in the composition and electronic properties of the silicide phase [11].

Recently we probed the lateral variation in the composition and electronic structure of complex Ni/Si interfaces. Using a SPE formation procedure involving RT deposition of 1-4 ML Ni on a 7x7-Si(111) surface under UHV conditions ($\sim 2*10^{-10}$ mbar) followed by annealing to 1100 K (above the solvus line of Ni in Si) and slow cooling, we obtained an interface consisting of large areas with a very low Ni content (less than 0.1 ML). This is the dominating phase and micron-sized 3D Ni-silicide islands covered less than 4% of the surface [18]. Thanks to the possibility to characterizing all phases, we identified two types of 3D silicide islands of composition and electronic structure close to NiSi and NiSi$_2$, respectively. The equal number of the two types of micron-sized islands contradicts the generally accepted sequence of reactions, based exclusively on structural characterization, where formation temperatures above 970 K for solid phase epitaxy should lead only to a NiSi$_2$ 3D phase [20, 21]. We explained this unexpected finding suggesting that the identified NiSi phase is structurally similar to NiSi$_2$ and should be considered as an intermediate transition reaction state. Using the same formation procedure but depositing Ni in an oxygen ambient ($\sim 5*10^{-9}$ mbar), where Ni acted as a promoter for Si oxidation, we were able to mimic the behaviour of the Si-Ni-O ternary system. In this system, where Si is actually the main component, the thermodynamical equilibrium corresponds to segregation into Si oxide, Ni silicide and Si phases.

After RT deposition the chemical maps and PE spectra of the formed Ni/SiO$_x$/Si system showed rather uniform distribution of Ni on top of a Si oxide film. A typical spectrum of the RT Ni/SiO$_x$/Si interface is shown in Fig. 3(a). The energy positions and the intensities of the Ni 3s and 3p peaks in the PE spectra correspond to a metallic Ni layer of $\sim$ 3 ML. From the deconvoluted Si 2p spectra we determined an oxide thickness, d_{oxide}, of $\sim$ 8Å, using the integrated Si 2p peak intensities of elemental Si^0, I$_0$, and oxidized Si states, ΣI_{x+}, following the relationship: $d_{oxide}= \lambda_{oxide}.cos(\alpha) \ln [1+(I_{x+}/ I_0)/c]$ [22]. The Si 2p photoelectron mean-free path, λ, in our experimental set-up is 4.7 Å in SiO$_2$ and the intensity ratio of pure SiO$_2$ and Si, c, is 0.7 [22]. The shape and the width of the chemically shifted Si 2p oxide component, (see Fig. 4(b)) indicated that in addition to the dominating Si^{4+} oxidation state there is a distinctive contribution of lower Si oxidation states. The weight of the Si^{4+} and other oxidation states changed negligibly from spot to spot which indicates that the size of the areas with different stoichiometries is smaller than the microprobe.

Figure 3 shows typical PE spectra and chemical maps of the Ni/SiO$_x$/Si interface obtained after annealing of the sample to $\sim$ 1100 K (this temperature is below the onset of Si oxide desorption). The maps displayed in Fig. 3(b) were obtained by collecting the Ni 3p and Si 2p photoelectrons. In the latter case, the analyser was tuned separately to the Si 2p photoelectrons emitted from the Si^0 and Si^{4+} states, respectively. The contrast of the maps corresponds to lateral variations in the Ni, Si and Si oxide content and reveals development of micron-sized Ni-rich areas. The grey scale of the Si^{4+} maps shows that the oxide film thickness changes in correlation with the local Ni content, becoming thinner onto the Ni-rich areas. This is confirmed by the PE spectra taken from these two distinct areas and displayed in Figs. 3(a) and 4.

(a)   **(b)**

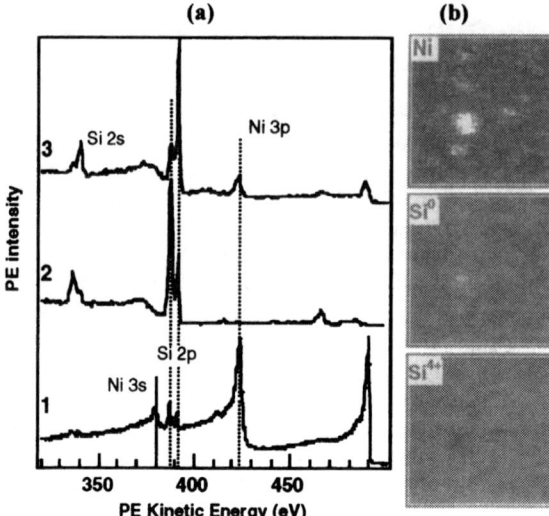

Fig. 3. (a) PE spectra: spectrum 1 is taken from the Ni/SiO$_x$/Si interface after RT Ni deposition in oxygen ambient, spectrum 2 is taken from out-of island areas after annealing and spectrum 3 from the Ni-rich islands, formed upon annealing. Photon energy: 492 eV.

(b) Chemical maps (12x12 μm^2) of the Ni/SiO$_x$/Si interface taken after annealing to 1100 K. From top to bottom the maps show the lateral changes of the Ni 3p, Si^0 and Si^{4+} 2p photoelectron intensities.

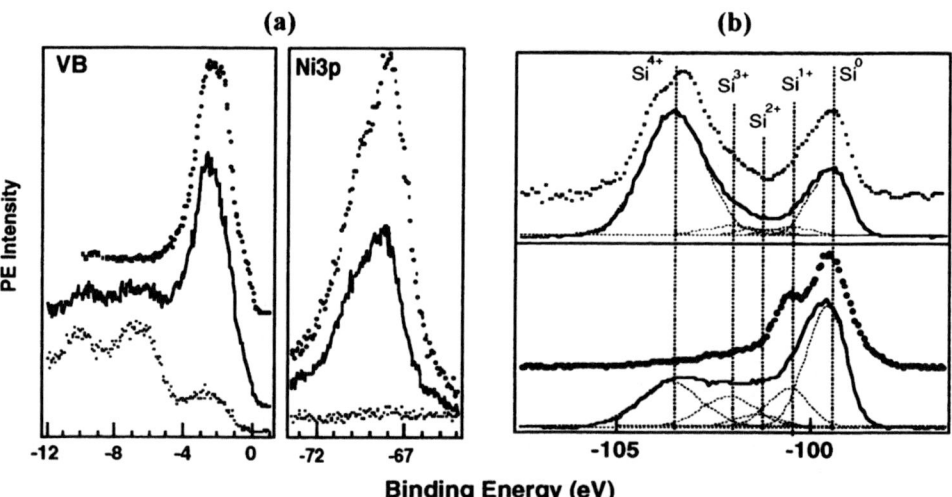

Binding Energy (eV)

Fig. 4 (a) Valence band (VB) and Ni 3p spectra taken from out-of-islands areas (dots), Ni-rich islands (line) and NiSi islands formed on a Ni/Si interface (markers). (b) the Si 2p core levels and the fitting components corresponding to different oxidation states from out-of-islands areas (top panel) and from Ni-rich islands (bottom panel). For the sake of comparison with dotted line the Si 2p spectra taken from the Ni/SiO$_x$/Si before annealing are shown in the top panel. The Si 2p spectra plotted with markers in the bottom panel are measured before from NiSi islands formed on a Ni/Si interface.

More detailed information about the composition of the Ni-rich islands and the out-of-islands areas can be extracted from the corresponding valence band (VB), Ni 3p and Si 2p core level spectra. Figure 4 shows the VB spectra, Ni 3p (a) and the Si 2p spectra (b) which reveal distinctive differences in the local chemical interactions and electronic structures. The spectra taken from the out-of-islands areas reproduce well those reported for SiO_2 thin films on Si [22, 23]. The Ni content in these SiO_2 areas was below the detection level of the microscope (~ 0.01 ML when probing the Ni 3p level with 500 eV photons). The VB spectra from all Ni-rich islands resemble closely the VB spectra of the NiSi-like islands formed at O-free Ni/Si interfaces (top spectrum in Fig. 4(a)): the Ni 3d-derived levels induce unresolved double structure at ~ -1.8 and -2.7 eV, whereas in the case of the $NiSi_2$ phase they develop a dominant sharp peak at ~ 3.1 eV [18]. However, compared to the spectra of the NiSi islands from a Ni/Si system (displayed with markers) in the VB spectra of the $Ni/SiO_x/Si$ system the relative contribution of the higher energy features is strongly enhanced. This enhancement is due to a mixture of Si 3s and 3p and O 2p states indicating presence of a detectable Si oxide film onto the islands in accordance with the Si 2p spectra. The Si 2p spectra taken from the islands and out-of-islands areas were deconvoluted using the established energy positions, relative widths and cross sections of the Si 2p corresponding to the four Si oxidation states, Si^{1+}, S^{2+}, Si^{3+} and Si^{4+} [22], and the parameters for unreacted Si^0, and reacted Si in the NiSi phase established in our previous study [18]. It was found that after annealing, the thickness of the Si oxide layer of the areas out-of the Ni-rich islands remains practically unchanged but the relative weight of the intermediate Si^{3+} oxidation state is reduced: the deconvoluted Si 2p spectra manifest dominance of the final Si^{4+} oxidation state. The deconvoluted Si 2p spectra from the Ni-rich islands clearly show that more than half of the intensity (60%) is due to by the Si in varying oxidation states. Because of the very small Si 2p chemical shift for NiSi (~ 0.1 eV), here we used only one component which accounts both for the silicide and unreacted Si. Our recent high energy resolution study have shown that the NiSi-like islands formed on a similar O-free Ni/Si interfaces are Si-terminated and a notable feature of these NiSi islands is that the out-diffused Si film is very thin (< 1Å). The ratio between the Ni 3p/Si 2p intensities for the NiSi-like islands formed in an O-free Ni/Si system, R, is ~ 2, where the sum of the reacted (silicide) and unreacted Si 2p components in the Si 2p spectra are considered. In the present case, the R value is only ~ 0.4. This low R value is consistent assuming that the Ni 3p signal is damped by the segregated Si oxide film, which according to the intensity of the Si oxide components in the Si 2p spectra should be ~ 5 Å thick. Similar dependence of the R value on the thickness of the segregated Si film was observed for the $NiSi_2$ islands on an O-free Ni/Si where the out-diffused Si overlayer reached ~ 4 Å [18, 19].

The peculiarity in the present data is the composition of the oxide cover of the islands: it contains comparable amounts of all Si oxidation states in contrast to out-of island areas which can be described as a rather homogeneous ~ 8 Å thick SiO_2 film, where the weight of the lower Si oxidation states is very small. This result is in accordance with the belief that the breakdown of the dielectric layer in MOS devices, due to diffusion of the metal through the oxide barrier, is a local phenomenon occurring at imperfections in the oxide phase. In the present model system we clearly see the correlation between the stoichiometric imperfections (distribution of the oxidation states) and the mass transport of Ni through the oxide and intermixing with Si: the NiSi-like islands form preferentially at areas where more intermediate-oxidation states are present. Note that the initial oxide layer before annealing shows an enhanced presence of lower oxidation states, which for an oxide film formed on a Si(111) substrate is viewed as kinds of dislocations [22]. The sizes of these dislocations are at nanoscales unresolved with our microscope. Another feature of the $Ni/SiO_x/Si$ system is that only NiSi-like islands are formed in contrast to the case of Ni/Si interfaces, where under the same temperature treatment $NiSi_2$

islands are formed as well [18]. Prevailing NiSi stoichiometry for the silicide phase was also reported for almost the same $Ni/SiO_x/Si$ system studied by a laterally averaging technique [24]. The suppressed formation of $NiSi_2$ islands can be attributed to the fact that the nucleation of $NiSi_2$ islands requires a sufficient number of mobile Si surface atoms [18-20]. In contrast to the case of a Ni/Si interface the NiSi-like islands here are encapsulated below a rather thick SiO_x cover which imposes a barrier to the mass transport of Si. The composition and evolution of the $Ni/SiO_x/Si$ system described here indicates that diffusion of the metal through imperfections of the oxide barrier should lead to local formation of metal-rich silicide phases.

Strained Ge/Si(100) interfaces

Ge/Si heterostructure has attracted both academic and technological interest as a novel band structure and in terms of its application in high-speed transistors and highly efficient photodiodes. Because of the 4% lattice mismatch between Si and Ge the epitaxial layers on Si can be considered as a classic strained-layer system: it obeys a Stranski-Krastanov-like growth mode, formation of 3D island on wetting Ge strained layer [25]. The transition from a 2D to 3D phase occurs at 3-4 ML coverage of Ge. The primary concern in formation of the Ge/Si interfaces is the control of their morphology because of the influence of lattice strain on the semiconductor energy gap.

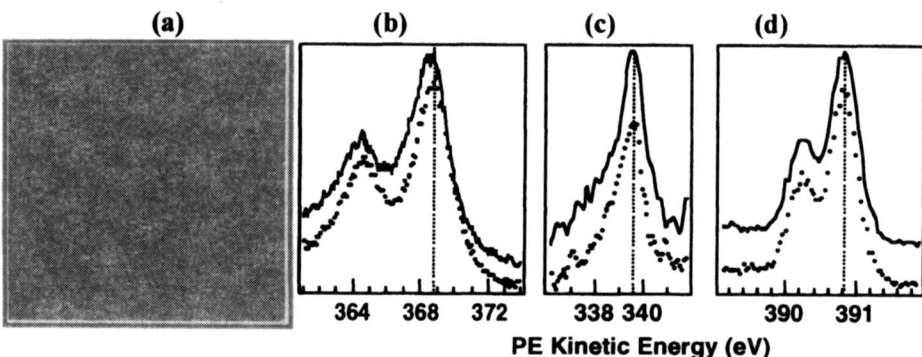

Fig. 5 (a) $6.5 \times 6.5 \ \mu m^2$ Ge $3p_{3/2}$ concentration map; (b) Ge 3p spectra from selected bright (dots) and dark region (line) from the map; (c) and (d) Si 2s and Si 3p core level spectra the same regions as (b). Photon energy: 490.5 eV.

For characterizing the lateral variation in the morphology of the Ge layers by SPEM we used PE core level spectromicroscopy to measure the strain-induced effects on the band discontinuities. We studied interfaces produced by deposition of more than 3-4 ML of Ge at temperatures in the range 700-800 K (epitaxial growth is reported to occur). The deposited Ge coverages are far beyond the Ge coverages (~ 0.2 ML) when the band bending effects occur and Ge-Si dimers are formed [26]. Thus the measured differences in the kinetic energies of the Si 2p and Ge 3p photoelectrons, ΔE_{Ge-Si}, collected from different regions can be used as a fingerprint for the local variation in the lattice distortion. Figure 5(a) shows Ge $3p_{3/2}$ concentration maps taken from layers prepared by Ge deposition at 850 K. They clearly reveal submicrometer-sized non-equivalent regions. From the intensity of the Ge 3p or 3d and Si 2s or 2p lines from the different regions we concluded that the brighter regions can be identified as agglomerates where the Ge coverage is above the critical thickness of 3-4 ML. The intensity analysis of Si and Ge core levels also indicated some Si out-diffusion for layers grown at higher temperatures. One set of the PE spectra taken from these non-equivalent regions of the Ge/Si(100) interface is shown in Fig 5 (b-d). The spectra reveal that the energy positions of the Si 2p and Si 2s core levels remain the same, whereas the Ge 3p energy position varies with the probed region. Comparison of the spectra from brighter and darker regions showed detectable ΔE_{Ge-Si} values of at least 0.1 eV. The band discontinuity variations between 0.6 and 0.7 eV can be estimated using theoretically calculated deformation parameters [27]. Inspecting the spectra taken from several very bright regions, we found that they had practically the same valence band discontinuity between Si and Ge as manifested by the ΔE_{Ge-Si} amounts. This is consistent with the expectations that the strain should be partially relaxed in the regions where Ge layer thickness exceeds the critical ~ 4 ML, but it obviously still exists in the darker low Ge coverage regions. The open question that remains is whether the observed differences in the local electronic structure are not correlated with a different density of dislocations which cannot be detected with our resolution.

Mass transport of the active phase during formation of supported catalysts

The active phase of the industrially used supported catalysts are metal or metal oxide particles deposited on oxide supports (SiO_2, Al_3O_2, TiO_2 etc.). The dimensions of these dispersed particles range between a few to a few hundreds nm. The local surface composition of the active catalytic phase and the support which determines the performance of the catalyst changes under the reaction conditions. At high temperatures and in the presence of reactive gases, e.g. under catalytic conditions, the active phase can develop surface mobility, which may result in sintering of the active phase (catalyst deactivation) or in wetting and spreading across the support surface (activation or reactivation) [28].

Our first study concerned spreading of MoO_3 catalyst particles on Al_3O_2 and TiO_2 supports. The experiments were performed using model supports, namely thin Al_2O_3 and TiO_2 films on Al or Ti foils. Using these supports we reduced the possible effects of the support polydispersity and expect to get direct experimental evidence for the molecular nature of the MoO_3 spreading onto the support oxides. Small molybdena particles were placed on the support foil using a suspension of MoO_3 micro-crystals in methanol. The spreading of MoO_3 micro-crystals was investigated by characterizing the samples before and after heat treatment in air (660 - 800 K for 5 - 7 h). High lateral resolution and analytical capabilities of SPEM allowed us to image this spreading process and to examine the variations in the local surface composition and morphology of the spread layer.

The Mo 3d and Al 2p maps and the core level spectra taken onto the support before and after catalyst conditioning confirmed the suggestion that during heating in air Mo-oxide species are released from the active phase and spreads over the support. Before annealing, the spectra from the MoO_3 crystals contained only Mo and oxygen peaks, whereas the spectra taken from the uncovered support surface only Al and O peaks, which proved that during deposition procedure no Mo oxide species dispersed away from the MoO_3 crystals.

After annealing, the Mo was detected all over the support. Since Mo appears in different oxidation states in the spread phases and large molybdena particles, the spread phase and the particles phase cannot be imaged simultaneously by tuning to the Mo 3d core levels, because the chemical shifts are ≥ 2 eV. However, quantitative information on the surface morphology of the surface can be easily obtained from the contrast of the Al 2p maps. Figure 5 shows an Al 2p map taken after annealing at 720 K for 6h, where the MoO_3 particle is distinguished as a topographic feature with a shadow. The contrast of the Al 2p maps reflect the lateral inhomogeneity of the spread phase: it varies with the local thickness of the spread molybdena phase, i.e. the darker areas correspond to stronger attenuation of the Al 2p emission due to local enrichment with Mo-oxide species forming a finite film covering the support foil. The PE spectra shown in Fig. 6 (b, c) confirm that indeed Mo can be found everywhere on the support and that the local concentration of the Mo-oxide species is higher in the darker spots of the Al 2p maps. From the attenuation of the Al 2p signal one can evaluate that the thickness of the spread Mo-oxide film is ~ a monolayer onto the dominating bright regions and several layers on the darkest spots, which can be described as islands. Another important feature is that the intensity of the Mo 3d signal did not show any dependence on the distance from the crystal as can be seen comparing the intensity of the Mo 3d spectra in Fig. 6(c) taken from bright and dark regions in the vicinity and away from the crystal. The two important findings from the maps and the spectra taken at different spots can be summarized as follows: (i) absence of concentration gradients near the MoO₃ particles, which questions the role of the "unrolling carpet" transport mechanism and favors diffusion and/or gas phase transport mechanisms; and (ii) lateral inhomogeneity of the spread Mo oxide phase: formation of islands of higher Mo oxide concentration of a size up to a few microns. These islands are almost homogeneously distributed with an island-island distance of 1 - 10 microns. These rather large island-island distances suggests fairly mobile surface species or/and very unstable nuclei, i.e. the Mo oxide species travel quite long on the surface before being trapped in an island.

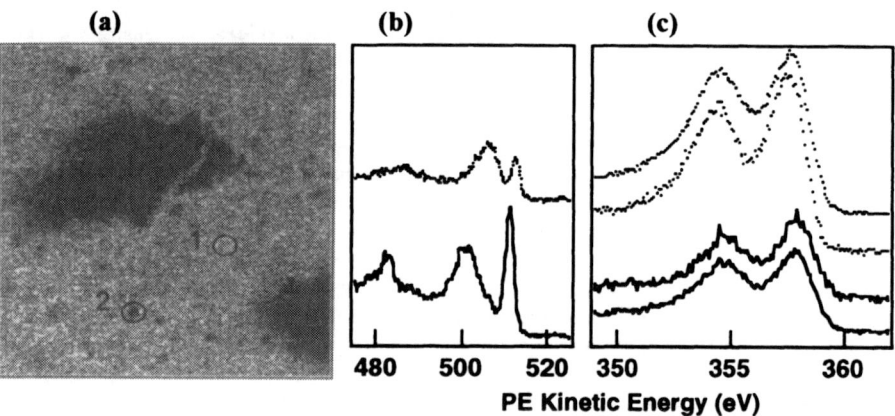

Fig. 6 (a) 50x50 μm² Al 2p map of a MoO₃ conglomerate on an Al₂O₃/Al support after annealing at 720 K for 6 h. 1 and 2 indicate the typical areas from where spectra are taken. (b) typical PE spectra of the Al 2p region taken from bright areas (lines) and dark spots (dots) in the Al 2p map. (c) and Mo 3d spectra taken from bright areas (dots) and dark spot (lines) close to and further away from the MoO crystal illustrating the lack of concentration gradient near the MoO₃ particles.

CONCLUSIONS AND OUTLOOK

The selected results presented here clearly manifest the benefits of using synchrotron radiation ESCA microscopy in applied research to detect morphological and chemical inhomogeneities by probing submicrometer fractions of the samples and processes at a microscopic scales that control the formation stages of interfaces. It should be noted that thanks to the high flux intensity of the focused beam, micrographs of 128*128 pixels can be acquired for a few minutes which is very important in characterization of more reactive interfaces with a finite lifetime, where the measurement times should be as short as possible. It should also be noted that the scanning geometry provides a more flexible and adaptable form of x-ray microscope. Both the transmitted signal (for thin samples) and any secondary emitted electrons or photons may be collected for each point adding appropriate analysers, so that a range of complementary data about the sample's structure and composition may be obtained. This classifies SR-SPEM as an easy-to-use, efficient tool for microcharacterization. Some of the near-future upgrades for improving the performance of SPEM at ELETTRA include increasing the lateral resolution using better optical elements and adding a sample cooling system in order to probe samples which have finite lifetime at room temperature or change their properties at low temperatures.

ACKNOWLEDGEMENTS

The present performance of the beamline is a result of an intense effort of many colleagues from ELETTRA (A. Abrami, F. DeBona, A. Gambitta, C. Fava, W. Jark, G. Loda, F. Mazzolini, R. Krempaska, P. Melpignano, D. Morris, R. Pugliese, F. Radovcic, R. Rosei, A. Savoia, F.-Q. Wei etc), ENIRicerche (S. Contarini, L. DeAngelis, C. Gariazzo, P. Nataletti, N. Minnaja and G. Perego) and IESS-CNR (E. Di Fabrizio, M. Baciocchi and M. Gentili) who produced the zone plates used in these experiments. Special thanks are due to D. Lonza and G. Sandrin for the design and construction of many SPEM parts and for their excellent technical assistance during the experiments, to G. Morrison from King's College for his help and illuminating discussions during the first tests of the microscope, to G. Margaritondo for his stimulating support of microscopy projects on ELETTRA and to A.J. Nelson for critical reading of the manuscript. The work was supported by an EC grant under contract ERBCHGECT920013 and by Sincrotrone Trieste SCpA.

REFERENCES

1. L. Casalis, W. Jark, M. Kiskinova, P. Melpignano, D. Morris, R. Rosei, A. Savoia, S. Contarini, L. DeAngelis, C. Gariazzo, P. Nataletti, G. Morrison, Rev. Sci. Instr. **66**, p. 4870 (1995).
2. J. Kirz, Ch. Jacobsen, M. Howells. Quart. Rev. of Biophysics **28**, p. 33 (1995) and references therein.
3. M. Marsi, L. Casalis, L. Gregoratti, S. Günther, A. Kolmakov, J. Kovac, D. Lonza, M. Kiskinova, J. Electr. Spectr. Rel. Phenom. **84**, p. 73 (1997).
4. D. Briggs, M.P. Seah, Practical Surface Analysis, John Wiley&Sons, Chichester (1990).
5. M. Kiskinova, Int. Journ. Imag. Syst. Technol. **8**, p. 462 (1997).
6. H.H. Rotermund, S. Jakubith, S. Kubala, A. von Oertzen, G. Ertl, J. Electr. Spectr. Rel. Phenom. **52** (1990) 811.
7. R. Imbihl, G. Ertl, Chem. Rev. 95 (1995) 697.
8. L. Casalis, L Gregoratti, M. Kiskinova, G. Margaritondo, F. M. Fernandez, G. Morrison, A.W. Potts, Surf. Interf. Anal. **25**, p. 374 (1997).
9. A. Potts, G.R. Morrison, S. Günther, M. Kiskinova, M. Marsi, Chem. Phys. Lett. (1998) in press.
10. A. Potts et al, in preparation.

11. M.-A. Nicolet, S.S. Lau, VLSI Electronics: Microstructure Science, Vol. 6, 1983, pp. 329-463.
12. A. Endo and S. Ino, Surf. Sci. **293**, p. 165 (1993) and references therein.
13. L. Calliari, M. Sancrotti, L. Braicovich, Phys. Rev. **B30** p. 4885 (1984).
14. Z. Suo, W. Wang, M. Yang, Appl. Phys. Lett. **64** p. 1944 (1994).
15. H. Yasunaga, Surf. Sci. **242**, p. 171 (1991).
16. S. Günther, A. Kolmakov, J. Kovac*, M. Marsi and M. Kiskinova, Phys. Rev. **B56**, p. 5003 (1997).
17. A. Kolmakov, S. Günther, J. Kovac, M. Marsi, L. Casalis, K. Kaznacheev, M. Kiskinova, Surf. Sci. **389**, p. 241 (1997).
18. L. Gregoratti, S. Günther, J. Kovac, L. Casalis, M. Marsi, M. Kiskinova, Phys. Rev. **B57** (1998) in press.
19. L. Gregoratti et al, in preparation.
20. S.A. Parikh, M.Y. Lee, P.A. Bennett, J. Vac. Sci. Technol. **A13**, p. 1589 (1995); Surf. Sci. **356** p. 53 (1996).
21. A.E. Dolbak, B.Z. Olshanetsky, S.I. Stenin, S.A. Teys, T.A. Gavrilova, Surf. Sci. **247**, p. 32 (1991).
22. F.J. Himpsel, F. R. McFeely, A. Taleb-Ibrahimi, J. Jarmoff, G. Hollinger, Phys. Rev. **B38**, p. 6084 (1988).
23. J.L. Alay, M. Hirose, J.Appl. Phys. **81**, p. 1006 (1997).
24. T. J. Sarapatka, J. Electr. Spectr. Rel. Phenom. **62**, p. 335 (1993).
25. Y.-M. Mo, M. G. Legally, J. Cryst. Growth **11**, p. 876 (1991).
26. K.-H. Huang, T.-S. Ku, D-S. Lin, Phys. Rev. **B56**, p. 4878 (1997).
27. L. Colombo, R. Resta, S. Baroni, Phys. Rev. **B44**, p. 5572 (1991).
28. H. Knözinger, E. Taglauer, Handbook of Heterogeneous Catalysis, , J. VCH-Verlag, vol. **1**, p. 216, (1997).
29. S. Günther, M. Marsi, A. Kolmakov, M. Kiskinova, M. Noeske, E. Taglauer, U.A. Schubert, G. Mestl, H. Knözinger, J. Phys. Chem. **101**, p.10004 (1997).

MAXIMUM AT ALS: A POWERFUL TOOL TO INVESTIGATE OPEN PROBLEMS IN MICRO AND OPTOELECTRONICS

G. F. LORUSSO*, H. SOLAK*, S. SINGH*, P. J. BATSON**,
J. H. UNDERWOOD**, and F. CERRINA*
*Center of X-ray Lithography, University of Wisconsin, Madison, Wisconsin
*Center of X-ray Optics, Lawrence National Berkeley Laboratory, Berkeley, California

ABSTRACT

We present recent results obtained by MAXIMUM at the Advanced Light Source (ALS), at the Lawrence Berkeley National Laboratory. MAXIMUM is a scanning photoemission microscope, based on a multilayer coated Schwarzschild objective. An electron energy analyzer collects the emitted photoelectrons to form an image as the sample itself is scanned. The microscope has been purposely designed to take advantage of the high brightness of the third generation synchrotron radiation sources, and it installation at ALS has been recently completed. The spatial resolution of 100 nm and the spectral resolution of 200 meV make our instrument an extremely interesting tool to investigate current problems in opto- and microelectronics. In order to illustrate the potential of MAXIMUM in these fields, we report new results obtained by studying the electromigration in Al-Cu lines and the Al segregation in AlGaN thin films.

INTRODUCTION

The term x-rays spectromicroscopy identifies a collection of experimental techniques combining conventional spectroscopy methods and high spatial resolution [1]. The low signal-to-noise level has originally limited the development of the photoemission spectromicroscopy approach. This limitation has been only recently overcome by the development of new high-brightness synchrotron radiation sources.

In 1987, The University of Wisconsin, in collaboration with the Advanced Light Source, begun the development of an x-ray microscope system of the scanning type with the goal of reaching a spatial resolution better than 0.1 μm, and a spectral resolution better than 200 meV, working with a base pressure better than 10^{-10} torr. All these goals were achieved in 1992, after a period of development of about 5 years. Several breakthroughs had to be achieved in order to deliver the required performances. In particular, MAXIMUM (Multiple Application X-ray Imaging Undulator Microscope) [2] mounted the first UHV-compatible Schwarzschild objective with in-situ alignment, using for the first time at-wavelength Foucalt test in extreme ultraviolet (EUV) and demonstrating the high efficiency of multilayer optics near the silicon edge.

In the case of MAXIMUM, radiation from the synchrotron source is monochromatized and focused by a Kirkpatrick-Baez (KB) system to illuminate a pinhole, which serve as source for the microscope optics. A multilayer coated (Ru-B$_4$C at 130 eV) Schwarzschild Objective (SO) produce an image of the pinhole with a 20x demagnification. When a sample is placed at the focus, photoelectrons are collected by a cylindrical mirror analyzed (CMA) electron spectrometer. The sample is mounted on a scanning stage, and by rastering the sample it is possible to produce a 2-d photoemission image. The schematic of MAXIMUM is reported in Fig. 1.

The photoemission microscope MAXIMUM was originally installed at the Synchrotron Radiation Center (SRC) in Madison, WI. During the period from 1992 to 1995, the microscope was successfully used to study semiconductor surfaces, interfaces, biological samples and organic particles. However, during the operation of the microscope, it become apparent that the microscope performance was severely hampered by the relatively low brightness of Aladdin. This limited the available flux at the microscope's focus and, consequently, the achievable spatial resolution because of signal-to-noise consideration. As a consequence, the experiments at SRC were often forced to operate at a reduced resolution in order to obtain realistic counting rates.

The installation of the microscope on a third-generation light source clearly appeared the only possible solution to this problem. A Participating Research Team (PRT) was formed in order to move MAXIMUM from Wisconsin to the Advanced Light Source (ALS). The microscope was moved to the ALS in April 1995, and it was preliminarily installed on the bend magnet beamline 6.3.2, were the preliminary tests were performed. In the present paper we report on the successful installation of MAXIMUM on its final location (beamline 12.0 at the ALS) in 1997. The present undulator beamline allows a flux $\Phi = 1.6 \times 10^{14}$ ph/s at 130 eV and 400 mA.

In this paper we report some recent results obtained by MAXIMUM investigating two open problems in microelectronics and optoelectronics: electromigration and AlGaN thin films.

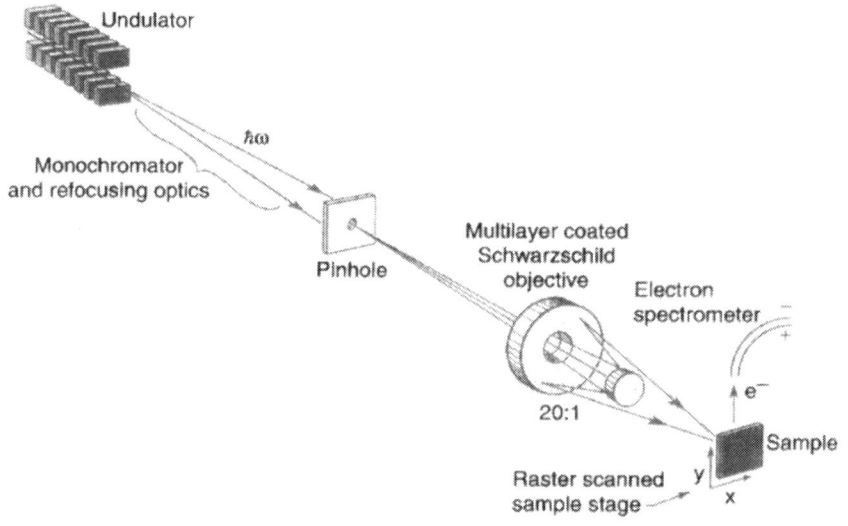

Fig. 1: Layout of MAXIMUM at beamline 12 at the ALS.

ELECTROMIGRATION

Electromigration (EM) is the movement of atoms in a conductor under the influence of an electric current. This process is one of the major reliability concerns in microelectronics industry because of its damaging effects on metal interconnect lines. Al, which has been the industrial choice for interconnect metallization, is especially susceptible to this damage mechanism [3]. Addition of small amounts of Cu (0.5-4%) to Al has been found to increase the lifetimes of interconnect lines against electromigration damage significantly and it is practiced commonly in industry [4].

In order to investigate EM, it is essential to obtain information on the Cu content of grain boundaries, on its chemical state and on the dynamics of Cu distribution between grain boundaries and Al_2Cu precipitates. MAXIMUM is an excellent tool to study this problem [5]. It can map the distribution of an element and can differentiate chemical states of the same element (like Al or Cu). Furthermore, it is especially suitable in examining the dynamics of the surface electromigration process thanks to its surface sensitivity. In order to test EM *in situ*, to preserve the chemical states of newly EM created surfaces, a special stage and a sample holder were constructed for in-situ testing   in the UHV chamber of the microscope. It allows heating of the sample above 300°C and four separate electrical contact.

For the EM test samples, 600nm of Al-Cu alloy was sputter deposited onto thermally oxidized Si wafers. Three different Cu concentrations, 0.5, 2 and 4% (weight), were chosen to observe the effect of Cu content. Lines were patterned using photolithography and wet etching processes. After patterning the wafers were annealed at 450°C for 30 minutes in forming gas to set the microstructure. No passivation was deposited onto the samples. Samples were slightly etched before introducing into the microscope in a solution of Ammonium-oxalate-monohydrate in ammonium hydroxide to reveal the Cu precipitates [6].

Samples were characterized before and after EM testing using spectromicroscopy. This is done by acquiring images of the sample at interesting photoelectron energies, as well as photoelectron spectra (PES, also called Energy Distribution Curves, EDC) on interesting spatial features on the sample. In this way we were able to compare the images and PES before and after the EM process and evaluate the changes that occurred.

Fig. 2 (left) shows an image acquired in an area of sample after testing for EM. The sample in that case is a 5 μm wide Al-4%Cu line. The image is acquired by collecting the secondary electrons of 5eV kinetic energy (KE); contrast at this energy arises from sample topography as well as elemental inhomogeneity, and the images are very similar to those obtained by SEM. Fig. 2 (right) shows three EDC's acquired at points A (hillock) B (small hillock) and C (ordinary), as marked in Fig. 1a. The micro-EDC analysis on the labeled hillock acquired at point A marked in Fig. 2 (left) indicated a narrow peak at 51.6 eV, due to metallic Al 2p core level. This is fresh Al that was carried here by the EM flux. The UHV environment of the microscope chamber slows oxidation of Al in that hillocks down. This area (A) appears bright in images acquired at all phototelectron energies due to high emission of clean Al surface. The fact that we observe metallic Al only in the hillocks and not anywhere else on the line indicates that surface electromigration did not take place in our experiment, even under favorable conditions of the UHV environment. In other words the EM was strictly limited to the grain boundary network.

The spectromicroscopic analysis of EM is also evidencing grain boundaries, the distribution of Cu in the sample, a 0.45eV shift towards higher kinetic energy of Al 2p core level in the Cu rich areas, and a charging due to the electrical isolation of the left over Al in the voids from the rest of the line. Furthermore, some of the same grain boundaries appear as bright regions in the 50 eV and 122 eV images, thus suggesting an increased presence of Cu in the grain boundaries after EM. In conclusion, the first photoemission spectromicroscopy study of electromigration in Al-Cu lines clearly evidenced the high potential of this technique in investigating interconnects in microelectronics.

 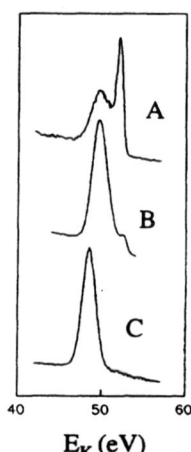

E_K (eV)

Fig. 2: X-ray image of a 5 μm wide Al-4%Cu line at 5 eV after electromigration (left), and micro electron distribution curves, showing the formation of clean localized Al after electromigration on the sample surface (right). See text for the positions of A, B, and C.

AlGaN THIN FILMS

Gallium nitride and related wide band gap semiconductors are an important class of electronic materials because of potential use in optoelectronic devices operating in the blue range [7]. As a consequence, many efforts have been devoted to investigate the electronic structure of such nitrides in the last few years, and x-ray photoemission spectroscopy (XPS) has been widely used to study GaN [8]. However, no investigation using spectromicroscopy has been reported up to now. As compared to conventional XPS, spectromicroscopy [1] can provide spatially resolved information on the chemical composition of the sample surface, as well as standard morphological and chemical analysis.

We investigate here for the first time AlGaN films by using x-ray photoemission spectromicroscopy. This preliminary analysis clearly indicates the great potential of spectromicroscopy in investigating chemical inhomogeneity, impurities and localization in AlGaN films. It also provides detailed information on the chemistry and on the morphology of the investigated systems in the submicron range.

In Fig. 3, we report three XPS microimages of sample B at different kinetic energy. The energies are 5 eV (secondary electrons), 21 eV (Ga3p), and 50 eV (Al2p), respectively. The size is 100 x 100 μm. In Fig 3 (left), the secondary electron images shows the morphology of the sample surface. A rhombohedral feature (dashed line) is evident in the upper part of Fig. 3 (left). The diagonal line (arrow) is a crystallographic plane at 60 degree with respect to the feature side. The chemical mapping of Ga and Al [Figs. 3 (center) and 3 (right), respectively] clearly indicate an Al excess and a Ga deficiency in the crystallographic plane.

The analysis also indicates the ability of spectromicroscopy to perform simultaneously morphological and chemical analysis *in situ*, as well as microscopic elemental mapping. In fact the images at 5 eV (secondary electrons) clearly evidenced the grain structure of the surface morphology and is in very good agreement with the results of atomic force microscope (AFM) analysis. Our results suggest future promising applications of this technique in investigating chemical and morphological inhomogeneity in AlGaN.

Fig. 3: XPS microimages of sample B at (left) 5 eV (secondary electrons), (center) 21 eV (Ga3p), and (right) 50 eV (Al2p). A rhombohedral feature (dashed line) and a crystallographic plane (arrow) are evident.

REFERENCES

1. G. Margaritondo and F. Cerrina, Nucl. Instr. and Meth. A291, 26 (1990).
2. C. Capasso, A. K. Ray-Chaudhuri, W. Ng, S. Liang, R. K. Cole, J. Wallace, F. Cerrina, G. Margaritondo, J. H. Underwood, J. K. Kortright, and R. C. C. Perera, J. Vac. Sci. Technol. A9, 1248 (1991).
3. Tom Seidel and Bin Zhao, Mat. Res. Soc. Sym. Proc. 427, 3 (1996)
4. I. Ames, F.M. d'Heurle, and R. Horstman, IBM J. of Res. Develop. 4, 461 (1970)
5. W. Ng, A.Ray-Chaudhuri, S. Liang, S. Singh, H. Solak, J. Welnak, F.Cerrina, G. Margaritondo, J. Underwood, J. Kortright, and R. Perera, Nucl. Instr. and Meth. A347, 422 (1994).
6. E.G. Colgan, and K.P. Rodbell, J. Appl. Phys., 75, 3423, (1994).
7. S. Nakamura, M Senoh, N. Iwasa, Jpn. J. Appl. Phys. 34, L797 (1995); S. Nakamura, M Senoh, S. Nagahama, N. Iwasa, T. Yamada, T. Matsushita, K. Kiyoku, and W. Sugimoto, Jpn. J. Appl. Phys. 35, L74 (1996). G. Martin, S. Strite, A. Botchkarev, A. Agarwal, H. Morkoc, W. R. Lambrecht, and B. Segall, Appl. Phys. Lett. 65, 610 (1994).
8. G. Martin, S. Strite, A. Botchkarev, A. Agarwal, H. Morkoc, W. R. Lambrecht, and B. Segall, Appl. Phys. Lett. 65, 610 (1994).

STATE-OF-THE-ART X-RAY PHOTOELECTRON SPECTROSCOPY (XPS):

CONVENTIONAL AND SYNCHROTRON X-RAY SOURCES FOR MICRO-XPS

E.L. Principe*, R.W. Odom*, A.L. Johnson**, G.D. Ackermann**, Z. Hussain**
and H. Padmore**
* Charles Evans & Associates, Redwood City, CA 94063
** Advanced Light Source, Lawrence Berkeley National Laboratory, Berkeley, CA 94720

ABSTRACT

This paper presents preliminary data on analyses of selected materials using two state-of-the-art XPS systems: the Physical Electronics Inc. (PHI, Eden Prairie, MN) Quantum 2000 instrument and the microXPS beamline (7.3.2.1) at the Advanced Light Source (ALS). This research compares and contrasts relevant performance characteristics of the two systems including elemental and chemical state detection sensitivity, imaging capabilities including lateral resolution and useful image fields, role of X-ray dose damage to surface, analysis speed as well as analytical throughput.

INTRODUCTION

X-ray photoelectron spectroscopy (XPS) is a mature, surface analysis technique which provides both elemental and chemical state composition of the top 10 nanometers of solids[1]. XPS has found extensive application to a wide range of materials including semiconductors, metals, polymers and glasses. XPS elemental detection limits are typically 0.1% atomic and, thus, it is suitable for analysis of major and minor constituents. There is an ever growing demand for elemental and molecular analysis within the near-surface region of a solid at micrometer (μm)and sub-μm lateral resolutions[2]. Microanalysis by XPS and other materials analysis techniques is driven primarily by manufacturers of semiconductors and medical devices as well as the medical diagnostic sector of biotechnology. There are several methods for achieving micro-XPS capabilities including development of small spot, conventional X-ray sources as well as the utilization of microfocused X-ray beams produced in synchrotrons. This paper presents preliminary results on a baseline study of two state-of-the-art XPS systems: the Physical Electronics Inc. Quantum 2000 which utilizes a newly designed, small spot conventional X-ray and the microXPS (μ-XPS) beamline at the Advanced Light Source.

EXPERIMENTAL

The PHI Quantum 2000 is a state-of-the-art XPS instrument utilizing a new X-ray source design to produce ~10 μm beam spot sizes. This system is equipped with an X-ray monochromator, Al and Mg anodes, an auxiliary ion beam for sputter depth profiling and charge neutralization as well as an electron flood gun for charge neutralization.

The μ-XPS beamline (7.3.2.1) at ALS utilizes a bending magnet design capable of delivering approximately 10^{10} photons/s in the soft X-ray region into a 2 μm diameter spot. The details of the beamline and the XPS instrument are described elsewhere[3].

The analytical program pursued in these baseline experiments included

1. Characterization of the X-ray sources for the two instruments and

2. Analysis of selected samples and comparison of the results.

Analyses were performed on ion beam sputterd Ag, a 100 μm diameter pin-hole, several regions on metallized regions of a Si device and etch pits in Al-23W and Al-9Mo.

RESULTS

Table 1 summarizes the salient capabilities of the two XPS systems which produce essentially equal photoelectron intensities for the Ag $3d^{5/2}$ transition. However, the μ-XPS system produces 80,000 counts per second (80kcps) photoelectron intensity at a 3μm spot size while the Quantum 2000 produced 90kcps at a 100μm diameter spot. The table also illustrates the tunability of the μ-XPS X-rays while the Quantum 2000 has fixed X-ray frequencies (in this study, the Al Kα transition at 1486.6eV). The energy resolution of the spectrometers are comparable and spatial resolutions are 2 and 10μm for the μ-XPS and Quantum 2000, respectively. The spectral reproducibility of the systems are comparable although beam stability in the μ-XPS system can have noticeable effects on spectral reproducibility while the X-ray source on the Quantum 2000 contributes minor instabilities to typical XPS signals.

Table 1. Comparison of the ALS μ-XPS Beamline and a PHI Quantum 2000 XPS

Characteristic	μ-XPS		Quantum 2000	
Intensity[1]	28 pA photocurrent	3μm beam	50 pA photocurrent	100μm beam
	80kcps	870eV photons	90kcps	1486.6eV photons
	23.5eV Pass Energy		23.5eV Pass Energy	
Spectral Resolution[1]	0.67eV @500eV photons		0.64eV @1486.6eV photons	
	1.08eV @870eV photons		23.5eV Pass Energy	
	1.68eV @1250eV photons		0.47eV @ 2.9eV Pass Energy	
	23.5eV Pass Energy			
Spectral Reproducibility	Determined by beam damage, counting statistics, beam stability		Determined by beam damage, counting statistics	
Spatial Resolution	2 μm		8 μm	
Spatial Reproducibility	<10μm over 40x40 mm^2		~10μm over 75x75 mm^2	

[1]Data taken from sputtered Ag foil, $3d^{5/2}$ transition

Analytical criteria most important to XPS microanalysis include the ultimate spatial resolution, photoelectron intensities and energy resolution, X-ray induced damage to the surface (especially important for organic surface analysis), the possibility of sample charging during analysis and the ability to routinely and reliably find the analytical area of interest. Minimum spatial resolutions for the two systems are determined primarily by the X-ray optics and are essentially fixed. Photoelectron energy resolutions are determined by the energy spectrometers which are essentially the same for the two instruments. X-ray induced surface damage depends on X-ray flux and the composition of the sample. The higher brightness of the ALS X-ray source could lead to more rapid surface damage; however, all analyses reported here were completed before significant damage was incurred. Sample charging currently poses more serious problems for the μ-XPS because of its higher source brightness and smaller spot sizes. The Quantum 2000 is equipped with both low energy positive ion and electron flood guns for charge neutralization in small spot analyses. The development of similar capabilities for the μ-XPS is currently in progress.

Microareas on samples are located with the Quantum 2000 using a photograph of the sample platen to locate large areas followed by localization using the relatively high accuracy PHI sample positioning system (± 15μm). The highest spatial resolution and, hence localization, of regions of interest on a sample is achieved using X-ray induced secondary electron imaging (referred to as SXI imaging). Microareas on the μ-XPS system are located using an accurate laser marking system (developed by MicroTherm, Minneapolis, mN), accurate stage translators and SXI imaging.

Figure 1 illustrates imaging of 100 μm diameter pinholes in Ni sheet for both the Quantum 2000 and the μ-XPS. Both sets of images include an optical micrograph, a secondary electron or SXI image as well as a 3D rendering of the two dimensional image. As expected, the μ-XPS has higher spatial resolution. Image resolutions in the SXI images are approximately 2 μm for the μ-XPS and 10 μm for the Quantum 2000. The curved surface in the Quantum 2000 3D image reflects intensity variations in the secondary emission due to variations in X-ray intensity. This curvature is essentially absent in the μ-XPS image illustrating its uniform X-ray intensity.

PHI Quantum 2000

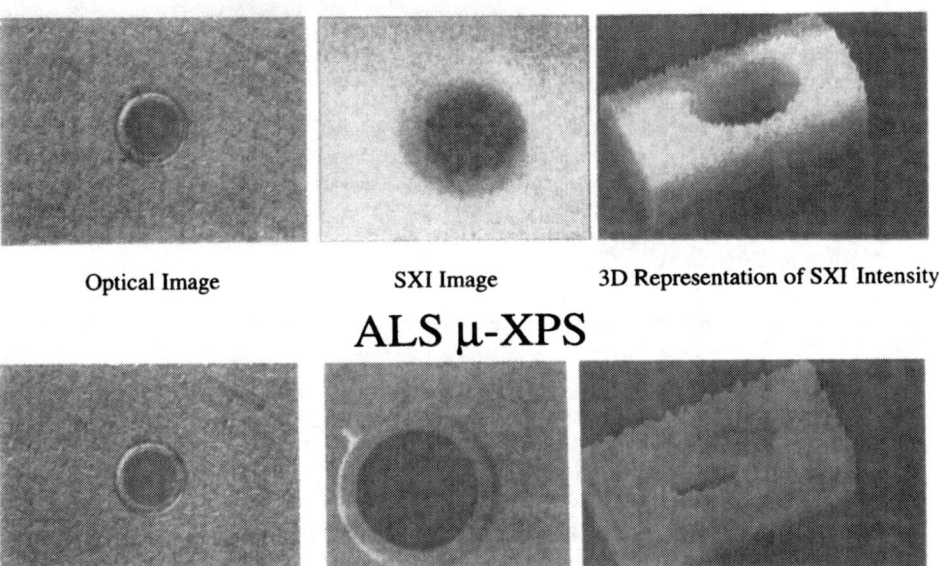

Optical Image   SXI Image   3D Representation of SXI Intensity

ALS μ-XPS

Figure 1. Optical and Secondary Electron Images from a 100 μm Pinhole in Ni Foil

Chemical microanalysis using both of these techniques is illustrated in Figures 2 and 3. The structure in this analysis was a patterned Si device in second metal layer (Metal 2) layer was exposed. The optical image in Figure 2 illustrates a number of features including an Al bond pad (100μm x100μm), field oxide (SiO₂) regions, vias and interconnect lines. Three XPS spectra were acquired from 3 distinct regions using the Quantum 2000 focused to a beam spot size of approximately 8 μm. The pass energy in these analyses was 187eV and the acquisition time was 30 minutes. The spectrum of the field oxide contains Si, O and weak C signals while the spectrum of the bond pad contains Al, O along with low levels of F contamination. The spectrum around the via region is dominated by oxygen photoelectrons and also contains readily detectable F, Si, Al and C signals. The image resolution in this analysis was approximately 10 μm.

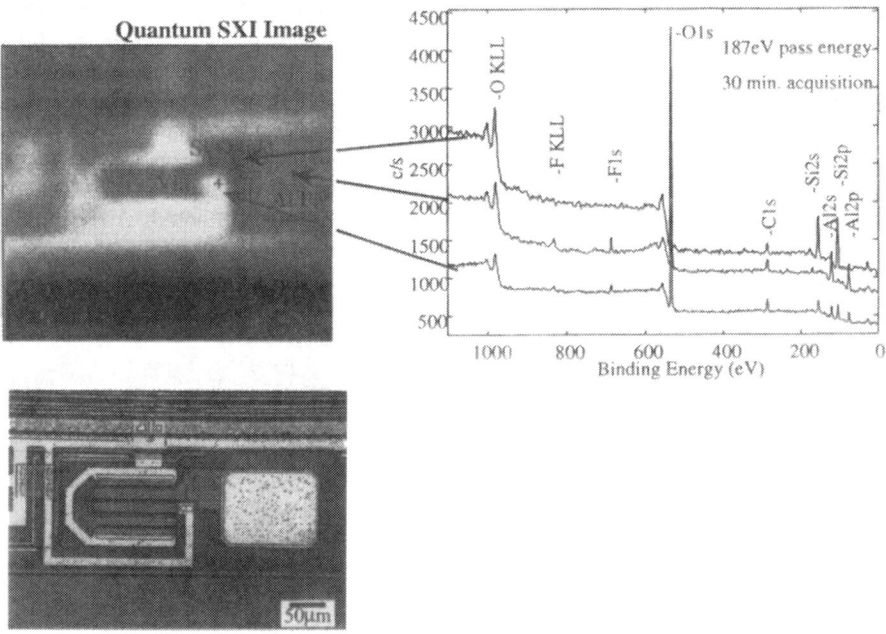

Figure 2. Secondary Electron Images and XPS Spectra from a Si Device using the Quantum 2000

An identical sample was analyzed on the μ-XPS systems operating at an image resolution of approximately 2 μm. The data in Figure 3 illustrates the high SXI image contrast and resolution of this system. A field oxide and Al run were analyzed with the μ-XPS beam and the XPS spectra show very distinctive peaks for Al, O, C and F in the metal region. The oxide spectrum provides lower intensity but distinctive signals for Si, O and C. One very distinctive characteristic of the μ-XPS system is high X-ray intensity leading to high XPS count rates. The spectrum shown in Figure 3 was acquired in 30 seconds using a band pass of 117eV. Peak signal intensities in this analysis were typically 2 orders of magnitude higher than those produced by the conventional X-ray source on the Quantum 2000.

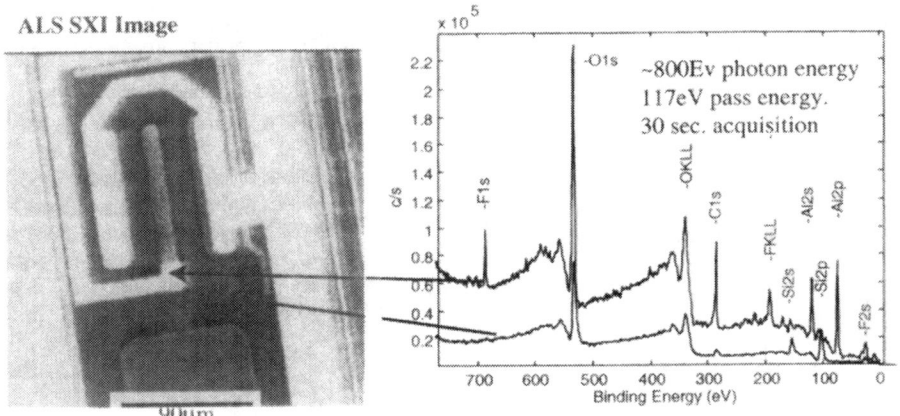

Figure 3. Secondary Electron Images and XPS Spectra from a Si Device using the ALS μ-XPS

CONCLUSIONS

The μ-XPS beamline developed in a relatively short time at the ALS demonstrates tremendous potential for XPS analysis at spatial resolutions at or below 2 μm. The high X-ray intensity of this system coupled with its tunability and small spot size will find extensive application in critical materials analyses in high technology areas such as microelectronics and biotechnology. There are several areas in which the μ-XPS system needs further development in order to have the ease of operation and range of sample types (conducting and insulating) of conventional microXPS techniques as exemplified by the PHI Quantum 2000 instrument. Necessary research and development on the μ-XPS beamline includes developing quantification techniques, incorporating effective charge neutralization of analytical surfaces, depth profiling by ion beam sputtering and data reduction and advanced image processing techniques to enhance high resolution XPS/NEXAFS data. All of these activities are in various phases of development and their successful implementation will provide material scientists with a very powerful microanalytical tool.

ACKNOWLEDGMENTS

We wish to acknowledge the assistance and encouragement of the following people along with their affiliations:

Ramone Ynzunza, Baylor Triplett and Fabia Gozzo—Intel Corporation

Pat Kinney and Yuri Uritsky—Applied Materials

REFERENCES

1. "Practical Surface Analysis, Eds. D. Briggs and M.P. Sean, 2nd edition, John Wiley & Sons, New York, 1990.

2. *Microscopic and Spectroscopic Imaging of the Chemical State*, Ed. M.D. Morris, Marcel Dekker, New York, 1993.

3. G.D. Ackermann, R. Duarte, K. Franck, M.R. Howells, Z. Hussain, S. Irick, A. Johnson, G. Morrison, H. A. Padmore, S.Y. Rah, T.R. Renner, B. Sheridan, W.Steele, C.Ayre, H. Fujimoto, F. Gozzo, B. Triplett, , R. Ynzunza, P.D. Kinney, Y.S. Uritsky, presented at the 1998 MRS Spring Meeting, San Francisco, CA

CHEMICAL ANALYSIS OF PARTICLES AND SEMICONDUCTOR MICROSTRUCTURES BY SYNCHROTRON RADIATION SOFT X-RAYS PHOTOEMISSION SPECTROMICROSCOPY

F. Gozzo*, B. Triplett*, H. Fujimoto*, R. Ynzunza[+], P. Coon[+], C. Ayre[+], P.D. Kinney[$], Y.S.Uritsky[@], G. Ackermann[#], A. Johnson[#], H. Padmore[#], T. Renner[#], B. Sheridan[#], W. Steele[#], Z. Hussain[#]

*Intel Corp., Dept. of Components Research, Santa Clara, CA 95052
[+]Intel Corp., Dept. of Materials and Technology, Santa Clara, CA 95052
[$]MicroTherm LLC, Minneapolis, MN 55413
[@]Applied Materials Inc., Dept. of Core Technologies, Santa Clara, CA 95054
[#]Lawrence Berkeley National Laboratory, Dept. of Adv. Light Source, Berkeley, CA 94620

ABSTRACT

Chemical analysis on a microscopic scale was performed on a TiN particle sample on silicon and on two patterned samples using a synchrotron source scanning photoemission microscope. For all the experiments, we exploit the ability, developed in our experimental system, to reach specific locations on the wafer and analyze the local chemical state.

INTRODUCTION

The rapid decrease of the device dimensions is defining ever smaller critical defect size for semiconductor microstructures. The aggressive trend towards decreasing device features requires the development of new analytical technologies with higher lateral resolution and microscopic chemical analysis capabilities.

A new bending magnet scanning photoemission spectromicroscope has been recently developed at the Advanced Light Source in Berkeley which can handle 50x50 mm wafer sections and is able to perform XPS analysis at one micron spot size as well as chemical bidimensional mapping. The combination of an *in situ* optical microscope with a submicron precision XY stage allows accurate navigation over the wafer and perform microscopic chemical analysis on specific features of interest as well as particle analysis on blanket wafers. Being able to perform XPS analysis on small particles is very important since the elemental and chemical analyses provide critical information on the origin of contaminants. The understanding of the local chemistry is crucial also when applied to patterned samples. One example is the influence of fluorine on aluminum bond pads surface chemistry in different manufacturing processes. The fluorine contamination of Al bond pads can compromise the bondability of the pads and is a serious problem. Poor bondability can arise from prolonged etching of the SiO_2/SiN passivation layer with a consequent excessive polymer formation on the pad region [1]. It may also be associated with an incomplete removal of the polyimide often used as a scratch protection. Very often, however, the fluorine contamination comes from unknown sources and the understanding of the surface chemistry involved can help in understanding the origin of the problem.

EXPERIMENT

All the experiments discussed in this article were performed at the Advanced Light Source in Berkeley using a new bending magnet scanning photoemission spectromicroscope which delivers 10^{10} photon/s in less than 2 microns spot beam in the soft X rays energy range. An *in situ* optical microscope allows one to easily locate the features of interest for subsequent analysis with the X ray beam. The details of the instrument are described elsewhere [2].

The particle sample was prepared by dispersing commercial particles in a dry powder form on a clean silicon wafer. With the help of an optical microscope, particles of different size were, then, selected for analysis. A new technique called Mark-Assisted Defect Analysis (MADA), recently developed by Microtherm and Applied Materials, was used to locate the particles of interest on the unpatterned wafers [3]. With the MADA technique, after creating a defect map with a standard optical defect detection system (TENCOR), an automated wafer marking system (MicroMark 5000, MicroTherm, LLC, Minneapolis, MN) relocates the defects with a better accuracy and produces small laser or indentation marks around each defect and in specific locations to assist the positioning of the defects in subsequent analysis tools, including the ALS μXPS spectromicroscope.

The two patterned samples discussed here are demonstration samples and were stripped to remove the passivation layer on the chip surface. (This procedure is usually not the one followed in normal chip manufacturing since the passivation layer was removed not only from the bonding pads surface, but also from the other features on the chip in order to perform chemical analysis and mapping in those locations). Thus, the analysis performed on these samples is meant to be mostly an example of microXPS analysis and chemical mapping using the ALS μXPS spectromicroscope.

RESULTS

Figure 1 shows a 20x20 μm photoelectron micrograph of an approximately 5 μm TiN cluster on silicon obtained with the ALS spectromicroscope by scanning the sample in x and y at 870 eV photon energy while detecting the photoemission signal from the Ti2p states. Energy distribution curves (EDC) on a microscopic scale taken on the cluster revealed, indeed, the presence of titanium in the locations suggested by the micrograph.

Figures 2a and 2b are survey spectra taken in regions A and B. Oxygen, titanium, silicon and carbon in the adventitious form but no nitrogen were detected on the cluster suggesting the Ti particles were oxidized on the surface. Also the presence of adventitious carbon justified the small Ti signal detected on the cluster. The Si2p signal is the combination of elemental silicon and silicon oxide. The silicon signal should come mostly from the empty spaces between the particles of the cluster and only in a small part from beam tail and scattered light, whose contribution has been previously estimated to be only the 20-25 % of the total signal [2]. The curve fitting performed on a higher resolution Ti2p energy region (not shown here) shows the presence of three components at 459, 456.4 and 455.3 eV identified as Ti2p in TiO₂, TiO and TiN. Approximately 70% of the total signal intensity is in the TiO₂ peak, 25% in TiO and only 5% in TiN. The cluster region was subsequently sputtered in order to remove the TiO₂ oxygen layer on the particles and the adventitious carbon lying on the surface. Figure 2c

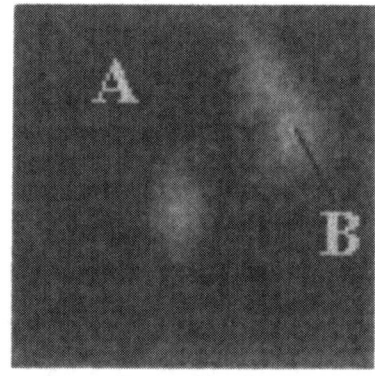

Fig. 1. 20x20 μm scanning photoemission micrograph of a TiN particle on clean Si, taken at a photon energy of 870 eV, detecting photoelectrons corresponding to the Ti2p states.

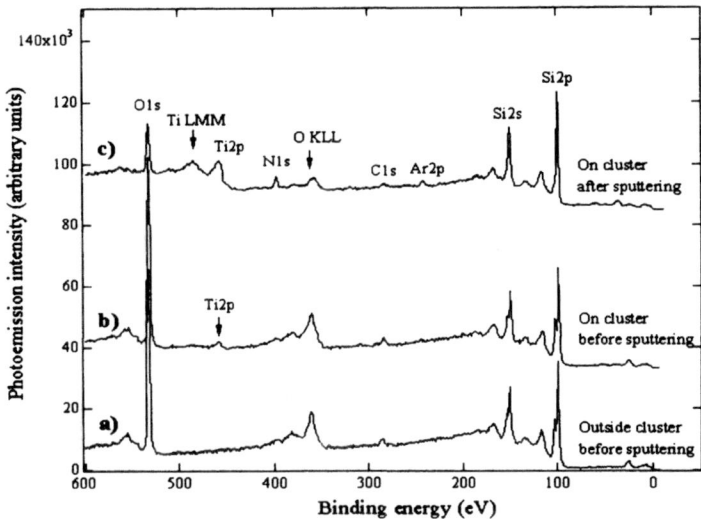

Fig. 2. XPS surveys outside the TiN cluster before sputtering (a), on the cluster before sputtering (b) and on the cluster after sputtering (c).

shows the survey taken in the same location after the sputtering. The N1s signal is now clearly detected, the O1s is strongly reduced while the Si2p signal is now mostly elemental silicon. C1s is at a lower binding energy value (283.6 eV), which is consistent with carbide, and the Ti2p core level is now a very complex structure suggesting the coexistence of TiN and different oxidation states close in binding energy.

The image in Fig. 3 shows the optical image of a feature of interest on a patterned sample (a) and the corresponding X rays photoemission micrograph (b) taken at the kinetic energy of the Al2p states. We refer to the selected Al bond pad feature as "spider bond pad" because of its peculiar

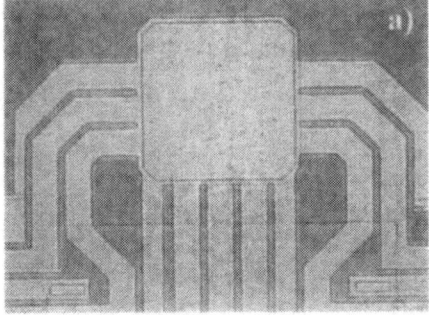

Fig. 3. 300x400 µm optical image of a patterned sample (a) and corresponding X ray photoemission micrograph taken at the energy of the Al2p states (b).

shape. Figure 4 shows overview XPS spectra on the spider pad and outside it. Fluorine, oxygen, carbon and aluminum were detected on the bond pad and essentially oxygen, silicon and silicon oxide with some fluorine traces outside the pad.

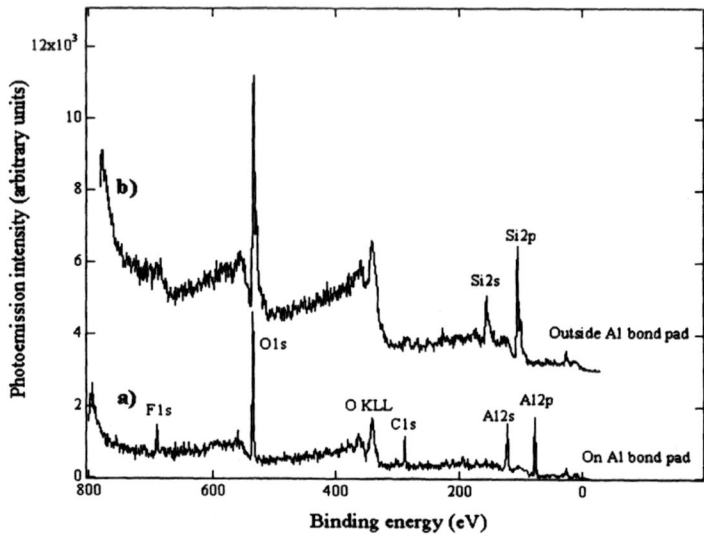

Fig. 4. XPS surveys on the Al bond pad (a) and outside the Al bond pad (b).

The curve fitting of the F1s and Al2p XPS spectra on the pad in Figs. 5a and 5b shows that the aluminum is mostly involved in fluorine bonding at the surface. The Al2p core level peak can be deconvolved into three components at approximately 76, 75.5 and 73 eV binding energy which are identified as being Al2p in aluminum fluoride, aluminum oxyfluoride and metallic aluminum. The corresponding F1s peak shows the presence of two components at approximately

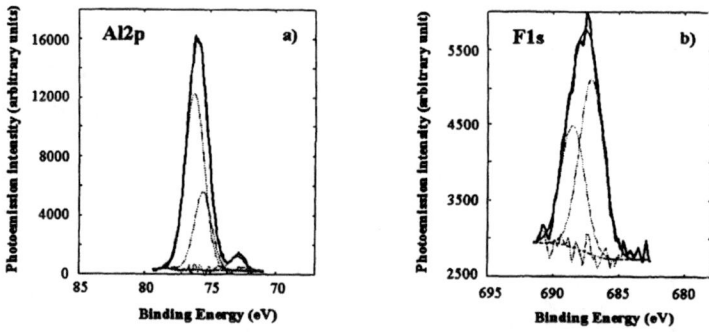

Fig. 5. XPS spectra of the Al2p (a) and F1s core levels (b) from the "spider bond pad" surface.

687 and 688.5 eV (see Fig. 5b) identified as AlFx and Al(OF)$_x$. The assignment of the different components is consistent with the previous work by Grman *et al* on surface analysis of fluorine contaminated bond pads [1]. No evidence for CF$_x$ bonding was found. Carbon is mostly involved in oxygen bonding.

Figures 6a and 6b show photoelectron micrographs of the second Al bond pad sample taken at the kinetic energy of the secondary electron tail (a) and the F1s states (b). The contrast in image b suggests a stronger reaction of fluorine in the bond pad region during the stripping process.

Fig. 6. 270x250 µm scanning photoemission micrographs of a patterned sample taken at the kinetic energy of the secondary electron tail (a) and the F1s states (b).

The overview spectra in Fig.7 confirm the chemical contrast observed in the image. Fluorine, oxygen, carbon, aluminum and traces of silicon oxide were detected on the bond pad and fluorine, oxygen, silicon and silicon oxide and traces of aluminum outside the pad. Higher energy resolution spectra of fluorine and carbon outside the pad clearly show the presence of higher binding energy components in both F1s and C1s, suggesting the formation of C-F

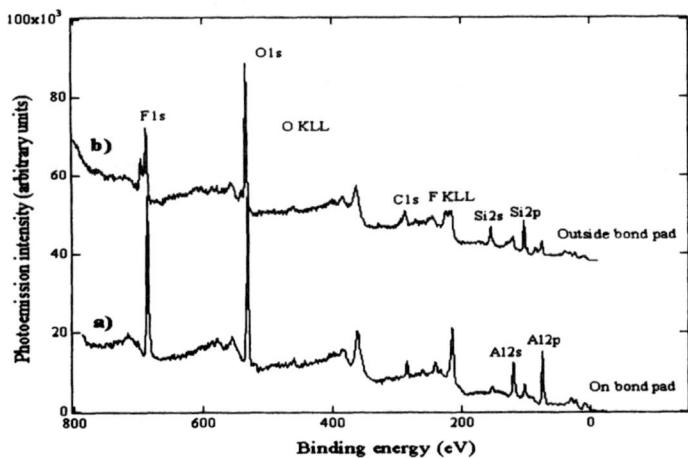

Fig. 7. XPS surveys on the Al bond pad (a) and outside the Al bond pad (b).

231

covalent bonding. Previous work by Tressaud *et al* on C-F bonds in fluorinated carbon materials assigned these higher binding energy components to perfluorinated groups or plasmon effects [4].

CONCLUSIONS

Successful TiN particle analysis was performed using synchrotron radiation and a scanning photoemission microscope which was able to detect the particle and its chemical changes after the removal of an oxide layer by sputter cleaning. The capabilities of the ALS spectromicroscope to perform chemical analysis on a microscopic scale were also tested on two patterned bond pad samples. Chemical mapping was performed as well as XPS spectroscopy on a microscopic scale and examples of identification of different chemical status in different location was also discussed.

ACKNOWLEDGMENTS

We acknowledge Charles Evans & Associates for providing one of the two patterned samples for analysis. We are also grateful to Judith Denes at Intel Corp. for preparing the samples for analysis with the competence that always characterizes her work. This work was partially supported by the Director, Office of Energy Research, Office of Basic Energy Sciences, Materials Sciences Division of the U.S. Department of Energy, under Contract No.DE-AC03-76SF00098.

REFERENCES

1. D. Grman, R. Hauert, E. Hollander and M. Amstutz, Solid State Technol., **35** (2), p.43-47, (1992).

2. G.D. Ackermann, R. Duarte, K. Franck, M.R. Howells, Z. Hussain, S.Irick, A. Johnson, G. Morrison, H.A. Padmore, S.-Y. Rah, T.R. Renner, B.Sheridan, W. Steele, C. Ayre, H. Fujimoto, F. Gozzo, B.B. Triplett, R.X. Ynzunza, P.D. Kinney, Y.S. Uritsky, presented at the 1998 MRS Spring Meeting, San Francisco, CA, 1998.

3. Y.S. Uritsky, P.D. Kinney, E.L. Principe, I. Mowat and L. McCaig, presented at the 1998 MRS Spring Meeting, San Francisco, CA, 1998.

4. Tressaud, F. Moguet, S. Flandrois, M. Chambons, C. Guimon, G. Nanse, E. Papirer, V. Gupta and O.P Bahl, J. Phys. Chem Solids, **57** (6-8), pp.745-751 (1996).

X-RAY FLUORESCENCE MICROTOMOGRAPHY ON A
SiC NUCLEAR FUEL SHELL

M. Naghedolfeizi; J.-S. Chung and G. E. Ice, Oak Ridge National Laboratory, Metals and Ceramics Div., Oak Ridge, TN; W. B. Yun, Z. Cai, B. Lai, Argonne National Laboratory, Adv. Photon Source, Argonne, IL

ABSTRACT

TRISO fuel particles contain a small kernel of nuclear fuel encapsulated by alternating layers of C and a barrier layer of SiC. The TRISO fuel particle is used in an advanced nuclear fuel where the SiC shell provides the primary barrier for radioactive elements in the kernel. The performance of this barrier is key to containment. We have used x-ray fluorescence microtomography to measure the trace element distribution in a SiC shell. Prior to our measurements the nuclear fuel and C layers were leached from the particle. The shell was then encapsulated by kapton tape to simplify handling. The shell was mounted on a glass fiber and measurements were made with an ~1x3 μm^2 x-ray probe on beamline 2-ID at the APS. The distribution of trace elements in the SiC shell was reconstructed after correcting the data for artifacts arising from absorption and scattering off the kapton tape. The observed trace elements are distributed in small <1μm regions through the SiC shell. The trace elements can be attributed to radiation enhanced diffusion of elements in the kernel or to trace elements introduced during fabrication. X-ray fluorescence microtomography is an ideal tool for this work because it is a penetrating nondestructive probe sensitive to trace elements in a low Z matrix and because it provides a picture of the elemental distribution in the shell.

INTRODUCTION

TRISO fuel particles, used in High-Temperature Gas Cooled Reactors(HTGR) are composite structures with a nuclear fuel kernel surrounded by alternating layers designed to contain fission products and compensate for radiation damage. As shown in Fig 1, a typical fuel particle contains an inner kernel of nuclear fuel, a low density buffer layer of pyrocarbon, dense layer of pyrocarbon coating, an interlayer of SiC and an outer dense layer of pyrocarbon. Depending on the type of reactor core design, the fuel kernel is chosen from UCO, UC_2, ThO_2, or UO_2. In addition, fuel kernel size, the thickness of the various layers, and the overall size of the TRISO fuel particle can vary with the type of fuel kernel[1]. TRISO coated fuel particles are compacted into fuel rods designed for passive containment of the radioactive isotopes. The SiC layer provides the primary barrier for both radioactive elements in the kernel and gaseous and metallic fission products. The effectiveness of this barrier layer under adverse conditions is critical to containment and has been the subject of previous studies.[1,2]

We report on measurements of the elemental *distribution* in a SiC shell after exposure to a fluence of ~10^{25} (neutrons/m^2). X-ray fluorescence microtomography is an ideal tool for this work because it is nondestructive, it is sensitive to heavy elements in a low Z matrix, and because it can provide a 3-D picture of the elemental distribution; the observed elemental distribution can be correlated with flaws and defects in the SiC shell.

Previous Studies and Sample Preparation

The shell examined came from a previous study of diffusion through SiC shells.[1] In the previous study, a total of 19 shells were exposed to varying fluences (Table 1). The C buffer layers and nuclear kernels of the TRISO particles were removed by laser drilling through the SiC and then leaching the particle in acid. The shells were repeatedly leached until a constant activity in the fission product was sensed[1]. At this point it was assumed that any remaining activity was due to daughter products which had migrated into the SiC shell. The shells were then analyzed to determine the total number of the various daughter products in each shell (Table 1). This method provides an accurate absolute measurement of the total loading of radioactive elements but gives no information about the distribution in the shell.

233

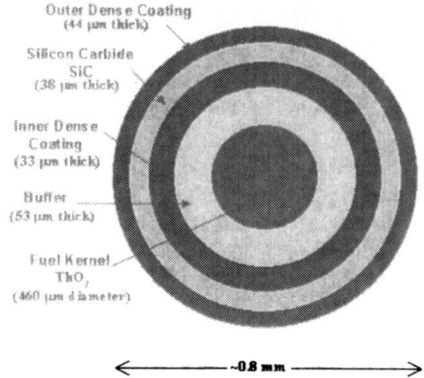

Outer Dense Coating
(44 μm thick)

Silicon Carbide
SiC
(38 μm thick)

Inner Dense
Coating
(33 μm thick)

Buffer
(53 μm thick)

Fuel Kernel
ThO₂
(460 μm diameter)

←————— ~0.8 mm —————→

Fig. 1 Schematic of TRISO fuel element.

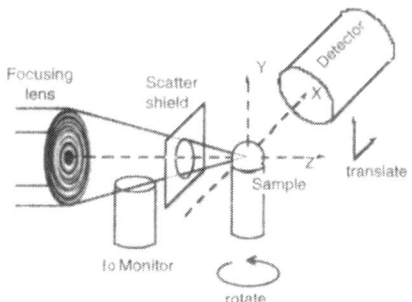

Focusing lens   Scatter shield   Y   Detector   X

Sample   translate   Z

to Monitor   rotate

Fig. 2. Key elements of an x-ray fluorescence tomography experiment. The x-ray fluorescence is monitored while the sample is rotated and translated by a precision stage. Both the step size of the stage motions and the focal spot size of the probe beam determine the spatial resolution in the reconstructed image.

Table 1. Selected isotopes and irradiation conditions for irradiated TRISO particles.

Ball ID	Kernel Type	Fluence (10^{25} n/m²)	Atoms in particles		
			Cs-137	Ce-144	Eu-154
6151-23-020-C	UC₂	6.4	9.91×10^{15}	4.95×10^{15}	1.04×10^{14}
6151-23-020-D	UC₂	6.4	1.01×10^{16}	5.58×10^{15}	1.14×10^{14}
6151-23-020-E	UC₂	3.5	7.69×10^{15}	4.37×10^{15}	1.02×10^{14}
6151-23-020-F	UC₂	3.5	7.82×10^{15}	$4.39 \times 10^{15})$	9.11×10^{13}
6157-08-020-A	UCO	5.8	9.05×10^{15}	4.36×10^{15}	1.19×10^{14}
6157-08-020-B	UCO	5.8	9.14×10^{15}	4.38×10^{15}	1.05×10^{14}
6157-11-020-A	UCO	6.1	9.52×10^{15}	4.60×10^{15}	1.10×10^{14}
6157-11-020-B	UCO	6.1	9.55×10^{15}	4.89×10^{15}	1.20×10^{14}
6252-12-COMP-A[a]	ThO₂	1.9	6.81×10^{14}	1.37×10^{14}	2.96×10^{12}
6252-12-COMP-B	ThO₂	1.9	6.56×10^{14}	1.36×10^{14}	3.56×10^{12}
7544-1-COMP-A	UC₂	1.9	2.79×10^{15}	5.71×10^{14}	2.20×10^{13}
7544-1-COMP-B	UC₂	1.9	3.22×10^{15}	6.30×10^{14}	2.16×10^{13}

X-ray fluorescence microtomography

The advent of intense synchrotron radiation sources has improved the resolution of tomography into the μm regime. Although most tomographic measurements use transmission tomography methods[3-5] experiments as early as 1985 demonstrated the elemental sensitivity of fluorescence tomography to high Z trace elements in a low Z matrix.[6] These pioneering measurements of the Fe distribution in a honey bee (*Apis mellifera*) found Fe concentrations at the surface and in the abdomen of the bee with a spatial resolution of ~150μm. Although transmission measurements are much faster, fluorescence measurements have better signal-to-noise for trace element detection.[7]

The experimental setup for an x-ray fluorescence tomography measurement is conceptually very simple (Fig. 2). The sample is placed on a stage which rotates and translates the sample. A detector is placed in the plane of the storage ring and at 90° to the incident beam. This detector geometry allows efficient measurement of the characteristic fluorescence from the trace elements while minimizing the x-ray elastic and Compton

---

[a] Used in this experiment

scattering.[7] The elemental distribution in a slice through the sample can be reconstructed after translating the sample through the x-ray beam and rotating the sample at least 180° for every x position. Finer resolution is achieved by decreasing both the translation and rotational step size.

Fluorescence tomography is very slow compared to transmission tomography because the number of volume elements (voxels) which can be resolved scales roughly with the number of data points collected. In contrast transmission tomography measures the attenuation through all translation positions simultaneously and therefore has an $\sim 10^6$ faster collection time. Fluorescence tomography is therefore most appropriate for use on materials where there is interest in the distribution of low concentrations of trace elements.

EXPERIMENT

The experiment was performed on beamline 2-ID of the Advanced Photon Source.[8] Beamline 2-ID uses a low bandpass x-ray mirror to define a beam axis, followed by a Si 111 perfect crystal monochromator and a hard x-ray zone plate. The x-ray energy was set at 10.5 keV. For this experiment a 40 cm focal length zone plate was used which produced a spot size of approximately $1x3\ \mu m^2$. The sample was epoxied to a glass fiber and sandwiched between 2 mil kapton tape simplify handling. The fiber was mounted on a small goniometer head which allowed the ball to be positioned at the center-of-rotation of the rotation stage. The ball was then centered on the x-ray beam so that the translation range of the measurement passed completely through the center of the ball.

Because a tomographic reconstruction requires consistent measurement conditions, the incident beam intensity was measured with an AMPTEC model XR-100T PIN Diode detector with 250 eV energy resolution.[9] The detector was placed at $\sim90°\ 2\theta$ to the beam but out of the plane of the x-ray ring (to optimize scattering efficiency). Scatter from the air between the sample and the zone plate was monitored. This crude incident beam monitor worked well after a backscatter shield was installed between the sample and the air volume viewed by the detector. In a previous fluorescence tomography attempt with the same sample, the measurements were rendered useless for reconstruction by large and unmonitored variations in the incident beam intensity.

The reconstruction was compromised by both limitations of the equipment and by limited counting time. For example, the stepping rate of the sample stage and the detector readout imposed an overhead of about 2 seconds for each measurement step. Because of the limited beam time available, a compromised data collection scheme with 8 μm translation steps (101 translation steps) and 3° rotation steps (101 rotational steps) was used. This step size was much larger than the probe beam size which complicated the data analysis. In addition, we note that 60° of rotation was inaccessible due to the design of the rotation stage.

Before the tomographic measurements were begun, the unfocused beam was centered on the fuel ball shell to determine the detectable trace elements. The fluorescence spectrum is shown in Fig. 3. Regions-of-interest ROI's were set around the dominant fluorescence lines. Because of software limitations only 10 (ROI's) could be stored. The 10 ROI's stored are listed in Table 2. Unfortunately, Cs, Ce and Eu L lines lie in the region from ~4.2-7 keV. This region is substantially masked by intense K fluorescence from Cr and Fe (Fig. 3). Therefore with the solid-state detector, the distribution of the radioactive elements was not measured.

A single line scan with 2 μm step size was then made to test the data collection software and hardware and to estimate typical feature sizes. For example, the measured linescan intensity of Zn is shown in Fig. 4. As can be seen, there are numerous small Zn features through the shell, some of which are smaller than the 2 μm step resolution. The origin of this Zn is not known.

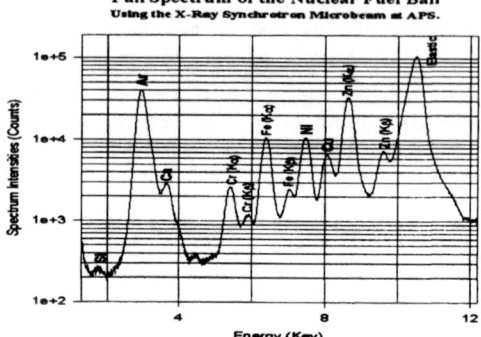

Fig. 3 Fluorescence spectrum from an unfocused beam on the fuel ball shell.

Table 2. Measured regions of interests (ROIs) in the fuel ball spectrum

ROI	Element	MCA Channel Number		Energy (KeV)
		Low	High	Center
1	Ar	248	289	2.943
2	Ca	398	475	3.68
3	Ba (lα)	656	698	4.48
4	Cr (kα)	780	845	5.41
5	Cr (kβ)	862	908	5.91
6	Fe	927	1000	6.39
7	Ni	1094	1160	7.482
8	Cu	1179	1242	8.046
9	Zn	1261	1343	8.637
10	Scattering	1556	1617	10.502

The absolute elemental concentrations were estimated by comparing the observed fluorescence signal, $S_{Fluorescence}$ to the x-ray elastic and Compton scattering signals. The total scattering cross section of SiC at 90° was estimated from Ref. 10. The beam polarization was estimated at ~5% and multiple scattering and absorption were assumed to be small. With these approximations, the factor $I_0\Omega$ was determined, where I_0 is the incident beam flux in photons/sec/μm^2 and Ω is the detector solid angle. The trace element concentrations were then estimated again assuming negligible absorption from the following equation.

$$S_{Fluorescence} \approx I_0\Omega \int\int\int (C(x,y,z)\sigma)dxdydz \qquad (1)$$

Here C is the elemental concentration of the trace element and σ is the fluorescence cross-section at the incident beam energy. Self absorption in the sample was later corrected during the tomographic reconstruction.

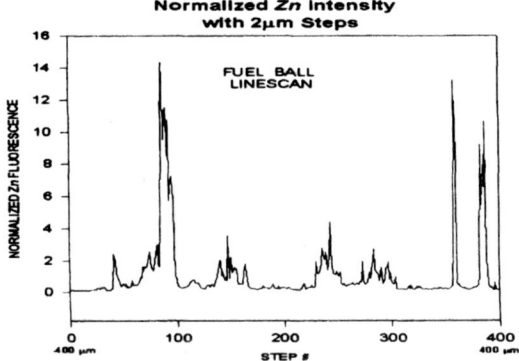

Fig. 4: Measured fluorescence intensity of Zn in a linescan experiment on TRISO fuel particle.

RESULTS AND EXPERIMENTAL ANALYSIS

Tomographic reconstructions were carried out, both with and without absorption correction, using the library suite RECLBL.[11] However, prior to the image reconstruction phase, it was necessary to correct the scanned projection data for artifacts. This is because over a small angular range, the kapton tape was at near glancing angle which deflected and absorbed the incident beam. These conditions produced large artifacts in the data which were removed by excluding the data points from a small angular range and averaging neighboring data points to estimate the fluorescence intensities in the missing region. Of the 10^4 data points for each ROI, ~25 were corrected for the effect of the kapton film. In addition, during the measurements, the detector was occasionally saturated by a huge fluorescence signal from an unusually high concentrations of one or more trace elements. Under these conditions, the ROI's from all the elements were effected. The elements responsible were identified by observing the angle/translation pattern of each element and comparing them to the intense signal position. Approximately 20 points for each ROI were corrected for saturation effects using average neighboring data points. After correcting the artifacts, the Fourier filtered backprojection routine of the library was incorporated into a computer program which reconstructed the 2-D images of the trace element distributions from the projection data on a Pentium PC. The mode of reconstruction was based on 360° rotations and the choice of filter was Hannig with a cutoff frequency of 0.25. As mentioned previously, the stage design limited rotations to ~300°. Because the reconstruction with 360° rotations appeared more robust, the missing 60° data was estimated from the 180° symmetric data.

The reconstructed spatial distributions of the monitored elements are presented in Fig. 5 and Fig. 6. Figure 5 shows the reconstructed distribution without absorption correction and Fig. 6 shows the distribution with absorption correction. Although the fuel ball shell is only ~38 μm thick, absorption corrections can be large, particularly for the low energy fluorescence lines. For example, ~97% of the Ba fluorescence (4.48 keV) can be absorbed by the SiC shell. As a result, the trace elements distributions are most easily detected near the surface of the SiC shell.

The analysis of the images indicates that the spatial distributions of the trace elements are mostly localized at the outer edge of the SiC shell. The reconstructed images with the absorption correction implies that a higher counting rate is required to improve the signal-to-noise ratio for low energy fluorescence lines. The quality of the images in reconstruction with absorption correction depends also on the selection of a proper number for intensity levels of object/background in the reconstruction routines. An elaborated trial and error procedure is required to obtain the proper number for a desired image quality. Though, at the present time, it is not possible to confidently identify the origin of the trace elements, they may have been originated from the impurities associated with the fuel, SiC, and/or fabrication processes. Measurements planned for non-irradiated fuel balls will settle this issue.

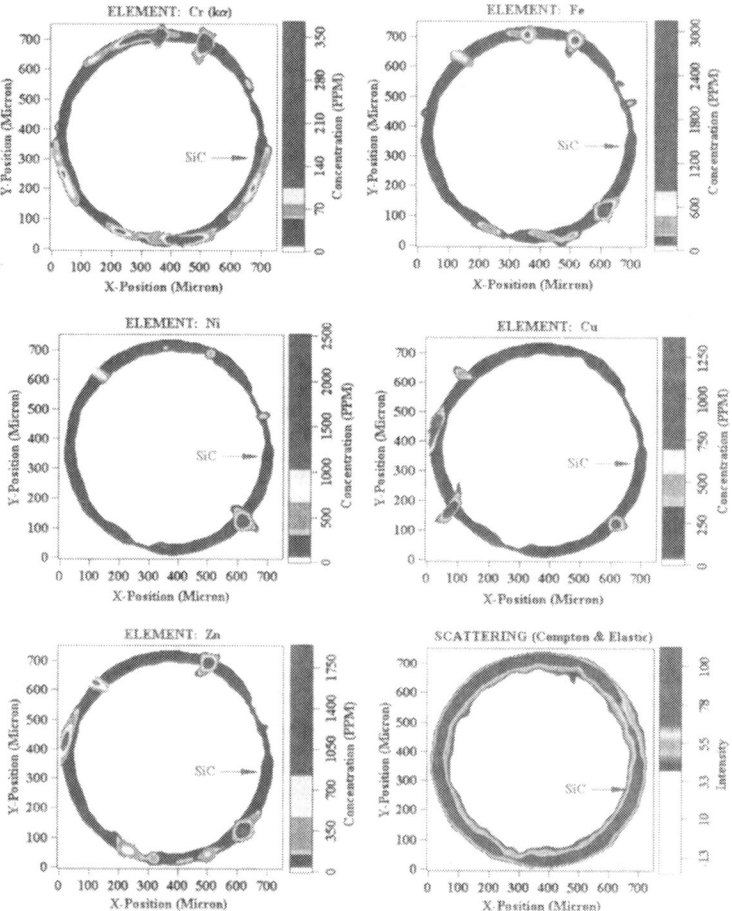

Fig .5: Reconstructed images of the spatial distribution of trace elements with no absorption correction. The reconstructed elemental distributions are superimposed on the reconstructed shell from the elastic/Compton scattering.

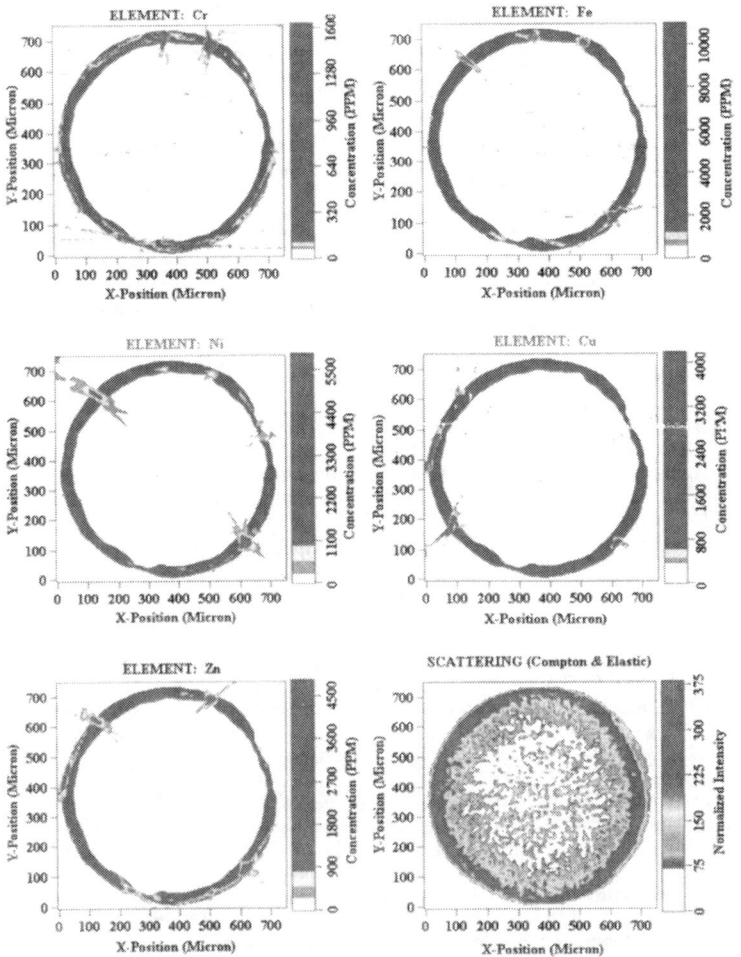

Fig. 6: Reconstructed images of the spatial distribution of trace elements with absorption correction.

CONCLUSIONS AND FUTURE WORK

X-ray microtomography is an emerging technique made practical by high brilliance x-ray sources, advanced x-ray focusing optics and high-performance x-ray detectors. We have demonstrated that fluorescence microtomography technique can be a powerful tool for investigating elemental distributions in materials. The technique is nondestructive and noninvasive with high spatial resolution. It is sensitive to high Z trace elements in a low Z matrix with PPM accuracy, and has a good signal-to-noise ratio. Fluorescence microtomography can be used to simultaneously identify the elemental distributions of many trace elements. The main drawback of fluorescence microtomography is the slow data collection rate. The results of this experiment point to several ways to greatly accelerate data collection. For example, with a broad-bandpass monochromator, the fluorescence signal can be increased by 2-3 orders of magnitude with no loss in spatial resolution. This increase in signal will allow 3-dimensional maps to be obtained. With such an intense probe, it will however be necessary to use wavelength dispersive spectrometers with integrating rather than single photon detectors. Such an arrangement will simultaneously increase spatial resolution and will greatly increase the sensitivity for minor trace elements of interest.

It should also be noted that in our experiment the beam energy was much lower than the K absorption edges of heavy elements such as Cs, Ce, Eu, and Ru. We therefore could not excite the K lines of these elements. The L lines are masked by overlapping K lines of less interesting trace elements. To positively measure the elemental distribution of Cs for example, either a crystal spectrometer or a high energy (>36 keV) x-ray probe are required. We intend to try both methods in future measurements.

ACKNOWLEDGMENT

Research sponsored in part by the Laboratory Directed Research and Development Program of the Oak Ridge National Laboratory and the Division of Materials Sciences, U.S. Dept. of Energy under contract DE-AC05-96OR22464 with Lockheed Martin Energy Research Corporation. Experimental measurements were made on beamline 2-ID-CD at the Advanced Photon Source ANL which is supported by the U.S. Dept. of Energy Basic Energy Science. The authors express their appreciation to Dr. E. D. Specht for his technical assistance and comments.

REFERENCES

1. B.F. Myers, F.C. Montgomery and K.E. Partain, "The Transport of Fission Products in SiC," Doc No. 909055, GA Technologies Inc., (1986).
2. P. Krautwasser, G. M. Begun, and Peter Angelini, *J. of American Ceramic Society*, **66** 424 (1983).
3. See for example, Q.C. Johnson, J.H. Kinney, U. Bonse and M.C. Nichols, *Proc. Mater. Soc. Symp.*, **69**, 203 (1986).
4. J.H. Kinney and M.C. Nichols, *Annu. Rev. Mater. Sci.*, **22** 121 (1992).
5. "Proceedings of the Workshop on High Resolution Computed Microtomography (CMT)," Lawrence Berkeley National Laboratory , U. of California, August 12-13, 1996.
6. Paul Boisseau, "Determination of Three Dimensional Trace Element Distributions by the Use of Monochromatic X-Ray Microbeams," Ph.D. Dissertation, (1986).
7. C.J. Sparks, "X-ray Fluorescence Microprobe for Chemical Analysis", pg 459-512 in Synchrotron Radiation Research, edited by H. Winick and S. Doniach Plenum Press, New York (1980).
8. W. Yun, B. Lai, D. Shu, A. Kounsary, Z. Cai, J. Barraza, D. Legnini, "Design of a Dedicated Beamline for X-ray Microfocusing and Coherence Based Techniques at the Advanced Photon Source".
9. A.C. Huber, J.A. Pantazis and V. Jordanov, *Nucl. Inst. Meth. B* **99** 665 (1995).
10. M. O. Krause,C.W. Nestor, Jr., C. J. Sparks, Jr., E. Ricci, "X-ray Fluorescence Cross Sections for K and L Rays of the Elements," Oak Ridge National Lab. Tech. Report ORNL-5399, (1978).
11. R. H. Huesman, G.T. Gullberg, W.L. Greenberg, T. F. Budinger, "RECLBL Library Users Maual, Donner Algorithms for Reconstruction Tomography," Pub. 214 Lawrence Berkeley Laboratory, U. of California, (1977).

APPLICATION OF SYNCHROTRON X-RAY FLUORESCENCE MICROSCOPY TO THE STUDY OF MULTI-METAL OXIDE CERAMICS

DALE L. PERRY,[1,2] SCOTT McHUGO,[1] ALBERT C. THOMPSON,[1] JOSEPH C. FARMER, [3] BART B. EBBINGHAUS,[3] RICHARD VAN KONYNENBURG,[3] WILLIAM A. BRUMMOND,[3] GUY ARMENTROUT, [3] THOMAS H. GOULD,[3] and NANCY YANG,[4] Lawrence Berkeley National Laboratory,[1] University of California, Berkeley, CA 94720, G. T. Seaborg Institute for Transactinium Science,[2] Lawrence Livermore National Laboratory, [3] University of California, P. O. Box 808, Livermore, CA 94551, and Sandia National Laboratory,[4] P. O. Box 696, Livermore, CA 94551. dlperry@lbl.gov

ABSTRACT

Synchrotron x-ray fluorescence microscopy has been used to study multi-metal oxide ceramics that have been designed to sequester radioactive actinide elements for long-term storage and disposal. X-ray fluorescent lines for the various elements have been used for lateral elemental mapping of the materials, and the heterogeneity of the samples is discussed with respect to the elements in the crystallographic phases that have previously been documented by other means of structural and chemical analyses.

INTRODUCTION

One of the most important aspects of nuclear technology is that of the long-term disposal of nuclear materials, both materials that are depleted such as spent nuclear fuel rods and materials that comprise stockpiles that, by mutual international agreement, must undergo permanent disposal. In the case of nuclear waste, for example, millions of gallons of nuclear waste are stored at U. S. Department of Energy facilities around the United States [1]. Research has been reported in the research literature dealing with several forms for use in the disposition of radioactive species, including glass [2-4] and ceramic [5-7] hosts.

The present study addresses the use of synchrotron-x-ray-induced fluorescence microprobe to study multi-metal oxide ceramic phases that have been designed to have radioactive, large-metal ions such as uranium and plutonium incorporated into them for purposes of long-term storage. The different crystallographic phases that comprise the ceramic materials are discussed, along with the heavy metals calcium, titanium, hafnium, gadolinium, and cerium that constitute the phase components. Lateral elemental mapping is shown, and the the elemental concentrations of the various metal ions are discussed with respect to the lack of homogeneity of the metals comprising the different crystallographic phases.

The work includes x-ray fluorescence microprobe data for several elements for each sample. In addition to their being highly important materials for storage applications, they also represent a wide array of metal ions that exhibit a varying degree of complexity in their x-ray fluorescence spectra and are very useful in proving the effectiveness of this technique in studying their chemical and structural details.

EXPERIMENTAL

Chemicals

Chemicals used for the syntheses of the ceramics were obtained commercially and used without further purification. Titanium dioxide, TiO_2, was purchased from Kronos (98 % purity), while HfO_2 (S Grade, 99.9 %) was bought from Wah-Chang/Allegheney-Teledyne, CeO_2 (99.5 %) from Cerac, Gd_2O_3 from Pacific Industrial Development Corporation (PIDC) (99 %), and $Ca(OH)_2$ (95 %) from GFS Chemical.

Synthesis of Ceramic Materials

The materials were synthesized by blending 37.40 g TiO_2, 23.75 g CeO_2, 13.78 g $Ca(OH)_2$, 11.11 g HfO_2, and 8.34 g Gd_2O_3 by wet milling, calcining the mixture at 750 $^{\circ}$C for 1 hr in an argon atmosphere, pressing the material into circular disks under ~ 12,000 psi, and finally sintering the disks at 1300 $^{\circ}$C in argon for 4 hrs.

Instrumentation

The films were analyzed using the x-ray fluorescence microprobe at the Advanced Light Source (ALS) at Lawrence Berkeley National Laboratory as previously described [8-10]. The x-ray microprobe uses a white radiation beam from a bending magnet on the synchrotron. A pair of multilayer-coated elliptical mirrors in a Kirkpatrick-Baez configuration are used to focus and monochromate (with 6 % bandpass) the beam. The detector was mounted perpendicular to the beam direction to minimize the scattered x-ray background. The spot size was varied from sample to sample in order to maximize the chance of observing heterogeneity of the sample. The samples were scanned in a series of long, one-dimensional scans of about 0.1 mm in which the step size between the points was 10 microns, and the counting time at each point was 1-2 minutes. At each point, the complete energy spectrum could be analyzed with a peak fitting program to find the net intensity in all of the elemental peaks.

RESULTS AND DISCUSSION

The elemental analyses and microstructure of the resulting ceramic, as studied by x-ray diffraction and high-resolution field emission scanning electron microscopy in conjunction with an electron microprobe equipped with a wave-length-dispersive x-ray (WDX) analyzer, showed the ceramic to consist of 95 % pyrochlore (with an elemental ratio of $Ca_{0.89}Gd_{0.22}Hf_{0.23}Ce_{0.66}Ti_2O_7$), 4 % rutile ($TiO_2$), and 1 % hafnia ($HfO_2$). Minute quantities of secondary phases containing calcium and titanium could also be detected using x-ray diffraction. Samples for investigation in the present study were synthesized as approximately one kilogram circular pucks from which smaller 1 cm x 1 cm x 2 mm sections were cut, yielding sample sizes that were manageable for use in the sample chamber. The sections used in this study are members of an almost endless number of "slices" that could be cut but are typically representative of the chemistry of the ceramic.

Elemental distribution by mapping within the ceramic material was found not to be uniform. A typical example of this distribution is found in Figure 1 in which distribution plots of the five different elements in a ceramic sample are shown. Each square image represents an area of 30 μ x 30 μ, which gives the researcher an opportunity to look at an area encompassing many average-size molecular unit cells, unit cells that have dimensions on the order of several Angstroms. One can see that in some areas, concentrations of some of the elements track very well. Gadolinium and hafnium, for example, are shown by the intensity maps to be in very roughly the same concentrations in comparing their respective, individual maps, especially in the lower left quadrant. However, there are concentration discrepancies in the upper right-hand

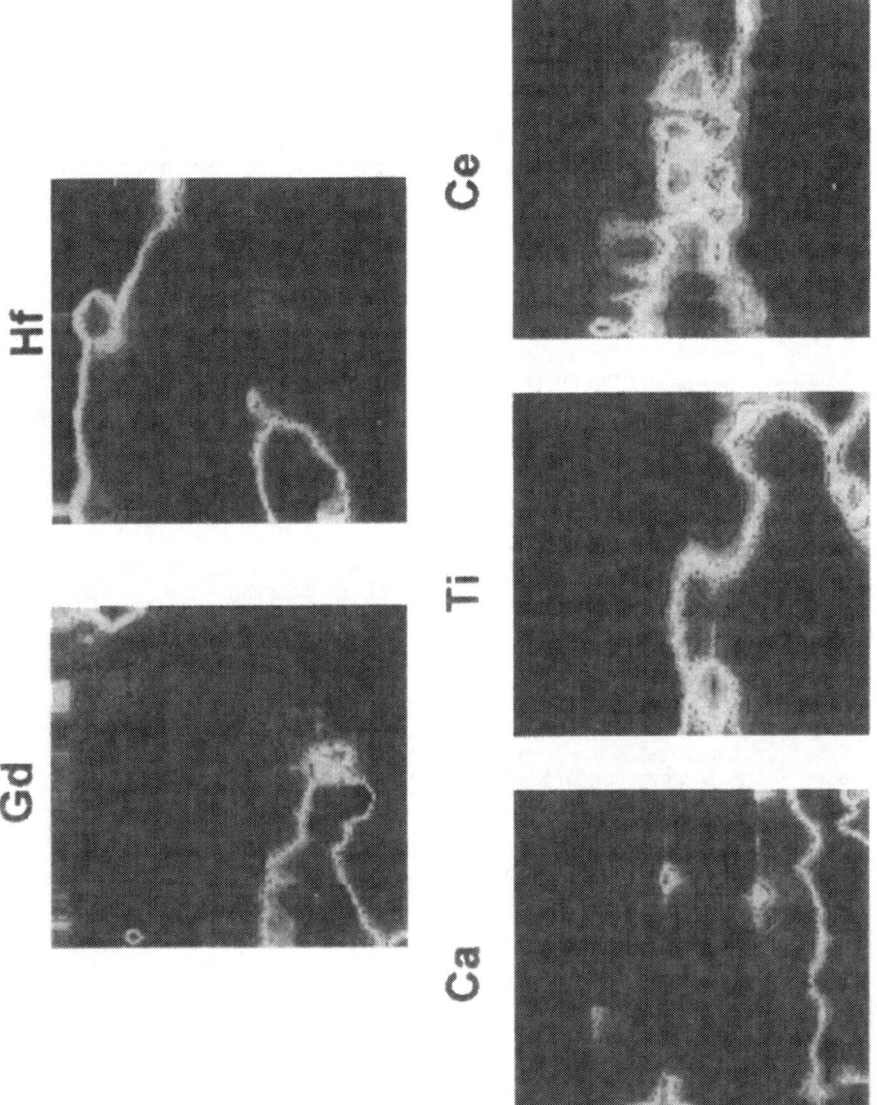

Figure 1. Elemental distribution maps for pyrochlore-rutile-hafnia ceramic samples.

portions of the maps in which hafnium is very clearly depleted with respect to the gadolinium. Another area of dissimilarity is that of hafnium and cerium in the lower right and left octants. Cerium and calcium track very well in the upper third of their maps but are in disagreement in the bottom third sectors. It must be remembered that calcium and titanium are represented in both major and minor phases in the ceramic.

The x-ray fluorescent lines of the elements in the present samples are easily differentiated in the spectra of the samples and thus provide for good elemental mapping. The sensitivity for the elements using an x-ray synchrotron facility such as the Advanced Light Source used here is many-fold that of the sensitivity of a traditional non-synchrotron instrument due to the brightness of the synchrotron x-ray source. In addition to the sensitivity, the mapping capability in terms of the resolution (down to ~ 1 μ) of the technique using the x-ray fluorescence microprobe at a synchrotron facility allows the researcher to observe much smaller areas for purposes of modeling aspects of the phases in ceramic materials such as the ones studied here in terms of elemental defects.

ACKNOWLEDGMENTS

This work was supported by the U. S. Department of Energy Fissile Materials Disposition Program and the Director, Office of Energy Research, Office of Basic Energy Sciences, Materials Sciences Division, under Contract Nos. W-7405-ENG-48 (LLNL) and DE-ACO3-76SF00098 (LBNL).

REFERENCES

1. *Closing the Circle on the Splitting of the Atom* (U. S. Department of Energy, Office of Environmental Management, Washington, D. C., 1995).
2. H. Matzke, Surf. Interface Anal., 22, 472(1994).
3. W. J. Weber, R. C. Ewing, C. A. Angell, G. W. Arnold, A. N. Cormack, J. M. Delaye, D. L. Griscom, L. W. Hobbs, A. Navrotsky, D. L. Price, A. M. Stoneham, and M. C. Weinberg, J. Mater. Res., 12, 1946(1997).
4. Radioactive Waste Forms for the Future, R. C. Ewing, Editor, North-Holland Amsterdam, 1988).
5. E. R. Vance, Mater. Res. Soc. Bull., 19, 28(1994).
6. A. G. Solomah, P. G. Richardson, A. K. McIlwain, J. Nucl. Mater., 148, 157(1987).
7. J. A. Cooper, D. R. Cousens, J. A. Hanna, R. A. Lewis, S. Myhra, R. L. Segall, R. S. C. Smart, P. S. Turner, and T. J. White, J. Am. Ceram. Soc., 69, 347(1986).
8. A. C. Thompson, K. L. Chapman, G. E. Ice, C. J. Sparks, W. Yun, B. Lai, D. Legnini, P. J. Vicarro, M. L. Rivers, D. H. Bilderback, and D. J. Thiel, Nucl. Instrumen. Meth. Phys. Res., Section A, 319, 320(1992).
9. A. C. Thompson, K. L. Chapman, and J. H. Underwood, *Optics for High- Brightness Synchrotron Beamlines,* (SPIE, Bellingham, Washington), 1740(1992).
10. D. L. Perry, A. C. Thompson, R. E. Russo, X. L.Mao, and K. L. Chapman, Appl. Spectrosc., 51, 1781(1997).

UPDATE ON SYNCHROTRON RADIATION TXRF: NEW RESULTS

S. BRENNAN*, P.PIANETTA*, S. GHOSH*, N. TAKAURA*, C WIEMER*,
A. FISCHER-COLBRIE**, S. LADERMAN**, A SHIMAZAKI†
A. WALDHAUER‡, M.A. ZAITZ***
*Stanford Synchrotron Radiation Laboratory, Stanford, CA 94309
**Hewlett-Packard Co., Palo Alto, CA 94301
†Toshiba Corp., Kawasaki, JAPAN
‡Applied Materials, 350 Bowers Ave., Santa Clara, CA 95051
***IBM Research Laboratory, Hopewell Junction, NY 12533

ABSTRACT

Synchrotron-based total-reflection x-ray fluorescence(SR-TXRF) has been developed as a leading technique for measuring wafer cleanliness. It holds advantages over other techniques in that it is non-destructive and allows mapping of the surface. The highest sensitivity observed thus far is $3x10^8$ atoms/cm^2 ($\approx$ 3fg) for 1000 second count time. Several applications of SR-TXRF are presented which take advantage of the energy tunability of the synchrotron source or the mapping capability.

INTRODUCTION

Integrated circuits continue to increase in complexity as the widths of the critical dimensions continue to shrink. With decreasing linewidths the acceptable level of metal contamination decreases as well. Residual metal contamination on the wafer surface can be a source of defects in the integrated circuit, if the contamination level is too high. Some of the defects known to result from metal contamination are increased leakage current, poor threshold voltage control and gate insulator breakdown. All of these can result in device failure or reliability problems. The ability to measure low levels of contamination is crucial for developing techniques to remove these contaminants. The Semiconductor Industry Association (SIA) 1997 National Technology Roadmap for Semiconductors (NTRS) projects that the required sensitivities for transition metals will be $3x10^8$ atoms/cm^2 in the year 2001[1]. For comparison, a monolayer of silicon atoms is 10^{15} atoms/cm^2. Thus the requirements for surface cleanliness are below the part-per-million (ppm) level of contaminants.

There are a variety of techniques which can be used to measure low levels of surface contamination on wafers. The two most prevalent are Wafer Surface Analysis (WSA)[2] (also known as Vapor Phase Decomposition, or VPD) and Total-reflection X-Ray Fluorescence (TXRF)[3,4]. The former uses a range of wet chemistry techniques to remove the residual metals on the surface by concentrating them into a droplet, which is then either studied using X-ray Fluorescence or inductively-coupled plasma mass spectroscopy (ICPMS). For many years, TXRF using a rotating anode has been available to the semiconductor industry. Some of these units are installed in the manufacturing facility, while others are in the analytical groups associated with the manufacturing process. The TXRF technique has several advantages over the WSA-ICPMS procedure. It is non-destructive, and can be used to map the distribution of contaminants on the wafer. The biggest disadvantage of the conventional TXRF technique is that it does not have the sensitivity of the WSA-ICPMS, which has demonstrated a sensitivity of 10^8 atoms/cm^2, whereas conventional TXRF has only recently demonstrated sensitivities of 5 x 10^9 atoms/cm^2. The TXRF technique consists of a beam of x-rays incident on the wafer surface at an angle of incidence below the critical angle for total external reflection[5]. With this requirement the x-ray penetration depth is approximately 30 Å, which reduces the scattered and fluorescence intensity from the substrate. For comparison, bulk metallic impurity levels are comparable to a surface contamination of $\sim 10^7$ atoms/cm^2.

Several years ago a collaboration of Hewlett-Packard, Intel and the Stanford Synchrotron Radiation Laboratory (SSRL) was formed to explore the possible utility of synchrotron radiation for TXRF. This collaboration is now coordinated by Sematech and involves a number of semiconductor manufacturers, including AMD, Applied Materials, IBM, Lucent, and Motorola. Although the actual list of metals known to cause problems includes low-Z materials (Al, Na), 3-D transition metals (Fe, Ni, Cu, Zn) and other metals (W), our group has emphasized the 3-D transition metals as the focus of our research. Because the fluorescence process is not strongly dependent on the incident photon energy as long as the energy is greater than that of the absorption edge, we have chosen to use multilayers rather than silicon single crystals for our monochromatization. This has resulted in a roughly 2 order-of-magnitude increase in flux[6,7].

EXPERIMENTAL

The data were collected on beam line 6-2 at SSRL using a focused beam from a 1-1 focusing bent cylindrical platinum coated fused silica mirror with a critical energy of 21 keV. Multilayers consisting of alternating layers of tungsten and B$_4$C with a d-spacing of 29 Å were installed in the monochromator and tuned to pass 11.1 keV radiation with a band pass of 280 eV. Because two multilayers are used in a +/- geometry it is a simple matter to change the photon energy entering the experiment. For most of the wafers 11.1 keV is a good energy because it positions the escape peak between the Zn K$_\alpha$ and Zn K$_\beta$ peaks. There are times when the energy is increased to excite a higher-Z material (e.g. As) or lowered to reduce the fluorescence from a film on the wafer (e.g. Cu, see below). The vacuum chamber[8] in which the measurements are made is equipped with a pair of translation stages for positioning any part of both 150 and 200 mm wafers in front of the single element Si(Li) detector. The vacuum chamber is enclosed in a "mini-environment" inside the experimental enclosure, which uses ULPA filters and is surrounded by clear plastic sheeting. With these precautions, wafers can be routinely measured without adding contamination. Wafers are loaded remotely using a robot arm. In the near future cassette-to-cassette loading will be achieved. The primary reason the measurements are performed under vacuum is to eliminate the air scatter and Ar fluorescence which would otherwise add to the total count rate in the detector. The entire vacuum chamber is mounted on a large rotary bearing and driven by a motorized micrometer to define the grazing angle of incidence of the photon beam on the wafer. The wafer is mounted vertically with its surface normal nearly perpendicular to the incident beam. This is done so that the fluorescence detector is positioned along the E-vector of the incident beam, which minimizes the scattered radiation into the detector (the incident beam is $\sim$ 95% linearly polarized in the horizontal plane). At each measurement position of the wafer (horizontally and vertically) an angle scan is performed monitoring the scattered light intensity vs. angle. These data are fit to a standard theory[5] and the angle of incidence is set based on that fit.

RESULTS

Of fundamental importance in TXRF is the minimum detection limit, or MDL[3]. For a given count time the instrument is incapable of detecting levels of contamination below this value. The typical counting time used for TXRF is 1000 seconds. Assuming stochastic noise, the MDL will improve with the square root of the count time. The MDL is determined by measuring the signal from a wafer with a known level of contamination (typically one cross-calibrated using other techniques such as ICP-MS). The MDL is calculated as:

$$MDL = \frac{3C\sqrt{I_B}}{I_p},\qquad(1)$$

where C is the known concentration of the sample being measured, I_B is the background under the fluorescence peak of interest, calculated by a linear fit to the background on either side of the peak, and I_p is the integrated peak intensity after the background has been subtracted. Figure 1 shows two spectra, one with a known concentration of $1x10^{11}$ atoms/cm^2 of three elements, Fe, Ni and Zn. The other spectrum is of a very clean wafer. The two spectra have been normalized so that their intensity is the same at $\approx$ 5 keV. The clean spectrum was measured for a total of 5741 seconds in order to increase the sensitivity. There are no discernable peaks in the region between iron and zinc, indicating that at the MDL of $3x10^8$ atoms/cm^2 there is no metal contamination. It should also be noted that the "clean" spectrum also indicates that there is no parasitic fluorescence in the Si(Li) detector down to this same MDL[8]. There are several peaks in the "clean" spectrum. Starting on the low energy side, they are the silicon fluorescence peak, followed by a chlorine peak. Because of the emphasis of this program on the transition metals a 25 μm teflon foil was placed over the detector. It reduces the silicon fluorescence by a factor of 500, and reduces the intensity of the Cl, S and Ar peaks. By the Ti fluorescence energy it transmits 67% of the intensity. The other significant peak in the clean spectrum is the "escape" peak, which is due to a small fraction of the elastically scattered photons absorbed by the detector which result in the emission of a Si K$_\alpha$ photon. Thus the escape peak is 1.74 keV below the elastic peak at 11 keV. This escape peak is observed with any peak of sufficient intensity.

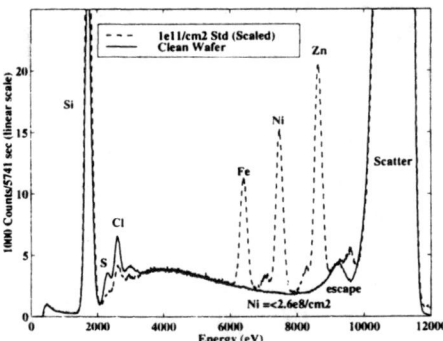

Figure 1: Fluorescent intensity from a silicon wafer intentionally contaminated with $1x10^{11}$ atoms/cm^2 of Fe, Ni and Zn. Also displayed is the spectrum from a clean wafer, showing no Fe, Ni, Cu or Zn contamination at a sensitivity of $3x10^8$ atoms/cm^2.

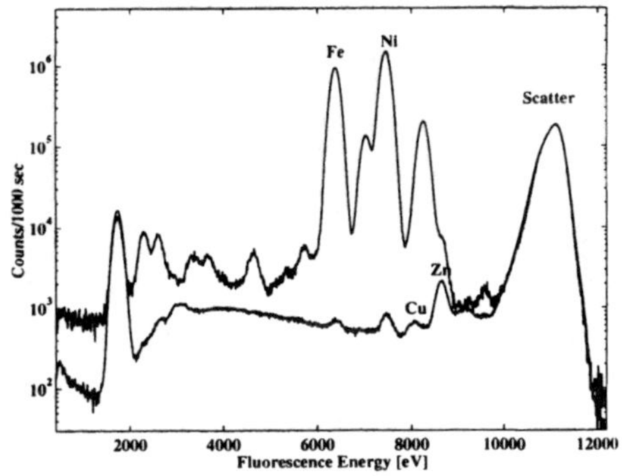

Figure 2: Fluorescent intensity from a droplet collected using the WSA technique and dried on the center of the wafer (upper curve). The lower curve shows the residual contamination not collected by the technique.

The WSA technique can also be used in conjunction with TXRF. By concentrating the contaminants in a droplet and drying the droplet on the wafer, there are several advantages. The mass of contaminant atoms in the view of the detector has increased by > 2000. An additional advantage is that the contaminant atoms consist of particles on the surface rather than mono-dispersed as they are under normal conditions. The particles can be excited by the incident beam even at a nominal x-ray angle of incidence of zero degrees, which greatly reduces the background signal from the silicon wafer. Figure 2 shows two spectra from a wafer on which WSA has been performed. The upper curve was measured on the residual droplet, and a concentration of 2×10^{13} atoms/cm^2 of Fe and Ni is observed. Because of the concentration factor, the entire surface was contaminated with 10^{11} atoms/cm^2 of those elements, which is a level observable with conventional TXRF. Although there are a variety of additional peaks in the region between 4000 and 8500 eV, they are either escape peaks or K$_\beta$ peaks from the Fe and Ni signal, with the exception of the shoulder at 8400 eV, which is probably due to Zn. This is more clearly observed in the lower spectrum, which was measured at a position away from the residual droplet. One can still see significant levels of contamination on the wafer. These levels are: Fe, 7×10^9, Ni, 1×10^{10}, Cu, 3×10^9, and Zn, 3×10^{10} atoms/cm^2. In other WSA wafers we have measured the residual contamination levels are much lower, although copper is often difficult to collect completely. Note also that as a result of concentration, the Fe and Ni have masked the presence of Cu on the surface, and the Zn concentration is difficult to quantify. For this figure the two spectra were normalized to have the same scattered beam intensity.

One of the strengths of the synchrotron radiation-based TXRF program is the ability to tune the incident beam energy to enhance the sensitivity to a particular element. Figure 3 shows the results of one study where one of the users was interested in determining whether there was bromine on the wafer. Bromine is a chemical analog of fluorine and chlorine, both of which are regularly used as the acids HCl and HF. Figure 3 has two spectra, one with the photon energy at the standard value near 11 keV. One can see small amounts of nickel and copper contamination on the wafer ($2\text{-}4\times10^9$ atoms/cm^2). However, with the incident photon energy set to 14 keV, a very clear Br peak is visible, with a concentration of 4×10^{11} atoms/cm^2. Note that the Cl peak at 2620 eV is also visible, and is 2.6×10^{12} atoms/cm^2. The teflon filter reduces the apparent intensity of the Cl peak. The 14 keV data were collected for 500 seconds, vs. 1000 seconds for the 11 keV data, so for the

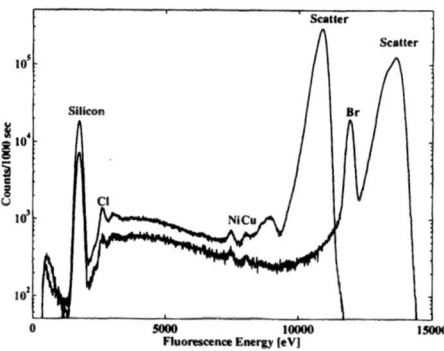

Figure 3: Fluorescent intensity from a silicon wafer using either 11 keV or 14 keV radiation to excite the fluorescence. The higher incident photon energy reveals the presence of bromine on the sample, probably a contaminant in the HCl used to clean the wafer. Low levels ($2\text{-}4\times10^9$ atoms/cm^2) of Ni and Cu are also present.

figure the 14 keV data are doubled. The lower incident intensity for the 14 keV spectrum is primarily due to the lower angle of incidence of the beam onto the multilayer monochromator. The incident beam overfills the vertical acceptance of the multilayers and the intensity decreases. The shoulder on the scatter peak for the 14 keV data is due to a weak diffraction peak from the silicon substrate.

There is increasing interest in copper films for metallization within the semiconductor industry. This poses a particular problem for conventional TXRF, as the typical excitation energy is grater than the Cu K-absorption edge, creating excessive Cu K$_\alpha$ radiation. Our preliminary work on Cu films has shown that one can see concentrations below

10^9 atoms/cm^2 of elements from Ti to Ni. This is made possible by setting the incident photon energy directly on the Ni K-absorption edge at 8.3 keV, which is below the K-absorption edge of copper. Due to higher order harmonics in the incident beam, there is still some Cu K$_\alpha$ fluorescence, but it is dramatically reduced. There is also a peak due to resonant Raman scattering from the Cu film[9]. However, by positioning the incident energy at the Ni edge, the Ni K$_\alpha$ peak can be observed if present, and the escape peak from the incident beam rests on the high-energy shoulder of the Fe K$_\alpha$ peak.

CONCLUSIONS

Synchrotron-based TXRF has been demonstrated to be a very effective technique for measuring a wide range of analytical problems of interest to members of the semiconductor processing industry. The combination of very high sensitivity, non-destructiveness, mapping and incident beam energy tunability makes this a unique research tool. It is in regular use by semiconductor industry groups at SSRL, being used for a variety of studies including materials and equipment qualification and measurement of process cleanliness. Although the emphasis has been on silicon and oxidized wafers, blanket films, ion implanted films and GaAs wafers have also been measured. We expect new applications of the technique to continue to occur.

ACKNOWLEDGMENTS

The authors would like to thank the staff of SSRL for their tireless and energetic support of this project. This work is supported by SSRL, which is funded by the Department of Energy, Office of Basic Energy Sciences under contract DE-AC03-76SF00515.

REFERENCES

1. A.C. Diebold and K. Monahan, Solid State Tech. **41**, 50 (1998)
2. A. Shimazaki, Electrochem. Soc. Syp. Proc. for Defects in Si II, Wash. D.C., May 5-10, 1991; ECS vol. 91-9, 47 (1991)
3. P. Bertin, *Principles of X-ray Spectrometric Analysis* (Plenum, New York, 1975)
4. R. Klockenkämper, *Total Reflection X-ray Fluorescence Analysis* (John Wiley & Sons, New York, 1997)
5. M. Born and E. Wolf, *Principles of Optics* (Pergamon, New York 1975)
6. S. Brennan, W. Tompkins, N. Takaura, P. Pianetta, S.S. Laderman, A Fischer-Colbrie, J.B. Kortright, M.C. Madden and D.C. Wherry, Nucl. Instrum. & Meth. **A347**, 417 (1994)
7. P. Pianetta, N. Takaura, S. Brennan, W. Tompkins, S.S. Laderman, A. Fischer-Colbrie, A. Shimazaki, K. Miyazaki, M. Madden, D.C. Wherry and J.B. Kortright, Rev. Sci. Instrum. **66**, 1293 (1995)
8. A. Fischer-Colbrie, S.S. Laderman, S. Brennan, S. Ghosh, N. Takaura, P. Pianetta, A. Shimazaki, A. Waldhauer, D. Wherry and S. Barkan, Mat. Res. Soc. Symp. Proc. **477**, 403 (1997)
9. See, for instance, *Resonant Anomalous X-ray Scattering: Theory and Applications* G. Materlik, C.J. Sparks and K. Fischer, ed.s, North-Holland (1994).

CHARACTERIZING TRACE METAL IMPURITIES IN OPTICAL WAVEGUIDE MATERIALS USING X-RAY ABSORPTION

P. H. CITRIN, P. A. NORTHRUP, R. M. ATKINS, P. F. GLODIS*, L. NIU[a], M. A. MARCUS[b], AND D. C. JACOBSON

Bell Laboratories, Lucent Technologies, Murray Hill, NJ 07974
*Bell Laboratories, Lucent Technologies, Norcross, GA 30071

ABSTRACT

X-ray absorption measurements are described for identifying metal impurities in silica preforms, the rod-like starting materials from which hair-like optical fibers are drawn. The results demonstrate the effectiveness of this approach as a non-destructive, quantitative, element-selective, position-sensitive, and chemical-state-specific means for characterizing transition metals in the concentration regime of parts per billion.

INTRODUCTION

Optical transmission loss is one of the most important factors in defining the utility of a waveguide material [1-3]. Improvements in optical fiber technology have resulted in silica fiber losses as low as ≤ 0.20 dB/km (the theoretical limit is 0.16 [4]), meaning that 0.01 dB/km is a very significant amount. A long suspected source of ~0.002 - 0.01 dB/km losses in the manufacturing of silica fiber has been the presence of trace metal impurities such as Fe. Supporting evidence for this has been only suggestive, however, because previous impurity measurements were either too insensitive, ill-defined, or far removed from the early stages of fabrication where identification of contaminants is most reliable and least costly. That is, characterizing those metal impurities that could contribute to transmission loss should ideally be performed *before* the optical fibers are drawn from the parent silica preforms, but this has proved to be too difficult because the metal concentrations in the preforms are so low.

In general, there are three distinct concentric regions of a preform: an inner core, a surrounding cladding, and an outer starting tube. (For the experimental preforms used in this study, the respective diameters of these regions are approximately 4, 12, and 24 mm.) A typical ~1-m-long preform is fed into a high temperature furnace where it is drawn into fiber. For a single mode fiber (either experimental or commercial), the diameters of the core and outermost region are ~6 - 9 μm and ~125 μm. The difference in refractive index between the cladding and core regions confines or guides the light pulses to travel within the core. Clearly the greatest concern is the existence of light-absorbing impurities within the core, but such impurities could originate from the outer cladding and/or shell regions and diffuse inwards. Therefore, it is important to analyze the preform as a function of position in order to determine (and ultimately eliminate) the source of impurities in the fiber.

A brief description of the manufacturing process is useful in understanding where possible contamination sources might arise. The modified chemical vapor deposition (MCVD) method used by Lucent Technologies [5] consists of torch heating the outside of a rotating, high-purity fused silica starting tube as gaseous halides of Si and of other dilute components (e.g.,Ge, P) flow through it in carrier gases of O_2 and Ar or He. Glassy SiO_2 and GeO_2 particles of the heat-reacted gases, along with selected dopants, are deposited inside the tube and downstream of the torch as it traverses the length of the tube. Varying the gas mixture with each pass of the torch thus builds up internal layers of different composition. Any dilute metal contaminants in the reactant gases, either from the original gas source or from the stainless steel gas-carrying tubes, are effectively left behind in the form of nonvolatile halides, but those that do manage to get carried along could be trapped in their own particular growth layer. (In a sense, the layers of a preform are like the rings of a tree, but here the newer layers grow inwards rather than outwards.) Subsequent heat treatment in the fiber drawing process might also promote diffusion of impurities across the different layers and regions of the preform, including the starting tube (supplied by an outside source and which may initially contain its own impurities as well).

In this work, we present x-ray absorption measurements which identify and quantify trace metal impurities as a function of position within experimental optical preforms prepared for this study. For certain transition metals, concentrations as low as ~10 ppb are measured. Direct correlations are established between the amount of optical loss and the degree of metal contamination. Implications for future work are discussed.

EXPERIMENT

The x-ray absorption measurements were performed at the National Synchrotron Light Source using the Bell Laboratories X15B beamline [6]. Extensive shielding and collimation using high purity Pb foil were required to reduce the large background fluorescence from the ubiquitous Fe-containing surroundings, enabling high detection sensitivity for most metal impurities within the preforms. As described in the next section, absolute concentrations were determined using samples of known implanted-metal doses. The existence and type of metals in the preforms were readily identified from their x-ray absorption spectra, while the metal oxidation states were determined by comparing their high-energy-resolution absorption-edge positions with those from prepared samples of known chemical state.

To generate glass samples with high enough concentrations of transition metals to observe excess fiber losses unambiguously, the preforms were made on a developmental fabrication lathe. With this facility, it is possible to vary the processing conditions and chemical purity in order to increase the likelihood of transition metal contamination. The ~24-mm-diameter experimental preforms were sliced into ~2-mm-thick sections, etched in HF, and rinsed in ethanol. Position sensitive x-ray fluorescence data were obtained using a ~0.5 x 0.5 mm^2 collimated x-ray beam incident along various points on the diameter of the sections. The collection region of the fluorescence detector was constructed of high purity, Fe-free components. Absorption filters of appropriate thickness and material were used to minimize collection of elastically scattered radiation.

RESULTS

The feasibility of the method was first tested using an experimental preform sectioned near the middle and near the end of the tube where differences in purity might be expected to exist (fibers had not been drawn from this preform, so optical loss measurements were not available). Fe K-edge absorption was also studied in these two slices as a function of radial position. Differences were indeed observed between the core regions, with greater Fe absorption found in the end slice. Differences were observed as well between the core and cladding region of the end slice, with greater absorption found in the core. Within experimental uncertainties, the middle slice showed indistinguishable absorption between the cladding and core regions. No attempts were made to identify the chemical state of Fe or to look for other metal impurities in these two preform slices.

To quantify the above results in atoms of Fe per cm^3, the x-ray absorption measured from the preform slices must be compared with that measured from the ion-implanted standards, whose known Fe concentrations (doses) are in atoms per cm^2. This requires knowledge of the effective sampling depth, x_{eff}, in the preform measurements. In general, an atom concentration C which varies as a function of depth x gives rise to a detected fluorescence intensity $I_{Fl} = \kappa \int C(x) \exp(-\mu_{tot} x) dx$, where κ is a proportionality constant. The total atomic absorption coefficient is $\mu_{tot} = (\mu_{in}/\sin\theta) + (\mu_{out}/\sin\phi)$, i.e., the sum of absorption of incident radiation with energy at the Fe K edge impinging on the sample at angle θ (measured from the surface), plus the absorption of outgoing Fe $K\alpha$ radiation emerging from the sample at angle $\phi (= 90 - \theta)$. In the standard samples, the typical range of atoms implanted at ≤ 5 MeV is only several μm or less, which is generally well below the penetration depth of energetic x-rays. This allows the depth distribution of that dose to be approximated as a delta function. With n_{surf} accordingly labeled as a "surface" dose, we have $C(x) = n_{surf}\delta(x)$, and thus $I_{Fl} = \kappa n_{surf}$. If this same dose of atoms were instead uniformly distributed in the bulk, i.e., $C(x) = C_{bulk}$, we would have $I_{Fl} = \kappa C_{bulk}/\mu_{tot}$, or $C_{bulk} = \mu_{tot} n_{surf} = n_{surf}/x_{eff}$.

Consider the case of Fe in SiO_2, with $\theta = \phi = 45°$. The calculated absorption coefficients [7] for incident Fe K-edge radiation and outgoing Fe $K\alpha$ radiation in vitreous silica ($\rho \approx 2.2$ gm/cm^3) are, respectively, $\mu_{in}(7.1$ keV$) \approx 96$ cm^{-1} and $\mu_{out}(6.4$ keV$) \approx 130$ cm^{-1}, giving a total absorption coefficient of $\mu_{tot} \approx (96 + 130)\sqrt{2} \approx 320$ cm^{-1}, or $x_{eff} \approx 31$ μm. A dose of $n_{surf} = 6.8 \times 10^{11}$ Fe atoms/cm^2 distributed uniformly in vitreous silica would therefore be equivalent to a bulk concentration of $C_{bulk} = 2.2 \times 10^{14}$ Fe atoms/cm^3, or 10 ppb. Since the x-ray absorption edge jump is directly proportional to atom concentration [8], it is straightforward to quantify the absorption measured in the preform slices by simply measuring the absorption in the standard samples and scaling accordingly. The small systematic uncertainties associated with the calculated absorption coefficients [7] and with the material densities are minimized by cross checking absorption data from a number of different material standards containing different doses.

Applying these calibration procedures to the above results for the preform end slice studied yields ~130 ppb Fe in the core region and ≤ 100 ppb in the cladding region. The middle slice shows ≤ 100 ppb in both core and cladding regions. Statistical uncertainties are ± 10 ppb. The significance of these values is discussed in the next section.

Two other experimental preform samples were studied which, though also not representative of typical manufactured material, had the important feature of having fibers of very different optical loss drawn from them. Absorption edge jump results from these samples are summarized in Fig. 1. Data from these two preforms yielding "more lossy" and "less lossy" fibers are indicated with filled and open shapes, respectively. The circles and triangles represent edge jump data from the Fe K and Cu K edges, with correspondingly different absolute concentration scales noted. Statistical uncertainties are reflected in the scatter of points taken at the same position. A dramatic concentration change with position is observed in the more lossy material, whereas no such change is seen in the less lossy

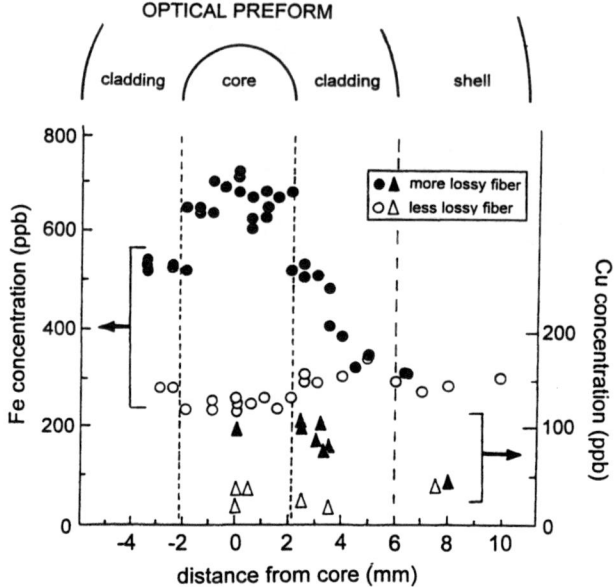

Fig. 1. X-ray absorption *K*-edge jump values for Fe (circles) and Cu (triangles). The data, calibrated against standards to give concentrations in ppb, are measured as a function of radial position within two experimental preform slices from which optical fibers of different loss were drawn. Note the striking position-dependent absorption of both Fe and Cu within the preform that yielded the "more lossy fiber".

sample. The core region of the more lossy sample exhibits the largest metal concentration, well above that in the outer starting tube, or "shell" region. The cladding region shows a monotonically decreasing concentration with increasing distance from the core. Similar trends are observed for Fe and Cu, albeit at much lower absolute concentrations and fewer data points for Cu. Attempts to study Cr and Ni were hampered by detection limits of about 200 ppb for Cr and 100 ppb for Ni (discussed in the next section). Presumably with better sensitivity, analogous trends would also be seen for these metals.

The striking spatial variation in metal concentration and the significantly different behavior of these variations in the "more" versus "less" lossy preforms clearly demonstrate the existence of a genuine correlation between metal contamination and optical loss. On the other hand, the absolute concentrations of Fe are so large — regardless of spatial position or even preform sample — as to raise serious concerns about the reliability of the calibration procedure. Calculations predict that as little as 1 ppb of Fe^{2+} would lead to absorption losses of almost 0.2 dB/km [3], so the Fe concentrations of > 200 ppb measured in these trial preforms would result in optical fibers that are essentially opaque. The key to understanding this dilemma is found in the chemical state of the Fe.

The top of Fig. 2 shows Fe *K*-edge data from two Fe-containing glass samples, one reduced in hydrogen to produce exclusively Fe^{2+} and the other oxidized to produce exclusively Fe^{3+}. The several volt shift in absorption edge position serves to distinguish between the different oxidation states. (The differences in shape above the edges reflects the different unfilled density of states with *p*-state symmetry [9].) Comparing the edge positions of these spectra from the standard samples with those of the normalized data from the two experimental preform slices, as indicated by the vertical line through their inflection points, clearly shows that the Fe in these preforms exists almost entirely in the +3 state. Significantly, this Fe^{3+} species is relatively transparent in the optical regions of interest, explaining how such large amounts of Fe are observed in preform slices from

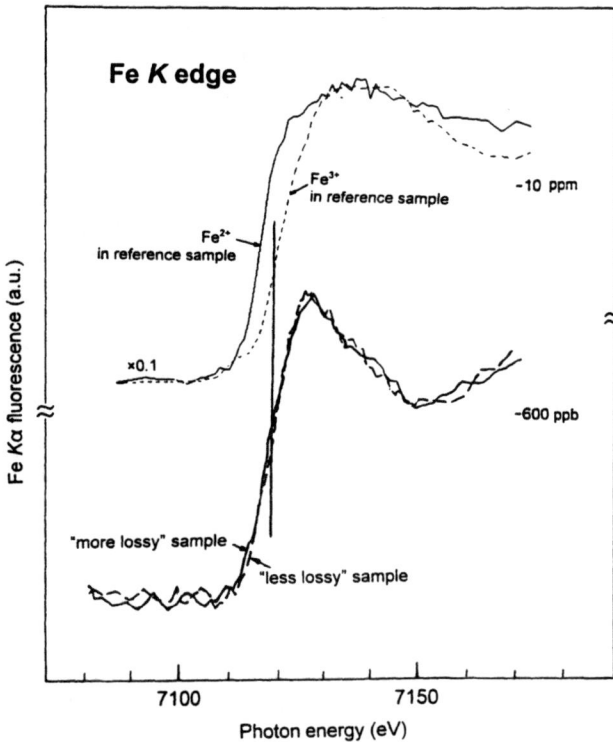

Fe K edge

Fe³⁺
in reference sample

-10 ppm

Fe²⁺
in reference sample

×0.1

-600 ppb

"more lossy" sample

"less lossy" sample

7100   7150

Photon energy (eV)

Fe Kα fluorescence (a.u.)

Fig. 2. Normalized Fe *K*-edge absorption data from two reference samples (top) and two experimental preform samples (bottom). The Fe in both preform samples is identified as being in the +3 state.

which functional optical fibers could, in principle, be drawn. Assuming that Fe in the more absorbing +2 state is indeed present, its concentration is substantially below that of the dominant Fe^{3+} species.

DISCUSSION

The immediate goal of this study was to establish a measurement procedure that is straightforward to interpret and sensitive enough to determine the identities and amounts of trace metal contaminants as a function of position within optical preforms. This goal was facilitated by using experimental rather than commercial preforms in order to insure that sufficient variations in metal concentrations (and in optical losses of subsequently drawn fibers) were studied. Some results of this work therefore apply only to these experimental samples, but several conclusions can be made which extend to future characterization of commercial grade preforms.

We first focus on the *relative* absorption edge jump results in the experimental preforms. The fact that no significant differences are seen between the shell and cladding regions in both the "more" and "less" lossy samples shows decidedly that the source of the metal contaminants does not originate from the starting tube. The monotonically increasing metal concentration with decreasing distance from the core observed in the cladding of the more lossy sample, along with the abrupt rise in the core region, suggests the following possible mechanism. Very small amounts of metal contamination, carried into the starting tube along with the gaseous chlorides of Si (mainly) and Ge, generally react with the (byproduct) Cl gas and harmlessly exit the tube in the form of chloride vapors. The success of this process requires the amount of Cl inside the tube to be sufficiently abundant, and this in turn depends on the flow rate of the reactant gases. For a variety of reasons, the gas flow rate, and thus the layer growth process, decrease as the cladding region approaches the core; these decrease further in the core region itself. Therefore, the observed trends correlate inversely with the amount of Cl gas available for reacting with the trace amounts of metal.

In discussing the *absolute* edge jump values measured in the experimental preforms, it is important to understand detection sensitivity. Even with the greatest of care taken to reduce artifactual background, detection sensitivities are determined largely by the energy regime in which a particular metal is being studied. In general, the lower the absorption-edge energy is for transition metals, the more difficult it is to discriminate that metal's characteristic fluorescence from elastically scattered radiation and ancillary fluorescence from the Z-1 filter material. Furthermore, the x-ray penetration, and thus effective sampling depth, is reduced. This is why the detection sensitivity for Cr is only about 200 ppb while that for Cu and Zn is less than 10 ppb (Ni is problematic due to an anomalous absorption feature in the x-ray monochromator crystals). We have studied samples in which there was "no" measurable absorption of Fe or Cu, i.e., the absolute concentrations in Fig. 1 are accurate. It is therefore significant that the "background" levels of absorption measured in, e.g., the shell regions, are nonzero. This implies that in these experimental preforms there is a constant source of trace metal contamination, which under certain conditions, increases and adversely affects the optical properties of the fibers drawn from them.

The next important step will be to characterize commercial preforms. This will test for background metal absorption and, to the degree that it exists, possible trends as a function of both radial and transverse (i.e., end versus middle) positioning within the preform. Such studies are eagerly anticipated.

ACKNOWLEDGMENTS

We thank L. F. Schneemeyer for the Fe-containing glass samples. The EXAFS experiments were performed at the NSLS, Brookhaven National Laboratory, which is supported by the DOE, Division of Materials Sciences and Division of Chemical Sciences.

REFERENCES

[a] Present address: Applied Materials, Inc., Santa Clara, CA 95054
[b] Present address: KLA-Tencor Corp., San Jose, CA 95134

[1]   Optical Fiber Telecommunications, edited by S. E. Miller and A. G. Chynoweth (Academic Press, Inc., New York, 1979), p. 705.

[2]   Optical Fiber Communications 1, edited by T. Li (Academic Press, Inc., New York, 1985), p. 363.

[3]   S. R. Nagel, IEEE Circ. Dev. Mag., March 1989, p. 36.

[4]   R. Csenscits, P. J. Lemaire W. A. Reed, D. S. Shenk, and K. L. Walker, Tech. Dig., Opt. Comm. Conf., Feb. 1984.

[5]   S. R. Nagel, J. B. MacChesney, and K. L. Walker, IEEE J. Quantum Electron. **18**, 459 (1982).

[6]   A. A. MacDowell, T. Hashizume, and P. H. Citrin, Rev. Sci. Instrum. **60**, 1901 (1989).

[7]   L. Gerward, Nucl. Instrum. Methods **181**, 11 (1981).

[8] P. A. Lee, P. H. Citrin, P. Eisenberger, and B. M. Kincaid, Rev. Mod. Phys. **53**, 769 (1981).

[9] For example, see S. Schuppler, D. L. Adler, L. N. Pfeiffer, K. W. West, E. E. Chaban, and P. H. Citrin, Phys. Rev. B **51**, 10527 (1995).

Part IV

Materials Characterization
With X-Ray Absorption

ANOMALOUS X-RAY SCATTERING STUDIES OF SHORT-, INTERMEDIATE- AND EXTENDED-RANGE ORDER IN GLASSES

DAVID L. PRICE*, MARIE-LOUISE SABOUNGI*, PASCALE ARMAND**, DAVID E. COX***
*Argonne National Laboratory, Argonne, IL 60439
**Argonne National Laboratory, Argonne, IL 60439; now at CNRS Montpellier, Place E. Bataillon Case 003, Montpellier, France
***Brookhaven National Laboratory, Upton, NY 11973

ABSTRACT

We present the formalism of anomalous x-ray scattering as applied to partial structure analysis of disordered materials, and give an example of how the technique has been applied, together with that of neutron diffraction, to investigate short-, intermediate- and extended-range order in vitreous germania and rubidium germanate.

INTRODUCTION

While short-range order (SRO) in network glasses is generally well understood, usually in terms of structural units identified thorough comparison with analogous crystalline compounds, the intermediate-range order (IRO) — the manner in which these units are organized to form a large random network — remains a controversial issue.[1] Further, the modification of such glasses by the addition of metal oxides produces an extended-range order (ERO) that is also not well understood.[2] Part of the problem stems from the fact that a single diffraction measurement of an n-component system yields a weighted average structure factor of the $n(n+1)/2$ separate partial structure factors:

$$S(Q) = \sum_{ab} W_a(Q) W_b(Q) S_{ab}(Q)$$

(1)

where $W_a(Q)$ is the weighting factor for element a and $S_{ab}(Q)$ is the partial structure factor for the element pair (a,b).[3] For a multicomponent system it is therefore difficult to extract reliable information about a specific atom pair from a single experiment. However, in recent years improvements in the radiation sources available have led to the exploitation of complementary techniques which allow a more detailed investigation of the structure. Spallation neutron sources and instruments dedicated to amorphous and liquid systems make it possible to carry out neutron diffraction (ND) measurements with sufficient resolution to resolve peaks in the radial distribution function from different atom pairs; in favorable cases, isotope substitution can also be used to vary the weighting factors in Eq. (1).[4] An especially important advance has been the advent of high-powered synchrotron x-ray sources which make it possible to obtain reliable difference measurements near the absorption edge of a particular element with anomalous x-ray scattering (AXS), varying the weighting factor for that element in Eq. (1) by changing the x-ray energy.[5]

In this review we present the formalism of AXS as applied to partial structure analysis of disordered materials, and then give an example of how the technique has been applied, together with that of neutron diffraction, to the case of vitreous germania and rubidium germanate.

FORMALISM

The scattering cross section in an x-ray diffraction measurement is given by a sum of products of scattering factors and phase factors over all pairs of atoms of the system:

261

$$\frac{d\sigma^s}{d\Omega} = \frac{1}{N}\sum_{ij} f_i(\mathbf{Q})f_j^*(\mathbf{Q})\exp\left[i\mathbf{Q}\cdot\left(\mathbf{r}_i - \mathbf{r}_j\right)\right]$$

(2)

where $f_i(\mathbf{Q})$ is the form factor of atom i evaluated at the scattering vector $\mathbf{Q}$ and $\mathbf{r}_i$ its instantaneous position. It is convenient to reformulate Eq. (2) as the sum of three terms which, in the Faber-Ziman formulation[6], becomes

$$\frac{d\sigma^s}{d\Omega} = I(\mathbf{Q}) + \left[\left\langle\left|f(\mathbf{Q})\right|^2\right\rangle - \left|\left\langle f(\mathbf{Q})\right\rangle\right|^2\right] + N\delta_{\mathbf{Q}0}\left|\langle f(0)\rangle\right|^2$$

(3)

where the angular brackets represent averages over all atoms in the sample. The first term in Eq. (3) is the *interference scattering* which contains the details of the atomic structure, the second is the *Laue diffuse scattering* which varies slowly with $\mathbf{Q}$, and the third is the *forward scattering* which is singular in the small-angle limit for a homogeneous system. Clearly the first term is the important one in the present context. In a multicomponent system, it can be expressed as a weighted sum over element pairs:

$$I(\mathbf{Q}) = \sum_{ab} c_a c_b f_a(\mathbf{Q})f_b^*(\mathbf{Q})S_{ab}(\mathbf{Q})$$

(4)

where the *partial structure factor* (psf)$S_{ab}(\mathbf{Q})$ represents the structure associated with a given pair of elements (a,b) and is given by

$$S_{ab}(\mathbf{Q}) = \frac{N}{N_a N_b}\sum_{i\in a, j\in b}\exp\left[i\mathbf{Q}\cdot\left(\mathbf{r}_i - \mathbf{r}_j\right)\right] - \frac{N}{N_a}\delta_{ab} + 1 - N\delta_{\mathbf{Q}0}$$

(5)

It is often convenient to define an *average structure factor* $S(\mathbf{Q})$:

$$S(\mathbf{Q}) = \frac{I(\mathbf{Q})}{\left|\langle f(\mathbf{Q})\rangle\right|^2}$$
$$= \sum_{ab} c_a c_b \frac{f_a(\mathbf{Q})f_b^*(\mathbf{Q})}{\left|\langle f(\mathbf{Q})\rangle\right|^2}S_{ab}(\mathbf{Q})$$

(6)

In the Faber-Ziman formalism used here, both $S(\mathbf{Q})$ and $S_{ab}(\mathbf{Q}) \to 1$ as $Q \to \infty$.

To obtain real-space information, pair correlation functions are obtained by Fourier transformation of the appropriate structure factors, *e.g.* :

$$g(\mathbf{r}) - 1 = \frac{1}{8\pi^3\rho}\int[S(\mathbf{Q}) - 1]\exp(-i\mathbf{Q}\cdot\mathbf{r})d\mathbf{Q}$$

(7)

where $g(\mathbf{r})d\mathbf{r}$ has a direct physical interpretation as the number of atoms in a volume element $d\mathbf{r}$ at a distance $\mathbf{r}$ from a reference atom at the origin.

We now treat the case of special case of anomalous scattering. Suppose that the incident energy is near an absorption edge for element A and we make a small change ΔE in the incident energy, which we assume has a significant effect on f_A but not on the other f_a. From Eq. (4), the corresponding change in $I(\mathbf{Q})$ is given by

$$\Delta I(\mathbf{Q}) = c_A\Delta f_A\sum_b c_b f_b^*(\mathbf{Q})S_{Ab}(\mathbf{Q}) + c.c.$$

(8)

In general, and especially near an absorption edge of element A, the form factor contains anomalous energy-dependent terms as well as the regular term:

$$f(\mathbf{Q}) = f_0(\mathbf{Q}) + f'(E) + if''(E) \tag{9}$$

Substitution of Eq. (9) into Eq.(8) gives, after some rearrangement,

$$\Delta I(\mathbf{Q}) = 2c_A \Delta f_A' \sum_b c_b \left[\left(f_{b0}(\mathbf{Q}) + f_b' \right) S_{Ab}' + f_b'' S_{Ab}'' \right] + 2c_A \Delta f_a'' \sum_b c_b \left[f_b'' S_{Ab}' - \left(f_{b0}(\mathbf{Q}) + f_b' \right) S_{Ab}'' \right] \tag{10}$$

By analogy with Eq. (6), we can define[7] a *difference structure factor* for element A:

$$S_A(\mathbf{Q}) = \frac{\Delta I(\mathbf{Q})}{2c_A \Delta f_A' \langle f_0(\mathbf{Q}) + f' \rangle + 2c_A \Delta f_a'' \langle f'' \rangle}$$

$$= \frac{2c_A \Delta f_A' \sum_b c_b \left[\left(f_{b0}(\mathbf{Q}) + f_b' \right) S_{Ab}' + f_b'' S_{Ab}'' \right] + 2c_A \Delta f_a'' \sum_b c_b \left[f_b'' S_{Ab}' - \left(f_{b0}(\mathbf{Q}) + f_b' \right) S_{Ab}'' \right]}{2c_A \Delta f_A' \langle f_0(\mathbf{Q}) + f' \rangle + 2c_A \Delta f_a'' \langle f'' \rangle} \tag{11}$$

which also $\to 1$ as $Q \to \infty$. Notice that $S_A(\mathbf{Q})$ as defined in Eq. (11) implicitly depends on both E and ΔE. The ΔE dependence could be removed by taking the limit of Eq. (11) as $\Delta E \to 0$. However, the error in the measured value of $S_A(\mathbf{Q})$ obviously becomes very large as ΔE is made small. In practice we try to choose a compromise value for ΔE that is large enough to reduce this error and small enough so that we can neglect the changes in f_b, $b \neq A$. The values of the f_b in Eq. (11) are then evaluated at $E + 1/2 \Delta E$. The choice of the distance of E from the absorption edge of element A is also a compromise between large values for the anomalous terms in f_A versus an accurate knowledge of Δf_A.

Eq. (11) can be simplified with the help of two approximations. First, a good approximation below the edge of element A is

$$\Delta f_A'' \ll \Delta f_A'. \tag{12}$$

E.g., for Ge at E_{Ge}-17 eV, $f_{Ge}' = -6.169$, $f_{Ge}'' = 0.494$; at E_{Ge}-200 eV, $f_{Ge}' = -3.987$, $f_{Ge}'' = 0.504$. With this approximation Eq. (11) simplifies to

$$S_A(\mathbf{Q}) = \frac{\sum_b c_b \left[\left(f_{b0}(\mathbf{Q}) + f_b' \right) S_{Ab}' + f_b'' S_{Ab}'' \right]}{\langle f_0(\mathbf{Q}) + f' \rangle} \tag{13}$$

Second, if the incident energies are far removed from the absorption edges of the elements $b \neq A$, a reasonable approximation is

$$f_b'' \ll f_{b0}, \quad b \neq A. \tag{14}$$

In this case, taking account of the fact that S_{AA} has to be real, Eq. (13) simplifies finally to

$$S_A(\mathbf{Q}) = \sum_b c_b \frac{\left(f_{b0}(\mathbf{Q}) + f_b' \right)}{\langle f_0(\mathbf{Q}) + f' \rangle} S_{Ab}' \tag{15}$$

To obtain the corresponding real-space information, we can use the equivalent of Eq. (7) to obtain the average real-space environment about an atom of element A.

EXPERIMENTS

We have addressed the issues of the SRO, IRO and ERO in a series of combined AXS and ND experiments on experiments on germania and rubidium germanate glasses.[8] The AXS measurements were carried out at the X-7A beam line at the National Synchrotron Light Source (NSLS) at Brookhaven National Laboratory, and the ND measurments at the GLAD facility at the Intense Pulsed Neutron Source (IPNS) at Argonne National Laboratory.

Germania- rather than silica-based glasses, and rubidium as the modifier element, were chosen because the energies of the Ge and Rb edges were suitable for AXS at the X-7A beamline. The glasses were prepared in solid form and polished to give smooth (~50μ roughness) flat surfaces toward the x-ray beam. A series of AXS measurements were carried out at the Ge and Rb K edges. The Rb edge measurements in the rubidium germanate glass were complicated by the high level of Ge fluorescence. For the neutron measurments the glasses were crushed and loaded into thin-walled vanadium tubes.

RESULTS

1. Neutron Diffraction

Neutron structure factors obtained for germania and rubidium germanate glasses are shown in Fig. 1. The results for the binary are in good agreement with those of Desa *et al.*[9] The first feature in the structure factor of GeO_2 is the peak at $Q_1 = 1.54$ Å^{-1}, corresponding to a correlation length $L_1 = 2\pi/Q_1 \sim 4.1$ Å characteristic of IRO in oxide and chalcogenide glasses.[1]

On the addition of only 10 % of the Rb_2O modifier to the GeO_2 network, a new peak arises at $Q_0 \sim 0.95$ Å^{-1}, implying ERO on a length scale $L_0 \sim 6.6$ Å. Q_0 is unaffected by changes in Rb_2O concentration x, but the peak height increases up to $x = 0.2$ and then decreases for larger x. This dependence reflects the behavior of the other physical properties such as T_g and viscosity. In the modified glass, the FSDP observed in GeO_2 at 1.54 Å^{-1} is replaced by a second peak at larger Q. Both the intensity of this peak and its Q value increase with x. Its position corresponds to a length scale L_1 ranging from 3.1 to 3.6 Å, slightly smaller than the characteristic scale of IRO in binary glasses.

2. X-Ray Diffraction

X-ray structure factors for GeO_2 and $(Rb_2O)_{0.2}(GeO_2)_{0.8}$ are shown in Fig. 2. Comparison of neutron and x-ray diffraction data makes it possible to infer information about the nature and origins of these two features, since the appropriate weighting factors are generally quite different for a given element. The same features are observed as in the neutron structure factors; however, for GeO_2 the FSDP is stronger in $S^X(Q)$ than in $S^N(Q)$, indicating that the cation-cation correlations play an important role in its existence, as is generally the case in oxide and chalcogenide glasses.[10-13] For $(Rb_2O)_{0.2}(GeO_2)_{0.8}$, the first peak is much weaker, and the second much stronger, than in $S^N(Q)$. Since correlations involving two cations (Ge or Rb) are emphasized in S^X relative to S^N, while all correlations involving O are de-emphasized, the relative weakness of the first peak in $S^X(Q)$ shows that correlations involving oxygen atoms are important; cation-cation correlations that make a *negative* contribution to this peak may also be involved. On the other hand, the dominance of the second peak in $S^X(Q)$ shows that the cation-cation correlations play a crucial role in its origin, as in that of the FSDP in the binary.

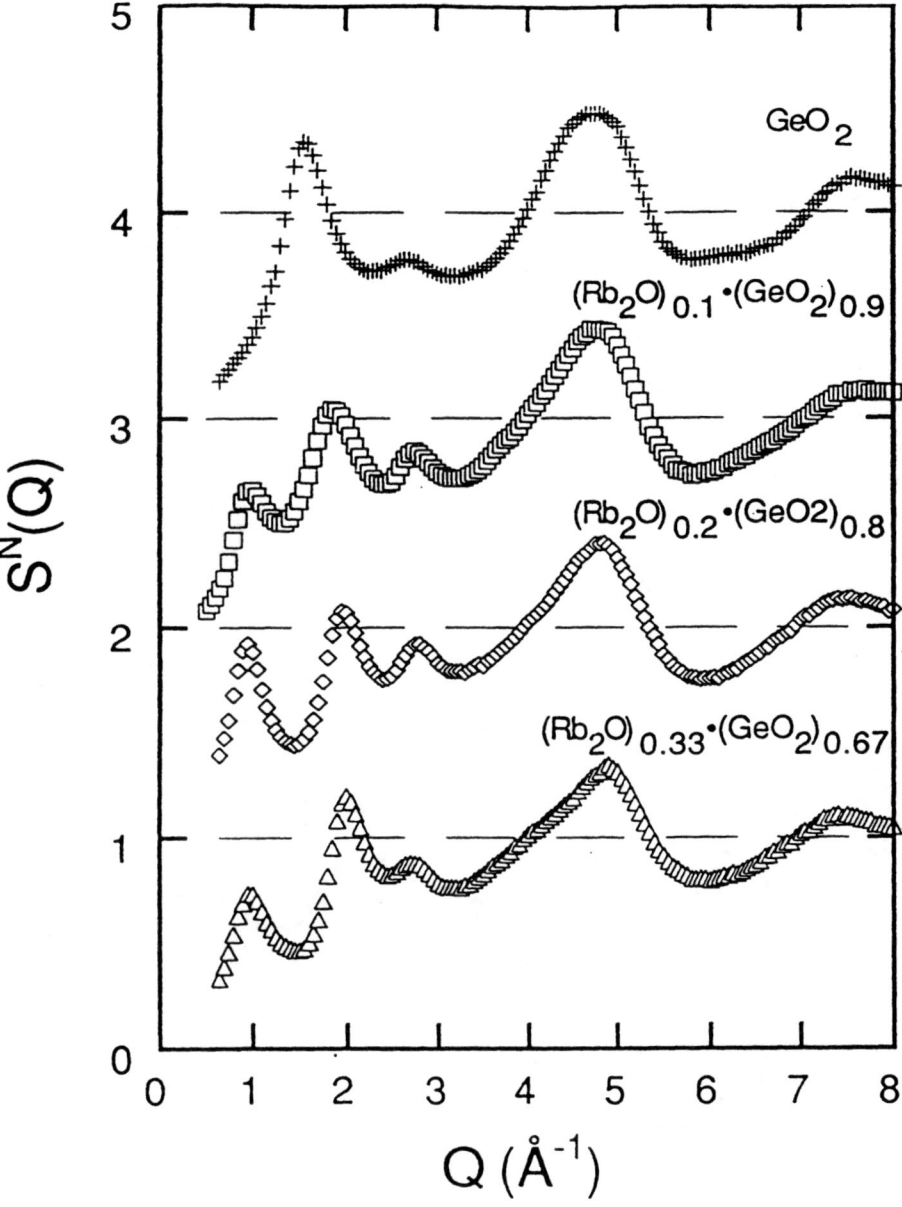

Fig. 1. Neutron structure factor $S^N(Q)$ of $(Rb_2O)_x(GeO_2)_{1-x}$ glasses; successive curves are displaced upward by one unit for clarity.

3. Anomalous X-Ray Diffraction

Further information about the origin of the IRO and ERO is obtained from the results of AXS measurements made near the Ge and Rb K-shell absorption edges. S^x_{Ge} is shown for GeO$_2$ and $(Rb_2O)_{0.2}(GeO_2)_{0.8}$ in Fig. 3, and S^x_{Rb} for $(Rb_2O)_{0.2}(GeO_2)_{0.8}$ in Fig. 4. For GeO$_2$, S^x_{Ge} has the same features as $S^x(Q)$, the main difference being that the peak at $Q_2 = 2.8$ Å^{-1}, attributed to chemical SRO of Ge and O, is considerably enhanced. In the molecular dynamics (MD) simulations of SiO$_2$ glass, S_{SiSi} and S_{OO} make positive contributions to this peak while S_{SiO} makes a negative contribution of almost identical magnitude.[10,13] In all three partials, the peak arises from Coulomb oscillations in the partial pair correlation function $g_{ij}(r)$ with period $L_2 \sim 2.2$ Å, in phase with $\sin(Q_2 r)$ for g_{SiSi} and g_{OO} but out of phase for g_{SiO}. By analogy with SiO$_2$ and taking into account differences in the appropriate weighing factors, this peak should be scarcely visible in $S^N(Q)$, stronger in $S^x(Q)$ and considerably enhanced in S^x_{Ge}, exactly the behavior observed.

For $(Rb_2O)_{0.2}(GeO_2)_{0.8}$, the intensity of the first peak is positively correlated with the scattering amplitude of Ge (Fig. 3), but *negatively* correlated with that of Rb (Fig. 4). This behavior can be explained in terms of a chemical ordering of Rb and Ge with a period $L_0 \sim 6.6$ Å, S_{RbRb} and S_{GeGe} being positive and S_{RbGe} negative at Q_0. The second peak has a strong positive correlation with both Rb and Ge scattering amplitudes, consistent with the crucial role of the cation-cation correlations discussed above. In molecular dynamics simulations[14] on the analogous silicate system $(Rb_2O)_{0.2}(SiO_2)_{0.8}$, the Rb-Rb partial structure factor shows positive peaks at 1.1 and 1.9 Å^{-1}, in agreement with this interpretation.

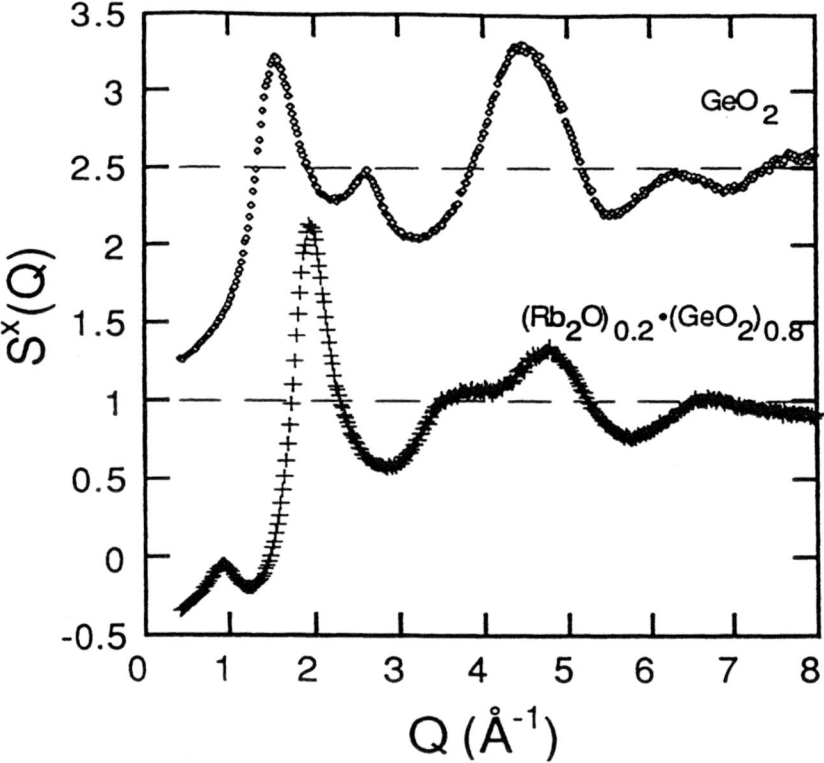

Fig. 2. X-ray structure factor $S^x(Q)$ of GeO$_2$ and $(Rb_2O)_{0.2}(GeO_2)_{0.8}$ glasses.

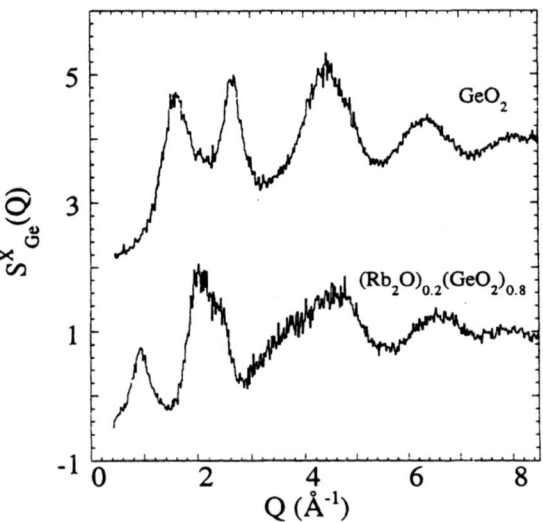

Fig. 3. Germanium difference structure factor $S^x_{Ge}(Q)$ of GeO_2 and $(Rb_2O)_{0.2}(GeO_2)_{0.8}$ glasses.

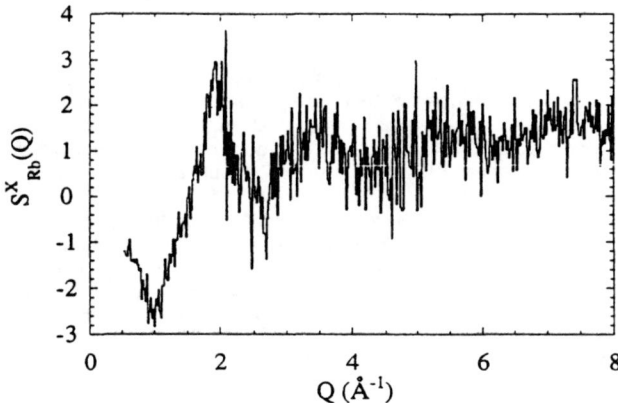

Fig. 4. Rb difference structure factor $S^x_{Rb}(Q)$ of $(Rb_2O)_{0.2}(GeO_2)_{0.8}$ glass.

A FULL PARTIAL STRUCTURE ANALYSIS OF GeO$_2$

Writing the ND, XRD and AXS data presented above as a vector $F(Q) = [S^N(Q), S^X_{13}(Q), S^X_{200}(Q)]$ and the required psf's as a vector $S(Q) = [S_{GeGe}(Q), S_{OO}(Q), S_{GeO}(Q)]$, the two vectors have a linear relationship

$$F = AS \tag{16}$$

where the elements of A are the weighting factors in Eqs. (6) and (15). Inserting the appropriate values for the nuclear and atomic parameters, it is found that degree of conditioning of A^{-1} is comparable to that obtained in many ND experiments with isotope substitution, so that reliable structure factors should be derived, provided the systematic and statistical errors are small.

The results for the psf's are shown as points in Fig. 5. The FSDP shows up as a positive peak in S_{GeGe} and S_{GeO} and a shallow negative peak in S_{OO}, supporting the assertion that cation correlations dominate the IRO. The peak at 2.6 Å^{-1} is strong and positive, with almost the same height, in S_{GeGe} and S_{OO}, and strong and negative, with almost the same magnitude, in S_{GeO}, confirming the chemical SRO nature of this peak. The topological SRO peak occurs at 4.4 Å^{-1}, predominantly in S_{GeGe}.

We are not aware of any published Molecular Dynamics (MD) computer simulations of GeO$_2$, but SiO$_2$ has been studied with classical MD by Vashishta et al.[10] and with ab initio MD by Sarnthein et al.[13] The ab initio MD results,[13,15] plotted as solid lines in Fig. 5 for comparison, are generally similar to the present ones for GeO$_2$ but displaced to higher Q due to the smaller length scale in silica. The main qualitative difference is that the FSDP and topological SRO peaks show up as weak positive features in S_{OO}.

An alternative formulation of the partial structure factors was derived by Bhatia and Thornton[16] in terms of number density (N) and concentration (C) fluctuations, shown in Fig. 6 for the present data and the MD results for SiO$_2$. A matter of some interest is the behavior of S_{CC} in the region of the FSDP, in view of the speculation of Elliott that the FSDP arises from concentration fluctuations in oxide and chalcogenide glasses.[17] In the present results for GeO$_2$, S_{CC} is in fact slightly negative in this region, so this explanation does not seem to apply here. As expected, the CSRO peak shows up most strongly in S_{CC} and the TSRO peak in S_{NN}. The MD results for SiO$_2$ are again qualitatively similar with a general shift to higher Q. For these, S_{CC} is close to zero in the region of the FSDP, again at variance with Ref. 17. A similar behavior was found was found in the classical MD result for SiO$_2$.[10]

DISCUSSION

This work provides results for the SRO and IRO of vitreous GeO$_2$ in terms of the contributions of correlations involving the three element pairs. The results are consistent with the identification of the first three peaks in the average neutron and x-ray structure factors with IRO, chemical SRO and topological SRO, respectively. Results for the FSDP indicate that in this glass IRO is associated with cation correlations as opposed to concentration-concentration correlations as has been suggested.

In the rubidium -modified glass, the strong oxygen correlations implied by the large weight of the first peak in the neutron diffraction pattern support the picture of Wright et al.[18] for network glasses modified with heavy alkali ions. In the binary, the FSDP reflects the periodicity arising from the succession of network cages;[19] in the modified glasses, the Rb$^+$ ions enter the larger cages and displace the oxygen atoms on their boundaries. This makes the boundaries of the large cages more spherical and leads to a strong diffraction peak at low Q. Clearly, this peak will be associated with strong oxygen correlations and with a chemical order due to the alternation of Ge and Rb ions. The cages without the Rb$^+$ ions will be correspondingly contracted, causing a shift of the second diffraction peak to Q values higher than the FSDP of the GeO$_2$.

BEYOND AXS?

The AXS technique makes use of the changes in the real part of $f_A(Q)$ on going through an absorption edge of element A. The question arises as to whether, if the high absorption for x-ray

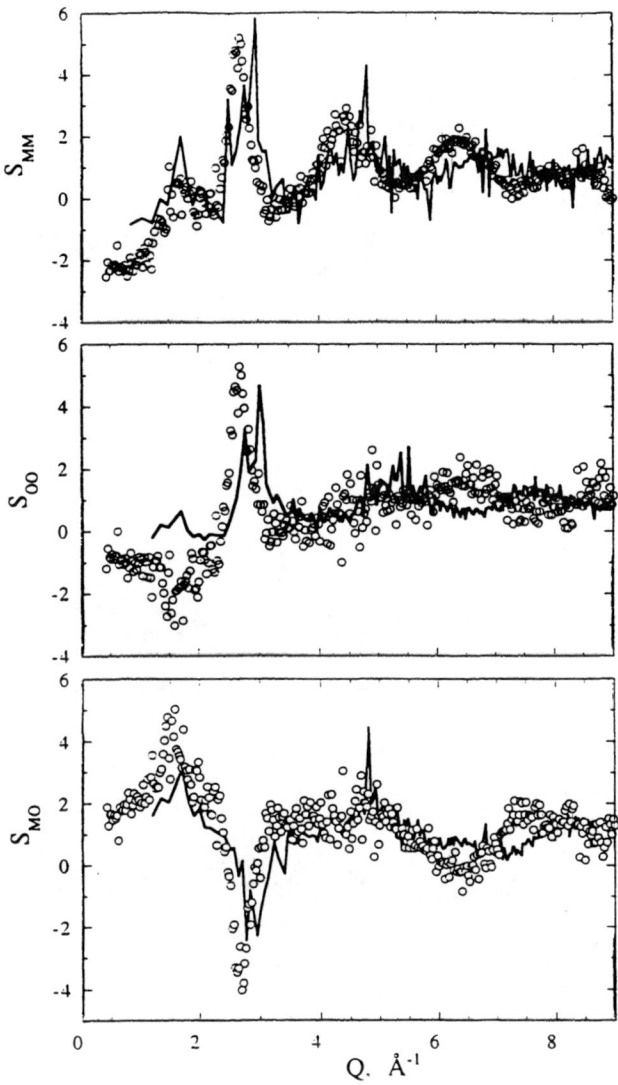

Fig. 5. Measured partial structure factors (Faber-Ziman definition) for the three atom pairs in vitreous GeO_2 (M = Ge, points), together with corresponding results from *ab initio* MD computer simulation[13,15] of vitreous silica (M = Si, lines).

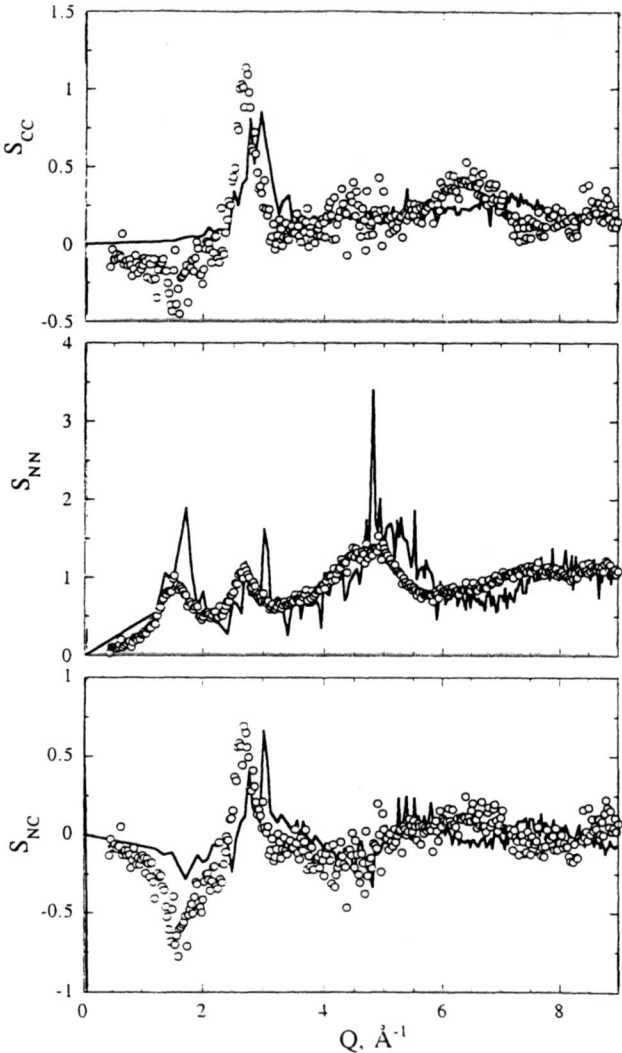

Fig. 6. Measured partial structure factors (Bhatia-Thornton definition[16]) in vitreous GeO_2 (points), together with corresponding results from *ab initio* MD computer simulation[13,15] of vitreous silica (lines).

energies above the edge can be tolerated, changes in the imaginary part of $f_A(Q)$ can be exploited as well. Inserting Eq. (9) into Eq. (6), the result can be written

$$I(\mathbf{Q}) = A(\mathbf{Q}) + 2f'_A(E)B(\mathbf{Q}) + \left[f'_A(E)^2 + f''_A(E)^2\right]C(\mathbf{Q}) \tag{17}$$

where

$$A(\mathbf{Q}) = \sum_{ab} c_a c_b f_{a0}(\mathbf{Q}) f^*_{b0}(\mathbf{Q}) S_{ab}(\mathbf{Q}) \tag{18a}$$

$$B(\mathbf{Q}) = c_A \sum_{ab} c_b f_{b0}\mathbf{Q}) S'_{Ab}(\mathbf{Q}) \tag{18b}$$

$$C(\mathbf{Q}) = c_A^2 S_{AA}(\mathbf{Q}) \tag{18c}$$

Since the coefficients in Eq. (17) are linearly independent functions of E, $A(Q)$, $B(Q)$, $C(Q)$ can in principle be determined independently, allowing a full partial analysis on the basis of the x-ray data alone. Furthermore, f_A' and f_A'' are also determined in the process for the sample under study, and will be quite different for a condensed sample compared with a free atom, a fact that has of course been exploited for many years in structural studies with absorption spectroscopy. In a diffraction context this leads to the "diffraction anomalous fine-structure" (DAFS) technique which has been widely used in the analysis of mixed or mixed-valence crystalline materials.[20] In the context of glass structure, this approach would yield information about structural irregularities in the glass as long as the scale for these is large compared with the range of the photoelectron. Such may well be the case for intermediate or extended range order in inorganic glasses.

CONCLUSION

This example makes it clear that AXS is a powerful technique for obtaining partial structure information in complex disordered materials, and in particular can probe the structure of minority species in a majority host. With the more powerful sources now available, and a corresponding development in experimental methods and analysis techniques, we can expect that before too long full partial structure analysis, with complete derivation of all partial structure factors, will become routine. Further, we may expect that DAFS type information may become available as a probe of structural inhomogeneities in glasses and other disordered materials.

ACKNOWLEDGMENTS.

The work was supported by the U.S. Department of Energy (DOE), Division of Materials Sciences, Office of Basic Energy Sciences, under Contracts W-31-109-ENG-38 at Argonne and DE-AC02-98CH10886 at Brookhaven. IPNS is supported by the Division of Materials Sciences (DOE) at Argonne National Laboratory. NSLS is supported by the Divisions of Materials and Chemical Sciences (DOE) at Brookhaven National Laboratory.

REFERENCES

1. S. C. Moss and D. L. Price, in *Physics of Disordered Materials*, ed. D. Adler, H. Fritzsche and S. R. Ovshinsky (Plenum, New York, 1985), p. 77; D. L. Price, S. C. Moss, R. Reijers, M.-L. Saboungi and S. Susman, J. Phys.: Condens. Matter 1, 1005 (1989); S. R. Elliott, Nature **354**, 445 (1991).
2. P. Armand, M. Beno, A. J. G. Ellison, G. S. Knapp, D. L. Price and M.-L. Saboungi, Europhys. Lett. **29**, 549 (1995).
3. Y. Waseda, *The Structure of Non-Crystalline Materials*. (McGraw-Hill, New York, 1980); *Methods in the Determination of Partial Structure Factors of Disordered Matter by Neutron and*

Anomalous X-Ray Diffraction, Ed. J. B. Suck, P. Chieux, D. Raoux and C. Riekel (World Scientific, Singapore, 1993).

4. D. L. Price, Ed., *Research Opportunities In Amorphous Solids With Pulsed Neutron Sources,* J. Non-Cryst. Solids **76**, No. 1 (1985)

5. P. H. Fuoss, P. Eisenberger, W. K. Warburton and A. Bienenstock, Phys. Rev. Lett. **46**, 1537 (1991).

6. T. E. Faber and J. M. Ziman, Phil. Mag. **11**, 153 (1965).

7. D. L. Price and M.-L. Saboungi, in Local Structure from Diffraction, Ed. S. J. L. Billinge and M. F. Thorpe (Plenum, NY, 1998), in press.

8. D. L. Price, A. J. G. Ellison, M.-L. Saboungi, R.-Z. Hu, T. Egami and W. S. Howells, Phys. Rev. B **55**, 11249 (1997).

9. J. A. E. Desa, A. C. Wright and R. Sinclair, J. Non-Cryst. Solids **99**, 276-288.

10. P. Vashishta, R. K. Kalia, J. P. Rino and I. Ebbsjö, Phys. Rev. B**41**,12197-12208 (1990).

11. I. T. Penfold and P. S. Salmon, Phys. Rev. Lett. **67**, 97-100 (1991).

12. P. Vashishta, R. K. Kalia, G. A.. Antonio and I. Ebbsjö, Phys. Rev. Lett. **62**,1651-1654 (1989).

13. J. Sarnthein, A. Pasquarello and R. Car, Phys. Rev. Lett. **74**, 4682 (1995); Phys. Rev. B **52**, 12690 (1995).

14. J. Kieffer and D. Nekhayev, private communication

15. A. Pasquarello, private communication (1998)

16. A. Bhatia and D. Thornton, Phys. Rev. B **2**, 3004 (1970)

17. S. R. Elliott, Phys. Rev. Lett. **67**, 711 (1991)

18. A. C. Wright, A. G. Clare, B. Bachra, R. N. Sinclair, A. C. Hannon and B. Vessal, Trans. ACA **27**, 239-254 (1991).

19. A. C. Wright, in *Proc. Symp. on Glass Structure, Amer. Cer. Soc. Mtg. Honolulu, HI, Nov. 1993,* J. Non-Cryst. Solids **179**, 84-115 (1994)

20. I. J. Pickering et al., J. Am. Chem. Soc. **115**, 6302 (1993), and references therein.

X-RAY ABSORPTION FINE STRUCTURE (XAFS) STUDIES OF COBALT SILICIDE THIN FILMS

S. J. Naftel, I. Coulthard, Y. Hu, T. K. Sham and M. Zinke-Allmang
Departments of Chemistry and Physics
University of Western Ontario, London, Canada.

ABSTRACT

Cobalt silicide thin films, prepared on Si(100) wafers, have been studied by X-ray absorption near edge structures (XANES) at the Si K-, $L_{2,3}$- and Co K-edges utilizing both total electron (TEY) and fluorescence yield (FLY) detection as well as extended X-ray absorption fine structure (EXAFS) at the Co K-edge. Samples made using DC sputter deposition on clean Si surfaces and MBE were studied along with a bulk $CoSi_2$ sample. XANES and EXAFS provide information about the electronic structure and morphology of the films. It was found that the films studied have essentially the same structure as bulk $CoSi_2$. Both the spectroscopy and materials characterization aspects of XAFS (X-ray absorption fine structures) are discussed.

INTRODUCTION

$CoSi_2$ and other silicides have found applications in the microelectronics industry. As the device technology approaches submicron size, research into the processes occurring at these unusual dimensions becomes important. One of the issues concerning device performance is the chemical uniformity and structure of small silicide features compared to that of the bulk silicides.

It has been found that the morphology of a sample depends on the preparation conditions. We have undertaken a study of the morphology differences of a few $CoSi_2$ thin film samples prepared under different conditions by utilizing the element and depth sensitivity afforded by XAFS allowing us to probe the surface and bulk spectra separately and identify changes in the composition of the films at various depths. This use of multi-core and multi-detection techniques allows us to characterize the differences in chemical uniformity of the films at various depths in a non-destructive approach.

The fine structures in X-ray absorption spectra originate from the interference of the outgoing and backscattered photoelectrons at the absorbing atom. The XAFS spectrum has been conveniently divided into two parts, the X-ray absorption near edge structure (XANES) which is the region up to about 50 eV above the edge and the extended X-ray absorption fine structure (EXAFS) which extends from the XANES up to 1000 eV or so above the edge.

In XANES the electron excited from the core is either bounded or of low kinetic energy (quasi bound) thus the electronic structure of the conduction states and multiple scattering are important in XANES. Thus XANES provides information about the local structure and bonding (symmetry, empty densities of states, etc.) about the absorbing atom. In EXAFS (photoelectron has a high kinetic energy) the core electrons of the other atoms are primarily responsible for the scattering and single scattering predominates.

For this study the sampling depths at the edges are of primary importance. The one absorption length (1/e attenuation of the photon) of the incident X-rays are ~50 nm, ~1.3 μm and ~300 μm for the Si $L_{3,2}$- (99 eV), Si K- (1839 eV) and Co K-edges (7709 eV) respectively as estimated from the absorption coefficients. It is immediately apparent that the X-rays at the Si $L_{3,2}$-edge energy have a penetration depth comparable to the thickness of the films thus

273

minimizing any contribution from the substrate silicon regardless of the detection technique. In addition the total electron yield technique is sensitive to only the first few nm of the sample, at the Si $L_{3,2}$-edge, as electrons have a short escape depth from the solid. The fluorescence yield has a larger (10's of nm) [1] sampling depth as the attenuation of the fluorescence x-rays is similar to the attenuation of the incident photons. At the Si K-edge the sampling depths are still different between TEY and FLY. TEY has a sampling depth of ~ 70 nm while for FLY the sampling depth is in the 100's of nm [1] range making both techniques relatively "bulk" sensitive compared to the Si L-edge. Thus a combined TEY and FLY study of silicon films at the appropriate edges can reveal, non-destructively, the electronic structure of both the ambient oxide layer, the silicide underneath and interfacial reactions between the silicide and substrate [2] without interference from the underlying Si substrate.

EXPERIMENT

$CoSi_2$ ingot was prepared by the materials preparation group at McMaster University using standard procedures. The films were prepared at Northern Telecom using a DC sputtering technique. 55.6 nm of Co were deposited on clean Si(100) wafers followed by a two step thermal annealing process in a nitrogen ambient as described previously [3]. A fourth film was prepared at the University of Western Ontario using an MBE procedure. Table I lists the preparation conditions of all samples used in this study. No attempt was made to remove native oxide from the films before spectra were taken. However, the bulk sample was mechanically scrapped with a diamond file to remove the majority of the oxide before introduction to the experimental chamber.

The Si L-edge spectra were taken in Total Electron Yield (TEY) utilizing sample current and in Fluorescence Yield (FLY) utilizing a channel plate detector at the Canadian Grasshopper Monochromator beamline of the Canadian Synchrotron Radiation Facility (CSRF) located at the Synchrotron Radiation Centre (SRC) of the University of Wisconsin-Madison. Si K-edge measurements, using the same detection methods, were made on the soft X-ray double crystal monochromator beamline of the CSRF using InSb(111) crystals. Co K-edge measurements were taken at the National Synchrotron Light Source (NSLS) at Brookhaven National Laboratory on beamline X-11A. The TEY spectra taken at NSLS utilized a He amplified TEY detector while, for FLY spectra a Lytle detector was used.

Table I. Description of Samples and Preparation conditions.

Sample	Description
$CoSi_2$	Bulk Ingot
Co-Si (1)	Co (55.6 nm) film, not annealed.
Co-Si (2)	Co (55.6 nm) film, annealed 450 °C/60 sec then 690 °C/60 sec.
Co-Si (3)	Co (55.6 nm) film, annealed 550 °C/60 sec then 690 °C/60 sec.
Co-Si (4)	Co (100 nm) film, annealed 525 °C/30 min, deposition at room temp.

RESULTS AND DISCUSSION

Figure 1 shows the Si $L_{3,2}$-edge TEY spectra for $CoSi_2$ and the Co-Si films as well as a

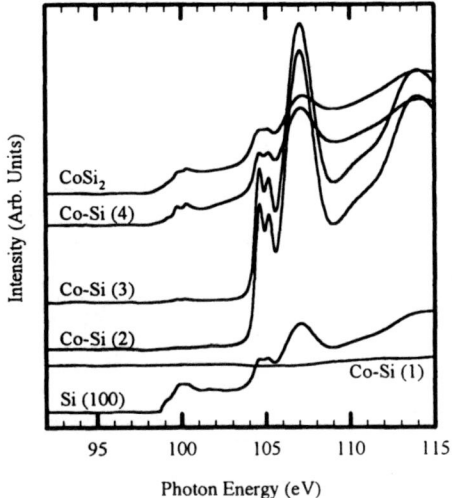

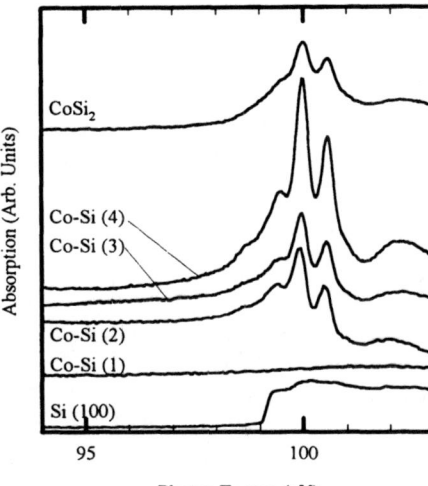

Figure 1. Si $L_{3,2}$-edge spectra taken in TEY mode. Spectra are shifted vertically for clarity.

Figure 2. Si $L_{3,2}$-edge spectra taken in FLY mode. Spectra are shifted vertically for clarity.

Si(100) crystal with the native oxide for comparison. All spectra have had a linear background removed.

The Si(100) spectrum clearly consists of an edge at ~100 eV which consists of a doublet with a separation of about 0.6 eV followed by a set of 3 peaks at ~ 104 eV which are attributed to SiO_2. The Si L-edge arises from *p-s* transitions. The spectra thus represent the densities of states above the Fermi level specifically of *s* character. The doublet is due to transitions to *s* like states starting from the $2p_{3/2}$ or $2p_{1/2}$ levels which have a separation of ~ 0.6 eV. It is clear that the high resolution (< 300 meV) at this edge and the large separation (~4 eV) between the crystalline Si edge and the oxide edge allows for a clear look at the Si edge without interference from the ambient oxide. The same is true for other silicides studied at the Si L-edge.

The spectra of the $CoSi_2$ bulk ingot also has two distinct edges, the first at ~ 99 eV is due to the transitions arising in the silicide, while that at ~ 104 eV is again the contribution of SiO_2 which was not removed in the scrapping procedure. The edge from the silicide also indicates the $2p_{3/2}$ - $2p_{1/2}$ spin orbit splitting between the two peaks visible in the spectra.

Now let us look at the TEY spectra of the films. The unannealed film, Co-Si (1), is a flat line showing no features at either the silicide edge or the oxide edge indicating that the Co film remains unreacted and that no silicon has migrated through the sample to the surface and that there has been no silicide formed at least on the surface of the film.

The spectra of the two DC sputtered films (Co-Si (2) and Co-Si (3)) that have been annealed however show no silicide edge but a clear SiO_2 edge indicating that the surface of the films is primarily SiO_2. The remaining film, Co-Si (4), also shows the presence of SiO_2 at the surface but this is clearly thinner (less than 5 nm) than that on the other films because we clearly see the presence of a silicide edge at 99 eV. This indicates that the MBE film is less sensitive to ambient oxidation than the other films.

Figure 2 shows the spectra for the same films and crystalline Si but taken in fluorescence yield mode and only showing the silicide region (the oxide absorption is significantly reduced in

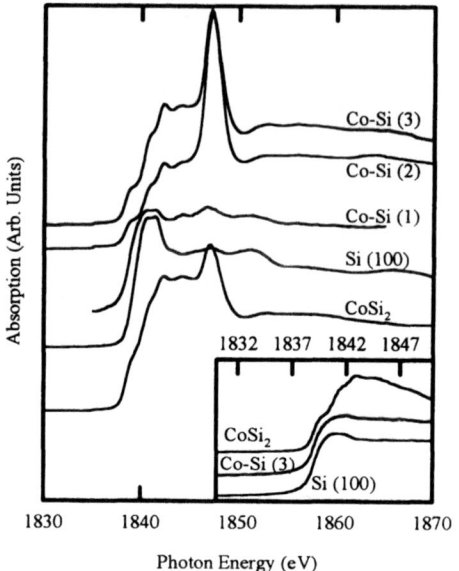

Figure 3. Si K-edge spectra taken in TEY mode. Inset shows spectra taken in FLY mode. Spectra are shifted vertically for clarity.

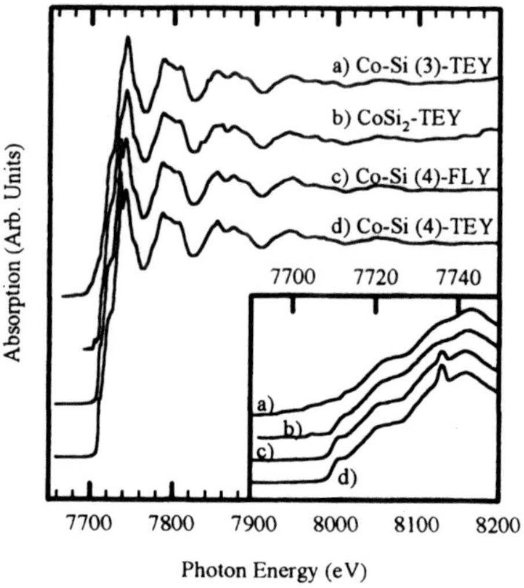

Figure 4. Co K-edge spectra for bulk CoSi$_2$ and two films. Spectra have been shifted vertically for clarity. Inset shows an expansion of the XANES region.

these spectra). As noted previously the sampling depth has increased thus we get a clearer look at the silicide spectra. The silicide spectra clearly consists of a shoulder followed by a spin orbit doublet. It is clear that all films consist of CoSi$_2$ as all the spectra are the same except for the unannealed film which still shows a flat spectrum here indicating that silicidation has not yet occurred. The spectra are all normalized to the photon flux thus the increase in the intensity of the film (4) spectra relative to the other spectra indicates again that the film (4) is not as sensitive to oxidation as the other films. This suggests that film (4) is the most uniform.

Figure 3 shows the Si K-edge TEY spectra for CoSi$_2$, the three DC sputtered films and Si(100). The Si K-edge spectra are more complex because as noted above the sampling depth of about a hundred nm means that the spectra consist of contributions from the native oxide, the silicide and the silicon substrate. Looking first at the unannealed film we see that it is similar to the spectrum for Si(100). However, three differences between the spectra are apparent; 1) the relative whiteline intensity is greatly reduced in the spectrum of the unannealed film, 2) there is a peak at ~1844 eV in the spectrum of the unannealed film and 3) the whiteline in the unannealed film is broader than in Si(100). Similar differences have been seen in the Ni-Si system [2] and indicate that some interaction occurs at the interface, most likely the formation of a Co rich silicide phase.

The spectra of the annealed films clearly resemble that of bulk CoSi$_2$ with a much larger peak at ~1847 eV. This peak is composed of a silicide peak and an SiO$_2$ peak [4] making it difficult to separate the contributions. The CoSi$_2$ spectra shown match previously published spectra and calculations [5] confirming that the bulk of the films is mainly CoSi$_2$ as found in the Si L$_{3,2}$-edge spectra.

The inset in figure 3 shows representative Si K-edge spectra taken in FLY mode. The sampling depth for FLY at the Si K-edge is several hundred nm [1]. With this sampling depth the contribution from the silicide and surface oxide are minimized and the spectra for all films resemble the Si(100) spectra.

Figure 4 shows the Co K-edge spectra of films (3) and (4) together with that of the CoSi$_2$ sample taken in TEY mode. The FLY spectrum for film (4) is also shown. All spectra were normalized to an edge jump of one. It is apparent that the EXAFS are identical in all the spectra shown indicating that the bulk of the films is CoSi$_2$, in agreement with the Si edge results. The Co K-edge spectra do not suffer from the complexities of the Si K-edge spectra because of the element specificity. The difference in the XANES region (see inset) of the film (4) spectra (a peak instead of a shoulder) may indicate a difference in the orientation of the sample to the detector or possibly, a difference in the crystallinity of the film as compared to the others.

CONCLUSIONS

We have reported some observations using multi-core multi-detection XANES spectra on a set of blanket CoSi$_2$ films. We found that all the films studied were composed of CoSi$_2$ except the unannealed film which was unreacted Co. The unannealed film gave an indication of some interfacial interaction between the Co and Si substrate. We clearly see that the sample grown by MBE was less sensitive to the surface oxidation in the ambient atmosphere than the other films.

The use of multi-core multi-detection techniques is a powerful means for non-destructive characterization of thin silicide films. Specifically, the Si L$_{3,2}$-edge is ideally suited for the study of thin silicide films (10 - 100 nm) since the sampling depth closely matches the film thickness.

ACKNOWLEDGMENTS

We wish to thank J. D. Garrett from the Materials preparation group at McMaster for the preparation of the bulk CoSi$_2$ ingot and M. Simara-Normandin of Nortel for providing the DC sputtered Co-Si films. Work at UWO is supported by an NSERC grant. SRC is supported under NSF award # DMR 95-31009.

REFERENCES

[1]   M. Kasrai, W. N. Lennard, R. W. Brunner, G. M. Bancroft, J. A. Bardwell and K. H. Tan, Appl. Surf. Phys. **99**, 303 (1996).

[2]   S. J. Naftel, I. Coulthard, T. K. Sham, D.-X. Xu, L. Erickson and S. R. Das, Thin Solid Films **308-309**, 580 (1997).

[3]   M. Saran, and A. Naem, J. Electrochemical Soc. **142**, 1688 (1995).

[4]   I. Coulthard, T. K. Sham, M. Simard-Normandin, M. Saran and J. D. Garrett, J. Phys. IV France 7, C2-1135 (1997).

[5]   P. Lerch, T. Jarlborg, V. Codazzi, G. Loupias and A. M. Flank, Phys. Rev. B **45**, 11481 (1992).

ELECTRONIC EFFECTS AT INTERFACES IN Cu   (Cr, Mo, W, Ta, Re)

MULTILAYERS

A. F. Bello, T. Van Buuren, J. E. Klepeis, T. W. Barbee, Jr.

Lawrence Livermore National Laboratory, Livermore, CA 94551

ABSTRACT

Interfacial electronic effects between Cu and the transition metals Cr, Mo, W, Ta, Re, are investigated by determining the strength of the white line absorption resonances on the $L_{3,2}$ edges of Cu in Cu_5/TM_5 multilayers. X-ray absorption (XAS) was performed to study the white lines, which are directly related to the unoccupied states of Cu in the multilayers. The metallic multilayers are 2 nm in period and 200 nm in total thickness. Each period contains 5 monolayers of Cu and 5 monolayers of the transition metal: $\approx$ 40% of the atoms are at interfaces. These material pairs form ideal structures for the investigation of interfacial electronic effects as they form no compounds and exhibit terminal solid solubility. Only weak white lines are observed on the $L_{3,2}$ edges of Cu since all the d-orbitals are filled. In the Cu/TM multilayers, however, we observed enhancement of the Cu white lines. We attribute this to the charge transfer from the "interfacial Cu atoms" d-orbital to the transition metal layers. Analysis of the white line enhancement enables calculation of the charge transfer calculation from the Cu to the transition metal. Cu shows a charge transfer of about 0.03 electrons/interfacial Cu atom in Cu/Cr, 0.064 in Cu/Mo, 0.35 in Cu/Ta, 0.17 in Cu/W , and 0.23 in Cu/Re. This charge transfer is consistent with the enhanced absorption energy of Cu on these materials as observed in thermal desorption experiments.

INTRODUCTION AND BACKGROUND

In this study we characterize electronic effects in short-period ($\approx$20 Å) metallic multilayer films in which 40% of the atoms are at an interface using near-edge ($L_{3,2}$) x-ray absorption. This study investigates Cu/TM where TM = Cr, Mo, W, Ta, Re. These immiscible elemental pairs are ideal to study as they form no compounds and exhibit terminal solid solubility. An interest in the charge transfer between elements in alloys and compounds has led to studies using x-ray absorption as described above. Near edge x-ray absorption fine structure (NEXAFS), a technique used for analyzing x-ray absorption near the absorption edge of the element, is especially suited to study the amount of unoccupied states in the conduction band of a metal. [1]

The d-metals spectra show large peaks at the absorption edges called "white lines." These are due to the unoccupied d-states just above the Fermi level in these metals. A study [2] of the white lines in the 3d metals show that as the d-band is increasingly occupied the white lines decrease in intensity. Starting with Ti ($3d^24s^2$), which has an almost empty d-band and shows strong white lines, the white-line intensities decrease across the Periodic Chart to Cu ($3d^{10}4s^1$), which has a full d-band and no white lines. Systematic measurement of the $L_{3,2}$ absorption spectra of bulk elemental Cu and Cu in the Cu/TM multilayers enabled measurement of the charge transfer.

NEXAFS on metallic multilayers has received less attention than alloys because of the difficulty in synthesizing multilayers with controllability up to the monolayer level and because there is little difference between the signal from the bulk and from longer period (> 30Å) multilayers [3]. For high-quality short period multilayers, however, the difference is clear. This is highlighted in a study of short period Co/Cu multilayers [4], where the electronic density of states of Cu in Cu/Co greatly differed from that of bulk Cu. The difference was attributed to both charge transfer and band structure changes of the interface atoms. Short period Cu/Fe was the subject of another NEXAFS study [4], where the signal from a periodic Cu(3 Å)/Fe(10 Å) multilayer was compared with that of a periodic Cu(10 Å)/Fe(3 Å) multilayer. The difference was attributed to the different structure of the Cu in each sample. Cu was BCC in one and was FCC in the other.

279

EXPERIMENT

Sample Preparation

All films in our experiments were fabricated by dc magnetron sputtering. Base pressures were typically 1×10^{-7} torr, argon gas was used to sputter at a pressure of about 2×10^{-4} torr, and the sputtering power was set to around 300-400KW. The set of samples, shown in Table I, are bilayers of Cu with another transition metal. In these samples, all layers were designed to have 5 monolayers (ML) each for a total period of about 21 Å. Monolayers were calculated using the bulk structure, listed in the table, of the constituent elements.

Table I. Bilayers of Cu With Other Transition Metals

Sample	bilayer thickness [Å]	bulk structure	Charge transferred from Cu atom (charge/interfacial atom)
Cu/Cu	10.435/10.435	fcc/fcc	
Cu/Cr	10.435/10.195	fcc/bcc	0.026
Cu/Mo	10.435/11.125	fcc/bcc	0.064
Cu/W	10.435/11.190	fcc/bcc	0.17
Cu/Ta	10.435/11.660	fcc/bcc	0.35
Cu/Re	10.435/11.145	fcc/hcp	0.23

Table I. Each layer is composed of 5 ML of the constituent elements or compounds. For example, 5 ML of fcc Cu in the [111] direction is 10.435 Å. The total thickness of each sample is about 2000 Å. The bulk structure for the constituent elements are listed for reference, but the film layers may not necessarily have that structure. The charge transfer between the interfacial atoms are also listed, and will be discussed in the later sections.

X-ray Absorption

Total Electron Yield (TEY) x-ray absorption experiments were conducted at Beamline 8-2 of the Stanford Synchrotron Radiation Laboratory (SSRL). In the range of 900 to 1100 eV, which is the region of interest for absorption in Cu, a resolution of 2000 was achieved. The experimental chamber vacuum was typically 1×10^{-9} torr. Samples were mounted so that the x rays were incident at 45°. A Au grid was placed in the incident x rays and its TEY signal was taken as proportional to the incident intensity. The TEY signal from both the sample and the Au grid were detected as current using a current amplifier. Data was acquired after annealing the sample to remove oxygen contamination on the surface. To anneal the sample, high current was passed through the film using Ta foil pressed against the outer edge of the sample. An optical pyrometer monitored the temperature, kept below 600 C to leave the film undamaged.

RESULTS AND DISCUSSION

X-ray Absorption of the Cu Standards

High-purity (99.999% pure) bulk Cu and a Cu_5/Cu_5 multilayers sputtered from a 99.999% Cu target were used as $L_{3,2}$ Cu edge standards for pure copper. TEY is surface sensitive because of the short emission depth of photoelectrons in solids, so any oxide contamination of a sample surface is can drastically change the near-edge spectrum. Cu $L_{3,2}$ edge results from the standards were compared to published Cu oxide spectra.[6] CuO has a resonance peak at 931.3 eV, which is just below the Cu L_3 edge of 932.5 eV, and Cu_2O has its absorption edge at 933.7 eV, which is just above the Cu L_3 edge. These experiments showed complete removal of the oxide phases from the bulk sample when it was annealed to a high temperature of roughly 700 C. The unannealed Cu_5/Cu_5 clearly showed CuO contamination — the shoulder just before the absorption peak in Figure 1, which disappeared after annealing to a temperature higher than 300 C. Figure 1 shows the data for Cu_5/Cu_5 before the anneal, after the anneal, and the annealed bulk Cu for comparison. An identical spectra from the high-temperature annealed bulk and the Cu_5/Cu_5 mean that the oxide in Cu/Cu has been completely removed at about 300 C.

Removal of the oxides is especially important since their signals are right at the L_3 Cu absorption edge, so their presence can lead to erroneous results in analyzing the near-edge absorption spectra. Figure 1 gives us confidence that on annealing the samples to 300C, the oxide phases vanish.

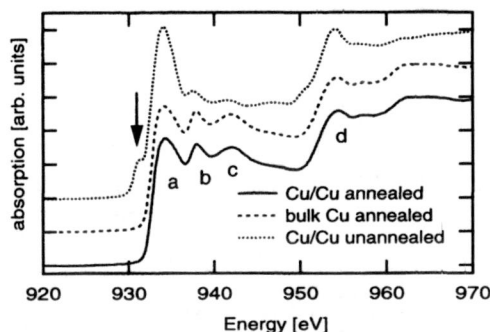

Figure 1. Absorption from the standards. The unannealed Cu5/Cu5 multilayer show oxide phases — the shoulder at the absorption edge as denoted by the arrow, and the large peak relative to the annealed sample L_3 peak (a). An identical spectra between the annealed Cu5/Cu5 and the annealed bulk Cu mean that the oxide from the multilayer has been completely removed at 300 C. The bulk sample was annealed to a high temperature of about 700 C to remove impurities, including oxides. The L_2 peak is labeled (d).

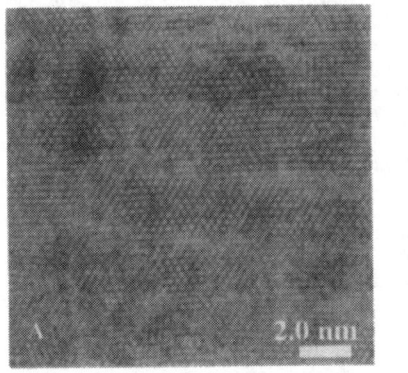 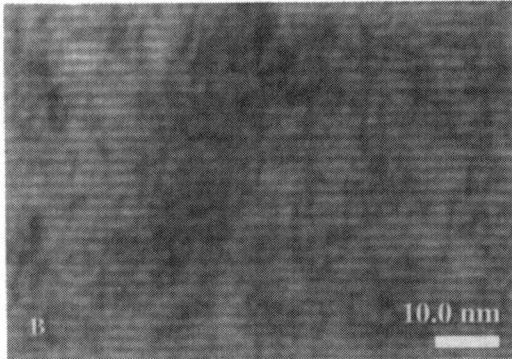

Figure 2. Copper molybdenum multilayer.

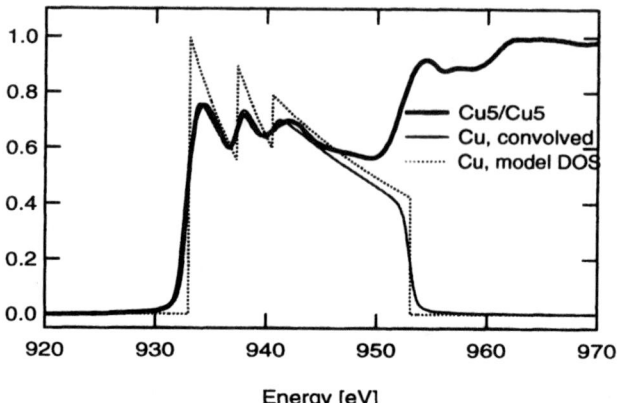

Figure 3. The Cu DOS is modeled by a set of exponentials, shown as the dotted curve. The model is convolved by the broadening function, shown by the thin curve, and a fit is made to the experiment data by optimizing the Cu DOS model.

Calculating the Charge Transfer from the White Lines

A method has been devised to calculate the charge transfer from the white lines. The basic idea is to fit the absorption spectrum by convolving a model of the unoccupied DOS with broadening functions. Detailed discussion will appear in another presentation [12], but a brief description is given here.

A complete model of the absorption spectra from the core level to the unoccupied states involve many factors. Ignoring some of these factors momentarily for a thought experiment, the absorption process is now examined. If there were no lifetime or instrumental broadening, no many-body (electrons) effects, and if the matrix elements were equal for all transitions, then the absorption spectra would equal the unoccupied DOS. A realistic model, however, must consider at least the largest of these effects. The lifetime and instrumental broadening are the two that are dominant in making the spectra different from the unoccupied DOS. The broadening can be described by the Voigt function [1], which is the convolution of a Gaussian (instrumental broadening) and a Lorentzian (lifetime broadening). To model the absorption, the Voigt function and the unoccupied DOS are convolved.

To calculate the charge transfer, the unoccupied states are modeled as a series of exponentials because of the similarity between the DOS features past the Fermi energy and exponential functions. Figure 3 shows the example for Cu, where the first peak simulates the tail of the 3d- band, and the following two peaks correspond to the higher DOS at the Brillouin zone boundaries (van Hove singularities). Also in the figure is the convolution of the Voigt function with the modeled DOS and the x-ray absorption data. The modeled unoccupied DOS is changed until a good fit between the absorption data and the convolved data is achieved. The similarity between the two curves in Figure 3 indicates that the modeled unoccupied DOS is a good approximation.

The same procedure is followed for all the Cu/X multilayers to obtain an approximation of the unoccupied states, with the results and experiment data for the series of samples shown in Figure 3. Again, the good fit between the convolved DOS and the experiment data suggests that the model for the DOS is a good approximation.

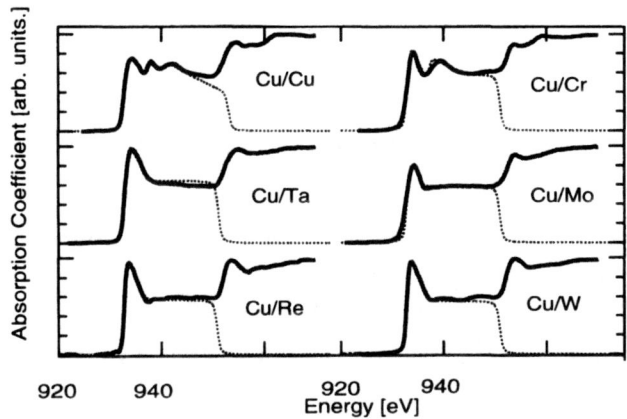

Figure 4. The absorption spectra and the fitted convolved data. Parameters for the modeled DOS are changed until there is a good fit of its convolution (with the broadening functions) to the x-ray absorption data.

Finally, the modeled DOS of the Cu/X multilayers are compared to the modeled DOS of bulk Cu. Since the number of unoccupied states in bulk Cu is known, then direct normalization gives the amount of unoccupied states of Cu in the Cu/X multilayers. Final results are in Table I. The charge transferred is presented as per interfacial atoms, a conclusion supported by DOS calculations for each monolayer[7]. Additionally, this allows these results to be related to interfaces in structures which contain bulk material and not limited dimension copper layers as used in this study.

SUMMARY

Charge transfer between Cu and the transition metals Cr, Mo, W, Ta, Re, were calculated by fitting the white line absorption resonances on the $L_{3,2}$ edges of Cu in multilayers to a phenomenological model. X-ray absorption (XAS) was performed to study the white lines, which are directly related to the unoccupied states of Cu in the multilayers. In the L-edge XAS of bulk Cu metal, only weak white lines are observed since all the d-orbitals are filled. In the L-edge XAS of Cu in the Cu/transition metal multilayers, however, enhanced Cu white lines are observed. This is attributed to charge transfer from the interfacial Cu d-orbital to the transition metal layers. Analysis of the white line enhancement enables calculation of the charge transfer calculation from the Cu to the transition metal. Cu in Cu/Cr shows a charge transfer of about 0.03 electrons/interfacial Cu atom, while Cu in Cu/Ta transfers about 0.35 electrons/interfacial Cu atom. Charge transfer in the remaining samples has values between these two.

ACKNOWLEDGMENTS

We would like to acknowledge L. Terminello for arranging the use of beam time at SSRL and ALS, and M. Wall for the TEM analysis. Technical support was provided by Jennifer Alameda and Don Hoffman.

This work was performed under the auspices of the U.S. Department of Energy (DOE) by Lawrence Livermore National Laboratory under Contract No. W-7405-Eng-48. TVB was supported by the Division of Materials Science, Office of Basic Energy Science, U.S. DOE. Part of this work was done at the ALS and at SSRL, which are supported by the DOE.

REFERENCES

[1] J. Stohr, *NEXAFS Spectroscopy*, Berlin: Springer-Verlag, 1992.

[2] D. H. Pearson, C. C. Ahn, B. Fultz, Phys. Rev. B. 47, 8471-8478 (1993).

[3] S. F. Cheng, A. N. Mansour, J. P. Teter, K. B. Hathaway, and L, T. Kabacoff, Phys. Rev. B 47, 206-216 (1993).

[4] A. Nilsson, J. Stohr, T. Wiell, M. Alden, P. Bennich, and N. Wassdahl, Phys. Rev. B. 54, 2917-2921 (1996).

[5] H. Ebert, J. Stohr, S. S. P. Parkin, M. Samant, and A. Nilsson, Phys. Rev. B. 53, 16067-16073 (1996).

[6] M. Grioni, J. F. Van Acker, M. T. Czyzyk, J. C. Fuggle, Phys. Rev. B. 45, 3309-3318 (1992).

[7] J. E. Klepeis, unpublished results (1997).

[8] M. Methfessel, Phys. Rev. B 38, 1537-1540 (1988).

[9] M. Methfessel, C. O. Rodriguez, and O. K. Andersen, Phys. Rev. B 40, 2009-2012 (1989).

[10] R. Hoffmann, Rev. Mod. Phys. 60, 601-628 (1988).

[11] A. K. McMahan, J. E. Klepeis, M. van Schilfgaarde, and M. Methfessel, Phys. Rev. B 50, 10742-10760 (1994).

[12] A. F. Bello, T. W. Barbee, Jr., unpublished results.

INTERFACIAL ELECTRONIC CHARGE TRANSFER AND DENSITY OF STATES IN SHORT PERIOD Cu/Cr MULTILAYERS

A. F. Bello, T. Van Buuren, J. E. Klepeis, T. W. Barbee, Jr.
Lawrence Livermore National Laboratory, Livermore, CA 94551.

ABSTRACT

Nanometer period metallic multilayers are ideal structures to investigate electronic phenomena at interfaces between metal films since interfacial atoms comprise a large atomic fraction of the samples. The multilayers studied were fabricated by magnetron sputtering and consist of bilayers from 1.9 nm to 3.3 nm. X-ray diffraction, cross-section TEM and plan-view TEM show the Cu layers to have a BCC structure Cu in contrast to its equilibrium FCC structure. The electronic structure of the Cu and the Cr layers in several samples of thin Cu/Cr multilayers were studied using x-ray absorption spectroscopy (XAS). Total electron yield was measured and used to study the white lines at the Cu L_2 and L_3 absorption edges. The white lines at the Cu absorption edges are strongly related to the unoccupied d-orbitals and are used to calculate the amount of charge transfer between the Cr and Cu atoms in interfaces. Analysis of the Cu white lines show a charge transfer of 0.026 electrons/interfacial Cu atom to the interfacial Cr atoms. In the Cu XAS spectra we also observe a van Hove singularity between the L_2 and L_3 absorption edges as expected from the structural analysis. The absorption spectra are compared to partial density of states obtained from a full-potential linear muffin-tin orbital calculation. The calculations confirm the presence of charge transfer and indicate that it is localized to the first two interfacial layers in both Cu and Cr.

INTRODUCTION AND BACKGROUND

In this work we have examined the electronic properties of Cu and Cr in Cu/Cr multilayers using x-ray absorption spectroscopy, with emphasis on the Cu/Cr interfaces. We characterized the structure of these short-period Cu/Cr multilayers using x-ray diffraction and transmission electron microscopy (TEM). The Cu/Cr binary pair is especially suited to study the interfaces in metals since these elements are mutually insoluble, thus eliminating mixing effects and compound formation and the lattice mismatch is very small. This allows the fabrication of high structural quality Cu/Cr multilayers that have a structure which can be approximated in calculations based on idealized atomic arrangements.

EXPERIMENT

Sample Preparation

All the Cu/Cr multilayer films studied were fabricated by DC magnetron sputtering. Base pressures were typically 10^{-7} torr; argon sputter gas at a pressure of about 2×10^{-4} torr was used. In these samples all layers were designed to have an integer number of monolayers (ML), where the thickness of a monolayer was calculated using bulk structures (FCC and BCC). Expected growth directions ([111]-Cu and [110]-Cr) of the constituent elements were also assumed. Table I presents the samples prepared for this study and their structural characteristics.

X-ray Diffraction and TEM Characterization

High and low angle θ--2θ x-ray diffraction scans were performed on all samples: high angle scans determine the interplanar spacings of the multilayer/constituent layers, while low angle scans characterize the multilayer period. Cu $K\alpha_{1,2}$ which has a average wavelength of 1.5418 Å was used. Cross-section TEM images of the sample Cu_5/Cr_5 (5 ML Cu and 5ML Cr) are shown in Figure 1. The uniformity of the layering is clearly shown in the lower magnification image reflecting sample quality. The higher magnification lattice image shows that a high degree of atomic order is present in this sample and that differentiation of the Cu and Cr layers is not possible. Additionally, the electron diffraction from this sample demonstrated that it is wholly BCC with a (110) interplanar spacing essentially equal to that of Cr as expected.

Mat. Res. Soc. Symp. Proc. Vol. 524 © 1998 Materials Research Society

Table I. Cu/Cr Bilayers of Varying Periods

Bilayer Thickness of Cu/Cr [ML]	Bilayer Thickness of Cu/Cr [Å]	Bilayer Period [Å]	% Cu in bilayer
		Cr (110)	0
3/6	6.25/12.89	19.14	32.6
6/8	12.53/16.31	28.84	43.4
3/3	6.25/6.44	12.69	49.2
6/6	12.53/12.89	25.42	49.3
8/8	16.66/16.31	32.97	50.5
5/5	10.43/10.19	20.63	50.57
8/6	16.66/12.89	29.55	56.4
6/3	12.53/6.44	18.97	66.1
		Cu (111)	100

Table I. Set of Cu/Cr samples studied. The samples were designed with the layer thicknesses shown and fabricated using independently measured Cu and Cr deposition rates.

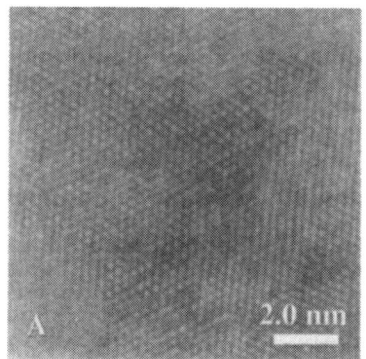 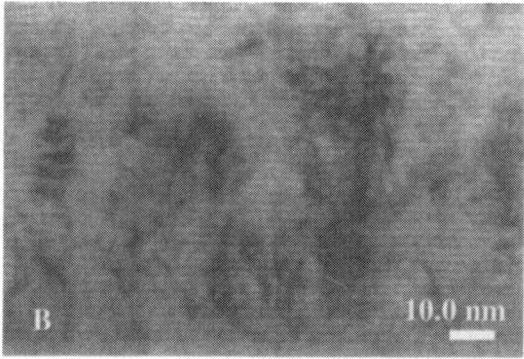

Figure 1. A) Cross-sectional high resolution lattice image of a Cu/Cr multilayer. B) Cross-sectional bright field image showing well defined layering of Cu and Cr in a columnar structure.

X-ray Absorption

Total Electron Yield (TEY) x-ray absorption experiments were conducted at Beamline 8-2 of the Stanford Synchrotron Radiation Laboratory (SSRL). Data was acquired after annealing the sample at progressively higher temperatures in order to remove oxygen contamination from the sample surface. An optical pyrometer monitored the annealing temperature which was kept below 600 C. Details of the experiment setup are presented in another paper[1] presented in this Symposia.

RESULTS AND DISCUSSION

Figure 2 shows θ-2θ x-ray diffraction scans from the samples with the highest and lowest fractional (by layer thickness) content of Cu. The Cr-rich sample, Cu_3/Cr_6, show significantly shorter plane spacing than the Cu-rich Cu_6/Cr_3 sample, which is the expected behavior since bulk Cr (110) has a plane spacings of 2.039 Å and Cu (111), 2.087 Å.

Figure 3 plots the plane-spacing calculated from the high-angle scans. The plane spacing for bulk Cu and Cr are also shown for reference. Samples with 50% fractional content in monolayers (Cu_3/Cr_3, Cu_6/Cr_6, Cu_8/Cr_8) favor the Cr bulk value. The full-width-half-max

(FWHM) of the diffraction peaks were measured and the Scherrer equation[2] applied to determine the microstructure grain size in the multilayers. Results show that the Cr-rich sample has a considerably larger grain size than the Cu-rich samples. Samples with 50% Cu vary in grain size, indicating that the number of interfaces, and not just the layer thickness fraction, influences the result.

The Cu layers in these thin period multilayers, especially the Cr-rich samples, are BCC because the high-angle scans support one peak, meaning one structure throughout the multilayer. The same behavior was observed[3] in co-sputtered films of Cu-Cr.

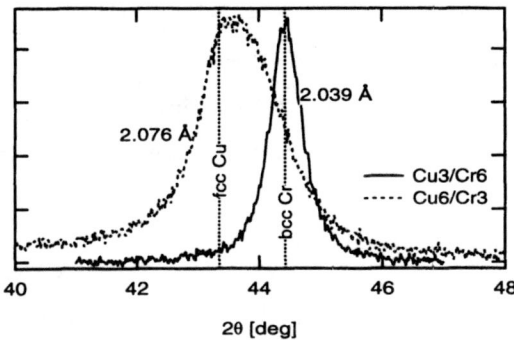

Figure 2. High-angle scans for two samples. The sample with 6 ML of Cr and 3 ML of Cu has a shorter lattice parameter of 2.039 Å than the sample with less Cu content, which show 2.076 Å.

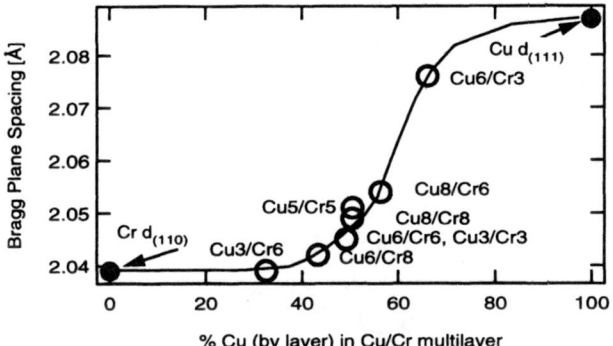

Figure 3. Plane spacing as a function of the layer fraction of Cu in a period. The lattice parameter changes gradually, as opposed to a sharp step. The lattice parameter for bulk Cu and Cr are taken from literature. The freely drawn line is used as a guide.

A major complicating factor in x-ray absorption is oxide contamination, which occurs when samples are in ambient atmosphere. Here, scans were taken after progressively higher temperature anneals to determine just when the oxide signals[4] disappeared. For Cu_5/Cr_5, the data taken at 300 C and higher temperature anneals removed the oxide signatures. At 300 C, features not directly related to the oxide were preserved. Higher temperature anneals distorted these features. Figure 4 shows how the Cu spectrum evolves upon annealing, starting with a slight shoulder due to the CuO and a relatively large absorption peak due to the Cu_2O. There is also one broad peak between the L_3 and L_2 absorption edges, indicative of a BCC symmetry point, where the DOS increases (van Hove singularity). The spectrum change is continued by the disappearance of the shoulder and a slight decrease in intensity of the absorption peak at about 300 C. Higher temperatures show the spectrum gradually becoming more like that for annealed Cu_5/Cu_5 or bulk Cu, and at the final annealing temperature of 500 C, the spectrum clearly shows the three peaks in the L_3 edge that is the distinct signature for FCC Cu.

On the DOS calculations, two methods were used.[5] The first is the full-potential linear muffin-tin orbital (FP) method[6,7], which employs a localized basis that provides a convenient framework for projecting out individual atom- and angular momentum-resolved partial densities of states (DOS). The partial DOS were calculated using a Mulliken decomposition as described by Hoffman[8] and McMahan[9]. The FP method makes no shape approximation to the crystal potential, whereas methods based on the atomic sphere approximation (ASA) assume that the potential is spherically symmetric about each atom. The second method, although less accurate, is the atomic sphere approximation, linear muffin-tin potential method, which works reasonably well for close-packed systems and provides an alternate means of obtaining the partial DOS. Therefore, ASA calculations were carried out to cross-check the FP results. The ASA method is the one used by Ebert, et. al.[10] in their study of $L_{2,3}$ x-ray absorption of Cu multilayers.

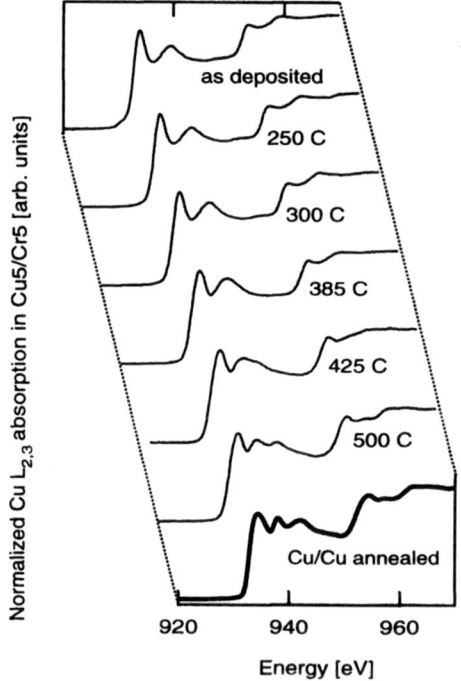

Figure 4. Cu spectra from Cu_5/Cr_5 as it is annealed progressively to higher temperatures. For clarity, the axes are shifted for each spectrum. For comparison, the annealed Cu_5/Cr_5 is shown at the bottom.

Integrating the DOS up to the Fermi energy gives 11 electrons, but integrating just the d-band DOS gives 9.89 electrons instead of the 10 electrons normally associated with the full d-band of Cu. This amount of unoccupied d-states make is possible to observe the white lines in the near-edge x-ray absorption of Cu.

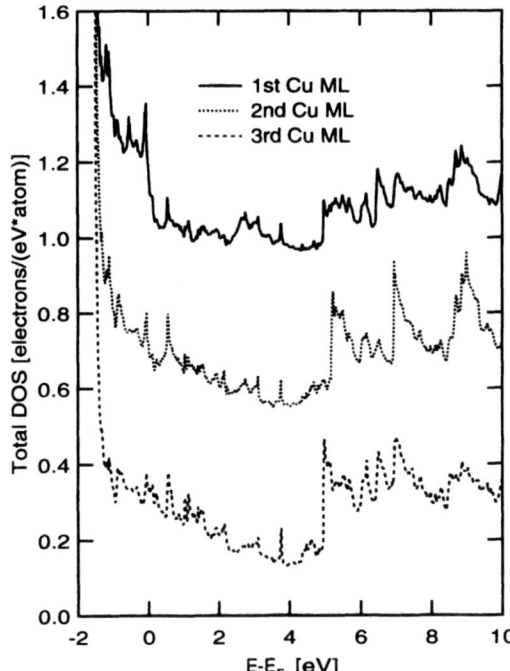

Figure 5. Total DOS of the individual Cu monolayers (plots are offset by 0.4 and 0.8 in the vertical axis for clarity.).

In addition to bulk Cu, the total DOS of the multilayer Cu_5/Cr_5 was calculated using the FP method, where the 5 Cr monolayers were taken to have the same structure and lattice parameter as bulk Cr (BCC, a=2.884 Å). The 5 Cu monolayers, however, were taken as BCC with a lattice parameter of 2.829 Å, leading to a 2% lattice mismatch at the interface. The x-ray diffraction data of Cu_5/Cr_5 other unpublished results[11] point to a BCC Cu structure and the specific lattice parameter value.

From the calculations, it is possible to separate the total DOS of individual monolayers in the multilayer in addition to computing the total DOS for the whole multilayer. Shown in Figure 5 are the total DOS for each of the Cu monolayers in the multilayer — the first Cu monolayer is the one right next to Cr, and the second Cu monolayer is the second closest monolayer to the Cr. Integrating the total DOS of the first Cu monolayer gives 10.85 electrons/atom, which is less than the full 11 electrons/atom expected for bulk Cu. The second Cu monolayer has 10.89 electrons/atom, which is also less than for bulk Cu, but the third Cu monolayer has 11 electrons/atom, which is identical to the value for bulk Cu. The same calculations are done for the Cr monolayers, and the results show that the Cr monolayer next to Cu has 6.17 electrons/atom, the next closest has 6.07 electrons/atom, and the third closest has 6.01 electrons/atom, which is identical to the 6 electrons/atom expected for bulk Cr.

The change in charge count relative to the bulk is greatest for the monolayers right at the interface, and the evidence point to charge transfer flow from Cu to Cr — the first Cu monolayer loses 0.15 electrons/atom, while the first Cr monolayer gains 0.17 electrons/atom. The second monolayers also show a net charge transfer — Cu gains 0.11 electrons/atom, and Cr loses 0.07 electrons/atom.

Figure 5 shows that in the first Cu monolayer, there is a sharp rise in the DOS at the Fermi energy. This is possibly due to electron states that characterize a bond introduced at the interface between Cu and Cr, and is where some of the charge transfer occurs. It is reasonable to expect that when there is charge transfer between Cu and Cr at an interface, the transferred charge gets localized close to the Fermi level since it is at the outermost electron levels where the interaction takes place. The same type of feature in the DOS of the first Cr monolayer is observed and is even more pronounced. In the region of the Fermi energy there is a sharp peak, which decreases for the second Cr layer, and finally disappears for the third Cr layer. The progressive weakening of the peak as the Cr monolayer gets farther from Cu indicates that it is associated with the interface. The figure also shows that in the unoccupied states from the Fermi level to about 2 eV above it, the general shape of the DOS remains the same. As previously discussed, the first Cu ML has a shoulder right below the Fermi energy. But with a typical average difference (absolute value, point-to-point) of 0.03, the curves are similar for all three monolayers in that energy region. In another study[1], we describe and use a technique to calculate the charge transfer from the x-ray absorption. The method uses the fact that the general shape of the DOS from the Fermi energy to about 2 eV above it is similar for the bulk and the monolayers. There, the calculated charge transfer was 0.026 electrons per interfacial atom.

SUMMARY

In unannealed samples of Cu/Cr of varying period and Cu composition, the x-ray diffraction scans show one reflection peak, indicating one structure throughout the multilayer. In the Cr rich samples, the BCC Cu has a lattice parameter nearly equal to that of Cr. In the Cu_5/Cr_5 sample, annealing to 500 C decomposes the multilayers into bulk-like Cu and Cr. TEM shows well-defined layering of Cu and Cr in a columnar structure.

Analysis of the XAS data show a charge transfer of about 0.03 electrons per interfacial atom, which we believe provides the enhanced bonding at the interface and promotes wetting of Cr by Cu.

ACKNOWLEDGMENTS

We would like to acknowledge L. Terminello for arranging the use of beam time at SSRL and ALS, and M. Wall for the TEM analysis. Technical support was provided by Jennifer Alameda and Don Hoffman.

This work was performed under the auspices of the U.S. Department of Energy (DOE) by Lawrence Livermore National Laboratory under Contract No. W-7405-Eng-48. TVB was supported by the Division of Materials Science, Office of Basic Energy Science, U.S. DOE. Part of this work was done at the ALS and at SSRL, which are supported by the DOE.

REFERENCES

[1] A. F. Bello, T. Van Buuren, T. W. Barbee, Jr., Conf. MRS Spring Meeting (1998).
[2] A. Guinier, *X-Ray Diffraction In Crystals, Imperfect Crystals, and Amorphous Bodies*, San Francisco: W. H. Freeman and Co., 1963.
[3] A. P. Payne and B. M. Clemens, J. Mat. Res. 7, 1370-1376 (1992).
[4] M. Grioni, J. F. Van Acker, M. T. Czyzyk, J. C. Fuggle, Phys. Rev. B. 45, 3309-3318 (1992).
[5] J. E. Klepeis, unpublished results (1997).
[6] M. Methfessel, Phys. Rev. B 38, 1537-1540 (1988).
[7] M. Methfessel, C. O. Rodriguez, and O. K. Andersen, Phys. Rev. B 40, 2009-2012 (1989).
[8] R. Hoffmann, Rev. Mod. Phys. 60, 601-628 (1988).
[9] A. K. McMahan, J. E. Klepeis, M. van Schilfgaarde, and M. Methfessel, Phys. Rev. B 50, 10742-10760 (1994).
[10] H. Ebert, J. Stohr, S. S. P. Parkin, M. Samant, and A. Nilsson, Phys. Rev. B. 53, 16067-16073 (1996).
[11] T. W. Barbee, Jr., unpublished results (1997).

SYNCHROTRON X-RAY ABSORPTION SPECTROSCOPY STUDIES OF Pt/Si SYSTEMS

I. Coulthard[a], S.J. Naftel[a], T.K. Sham[a]
[a]Department of Chemistry, University of Western Ontario, London, Canada.

ABSTRACT

Platinum was deposited onto porous silicon by a reductive deposition technique utilizing the inherent reducing power of porous silicon. The resulting deposits were studied by X-ray Absorption Near Edge Structure (XANES) at the Si-K, Pt-$M_{3,2}$, and Pt-$L_{3,2}$ edges. Samples of varying deposition concentrations were studied and were compared with untreated porous silicon and platinum silicides to determine the nature of the deposits and their effect upon the porous silicon substrate itself.

INTRODUCTION

The ability of porous silicon, a material created by the electrochemical etching of crystalline silicon, to produce intense visible luminescence at room temperature [1], is completely in contrast to its parent material crystalline silicon which is an indirect band gap semiconductor. For this reason, porous silicon has been investigated as a potential material for optoelectronic devices [2]. Thin layers of metal deposited onto the surface of porous silicon by methods such as chemical vapour deposition (CVD) [3] and electroplating [4] have been shown to chemically and physically stabilize the porous silicon layer making it more potentially suitable for these applications.

Another lesser known property of porous silicon is its ability to act as a reducing agent [5]. Any system with a positive half reduction potential with respect to hydrogen may be reduced by porous silicon. As a result, several noble metals such as Au, Cu, Ag, Pd, Rh, and Pt may be readily reduced from solution by porous silicon resulting in the deposition of the metal onto the surface of the porous silicon. Reductive deposition may provide an alternative method for producing the metal-porous silicon composites for optoelectronic applications. The metal deposits themselves may also be of interest. The nucleation and growth of metal deposits involved in the reductive deposition may produce metal deposits of interesting dimensions (i.e. nanoparticles) which could have electronic properties different from that of the bulk metals [6]. It may also be possible to encourage the formation of nanoscale metal silicides.

In this study, Pt was deposited onto porous silicon using reductive deposition. Deposition concentrations were varied to study the effect upon the deposition. The effect of the deposition upon the porous silicon substrate and the nature of the Pt deposits themselves were also studied.

Mat. Res. Soc. Symp. Proc. Vol. 524 © 1998 Materials Research Society

EXPERIMENTAL

Porous silicon films were prepared utilizing a method described previously [5]. Utilizing a p-type (B doped) Si (100) wafer with a resistivity of 1-10 Ohm.cm, a current density of 20 mA/cm^2, and an anodization time of 20 minutes, a typical porous silicon layer was produced. This layer approximately 10-15 microns thick is reddish brown in colour, produces an orange-red luminescence, and readily oxidizes in ambient conditions.

Pt was reductively deposited utilizing solutions of K$_2$[PtCl$_4$] in a general procedure described elsewhere [7]. Solution concentration was varied from 10^{-3} to 10^{-4} M. Deposition time was typically 2 minutes and was accomplished either by immersion of the porous silicon in solution or placement of a surface tension bubble of solution on the top of the porous silicon layer. The formation of bubbles of hydrogen during the reductive deposition is readily observable.

The Si-K edge and Pt-M$_{3,2}$ edge spectra were recorded on the Canadian Double Crystal Monochromator Beamline [8] at the Canadian Synchrotron Radiation Facility (CSRF) at the Synchrotron Radiation Centre (SRC) at the University of Wisconsin-Madison and were recorded in total electron yield (TEY) mode using sample current. Pt-L$_{3,2}$ edge spectra were recorded at beamline X11-A at the National Synchrotron Light Source (NSLS) 2.5 GeV ring at Brookhaven National Laboratories in TEY mode using a He amplified TEY detector.

Scanning electron microscope (SEM) images were recorded with a Hitachi S-4500 field emission SEM found at Surface Science Western. Platinum silicides were prepared by the Materials Preparation Group at McMaster University.

RESULTS AND DISCUSSION

Figure 1 shows the Si-K edge TEY spectra for : fresh porous silicon, ambient oxidized porous silicon, platinum deposited onto porous silicon (Pt/Si), and two platinum silicides. It can clearly be seen that porous silicon readily oxidizes in ambient conditions as seen by the large oxide peak at 1847 eV. The analysis is complicated however by the fact that silicides in general display a peak at 1847 eV resulting from silicide and surface oxide. This complicates the analysis of the Pt / Si spectrum. The spectra of Pt / Si and Pt$_2$Si are clearly very similar both showing an additional shoulder at 1845 eV. However this shoulder in the Pt / Si spectrum could be attributed to silicon suboxide which results from the partial chemical capping of the porous silicon layer by the platinum deposits. Other edges must be used to resolve this complicated situation.

Figure 2 shows the Pt-M$_3$ edge for two Pt / Si samples, the platinum silicides, and Pt metal. As expected the Pt-L$_3$ results are identical. These spectra clearly show that the Pt deposited onto porous silicon is in fact Pt metal and not a silicide. Referring back to Figure 1, this means that the peaks at 1845 eV and 1847 eV in the Pt / Si spectra are due to silicon oxides. The fact that the peak at 1847 eV lessens in intensity with respect to ambient oxidized porous silicon and the presence of a suboxide peak suggests that the Pt deposits in fact do to some extent chemically cap the porous silicon layer resulting in reduced exposure to ambient conditions. Complete capping of the layer may require vastly increased deposition times or may be

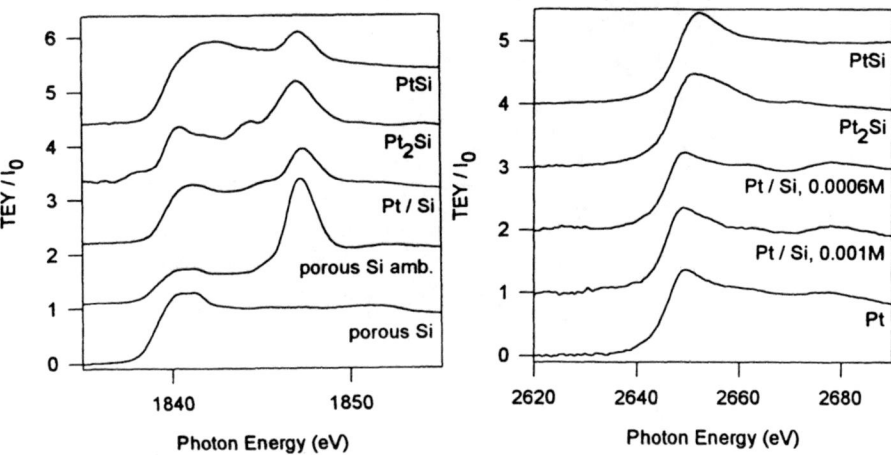

Figure 1   Si-K edge XANES comparison of porous silicon, platinum silicides and platinum deposited porous silicon.

Figure 2   Pt-M₃ edge XANES comparison of platinum silicides, Pt, and Pt / Si samples prepared at differing concentrations.

hindered by the adhesion of hydrogen bubbles to the layer surface during the reductive deposition and/or trapped air within the porous silicon layer.

Although Figure 2 clearly shows that the deposits on the porous silicon are in fact Pt metal, it also shows that the white line intensity of the spectra is greater than that of Pt metal. The white line is a measure of the unoccupied density of d states [9]. An increase in the white line intensity indicates an increase in the density of unoccupied d states. The cause for this increase in white line intensity is elucidated when the morphology of the Pt deposits on porous silicon are studied utilizing SEM. Figure 3 shows an SEM image of Pt deposits on porous silicon. It can clearly be seen that nanoclusters of platinum are found on the surface of the porous silicon. The nanoclusters are aggregates of smaller clusters on the order of 5-10 nm in diameter. The change in white line intensity results from the small size of the Pt nanoparticles. The vast increase in the number of surface and near surface atoms relative to the bulk results in the rehybridization at the surface atoms which results in a change in the density of unoccupied d states. It is expected that the smaller the cluster the smaller the average coordination number, the larger the 5d hole population since the atomic configuration is typically $6s^2 5d^8$ for the free atom and $6s^1 5d^9$ for bulk metal. This is consistent with the spectra in Figure 2 which shows a greater white line intensity when lesser concentrations of $[PtCl_4]^{2-}$ are utilized in the reductive deposition. It would be expected that lesser concentrations would result in a slower rate of reductive deposition which would result in smaller nanoparticles with a greater number of surface and near surface atoms. The effect of particle size upon the XANES spectra is even more clearly evident in the Pt-L₂ spectra of Pt / Si samples as seen in Figure 4. Pt metal displays no white line at the L₂ or M₂ edge [9]. However it can clearly be seen that when nanoparticles of platinum on porous silicon are studied that these nanoparticles do in fact display white lines at the L₂ edge unlike the bulk material. It is clear that the $d_{3/2}$

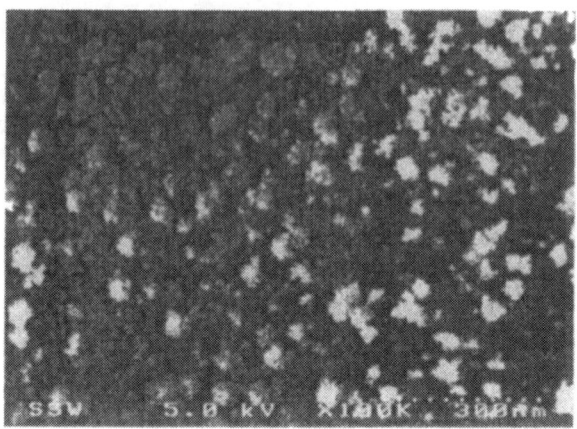

Figure 3   SEM image of Pt clusters
reductively deposited onto porous silicon.

state which was fully occupied in Pt metal resulting in the lack of white line in the L_2 edge for bulk Pt has become unoccupied in small nanoparticles of Pt.

Figure 5 shows the Pt-M_3 edge spectra for Pt metal, the two platinum silicides, and a Pt / Si sample which has been rapidly thermal annealed. It was hoped that the annealing process would result in some distinct change in the nature of the Pt deposits on porous silicon and perhaps would induce the formation of nanoscale platinum silicides. It can clearly be seen however that platinum silicides of any type were not produced and that the platinum deposits on porous silicon still remain as platinum metal. The only observable change due to the annealing is the fact that the Pt / Si sample no longer displays a greater white line intensity than the bulk Pt. Visual observation also reveals that after annealing the Pt deposits appear as a more uniform metallic layer where in the non-annealed samples, no metallic layer was observed. It would appear that the annealing has allowed the nanoparticles of platinum to diffuse on the surface and form a uniform layer of platinum on the surface of the porous silicon of a thickness such that it behaves in a identical fashion electronically to bulk Pt.

CONCLUSIONS

Pt can be readily deposited onto the surface of porous silicon utilizing a reductive deposition method. The resulting deposits of platinum are found to be nanoclusters of Pt which due to their dimensions exhibit an electronic structure (increasing $d_{5/2}$, $d_{3/2}$ hole population) different from bulk Pt. The deposits of platinum on the surface of the porous silicon also serve to partially chemically cap the porous silicon layer and reduce its exposure to the ambient conditions in which a porous silicon layer will readily oxidize. More detailed study will be

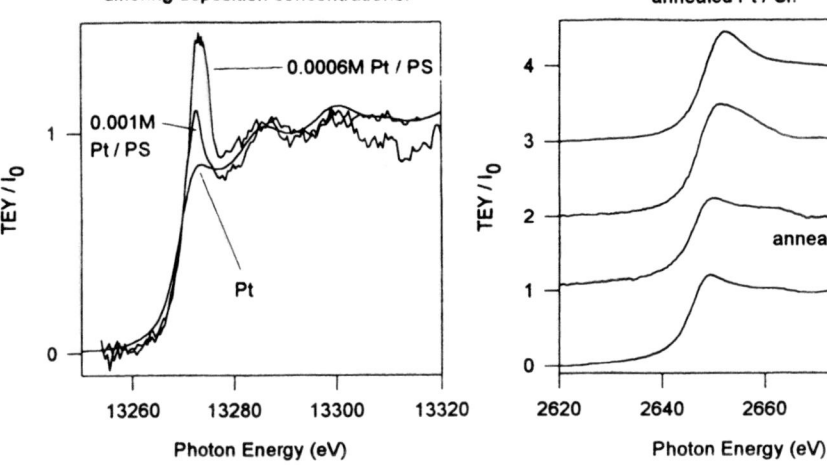

Figure 4   Pt-L_2 edge XANES comparison of Pt and Pt / Si samples prepared at differing deposition concentrations.

Figure 5   Pt-M_3 edge XANES comparison of Pt, platinum silicides, and thermally annealed Pt / Si.

required to produce a more complete chemical capping of a porous silicon layer using this method. During the experiments no indication of any platinum silicide formation was observed however rapid thermal annealing of a Pt / Si sample resulted in the production of a thin, uniform, metallic layer of Pt. Both the deposited nanoparticles themselves and the ability of the deposits to cap the porous silicon layer warrant further study using Pt and other suitable noble metals.

ACKNOWLEDGEMENTS

Research at the University of Western Ontario and CSRF is supported by the Natural Sciences and Engineering Research Council of Canada (NSERC). We thank S.R. Das of the National Research Council of Canada for the rapid thermal annealing of the Pt / Si sample.

REFERENCES

[1]   L.T. Canham, *Appl. Phys. Lett.* **57**, p. 1046 (1990).

[2]   F. Namavar, H.P. Marusha, and N.M. Kalkhoran, *Appl. Phys. Lett.* **60**, p. 2514 (1992).

[3]   A. Halimaoui, Y. Campidelli, P.A. Badoz, and D. Bensahel, *J. Appl. Phys.* **78**, p. 3428 (1995).

[4]   F. Ronkel, J.W. Schultze, and R. Arens-Fischer, *Thin Sol. Films* **276**, p. 40 (1996).

[5]   I. Coulthard, D.T. Jiang, J.W. Lorimer, and T.K. Sham, *Langmuir* **9**, p. 3441 (1993).

[6]   H. Gleiter, *Prog. Mater Sci.* **33(4)**, p. 223 (1989).

[7]   I. Coulthard, and T.K. Sham, *Appl. Surf. Sci.* (in press)

[8]   B.X. Yang, F.H. Middleton, B.G. Olsson, G.M. Bancroft, J.M. Chen, T.K. Sham, K. Tan, and D.J. Wallace, *Rev. Sci. Instrum.* **63**, p. 1355 (1992).

[9]   N.F. Mott, *Proc. Phys. Soc. London* **A62**, p. 416 (1949).

DIRECT CORRELATION OF SOLAR CELL PERFORMANCE WITH METAL IMPURITY DISTRIBUTIONS IN POLYCRYSTALLINE SILICON USING SYNCHROTRON-BASED X-RAY ANALYSIS

S.A. McHugo[1], A.C. Thompson[1], G. Lamble[1], A. MacDowell[1], R. Celestre[1], H. Padmore[1], M. Imaizumi[2], M. Yamaguchi[2], I. Perichaud[3], S. Martinuzzi[3], M. Werner[4], M. Rinio[5], H.J. Moller[5], B. Sopori[6], H. Hieslmair[7], C. Flink[7], A. Istratov[7], E.R. Weber[7]

[1]Lawrence Berkeley National Laboratory, Berkeley, CA, USA, samchugo@lbl.gov
[2]Toyota Technological Institute, Nagoya, JAPAN
[3]Lab. de Photoélectricité des Semi-Conducteurs, University of Marseille, FRANCE
[4]Max-Planck-Institute für Mikrostructur Physik, Halle, GERMANY
[5]University of Freiberg, GERMANY
[6]National Renewable Energy Laboratory, Golden, CO, USA
[7]University of California at Berkeley, Department of Materials Science, CA, USA

ABSTRACT

The work presented here directly measures metal impurity distributions and their chemical state in as-grown and fully processed polycrystalline silicon used for terrestrial-based solar cells. The goal was to determine if a correlation exists between poorly performing regions of solar cells and metal impurity distributions as well as to ascertain the chemical state of the impurities. Synchrotron-based x-ray fluorescence mapping and x-ray absorption spectroscopy, both with a spatial resolution of !μm, were used to measure impurity distributions and chemical state, respectively, in poorly performing regions of polycrystalline silicon. The Light Beam Induced Current method was used to measure minority carrier recombination in the material in order to identify poor performance regions. We have detected iron, chromium, nickel, gold and copper impurity precipitates and we have recognized a direct correlation between impurity distributions and poor performing regions in both as-grown and fully processed material. Furthermore, from x-ray absorption studies, we have initial results, indicating that the Fe in this material is in oxide form, not $FeSi_2$. These results provide a fundamental understanding into the efficiency-limiting factors of polycrystalline silicon solar cells as well as yielding insight for methods of solar cell improvement.

INTRODUCTION

Polycrystalline silicon can be used to fabricate solar cells with a moderate solar conversion efficiency and low fabrication costs. Although these cells are presently manufactured for terrestrial-based applications, an improvement in the efficiency of these cells would greatly increase their commercial viability. Presently, polycrystalline solar cells have efficiencies of 13-15% as compared to more expensive, single crystalline solar cells with efficiencies of 17-20% [1]. The primary cause for lowered efficiencies is localized regions of high minority carrier recombination, which possess high concentrations of dislocations [2-4]. However, it is known that minority carrier recombination at "clean" dislocations is relatively weak but greatly increases by decoration or precipitation of transition metal impurities [5-8]. This suggests dislocations in high recombination regions of polycrystalline silicon are decorated with transition metals. Past work, [9], has provided indirect evidence that metal impurities are precipitated in regions of high carrier recombination while other work, [10], has revealed metal impurity agglomerations at dislocations in polycrystalline silicon. However, no direct evidence has been provided to relate high minority carrier recombination with transition metals in this material. IN fact, carbon or oxygen may play an important role since these impurities are found in high concentrations in most polycrystalline silicon, $\approx 10^{18}$ atoms/cm^3.

This research seeks to determine whether a correlation between impurity distributions and regions of high minority carrier recombination exists in polycrystalline silicon solar cells. Additionally, we have attempted to ascertain the chemical state of impurities in the silicon, which would be the first identification of its kind in this material and could have serious implications in regards to strategies for material improvement.

EXPERIMENT

Boron doped polycrystalline silicon grown by an electromagnetic casting method [11] or the rapid growth on substrate (RGS) method, [12], were used in this study. Minority carrier recombination was mapped across the as-grown material with the Light Beam Induced Current (LBIC) method, using 880nm wavelength light. Aluminum layers were evaporated onto the frontside and backside of the sample, respectively, in order to provide electrical contacts for LBIC measurements. Prior to evaporations, the material was cleaned with a RCA cleaning process and silicon etching. The frontside of the samples were analyzed using synchrotron-based x-ray fluorescence (XRF) mapping in order to determine metal impurity content and distribution. Prior to XRF measurements, the Al layers were removed with a VLSI grade HCl and HNO$_3$ chemical etch and the surface of the material was cleaned with VLSI grade piranha (H$_2$SO$_4$:H$_2$O$_2$) etch and HF dips. The XRF equipment is located at the Center for X-ray Optics microprobe beamline 10.3.1 in the Advanced Light Source Center. It uses 12.5keV monochromatic radiation to excite elements in the sample with a spatial resolution of $1\mu m^2$ and a Si-Li detector to measure fluorescence x-rays from the sample, all in atmospheric conditions. The XRF microprobe sensitivity is impurity and matrix specific but, for example, the system can detect a single Fe or Cu precipitate/agglomerate in silicon greater than 10-20nm in radius. The sampling depth for 3d transition metal impurities in silicon is approximately $50\mu m$. It should be noted that the sensitivity of the microprobe drops considerably for elements with an atomic number < 16. Impurity concentrations are quantified with the use of standard samples of known impurity levels. X-ray absorption studies were carried out at the Center for X-ray Optics beamline 10.3.2 in the Advanced Light Source Center. A 4 crystal Si monochromator produces a tunable monochromatic x-ray beam, which is focussed to a $1\mu m^2$ spot with a pair of grazing incidence mirrors. A Si-Li detector was used to detect fluorescence x-rays and quantify absorption for specific elements. Detail of the absorption apparatus is given in [13]. Etch features on the sample surface, caused by slight etching of grain boundaries during the silicon etch prior to Al contact formation, were used as reference points to locate regions of interest in both fluorescence and absorption studies.

RESULTS AND DISCUSSION

LBIC mapping of minority carrier recombination across the polycrystalline silicon sample revealed localized regions of high carrier recombination. A typical LBIC map in a portion of the cast polysilicon is shown in Figure 1 where dark regions indicate areas of high carrier recombination. Note the regions of high recombination located approximately in the center of the LBIC scan area.

X-ray Fluorescence (XRF)

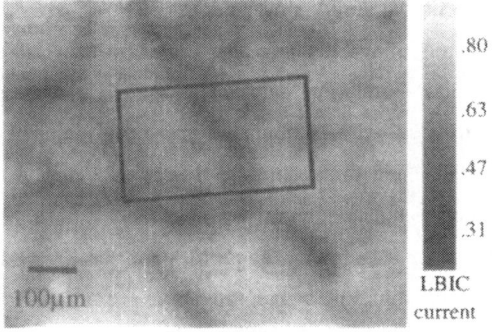

Figure 1: Light Beam Induced Current map of carrier recombination across a portion of multicrystalline silicon. Dark regions indicate high carrier recombination. The black box denotes the area analyzed with x-ray fluorescence.

spectra were taken at $1\mu m^2$ points in the region of Figure 1 as denoted by the black box. No impurity-generated x-ray fluorescent radiation emits in regions of low minority carrier recombination. However, x-ray fluorescent radiation associated with Cr, Fe and Ni emits from regions of high carrier recombination. Specifically, Fe Kα and Fe Kβ fluorescence radiation at

6.4 keV and 7.06 keV, respectively, are clearly discernable above background noise. The ratio of these spectral peak heights is as expected, 4:1 for Kα to Kβ, which provides direct evidence that Fe is present in this region of the material. Fluorescent radiation at 5.4 keV and 7.47 keV, associated with Cr Kα and Ni Kα, respectively, is also clearly distinguished above background noise, strongly indicating that Cr and Ni impurities are present in this region of the material.

Concentrations of impurities at each $1\mu m^2$ spot were calculated by data analysis of the collected spectra and comparison to standard samples with known concentrations of impurities. Impurity maps were produced in the region denoted by the black box in Figure 1. Figure 2 is an impurity map of Fe in this region. Maps of Ni and Cr revealed the exact same distribution as the Fe, indicating a preferred precipitation site exists for these impurities. Clearly there is a correlation between metal impurity distributions and minority carrier recombination. We have performed similar analysis on fully processed RGS polysilicon and have also found a correlation between

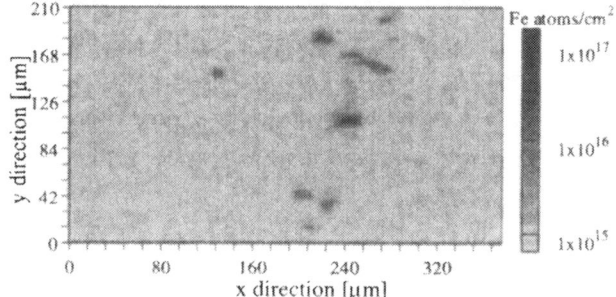

Figure 2: Fe distribution in polycrystalline silicon. The mapped area directly corresponds to the area in the black box of Figure 1. Note the correlation between metal impurity distributions and carrier recombination.

minority carrier recombination and metal impurity distributions. Since impurities remain after complete cell processing, the results also suggest that processing steps should be modified to ensure complete impurity removal and realization of maximum solar cell efficiency. These results indicate metal impurities are the cause for high carrier recombination regions in polycrystalline silicon and, therefore, play a significant role in the performance of the material as a solar cell.

At each high impurity content region of Figure 2, a single precipitate or a number of precipitates may reside in each $1\mu m^2$ spot. If only one precipitate is assumed present, the diameter can be calculated from the measured impurity concentrations and assuming the precipitate is located at or near the surface:

$$d_{@\,surface} = 2*\left[\left(\frac{3}{4\pi}\right)*\frac{dose*area}{density}\right]^{1/3} \qquad (1)$$

where d is the precipitate diameter, dose is the measured concentration (atoms/cm^2), area is the sampled area, i.e. the x-ray spot size, and density is the atomic density of the impurity in the precipitate. For instance, the measured concentration of 5×10^{16} atoms/cm^2 for Fe shown in Figure 2 would suggest one precipitate with a diameter of 288nm is present in that area, this is calculated with a 1μm^2 x-ray spot and the Fe atomic density of 4×10^{22} atoms/cm^3 in FeSi$_2$. Equation (1) assumes only one precipitate exists in the 1μm^2 measured area. Contrary to this assumption is past work, [10], which has shown Cu and Ni precipitates in polycrystalline silicon exist in nm-scale precipitates/agglomerates, which are dispersed over μm-scale regions. Furthermore, work of Shen etal., [14], has indicated that Fe forms a fine dispersion along dislocation lines as opposed to solitary clusters or precipitates. These results would suggest a fine dispersion of impurity precipitates exist in the XRF maps of Figure 2.

We have also performed x-ray absorption studies on electromagnetically cast polysilicon. Spectra were taken with a 1μm^2 spot on regions of high Fe content. A summation of spectral scans at the Fe K absorption edge taken in the Fe-rich region is shown in Figure 3. Additionally, for purpose of comparison, we present the summed spectrum of a standard sample with Fe in Fe

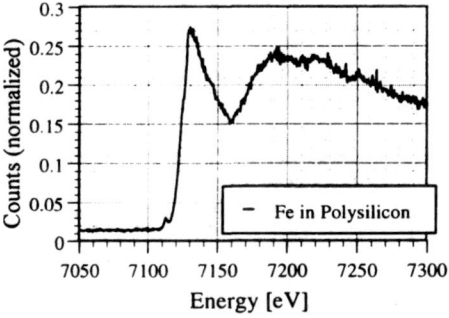

Figure 3: Summed spectrum of Fe in polycrystalline silicon. Note the similarity with Figure 4.

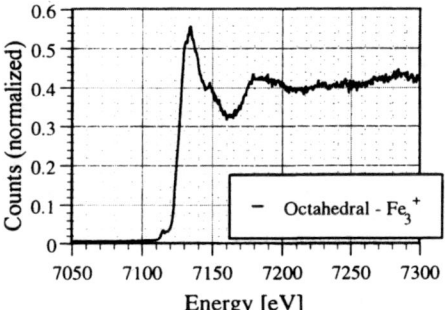

Figure 4: Summed spectrum of Fe$_3^+$ in octahedral coordination. Note the similarity with Figure 3.

oxide with an octahedral coordination in the +3 charge-state in Figure 4. The spectra are closely matched, especially in the near edge region, indicating the Fe in polysilicon is also in the +3 charge-state with octahedral coordination. Considering FeSi$_2$ is in either a tetragonal or orthorhombic coordination while Fe$_2$O$_3$ or Fe$_3$O$_4$ possesses an octahedral coordination, our results indicate the Fe in the polysilicon is not FeSi$_2$ but may be in a Fe$_2$O$_3$ or Fe$_3$O$_4$ state. It is important to note that the binding energy of a Fe iron to a Fe oxide phase is drastically higher than Fe to a Fe silicide phase, [16]. This could constitute a lower concentration of dissolved Fe impurities in the presence of a Fe oxide particle as compared to a Fe silicide particle. Therefore, the identification of Fe oxide in polycrystalline silicon has serious implications for removal of impurities from the silicon because the flux of Fe impurities out of the material is strongly dependent on the dissolved Fe concentration.

CONCLUSIONS

In summation, metal impurities were found in polycrystalline silicon used for solar cells. The distribution of impurities correlated directly with poor performing regions of the material. These results are the first direct proof that metal impurities significantly affect the performance of polycrystalline silicon solar cells. Furthermore, we have initial results, which indicate that the chemical state of Fe in this material is Fe oxide.

ACKNOWLEDGMENTS

This work was supported by the Director, Office of Energy Research, Office of Basic Energy Sciences, Materials Sciences Division, of the U.S. Department of Energy, under Contract No. DE-AC03-76SF00098.

REFERENCES

1. J. M. Gee, R. R. King and K. W. Mitchell, 25th IEEE Photovoltaic Specialists Conference, Washington D.C., pg. 409, (1996)
2. S. Pizzini, A. Sandrinelli, M. Beghi, D. Narducci, F. Allegretti, S. Torchio, G. Fabbri, G. P. Ottaviani, F. Demartin and A. Fusi, *J. Electrochem. Soc.* **135**, 155, (1988)
3. B. L. Sopori, L. Jastrzebski, T. Tan and S. Narayanan, 12th European Photovoltaic Solar Energy Conference, Netherlands, 1003, (1994)
4. S. A. McHugo, H. Hieslmair and E. R. Weber, *Appl. Phys. A.* **64**, 127, (1997)
5. C. Cabanel and J. Y. Laval, *J. Appl. Phys.* **67**, 1425, (1990)
6. T. S. Fell, P. R. Wilshaw and M. D. d. Coteau, *Phys. Stat. Sol. (a).* **138**, 695, (1993)
7. V. Higgs and M. Kittler, *Appl. Phys. Lett.* **63**, 2085, (1993)
8. M. Kittler, W. Seifert and V. Higgs, *Phys. Stat. Sol. (a).* **137**, 327, (1993)
9. L. Jastrzebski, W. Henley, D. Schielein and J. Lagowski, *J. Elec. Chem. Soc.* **142** 3869, (1995)
10. S. A. McHugo, *Appl. Phys. Lett.* **71**, 1984, (1997)
11. I. Périchaud, G. Dour, B. Pillin, F. Durand, D. Sarti and S. Martinuzzi, *Sol. State Phen.* **51-52**, pg. 473, (1996)
12. H. Lange and I. A. Schwirtlich, *J. Crys. Grow.* **104**, 108, (1990)
13. A.A. MacDowell, R. Celestre, C.H. Chang, K. Franck, M.R. Howells, S. Locklin, H.A. Padmore, J.R. Patel, and R. Sandler, SPIE Proceedings 3152, 126-135 (1998).
14. B. Shen, T. Sekiguchi, R. Zhang, Y. Shi, H. Shi, K. Yang, Y. Zheng and K. Sumino, *Jpn. J. Appl. Phys.* **35**, 3301, (1996)
15. B. Shen, J. Jablonski, T. Sekiguchi and K. Sumino, *Jpn. J. Appl. Phys.* **35**, 4187, (1996)

EXAFS CHARACTERISATION OF AMORPHOUS GaAs

M.C. Ridgway*, C.J. Glover*, G.J. Foran** and K.M. Yu***
*Department of Electronic Materials Engineering, Research School of Physical Sciences and Engineering, Australian National University, Canberra, Australia
**Australian Nuclear Science and Technology Organisation, Menai, Australia
***Materials Sciences Division, Lawrence Berkeley National Laboratory, Berkeley, USA

ABSTRACT

The structural parameters of stoichiometric, amorphous GaAs have been determined with extended x-ray absorption fine structure (EXAFS) measurements performed in transmission mode at 10K. Amorphous GaAs samples were fabricated with a combination of epitaxial growth, ion implantation and selective chemical etching. Relative to a crystalline sample, the nearest-neighbor bond length and Debye-Waller factor both increased for amorphous material. In contrast, the coordination numbers about both Ga and As atoms in the amorphous phase decreased to ~3.85 atoms from the crystalline value of four. All structural parameters were independent of implantation conditions and as a consequence, were considered representative of intrinsic, amorphous GaAs as opposed to an implantation-induced extrinsic structure.

INTRODUCTION

Amorphous GaAs has a multitude of potential technological applications and thus, a structural parameter determination is of considerable relevance. Accordingly, a variety of theoretical studies have been reported [1-6] though results vary with computational methodology. None-the-less, the amorphous phase has in general been considered as a combination of bulk and defective sites [1], the former analogous to that of the four-fold coordinated crystalline phase. Conversely, the possible defect configurations include three-fold coordinated atoms and like-atom bonding [1]. In amorphous GaAs, the establishment of the long-range order characteristic of the crystalline phase is inhibited through the uncorrelated connection of the basic tetrahedra to form a disordered, continuous random network comprised of both even- and odd-membered rings. The latter, absent in the crystalline phase, necessitate the presence of like-atom bonding. Furthermore, a distribution of coordination numbers is expected, as compared for the amorphous and crystalline phases in Figure 1.

Despite the theoretical efforts described above, an experimental structural determination of amorphous GaAs, at the atomic scale, has yet to be reported. Though the structures of amorphous Ga_xAs_{1-x} prepared by sputtering ($0.47 \leq x \leq 0.53$) or flash evaporation ($x = 0.48$-0.49) have previously been investigated using extended x-ray absorption fine structure (EXAFS) analysis [7-10], contradictory results were reported, as attributed to sample-preparation-specific artefacts [8]. For the present study, ion implantation has been utilised to produce stoichiometric, amorphous GaAs samples for EXAFS analysis and thereafter, the artefact-free, structural parameters of amorphous GaAs were measured for the first time.

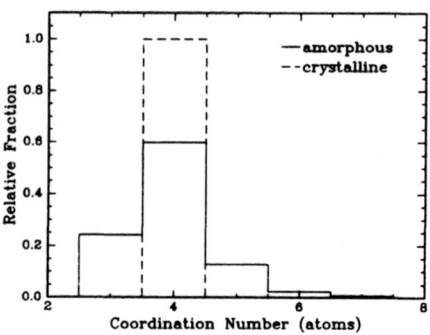

Figure One: Histogram of the distribution of coordination numbers in amorphous and crystalline GaAs. (Data for amorphous GaAs from reference 9.)

EXPERIMENTAL

To fabricate stoichiometric, amorphous GaAs samples for transmission EXAFS measurements, semiconductor device processing techniques were utilised. In brief, a GaAs/AlAs/GaAs heterostructure, grown by metal organic chemical vapour deposition, was amorphised to a depth beyond the AlAs/GaAs interface by ion implantation using an equal total dose of Ga and As ions to preserve stoichiometry. (To confirm the measured structural parameters were independent of implantation conditions, an ion dose range extending two orders of magnitude beyond that required for amorphisation was investigated.) The AlAs layer was then dissolved with a selective chemical etchant (HF:H$_2$O) to separate the amorphous GaAs surface layer, of thickness ~2.5 μm, from the GaAs substrate. The former, supported on adhesive Kapton film, were then stacked together to achieve μx = ~1.5 where μ is the energy-dependent attenuation coefficient and x is the total sample thickness. This technique, described in detail in references 11 and 12, yielded samples of superior stoichiometry, homogeneity, continuity and planarity relative to those fabricated by alternative techniques.

Transmission EXAFS measurements were performed at 10K on beamline 2-3 of the Stanford Synchrotron Radiation Laboratory. Absorption spectra were recorded at both the Ga and As K-edges (10.368 and 11.863 keV, respectively). The normalised EXAFS was Fourier transformed over a photoelectron momentum (k) range of 3-18 and 3-20 Å^{-1} for the Ga and As K-edges, respectively. Thereafter, the structural parameters of the nearest-neighbor shell were isolated by inverse transforming over a non-phase-corrected radial distance range of 1.5-2.7 Å for both Ga and As absorbers. Analysis was performed with the XFIT code [13], in the single-scattering approximation, with the phase and backscattering amplitude calculated *ab initio* with the FEFF4.0 code [14]. For each K-edge, the amplitude reduction factor (S_0^2) was determined from the crystalline sample with the coordination number (n) fixed at a value of 4. This edge-specific S_0^2 value was then used for the analysis of all amorphous sample spectra to facilitate a meaningful comparison of implantation-induced changes in n and the relative mean square deviation in bond length or equivalently, the Debye-Waller factor (σ^2). The ratio method [15] was also applied for the determination of n and σ^2 independent of S_0^2. Within experimental uncertainty, values of n and σ^2 determined with XFIT were consistent with those from the ratio method though the error in n was reduced from ~10 to ~4% with the latter.

RESULTS AND DISCUSSION

Figures 2(a) and (b) show EXAFS and Fourier transformed spectra, respectively, of crystalline and amorphous samples. For the crystalline sample, the structure apparent in the EXAFS spectrum (Fig. 2(a)) was the result of scattering from next-nearest neighbors and accordingly, second- and third-nearest neighbor contributions were readily apparent following Fourier transformation (Fig. 2(b)). In contrast, EXAFS spectra for all amorphous samples were of a single frequency indicative of scattering from a single shell. As a consequence, there was no evidence of next-nearest neighbors following Fourier transformation (Fig. 2(b)). The combination of both bond-length and bond-angle distortions in the amorphous material was evidently sufficient to damp out contributions from beyond the nearest neighbor.

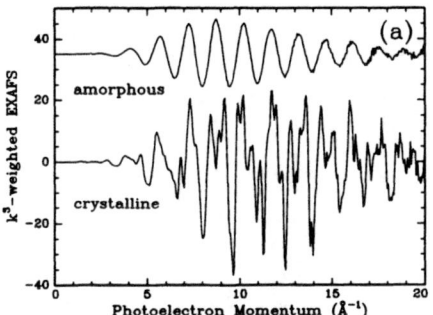 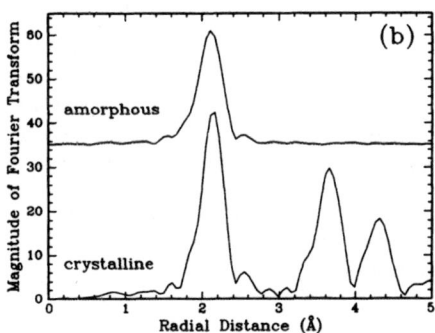

Figure Two: *Spectra of (a) k^3-weighted EXAFS as a function of photoelectron momentum and (b) non-phase-corrected Fourier Transform as a function of radial distance measured at the As K-edge and comparing crystalline and amorphous GaAs samples. Individual spectra have been offset vertically for clarity of presentation.*

Figures 3(a), (b) and (c) show nearest-neighbour bond length (r), σ^2 and n, respectively, of amorphous samples as a function of the energy deposited in vacancy production. (The latter parameter is proportional to implanted ion dose.) All figures include measurements for both Ga and As absorbers. From Fig. 3(a), it was evident that implantation-induced amorphisation yielded an increase in nearest-neighbor bond length, from 2.445 ± 0.003 Å in the crystalline sample to 2.451 ± 0.006 Å in amorphous material, the latter independent of the energy deposited in vacancy production. (Though the magnitude of this change was within the bounds of experimental uncertainty, the relative difference was consistently measured over the extent of the ion-dose range examined.) Note that an increase in bond length, relative to crystalline GaAs, has also been measured with EXAFS in amorphous $Ga_{0.49}As_{0.51}$ and $Ga_{0.48}As_{0.52}$ deposited by sputter deposition and flash evaporation, respectively [7-10].

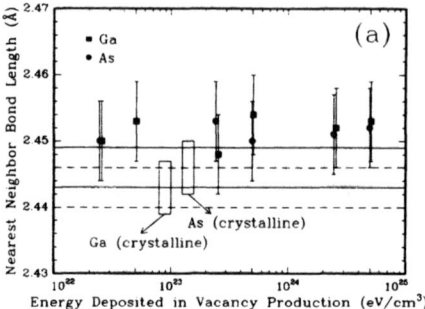

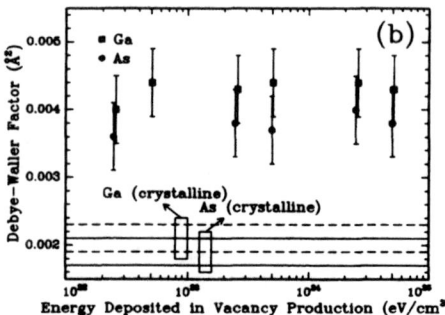

Figure Three: Experimental values of (a) nearest-neighbor bond length, (b) relative mean square deviation in nearest-neighbor bond length and (c) nearest-neighbor coordination number as a function of energy deposited in vacancy production and as measured at the Ga and As K-edges. Horizontal lines define the uncertainty limits of values measured in a crystalline sample at both edges.

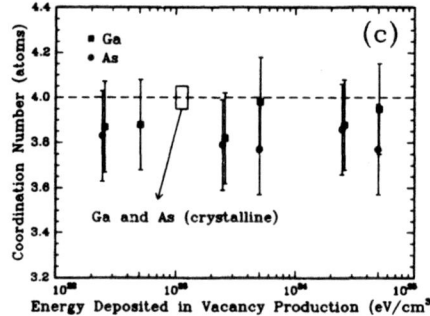

From Fig. 3(b), the σ^2 values for amorphised material were greater than that of the crystalline sample as anticipated given $\sigma^2_{total} = \sigma^2_{thermal\ disorder} + \sigma^2_{structural\ disorder}$ and the structural disorder inherent in an amorphous phase. As above, the Debye-Waller factor was independent of the energy deposited in vacancy production.

Figure 3(c) shows nearest-neighbor coordination numbers as a function of the energy deposited in vacancy production. As noted previously, experimental values of n were determined with the ratio method, an example of which is shown in Figure 4. For both Ga and As absorbers, a reduction in coordination number, from four to ~3.85 atoms, was observed following amorphisation, again independent of ion dose. (Despite the extent of experimental uncertainty (~4%), as above the relative change in coordination number was considered experimentally valid.) Note that such a measurement was in excellent agreement with theoretical predictions of 3.83-4.09 atoms [3-6]. The values of n derived herein were representative of an average, comprised of the superposition of coordination numbers predicted to range from three to seven atoms with a value of four the most common [3-6] as shown in previously in Fig. 1. The experimental results of Fig. 3(c) indicated that the majority of atoms in amorphous GaAs that were not four-fold coordinated were thus under-coordinated. This observation was in excellent agreement with the theoretical prediction that the predominant defect in stoichiometric, amorphous III-V semiconductors is a three-fold coordinated atom [1].

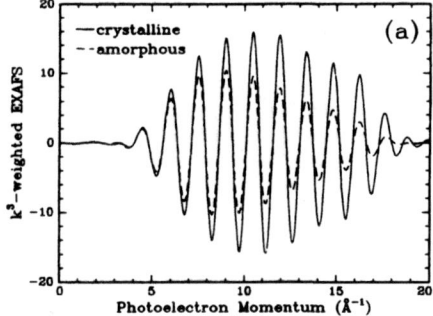 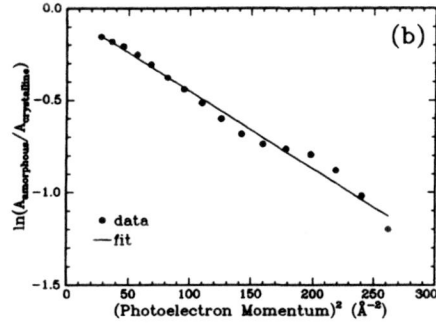

Figure Four: Spectra of (a) back transformed, k^3-weighted EXAFS as a function of photoelectron momentum for the first shell of crystalline and amorphous GaAs measured at the Ga K-edge and (b) ratio of the EXAFS amplitudes $A(k)_{amorphous}/(A(k)_{crystalline}$ as a function of the square of photoelectron momentum.

Following amorphisation, selected samples were annealed at a temperature of 200°C for 1 hr to investigate the potential structural relaxation of the amorphous phase. In addition, selected samples were implanted at room temperature, as opposed to ~-196°C, to examine the potential implant-temperature dependence of the structural parameters of amorphised GaAs. However, within experimental error, changes in r, σ^2 or n were not observed in either case. Furthermore, given the complete lack of ion dose dependence for the structural parameters as detailed above, it is suggested that the r, σ^2 and n values presented herein were representative of stoichiometric, amorphous GaAs. (Equivalently, though the crystalline-to-amorphous transformation was induced by ion implantation, the resulting structure exhibited no dependence on implantation parameters.)

CONCLUSIONS

In conclusion, the structural parameters of amorphous GaAs have been determined from transmission EXAFS measurements using artefact-free samples produced by ion implantation. In the amorphous phase, both the nearest-neighbor bond length and Debye-Waller factor increased, the latter consistent with the presence of structural disorder. Also, the average coordination number about each lattice constituent decreased from the crystalline value of four. All structural parameters were independent of the implantation conditions and thus, the experimental values of r, σ^2 and n were considered representative of the intrinsic structure of amorphous GaAs as opposed to an implantation-induced extrinsic structure.

ACKNOWLEDGMENTS

We thank H.H. Tan, A. Clark and C. Jagadish for MOCVD growths, R.G. Elliman and J.S. Williams for helpful discussions, P. Ellis and H. Freeman for assistance with the XFIT code, D. Creagh and W. Hutchinson for the provision of the Laue apparatus and the Stanford Synchrotron Radiation Laboratory for access to the facility. MCR, CJG and GJF were supported by the Australian Synchrotron Research Program which has been funded by the Commonwealth of Australia via the Major National Research Facilities Program. KMY is supported by the Director, Office of Energy Research, Office of Basic Energy Sciences, Materials Science Division of the U.S. Department of Energy. The Stanford Synchrotron Radiation Laboratory is supported by the Office of Basic Energy Sciences of the U.S. Department of Energy.

REFERENCES

[1]   E.P. O'Reilly and J. Robertson, Phys.Rev. **B34**, 8684 (1986).

[2]   J.P. Xanthakis, P. Katsooulakos and D. Georgiakos, J.Phys: Cond.Mat. **5**, 8677 (1993).

[3]   E. Fois, A. Selloni, G. Pastore, Q.M. Zhang and R. Car, Phys.Rev. **B45**, 13378 (1992).

[4]   C. Molteni, L. Colombo and L. Miglio, Phys.Rev. **B50**, 4371 (1994).

[5]   H. Seong and L.J. Lewis, Phys.Rev. **B53**, 4408 (1996).

[6]   N. Mosseau and L.J. Lewis, Phys.Rev.Lett. **78**, 1484 (1997).

[7]   J.A. Del Cueto and N.J. Shevchik, J.Phys. C: Sol.St.Phys. **11**, L829 (1978).

[8]   M.-L. Theye, A. Gheorghiu and H. Launois, J.Phys. C: Sol.St.Phys. **13**, 6569 (1980).

[9]   D. Udron, M.-L. Theye, D. Raoux, A.-M. Flank, P. Lagarde and J.-P. Gaspard, J.Non-Cryst.Sol. **137/138**, 131 (1991).

[10] S.H. Baker, M.I. Manssor, S.J. Gurman, S.C. Bayliss and E.A. Davis, J.Non-Cryst.Sol. **144**, 63 (1992).

[11] M.C. Ridgway, C.J. Glover, G.J. Foran and K.M.Yu, J.Appl.Phys., to appear May 1998.

[12] M.C. Ridgway, C.J. Glover, H.H. Tan, A. Clark, F. Karouta, G.J. Foran, T.W. Lee, Y. Moon, E. Yoon, J.L. Hansen, A. Nylandsted-Larsen, C. Clerc and J. Chaumont, to appear in these preceedings.

[13] P.J. Ellis and H. Freeman, J.Synch.Rad. **2**, 190 (1995).

[14] J.J. Rehrs and R.C. Albers, Phys.Rev. **B41**, 8139 (1990).

[15] E.A. Stern, D.E. Sayers and F.W. Lytle, Phys.Rev. **B11**, 4836 (1975).

AMORPHOUS SEMICONDUCTOR SAMPLE PREPARATION
FOR TRANSMISSION EXAFS MEASUREMENTS

M.C. RIDGWAY[1], C.J. GLOVER[1], H.H TAN[1], A. CLARK[1], F. KAROUTA[1], G.J FORAN[2], T.W. LEE[3], Y.MOON[3], E.YOON[3], J.L. HANSEN[4], A. NYLANDSTED-LARSEN[4], C. CLERC[5] AND J. CHAUMONT[5]

[1]Department of Electronic Materials Engineering, Australian National University, Canberra, Australia
[2]Australian Nuclear Science and Technology Organisation, Menai, Australia
[3]School of Materials Science and Engineering, Seoul National University, Seoul, Korea
[4]Institute of Physics and Astronomy, Aarhus University, Aarhus, Denmark
[5]Centre de Spectrometrie Nucleaire et de Spectrometrie de Masse, Centre National Recherche Scientifique, Orsay, France

ABSTRACT

A novel methodology has been developed for the preparation of amorphous semiconductor samples for use in transmission extended x-ray absorption fine structure (EXAFS) measurements. Epitaxial heterostructures were fabricated by metal organic chemical vapour deposition (group III-Vs) or molecular beam epitaxy (group IVs). An epitaxial layer of ~2 μm thickness was separated from the underlying substrate by selective chemical etching of an intermediate sacrificial layer. Ion implantation was utilised to amorphise the epitaxial layer either before or after selective chemical etching. The resulting samples were both stoichiometric and homogeneous in contrast to those produced by conventional techniques. The fabrication of amorphous GaAs, InP, $In_{0.53}Ga_{0.47}As$ and Si_xGe_{1-x} samples is described. Furthermore, EXAFS measurements comparing both fluorescence and transmission detection, and crystalline and amorphised GaAs, are shown.

INTRODUCTION

Quantification of the local disorder in amorphous semiconductors is of significance to both the materials science and technological communities. Disorder may be present in either chemical or structural forms as like-atom bonding (group III-Vs) or changes in bondlength, bond angle and/or co-ordination number (groups III-Vs and IVs), respectively. Techniques such as extended x-ray absorption fine structure (EXAFS), electron spin resonance and Raman, neutron and x-ray scattering are applicable for the measurement of short and/or medium range order in amorphous semiconductors. Similarly, a variety of computational methods can be utilised for the theoretical formulation of structural models.

EXAFS can yield accurate information on the local atomic environment of an x-ray absorbing atom surrounded by atomic scattering centres. Measurements can be performed in two modes, transmission or fluorescence, with semi-transparent or bulk samples, respectively. In general, transmission is the preferred means of analysis as it utilises all the incident photon flux and lacks the complexity and count-rate limitations inherent with a solid-state detector. However, transmission EXAFS measurements for amorphous GaAs [1,2] have yielded contradictory results where the sample preparation methodology was considered responsible for such differences [1]. Samples for these studies were prepared by flash evaporation or sputtering which, unfortunately, produced artefacts that included voids, inhomogeneity and/or non-stoichiometry. Similarly, theoretical calculations of the structure of amorphous GaAs have exhibited a model dependence [3,4,5]. The preparation of artefact-free samples is thus of necessity to unambiguously determine the structural parameters of amorphous semiconductors for comparison with theoretical calculations. Accordingly, semiconductor device processing techniques [6] have been utilised herein to produce amorphous semiconductor samples for transmission EXAFS measurements that, as described below, were of superior stoichiometry, homogeneity, continuity and planarity relative to those fabricated with conventional techniques.

EXPERIMENTAL

Fluorescence-Detected EXAFS Measurements

Bulk, (100) GaAs substrates were amorphised with ion implantation to a depth of ~2.5 μm. Multiple energy, multiple dose implants of both Ga and As ions were utilised to produce a near-constant energy deposition in vacancy production over the extent of the amorphous layer as shown in Figure 1. TRIM [7] was used to calculate the ion energies and minimum required doses based on a GaAs threshold damage density[1] of 4×10^{22} eV/cm^3 [8]. All implants were performed at ~-196°C with the incident ion beam oriented in a non-channeling direction and the ion flux minimised to yield a maximum areal power density of <1 W/cm^2.

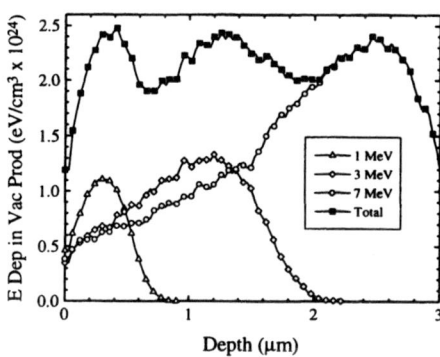

Figure 1. Depth distribution of the energy deposited in vacancy production for multiple-energy, mulitple-dose As implants. The total As dose was 8.3×10^{15} ions/cm^2.

Fluorescence-detected EXAFS measurements were performed at the Photon Factory, at room temperature, with a 10-element, solid-state Ge detector. A grazing-incidence orientation was utilised to inhibit x-ray penetration to depths beyond the amorphous layer. For both fluorescence and transmission measurements, the latter described below, structural parameters were determined with XFIT [11].

Transmission-Detected EXAFS Measurements

1. GaAs

Figure 2 shows a schematic diagram of the methodology used to fabricate amorphous GaAs samples for transmission EXAFS measurements. Crystalline GaAs/AlAs/GaAs heterostructures (2 μm / 0.02 μm / (100) substrate) were deposited by metal organic chemical vapor deposition (MOCVD). The heterostructure was then amorphised by ion implantation to a depth of ~2.5 μm as described above.

---

[1] The threshold damage density is the value of energy deposited in vacancy production sufficient to yield amorphisation.

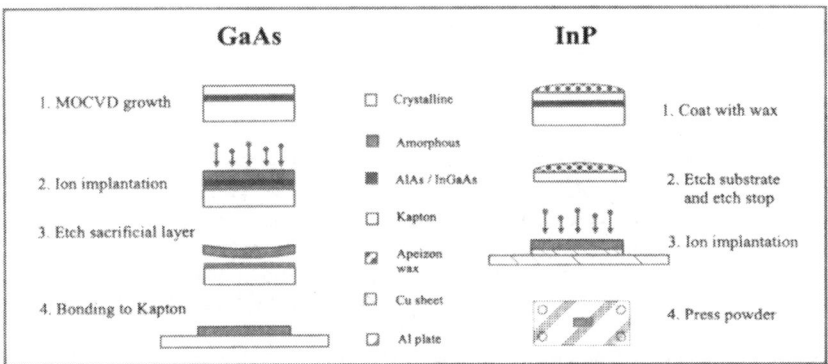

Figure 2. GaAs and InP amorphous sample preparation procedure for transmission EXAFS measurements.

Implanted samples were cleaved to dimensions of ~5x8 mm and immersed in a $HF:H_2O$ (1:10) solution, at room temperature, for ~48 hr to preferentially remove the AlAs sacrificial layer. After floating free of the substrate, the thin amorphised layer was transferred to a H_2O bath and retrieved from the liquid surface with adhesive, x-ray transparent Kapton tape on an Al support. The amorphous nature of the samples was confirmed with transmission Laue measurements.

The technique described herein necessitated an etchant with sufficient selectivity to remove the sacrificial AlAs layer whilst maintaining a constant thickness of the amorphised GaAs layer. The selectivity of $HF:H_2O$ (1:10) in the crystalline AlAs/GaAs system was ~$10^8/1$ [6] and no change in selectivity was measurable for the amorphous AlAs/GaAs system.

2. Ge_xSi_{1-x}

$Ge_xSi_{1-x}/Si/SiO_2/Si$ heterostructures (2 µm / 0.2 µm / 0.4 µm / (100) substrate) were fabricated by molecular beam epitaxy (MBE) of Ge_xSi_{1-x} layers on standard Si-on-Insulator substrates. Following deposition, samples were annealed at 700-900°C to minimise residual strain in the lattice mismatched Ge_xSi_{1-x} layer. The measured selectivity of the amorphous SiO_2/Si system in a $HF:H_2O$ (1:2) solution was insufficient for post-implant etching and thus, the Ge_xSi_{1-x}/Si layers were separated from the substrate in a crystalline state. The Ge_xSi_{1-x} layer was masked with Apiezon black wax (W-100) and the sacrificial SiO_2 layer was removed through immersion in a $HF:H_2O$ (1:2) solution under slow agitation for ~4 days. Note that the wax also served as a structural support for the Ge_xSi_{1-x}/Si layers and the strain inherent in the convex geometry assisted the etching process by separating the edges of the epitaxial layer from the substrate. Following selective etching, samples were removed from the solution and bonded to Cu tape with double-sided, conductive adhesive (3M #1182) sufficient for both electrical and thermal conduction. The black wax was removed with trichloroethylene (TCE) and the thin Ge_xSi_{1-x}/Si layer was then amorphised with multiple energy, multiple dose implants of Si and/or Ge ions as described above for GaAs. The adhesive was subsequently dissolved with TCE and acetone to separate the Ge_xSi_{1-x}/Si layers from the Cu tape.

Differences in thermal expansion coefficients and the presence of irradiation-induced strain prior to amorphisation resulted in slight cracking of the Ge_xSi_{1-x}/Si layers. To avoid errors resulting from 'thickness effects' [10] due to a lack of continuity and/or planarity, the amorphised material was thus crushed and packed in a BN binder for transmission EXAFS measurements. (Note that the presence of the thin, amorphised Si layer had no influence on EXAFS measurements performed at the Ge K-edge.)

3. InP, $In_{0.53}Ga_{0.47}As$

Lattice-matched $InP/In_{0.53}Ga_{0.47}As/InP$ heterostructures (2 μm / 0.2 μm / (100) substrate) were deposited by MOCVD. The selectivity of both the amorphous and crystalline $InP/In_{0.53}Ga_{0.47}As$ systems were insufficient for either of the etching procedures described above. As shown in Figure 2, the InP epitaxial layer was thus masked with black wax and the InP substrate was removed by etching in 37% HCl for ~1 hr, utilising the $In_{0.53}Ga_{0.47}As$ layer as an etch stop. The $In_{0.53}Ga_{0.47}As$ layer was then removed in a $H_2SO_4:H_2O_2:H_2O$ (1:1:10) solution for ~5 min. As above, the masked InP layer was mounted on Cu tape, amorphised with equal doses of In and P ions and then processed to yield powder transmission EXAFS samples. $In_{0.53}Ga_{0.47}As$ samples were fabricated in an identical fashion in the absence of an InP epitaxial layer.

Transmission-detected EXAFS measurements were performed at the Stanford Synchrotron Radiation Laboratory, at either room temperature or 10K, utilising conventional, gas-filled ion chambers and calibration foils.

RESULTS

Given the complexity of the sample preparation techniques for transmission EXAFS measurements, a comparison of the two detection methodologies was thus warranted. Spectra showing fluorescence and transmission EXAFS measurements performed at room temperature at the As K-edge on crystalline GaAs samples are presented in Figure 3. The superior signal-to-noise ratio and greater k range obtainable with transmission detection were readily apparent from a comparison of the k^3-weighted EXAFS spectra (Figure 3(a)), where k is the photoelectron momentum. Transmission measurements thus yielded greater detail (note the amplitude of the second nearest neighbour contribution) and lesser artefacts (note the non-zero components at non-physical distances) following Fourier transformation (Figure 3(b)). Similar results were obtained with amorphous samples though, as described below, second nearest neighbour contributions were not evident.

Figure 4 is a comparison of transmission EXAFS spectra performed at 10K at the As K-edge of crystalline and amorphous GaAs samples. In general, the optimum thickness (x) for transmission samples is calculable as $\mu x = $ ~1.5 where μ is the energy-dependent attenuation coefficient. (For GaAs, $x = $ ~15 μm at both the Ga and As K-edge absorption maxima.) For amorphous GaAs samples, the optimum total thickness was achieved by simply stacking samples together. For the crystalline sample, crushed GaAs powder with an equivalent thickness of ~15μm was dispersed in a BN binder to negate diffraction and polarisation effects.

Comparing spectra for crystalline samples derived from measurements at room temperature and 10K (Figures 3 and 4, respectively), significantly increased structure was apparent in both the k^3-weighted EXAFS and Fourier transformed spectra obtained at low temperature. As a consequence of the reduction in the thermal component of the Debye-Waller factor, the scattering contributions from second and third nearest neighbours were readily apparent following Fourier transformation (Figure 4(b)). In contrast, there was no evidence of scattering from next-nearest neighbours in the amorphous sample, as measured at 10K and as consistent with the anticipated lack of long range order. Amorphisation produced an increase in

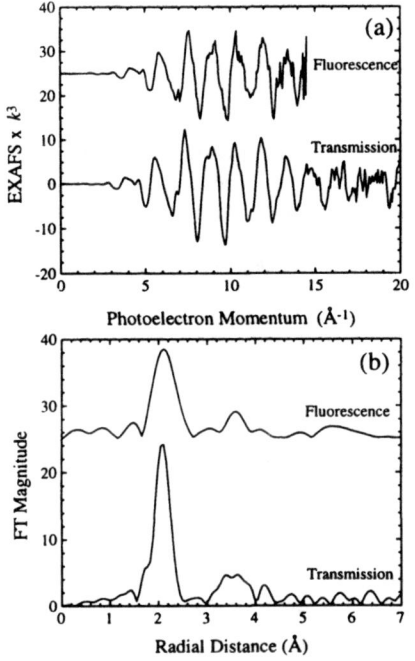

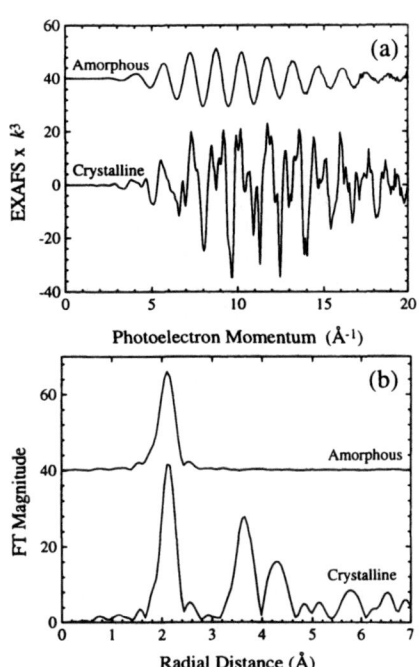

Figure 3. k^3-weighted EXAFS (a) and non phase corrected Fourier Transform (FT) magnitude (b) comparing transmission with fluorescence data, at the As K-edge, measured at room temperature.

Figure 4. k^3-weighted EXAFS (a) and non phase corrected Fourier Transform (FT) magnitude (b) comparing a-GaAs and c-GaAs, at the As K-edge, measured at 10 K.

nearest-neighbour bond length, from the crystalline value of 2.445±0.003 Å, to 2.451±0.006 Å. Also, the measured decrease of the average nearest neighbour co-ordination number from 4 (crystalline) to ~3.85±4% atoms (amorphous) was in excellent agreement with theoretical predictions [12]. A thorough analysis of implantation-induced changes in the structure of GaAs will be published elsewhere [13].

The ion implantation and selective chemical etching processes described above for GaAs and InP yielded stoichiometric and homogenous samples. Utilising MOCVD and ion implantation, the maximum deviation from stoichiometry was estimated to be less that 0.01%. In contrast, transmission samples produced by flash evaporation or sputter deposition were non-stoichiometric $Ga_{0.48}As_{0.52}$ [1] and $In_{0.40}P_{0.60}$ [9]. Such deviations from stoichiometry may well be expected to yield a two phase mixture and poor homogeneity has, in fact, been cited as a source of error in previous studies [1].

Transmission measurements are also subject to errors resulting from a lack of sample continuity and/or planarity. Flash evaporation and sputter deposition can yield discontinuities in the form of voids and pin holes whilst non-planarity can result as a consequence of evaporation from a point source with a fixed sample-source distance. Both effects lead to a reduction of the EXAFS amplitude [10]. Amorphous GaAs samples fabricated as above were continuous and pin-hole free with a non-uniformity in sample thickness estimated to be < 2% across the sample area.

CONCLUSIONS

A novel methodology for the preparation of stoichiometric amorphous semiconductor samples for use in transmission EXAFS measurements has been developed. Subsequent determination of the structural parameters of amorphous GaAs have yielded results consistent with theoretical predictions. The sample preparation procedure is applicable to a range of multi- and mono-elemental semiconductor systems, a variety of which have been described herein. Accordingly, both chemical and structural disorder can now be measured in stoichiometric, amorphous semiconductor samples in the absence of artefacts resulting from the sample preparation technique.

ACKNOWLEDGEMENTS

We thank the Australian Synchrotron Radiation Program for financial support, C. Jagadish for helpful discussions, G. Li and M. Phillips for additional technical assistance, D. Creagh and W. Hutchinson for the provision of the transmission Laue apparatus, P. Ellis and H. Freeman for assistance with the XFIT code and both the Photon Factory and Stanford Synchrotron Radiation Laboratory (SSRL) for providing access to these facilities. Work done (partially) at SSRL which is operated by the Department of Energy, Office of Basic Energy Sciences.

REFERENCES

1. M.L. Theye, A. Gheorghiu, and H. Launois, J. Phys. C: Solid St. Phys. **13**, 6569 (1980).
2. J.A. Del Cueto and N.J. Shevchik, J. Phys. C: Solid St. Phys. **11**, L829 (1978).
3. J.P. Xanthakis, P.Katsoulakos, and D. Georgiakos, J. Phys.: Condens. Matter **5**, 8677 (1993).
4. E.P. O'Reilly and J.Robertson, Phys. Rev. **B 34**, 8684 (1986).
5. N. Mousseau and L.J. Lewis, Phys. Rev. **B 56**, 9461 (1997).
6. N.M. Jokerst, in *Handbook of Compound Semiconductors*, edited by P.H. Holloway (Noyes Publications, New Jersey, 1995), p. 518.
7. J.P. Biersack and L.G. Haggmark, Nucl.Instrum.Meth. **174**, 257 (1980).
8. K.B. Belay, D.J. Llewellyn, and M.C. Ridgway, Appl. Phys. Lett. **69**, 2534 (1996).
9. D. Udron, A.M. Flank, P.Lagarde, D. Raoux, and M.L. Theye, J. Non-Cryst. Solids **150**, 361 (1992).
10. S.M. Heald, in *X-Ray Absorption; Vol. 92*, edited by D.C. Koningsberger (John Wiley and Sons, Brisbane, 1988), p. 87.
11. P.J. Ellis and H.C. Freeman, J. Synchrotron Rad. **2**, 190 (1995).
12. H. Seong and L.J. Lewis, Phys. Rev. **B 53**, 4408 (1996).
13. M.C. Ridgway, C.J. Glover, G.J. Foran and K.M. Yu, J. Appl. Phys. **83**, 4610 (1998).

KINETICS OF THE GROWTH OF COPPER CLUSTERS ON THE ALUMINA (0001) SURFACE : INFLUENCE OF SURFACE STRUCTURE

M. GAUTIER-SOYER*, S. GOTA*, L. DOUILLARD*, P. LE FEVRE**, H. MAGNAN*, J. P. DURAUD***
* CEA, DSM / DRECAM / SRSIM, CEA / SACLAY, 91191 - Gif sur Yvette Cedex, mgautiersoyer@cea.fr
** LURE, batiment 209 D, Université Paris Sud, 91405 Orsay Cedex
*** Laboratoire Pierre Süe, CEA-CNRS, CEA / SACLAY, 91191 - Gif sur Yvette Cedex

ABSTRACT

The kinetics of the growth of copper clusters on the alumina (0001) surface was studied as a function of surface structure, using EXAFS at the Cu K edge. Equivalent Cu coverages ranging from 0.5 to 4 equivalent monolayers were deposited in situ, at room temperature, on alumina (0001) surfaces exhibiting the (1x1) or the $(\sqrt{31} \times \sqrt{31})R9°$ reconstructed structure. The evolution of mean cluster size with deposition time was followed from the mean Cu coordination number in the clusters deduced from the EXAFS data. The increase of the mean cluster radius with deposition time is characteristic of a coalescence mechanism on both surfaces. The growth is quicker on the reconstructed surface, likely due to different surface diffusion properties of both surfaces.

INTRODUCTION

The growth of copper on α-Al_2O_3 has received renewed attention in the last years. The first motivation arises from the interest of the Cu / Al_2O_3 interface in catalytic reactions and in microelectronics applications. But in addition, from a fundamental point of view, the Cu / Al_2O_3 system can be considered as a model non-reactive metal / oxide system to investigate the physical mechanisms of the growth. Indeed the interaction between Cu and Al_2O_3 is very weak. No evidence of Cu oxidation or a Cu / Al reaction was observed on very thin Al_2O_3 films [1]. The (0001) surface of α-Al_2O_3 single crystals can exhibit different structures, from the non-reconstructed (1x1) to the reconstructed $(\sqrt{31} \times \sqrt{31})R9°$. Both structures have been recently resolved by using grazing incidence x-ray diffraction [2]. They are very different, both from the size of the unit cell, and from composition, the reconstructed surface being strongly aluminum enriched. So such surfaces are expected to behave differently when used as a substrate for copper growth. On both surfaces, three-dimensional clusters are formed in the initial stages of the growth, following the Volmer-Weber mode [3]. However, based on the observed Auger parameter of copper in x-ray photoelectron spectroscopy (XPS), we had shown that for a given Cu coverage, the cluster size was larger on the reconstructed surface than on the primitive one [3]. Such a behavior could be explained by at least two reasons : a difference in the nucleation stage, with less nucleation sites on the reconstructed surface, or a different kinetics of the growth on the two surfaces. More recently, we have studied the growth kinetics of copper clusters on a reconstructed $(\sqrt{31} \times \sqrt{31})R9°$ surface kept at room temperature [4]. We used

EXAFS (Extended X-ray Absorption Fine Structure) at the Cu K edge, which proved to be useful to study the specific properties of small metal clusters, like the reduction of the coordination number [5-7]. We found that the mean cluster radius increased with deposition time following a scaling law characteristic of a coalescence process. In order to determine the influence of the surface structure of alumina on the kinetics of the growth of clusters, we have performed new EXAFS experiments on the (1x1) surface kept at room temperature. These results are discussed in comparison with those previously obtained on the $(\sqrt{31} \times \sqrt{31})R9°$ surface, so as to evaluate if the observed difference in cluster size is mainly due to a difference in the nucleation stage or in the growth kinetics.

EXPERIMENT

Polished single crystals of alumina (0001) were heated at 1500 °C under air, and in-situ post-annealed under UHV at 1000 °C in the EXAFS chamber to obtain the (1x1) surface structure, as checked by LEED. Equivalent coverages ranging from 0.5 to 4 monolayers (ML) of copper, carefully calibrated with a quartz microbalance, were deposited at a rate of 1Å/mn. Evaporation was performed in the EXAFS chamber (10^{-10} Torr) with the substrate at room temperature. EXAFS measurements were carried out at 80 K. They were performed on the wiggler DW21 beamline of the storage ring DCI at the LURE facility (Orsay, France), using a Si (311) double crystal monochromator. The Cu K edge EXAFS spectra [8950 - 9650 eV] were recorded with the photon beam normal to the surface, in the fluorescence detection mode.

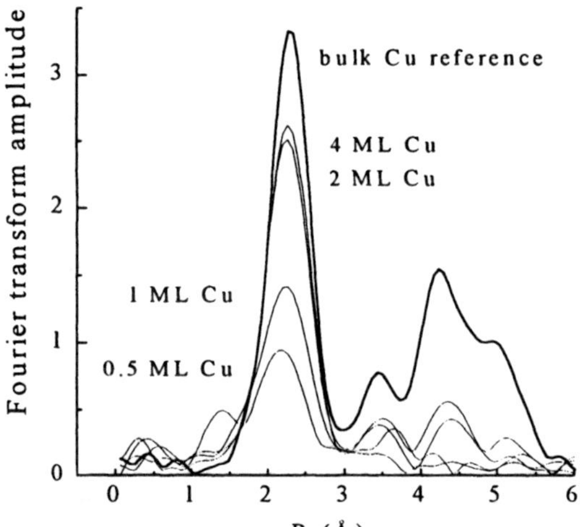

RESULTS

Data processing for extracting the EXAFS curve $\chi(k)$ from the raw data was done following standard data procedure [8]. The Fourier transform amplitude of $\chi(k)$, also called radial distribution function (RDF) was calculated between 3.4 and 12 $Å^{-1}$ using a Kaiser window (figure 1). To get the local structure around Cu in the clusters, we have isolated the main peak of the RDF using a window between 1.47 and 3.0 Å. The fitting procedure has been detailed elsewhere [5]. For ensuring the correctness of the results, the obtained Fourier back-transformed spectra have been fitted following two independent procedures : we have used the

Fig.1 : Fourier transform amplitude of $\chi(k)$ for the four Cu deposits on the (1x1) surface.

amplitude and the phase shifts computed by Mc Kale et al [9] and also we have used those extracted from a reference copper sample recorded at 80 K. We have obtained high quality fits for similar structural parameters from both procedures. The results are summarized in table 1. Only Cu-Cu bonds were considered, as adding Cu-Al or Cu-O bonds did not improve the fits at all.

deposit	surface density of Cu atoms deposited	d(Å) $\pm$ 0.02 Å	N $\pm$ 10%	$\Delta\sigma^2$ (Å^2) $\pm$ 0.5 10^{-3}
0.5 ML	9 10^{14} at.cm-^{2}	2.46	5	0.094
1 ML	1.8 10^{15} at.cm-^{2}	2.50	6.6	0.093
2 ML	4.2 10^{15} at.cm-^{2}	2.52	10.5	0.089
4 ML	8.4 10^{15} at.cm-^{2}	2.525	11	0.091
bulk Cu	------------------	2.54	12	------------------

Table 1 : Structural parameters of the first shell giving the best least square fits for the four deposits considered on the (1x1) surface. For simplicity, the deposits are denoted by 0.5, 1, 2 and 4 ML (which are approximate values). The exact Cu atom surface density, as measured by the quartz microbalance, is given in the second column. d is the mean Cu-Cu distance, N the average number of Cu first neighbors and $\Delta\sigma^2$ the mean square relative displacement.

DISCUSSION

The aim of this work was to compare the growth on both (1x1) and $(\sqrt{31} \times \sqrt{31})R9°$ alumina surfaces. Then the first step was to deduce, from the experimental Cu coordination numbers, an average cluster size. We used the same calculated curve as in ref [4], which gave the evolution of the mean Cu coordination number versus cluster radius (figure 2, solid line). This implicitly assumes that the Cu cluster distribution is monomodal and that the Cu clusters have the same fcc structure as the bulk [4]. It is worth mentioning that this approach is best suited to coordination numbers in the range [5.5 - 10.5], corresponding in our approach to cluster radius between 2.55 and 15 Å. For larger clusters, EXAFS is no longer sensitive to the cluster size, and for smaller ones, the fcc model does not apply any more. Indeed, the smallest fcc cluster which can be built contains 13 atoms, with a mean coordination number of 5.5 and a radius of 2.54 Å. In figure 2, the N values of table 1 ((1x1) substrate) are reported, together with the values we had obtained for the $(\sqrt{31} \times \sqrt{31})R9°$ substrate, for 0.5, 0.6 and 1 ML deposits [4]. For the 0.5 ML deposits on both surfaces, we are clearly at the lower limit of our technique, and we can only deduce that the mean cluster radius is smaller than 2.55 Å. For the 4

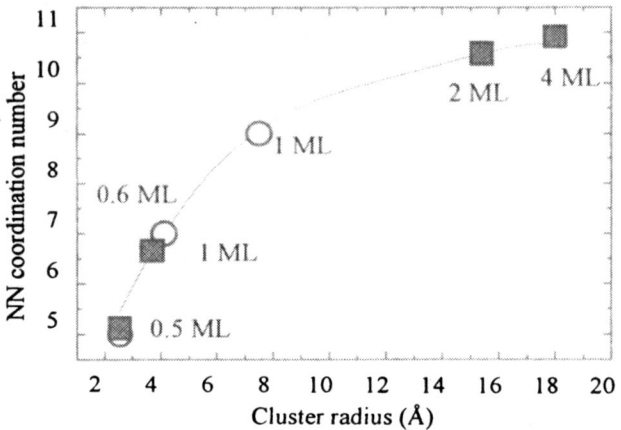

Fig. 2 : Calculated coordination number of the first neighbor shell for clusters of radius from 2.5 to 18 Å, built by adding successive shells around a central Cu atom, while keeping the fcc structure (solid line). The experimental N values extracted from EXAFS analysis for the deposits on the (1x1) surface (this work) are plotted by black squares and those obtained in ref [4] on the $(\sqrt{31} \times \sqrt{31})R9°$ surface by open circles.

ML deposit, the large value of the coordination number (11) indicates that the cluster size is larger than 18 Å (upper limit of the model). First examination of figure 2 confirms that, for a given Cu equivalent coverage, the cluster size is larger on the reconstructed surface. To have a more quantitative comparison of the growth kinetics on both surfaces at 300 K, we have plotted the mean cluster radius as a function of the number of copper atoms deposited (second column of table1) in a logarithmic scale (figure 3).

As concerns the points which are at the limit of our model, the mean radius of the 0.5 ML deposit has been set to 2.55 Å for both surfaces and that of the 4 ML deposit on the (1x1) to 18 Å. Then the evolution of cluster radius with deposition time (the deposition rate is constant) can be fitted by a power law t^x, with an x exponent close to 1 for the (1x1) surface, and close to 1.4 for the $(\sqrt{31} \times \sqrt{31})R9°$. At this point, it is important to mention that, as explained above, the mean radius of the 0.5 ML deposits may be overestimated, while that of 4 ML may be underestimated. This would lead however to a larger slope than deduced from figure 3. Therefore, in the [0.5 - 4 ML] range, an individual cluster growth, which would be characterized by exponents such as 1/3 or 1/4 [10], can be definitely discarded. The values close to 1 are characteristic of a coalescence regime of the clusters, as discussed for the submonolayer growth of copper on the $(\sqrt{31} \times \sqrt{31})R9°$ surface [4]. The slope is slightly larger for the $(\sqrt{31} \times \sqrt{31})R9°$ surface, indicating that the growth is quicker on the reconstructed surface than on the (1 x 1) one. This result show that, even if there is a difference in the nucleation sites on both surfaces [11], the difference in the kinetics is a major effect. Such a difference suggests different surface diffusion properties induced by the reconstruction of the alumina (0001) surface.

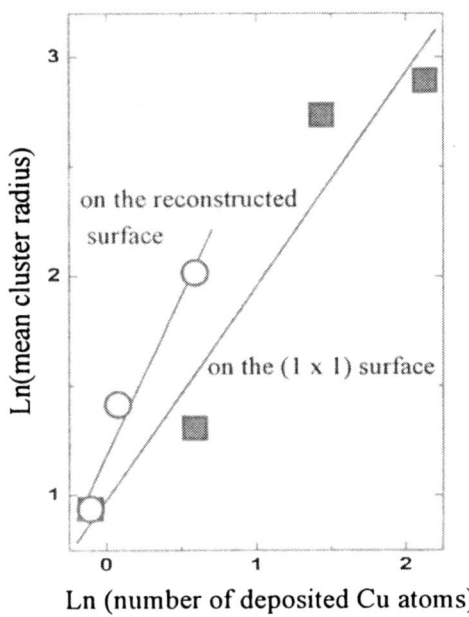

Fig. 3 : Evolution of the mean cluster radius (in Å) as a function of deposition time, measured as the deposited Cu atom density (see table 1, column 2). Black squares correspond to the (1x1) surface (this work), and open circles to the $(\sqrt{31} \times \sqrt{31})R9°$ (data of ref [4]). The scale is logarithmic.

CONCLUSION

By using EXAFS at the Cu K edge, we have studied the evolution of the mean cluster radius as a function of deposition time on two different Al_2O_3 (0001) surfaces, exhibiting the (1x1) structure or the $(\sqrt{31} \times \sqrt{31})R9°$ reconstruction. In the 0.5 - 4 monolayer range, the growth on both surfaces proceeds by the coalescence of three-dimensional clusters with nanometric size. However, the growth is quicker on the reconstructed surface, suggesting different surface diffusion properties.

ACKNOWLEDGMENTS

The authors are grateful to D. Ragonnet (LURE, Orsay) and the staff of LURE for their help in the EXAFS experiments. They thank A. Traverse (LURE, Orsay) for the Cu EXAFS reference data and A. Potier (SRSIM, CEA / Saclay) for his help in data processing.

REFERENCES

[1] Y. Wu, E. Garfunkel, T. E. Madey, J. Vac. Sci. Technol. A4 (1996) 1662, and references therein.

[2] G. Renaud, B. Villette, I. Vilfan, A. Bourret, Phys. Rev. Lett. 73 (1994) 1825 ; P. Guénard, G. Renaud, A. Barbier, M. Gautier-Soyer, Mat. Res. Soc. Symp. Proc. Vol 437 (1986) 15.

[3] M. Gautier, L. Pham Van, J. P. Duraud, Europhys. Lett. 18 (1992) 175 ; Surf. Sci. Lett. 249 (1991) L327.

[4] M. Gautier-Soyer, S. Gota, L. Douillard, J. P. Duraud, P. Le Fèvre, Phys. Rev. B 54 (1996) 10366 S. Gota, M. Gautier-Soyer, L. Douillard, J. P. Duraud, P. Le Fèvre, J. Phys. IV France 7 (1997) 675.

[5] S. Gota, M. Gautier, L. Douillard, N. Thromat, J. P. Duraud, P. Le Fèvre, Surf. Sci. 323 (1995) 163.

[6] P.A. Montano, G. K. Shenoy, E. E. Alp, W. Schulze, J. Urban, Phys. Rev. Lett. 56 (1986) 2076.

[7] A. Pinto, A. R. Pennisi, G. Faraci, G. D'Agostino, S. Mobilio, F. Boscherini, Phys. Rev. B 51 (1995) 5315.

[8] B. K. Teo, in «EXAFS : Basic principles and data analysis», Springer Verlag, Berlin, 1984.

[9] A. G. Mc Kale, B. W. Veal, A. P. Paulikas, S. K. Chan and G. S. Knapp, J. Am. Chem. Soc. 110 (1988) 3763.

[10] D. Beysens, C. M. Knobler, H. Schaffar, Phys. Rev. B 41 (1990) 9814.

[11] S. Gota, M. Gautier-Soyer, L. Douillard, J. P. Duraud, P. Le Fevre, Surf. Sci. 352-354 (1996) 1016.

Temperature-Driven Oxidation Behavior of Pure Iron Surface Investigated by Time-Resolved EXAFS Measurements

S.J.Doh[1], J.M.Lee[2], D.Y. Noh[3] and J.H.Je[1]

[1] Department of Materials Science & Engineering and [2] Pohang Accelerator Lab, POSTECH, Pohang, 790-784, Korea
[3] Department of Materials Science & Engineering, K-JIST, Kwangju, 506-712, Korea

ABSTRACT

The surface-front oxidation mechanism of iron was investigated by time-resolved, glancing-angle Fe K-edge fluorescence EXAFS measurements at various oxidation temperatures of 200-700 ℃. The glancing angle was chosen according to the depth of the oxide layer, roughly 1500-2000 Å. The oxidation behavior under rapid heating(up to 600 ℃ within 10 minutes) was compared with the slowly heated oxidation process using the Quick-EXAFS measurements. In the slowly heated process, Fe_3O_4 was the dominating phase at a relatively low temperature (300-400 ℃) initially. However, at a relatively high temperature (above 600 ℃), the Fe_2O_3 and FeO crystalline phases are gradually enriched as the successive oxidation process involving intrusive oxygen proceeded. Remarkably under a prolonged heat treatment above 600 ℃, the stable FeO phase that exists in a deep-lying interface structure and Fe_2O_3 phase eventually dominates the thick front-surface structure. In a quickly heated process, however, Fe_3O_4 phase is less dominating, which is contradictory to the commonly accepted oxidation models. The EXAFS results are discussed in conjunction with the x-ray diffraction features under the same heat treatment conditions.

INTRODUCTION

The oxidation behavior of pure iron has been studied for several decades.[1] Since oxygen atoms thermally diffuse into Fe when Fe is heated to a high temperature (about 1200 ℃), it has been proposed that the oxygen rich phases would be formed near the surface while the oxygen poor phases would be away from the surface. Accordingly the Fe_2O_3 (the richest oxygen content) phase would be placed on top of the Fe_3O_4, Fe_2O_3 and, FeO . This is commonly accepted picture of the oxidation. However, it has not been proven by experiments rigorously so far, although the early-stage surface oxidation state within several atomic layers at room temperature [2-4], and the bulk oxidation process involving the crystalline phase transition behaviors[5] were investigated. These studies are not directly linked to the depth profile of layered structure of the oxides. To observe such surface-front oxidation behaviors in real-time under in-situ heating

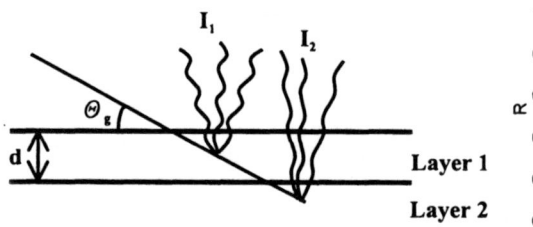

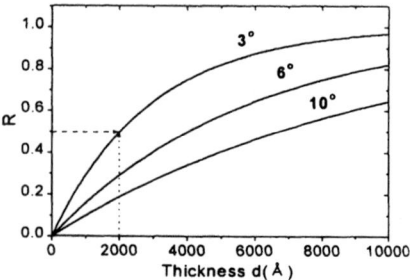

Fig. 1. The glancing-angle fluorescence EXAFS geometry and the EXAFS intensity from the thin film layer. I_1 represents the fluorescence intensity from the top layer (thickness d), and I_2 does the fluorescence intensity from the remaining extended region.

Fig. 2. Computed $R=I_1/(I_1+I_2)$ as a function of layer thickness (d) at various glancing angles (Θ_g). Here the approximate formula can be expressed as $R = 1 - \exp(-\mu d / \sin\Theta_g)$. When μ =0.18μm^{-1} and Θ_g =3°, d=2000 Å is obtained by choosing R=50%.

condition, we performed synchrotron x-ray scattering measurements and the glancing-angle fluorescence and reflection EXAFS measurements.

EXPERIMENTS

Pure Fe (99.999%) samples of 10x10 mm^2 were polished to obtain a flat, clean surface without excluding the initial natural oxide state. For in-situ experiments, samples were heated in air on a ceramic heating stage capable of heating in the range of 25-700℃. The XRD measurements were performed at the beamline 3C2 , and the glancing-angle EXAFS at the beamline 3C1 at PLS (Pohang Light Source). Time-resolved EXAFS experiments can be performed at the fast scan mode, i.e. under continuous scan motion. [6] Since nominal speed of the EXAFS scan reaches to one scan per minute, we may take time-resolved EXAFS data under quick temperature change as fast as 50℃/min. The temperature was resolvable to 50℃ per each scan.

The oxide depth of interest can be tuned by choosing an appropriate grazing incident angle. In figure 1, the experimental diagram illustrates the fluorescence signal intensities with I_1 (from the top layer) and I_2 (from the remaining extended region). In figure 2, at a chosen glancing angle (Θ_g), each curve for $I_1/(I_1+I_2)$ - computed as a function of layer thickness - represents the fluorescence contribution from the developed oxide layer. In these experiments, as shown for Θ_g= 3°, 2000 Å was the typical thickness of interest in probing. The fluorescence is measured using a silicon detector (Canberra PIPS), placed along the surface normal. In this fluorescence geometry, the XANES data are rather enhanced, but the EXAFS data are largely distorted by the amplitude reduction. In our reflection EXAFS measurements, the glancing angle was selected below the critical angle 0.5° to accommodate the Fe K-edge (E_o=7112eV) EXAFS scan ranges.

RESULTS AND DISCUSSION

Figure 3 shows the XANES data taken at different heating temperatures under the step-scan (a relatively slow scan) mode (a) and the QEXAFS(Quick-EXAFS) scan mode (b). As interpreted from the standard XANES data, the mid-edge-bump near E_o-threshold in Fe_3O_4 and Fe_2O_3 is more oscillatory compared to that of FeO phase. Heating below 500 ℃ leads the growth of the edge-bump oscillation, with increasing the oxide layer thickness. At 600 ℃, this edge-bump is rather smoothened by newly populated FeO phase. The successive heating at 650 ℃ makes this edge-bump recover to a Fe_2O_3-like bump. Since XANES data are not high energy-resolved, the Fe_2O_3 phase is hardly distinguished from the Fe_3O_4 phase.[7] On the other hand, a dramatic reduction of the fluorescence EXAFS yield (associated with the step size from the pre-edge background to the EXAFS signal level) took place between 400 ℃ and 500 ℃. This apparent decrease of the absorption coefficient results from a lower density of the oxide layer in our observation depth, typically 2000 Å.

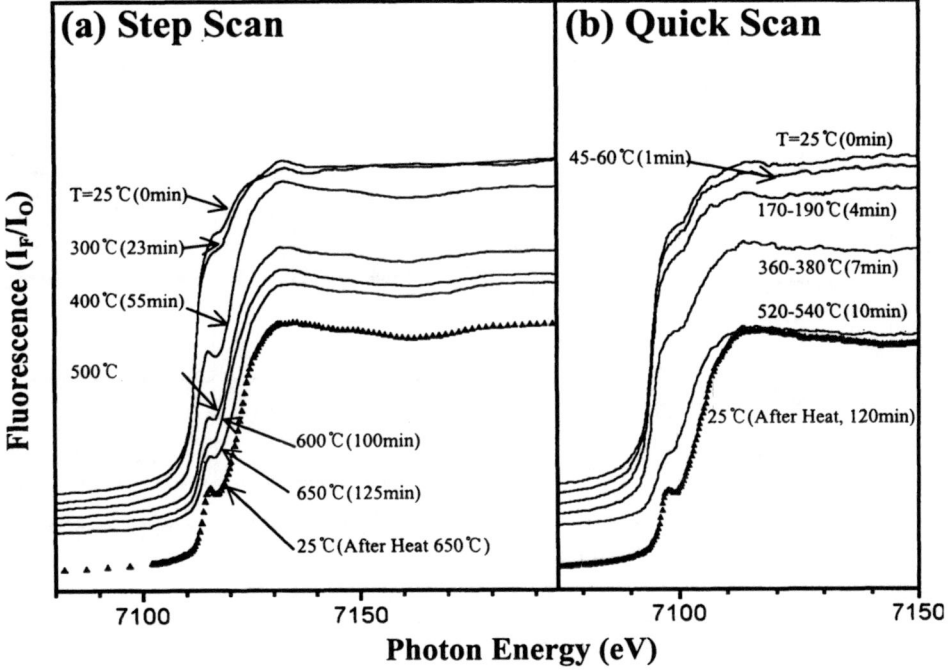

Fig. 3. (a) Step-scan EXAFS and (b) Quick-scan EXAFS measured through the glancing-angle Fe K-edge fluorescence yields during temperature-driven oxidation process. QEXAFS is performed to measure the quickly developed oxidation process within 10 minutes while the Fe is heated to 600 ℃.

At 400℃, the early-stage iron oxide phase is formed, as observed from the significant difference between 25℃ (pure iron) and 400℃ through FT r-space diagram in figure 4. The first shell in 1.5-2.5 Å range is attributed to the Fe-O bond(1.93-1.95 Å) from the Fe_2O_3 or the Fe_3O_4 crystalline phase.[8-9] The second shell featured from the dominant Fe-Fe bond represents that a large amount of pure iron phase remains near the surface volume. At this point, through EXAFS analysis, it was not easy to determine the phase between (Fe_3O_4 + Fe) and Fe_2O_3. More convincing results are available from our XRD data in figure 5, supporting the existence of the Fe_3O_4 rather than the Fe_2O_3. It is now conclusive that the Fe_3O_4 phase below 400℃ is initially dominant near surface zone.

At 500℃, the increasing amount of the Fe-O bond in the first shell and the decreasing amount of the Fe-Fe bond in the second shell - as observed through the amplitude reduction in figure 4 - indicate the growth of Fe_3O_4 phase as temperature increases. XRD data are consistently

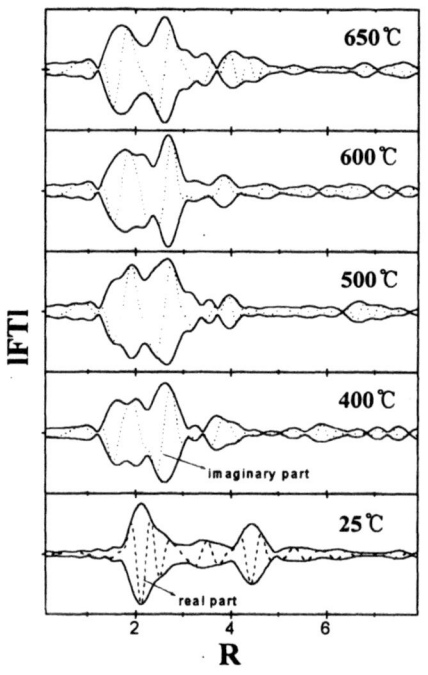

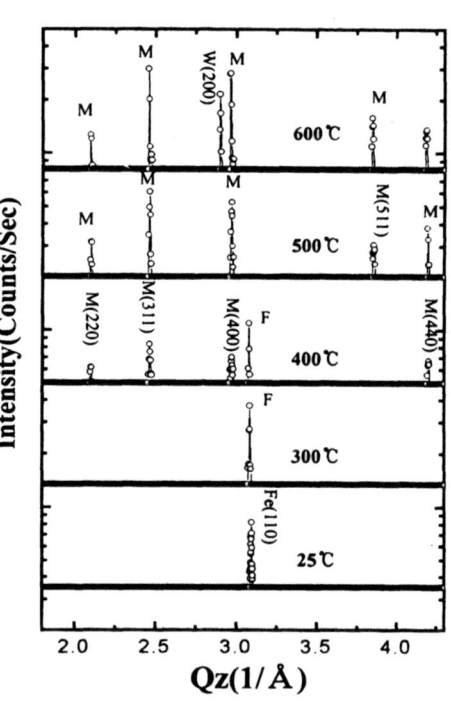

Fig. 4. The k-weighted Fourier transforms of the chi-data for Fe oxides at different heating temperatures. At 25℃, the real part (dashed) of FT denotes the Fe-Fe contribution from the iron crystalline. In 400-650℃, the imaginary part (dotted) of FT denotes the Fe-O contribution. The parameters chosen for this UWXAFS ··· ıputation are K_{min}=2.6, K_{max}=11.3 and dK=1.0.

Fig. 5. X-ray diffraction patterns at various oxidation in-situ temperatures. Fe peaks appear at 25℃ and 300℃, and Fe_3O_4 peaks at 400℃. Fe peaks are reduced until disappearance at 500℃, but Fe_3O_4 peaks increase. At 600℃, FeO peaks appear. After a prolonged heating at 650℃, Fe_2O_3 peaks appear.

324

interpreted as the growing Fe_3O_4 and the reducing Fe phase.

At 600℃, a dramatic change is observed at the second shell in the FT r-space, indicating the dominant Fe-O contribution – it is reasoned since the imaginary component plotted with dotted line becomes a main contributor to the second shell peak in |FT|. From this we might conclude that the FeO phase was dominant because the Fe-O bond (2.87 Å) exists only in the FeO phase. This FT r-space interpretation is consistent with emerging FeO crystalline phase in XRD. However, the Fe_2O_3 phase is not detected from both EXAFS and XRD.

At 650℃, the increasing amount of the Fe-O bond is observed from the first shell, and Fe-Fe (2.88 Å in Fe_2O_3) bond from the second shell. Thus it is conclusive that the surface-front layer (within 2000 Å) is almost covered with the Fe_2O_3. It is also understood that the complete coverage of the Fe_2O_3 layer requires the enough annealing time and temperature.

From other studies, of which detailed experimental data are not included in this paper, we can point out the following facts; (i) Our reflection EXAFS measurements probing the species of utmost-surface structure reflect that the more plausible utmost-surface oxidation phase is Fe_2O_3 rather than Fe_3O_4, even below 500℃. (ii) The relevant SEM picture, 2-D surface-front image, shows the phase separation between the faceted part (Fe_2O_3 island) and the smooth part (background Fe_3O_4 layer).

Below 600℃, the Fe_2O_3 phase is not observed significantly from both fluorescence EXAFS and XRD, but measurable through a reflection EXAFS. This implies that the surface-front Fe_2O_3 formation is associated with the island-to-layer crystal growth mechanism. Fe_2O_3 phase is initially growing in island mode above 500℃, and then gradually covers the whole surface until the stabilized oxide layer is formed with time and temperature. Recalling our previous Quick-EXAFS data, since the quickly heated surface-front layer is less populated with the Fe_2O_3 or the Fe_3O_4 phase in a short time scale, the Fe_2O_3-like XANES features are not apparently observed in figure 3(b). Although the XANES feature resembles the FeO phase, it is not reasonable to accept the stable FeO phase below 600℃. At this point, we may assert that the early phase is formed in FeO-like structure due to oxygen deficiency, but it may not be identifiable from XRD due to its small crystalline size. In the process of forming such nanocrystalline structure, it is likely that the oxygen diffusion velocity competes with the crystalline growth velocity at a given temperature.

In this study, it is better to conclude that such quickly developed oxidation mechanism is not fully understood so far. However, we note that we have demonstrated that our QEXAFS technique is well suited to study the temperature-driven oxidation process, because the angle-resolved x-ray diffraction experiment cannot compete with this time-resolved EXAFS in data collection speed.

CONCLUSIONS

In this study, the synchrotron based XRD, glancing-angle fluorescence EXAFS, and reflection EXAFS, were employed to investigate the surface-front oxidation mechanism of iron. Through time-resolved Quick-EXAFS measurements, the formation mechanism of iron oxide phases were monitored under a rapidly rising temperature up to 600℃ for 10 minutes. EXAFS results indicate that the FeO-like phase was prevailing in the microstructure even below 600℃, indicating that only small amount of the Fe_2O_3 and the Fe_3O_4 phase were formed under quick annealing. These results are somewhat contradictory to those of slowly heated oxidation process, in which the Fe_3O_4 phase was dominant below 500℃, and the Fe_2O_3 and the FeO phase were dominant above 600℃.

ACKNOWLEDGEMENTS

Experiments at PLS were supported in part by MOST and POSCO(1PD9603101).

REFERENCES

1. John M. West, Basic Corrosion and Oxidation, 2nd ed. John Wiley & Sons, New York, 1986, pp 202-203
2. Tie Guo and M.L.denBoer, Phy. Rev. B **38**, 3711(1988)
3. Franco Ciccacci, Lamberto Duo and Ezio Puppin, Surface Science, **269**, 533(1992)
4. Bong-Soo Kim, S. Hong and D.W.Lynch, Phys. Rev. B **41**, 12227(1990)
5. J. Kucera and M. Hajduga, Oxidation of Metals, 41, 1(1994)
6. J.M. Lee, N.-E. Sung, J.-K. Park, J.-G. Yoon, J.-H. Kim, M.-H. Choi and K.-B. Lee, J. Synchrotron Rad., **5**, xxx (1998) in press
7. Alison J. Davenport and Michael Sansone, J. Electrochem. Soc. , **142**, 725(1995)
8. N T Barrett, P N Gibson, G N Greaves, P Mackle, K J Roberts and M Sacchi, J. Phys. D. , **22**, 546(1989)
9. http://cars.uchicago.edu/~newville/adb/, the Atoms.inp Archive

THE CHEMICAL ENVIRONMENT OF Er^{3+} IN a-Si:Er:H AND a-Si:Er:O:H

LEANDRO R. TESSLER[*], CÍNTHIA PIAMONTEZE[*], ANA CAROLA IÑIGUEZ[*],
M. C. MARTINS ALVES[**] and H. TOLENTINO[**]
[*]Instituto de Física "Gleb Wataghin", UNICAMP, C. P. 6165, 13083-970 Campinas, SP, Brazil,
tessler@ifi.unicamp.br
[**]Laboratório Nacional de Luz Síncrotron, C. P. 6192, 13083-970, Campinas, SP, Brazil

ABSTRACT

We have measured extended x-ray absorption fine structure (EXAFS) of the Er L_{III} edge in a-Si:Er:O:H with different concentrations of Er and O. The samples were prepared by reactive RF co-sputtering from a silicon target partially covered with metallic erbium platelets. They present the characteristic Er^{3+} photoluminescence at 1.54 μm as deposited. The FFT of the Er EXAFS provides two well separated peaks. The characteristics of the first peak resemble those of Er$_2$O$_3$. We associate the first neighbor shell with oxygen atoms, even in non intentionally oxygenated samples. The average coordination and Er-O separation are significantly smaller than in Er$_2$O$_3$. This may be the reason why Er^{3+} luminescence in a-Si:H presents small temperature quenching. The second shell is interpreted as being composed of silicon atoms.

INTRODUCTION

The study of triply ionized erbium (Er^{3+}) luminescence in semiconductors has been an intensive area of research over the last few years because of possible photonic applications[1]. The Er^{3+} ions exhibit atomic-like luminescence that arises from electronic transitions within its incomplete internal 4f shell. In particular the $^4I_{13/2} \rightarrow {}^4I_{15/2}$ transition from the first excited to the fundamental state emits photons with a wavelength of ~ 1.54 μm. Because of the shielding provided by the outer filled 5s^2 and 5p^6 shells the transition wavelength depends very weakly on the details of the host. This wavelength is especially important because it corresponds to the window of minimum attenuation in the conventional silica based optical fibers currently used in optical communications. Although f-f transitions are dipole forbidden in the free ion, they become at least partially allowed in solid hosts due to the admixture of other angular momentum states to the f-wavefunction. Among the various semiconductor hosts available, silicon is especially interesting because of the possibility of photonic-electronic integration. Erbium luminescence has been reported both in crystalline [2] and hydrogenated amorphous silicon (a-Si:H) hosts. Roughly speaking, luminescent Er-containing a-Si:H has been prepared either by erbium ion implantation [3] or co-deposition [4,5].

The details of the chemical environment of erbium in crystalline silicon have been investigated by EXAFS [6,7] and in a-Si:H by emission Mössbauer spectroscopy [8]. In the present work we present the first EXAFS studies of Er containing a-Si:H. All samples studied turned out to contain oxygen, either intentional or as a contaminant. We confirmed that erbium acts as a strong getter for oxygen in silicon. Moreover, although the erbium chemical environment is similar to that found in Er$_2$O$_3$, the average coordination is normally smaller and the Er-O distance decreases as the oxygen concentration decreases.

EXPERIMENT

Sample Preparation

Erbium doped a-Si:H and a-Si:O:H samples were prepared by RF co-sputtering following the procedure described elsewhere [9]. Briefly, we used a 3" silicon target (99.999%) partially covered by small metallic erbium platelets (99.9%). The crystalline silicon substrates were kept at 200°C during the depositions. The base pressure was always better than 2×10^{-6} torr. The sputtering gas consisted of 1 mTorr of H_2 (99.999%) and 9 mTorr of Ar (99.999%). A partial pressure of O_2 (99.999%) between 3×10^{-6} and 10^{-4} was added in order to obtain oxidized samples. In an effort to obtain a sample as free of oxygen as possible, after pumping the deposition system overnight the target was pre-sputtered for 80 minutes before the depositions.

Erbium and oxygen concentrations were determined by RBS. Erbium to silicon concentrations for all samples is below 1 at. %. All samples studied present room temperature photoluminescence as deposited when excited by the 514.5 nm line of an Ar^+ laser. We did not perform any post-deposition thermal treatments in our samples.

EXAFS Measurements

Erbium L_{III}-edge EXAFS was measured with total secondary electron yield detection at the XAFS beamline of the Laboratório Nacional de Luz Síncrotron (LNLS) in Campinas, Brazil[10]. Care was taken in order to eliminate the Bragg peaks from the crystalline silicon substrates. The samples were at room temperature during the measurements. Reference samples of bulk Er_2O_3 and a thin film of $ErSi_2$ were used as reference standards for determining coordination numbers and bond lengths. EXAFS data were processed following the recommendations of the international committee [11]. Data treatment and simulations were done using the WINXAS analysis software [12].

RESULTS

In Figure 1 we show the raw EXAFS data for a representative sample and for the model compounds. The corresponding $\chi(k)$ are represented in Figure 2. It is clear from these figure that Er_2O_3-like oscillations are present in a-Si:Er:O:H . This pattern occurred for all the samples studied. Fourier transforms of $\chi(k) \cdot k^2$ are shown in Figure 3. From the experimental curves it is clear that the first neighbor shell of Er in a-Si:H is similar to that in Er_2O_3, indicating that in average the first neighbors of Er are oxygen atoms. The second peak is closer to the Er-Si separation in $ErSi_2$. In principle, this peak could be due to the presence of a different erbium site or to the second neighbor shell in erbium atoms surrounded by oxygen. In order to check these possibilities we performed backtransforms of the filtered first and second peaks, and compared them with simulations using the amplitudes and phases obtained from the Er_2O_3 and $ErSi_2$ standards. The same systematic procedure was repeated for all the samples: atomic separation r_0 coordination N_O and Debye-Waller factor σ_O^2 for the first shell were obtained assuming an oxygen first shell. The results obtained for different oxygen concentrations are represented in Table I. Then, with these parameters fixed, the second shell (assumed to be silicon) was included in the simulation, yielding the atomic separation r_{Si}, coordination N_{Si} and Debye-Waller factor σ_{Si}^2. The result of one of such simulations is represented in Figure 4. It turned out that except for improvement in the signal-to-noise ratio, the results did not depend on the erbium concentration in the range studied.

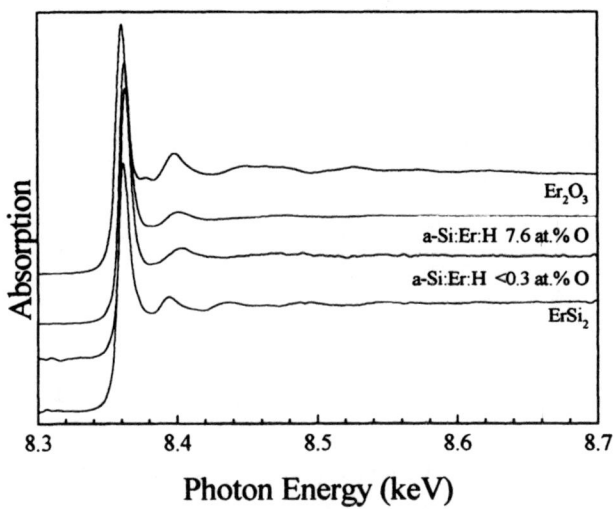

Figure 1. Raw Er L_{III}-edge x-ray absorption data from reference samples and erbium-doped a-Si:O:H. The oxygen concentrations are indicated.

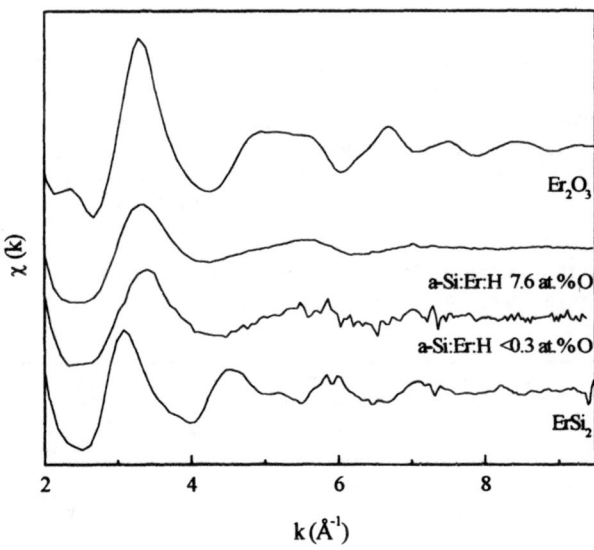

Figure 2. EXAFS signal $\chi(k)$ obtained from Figure 1

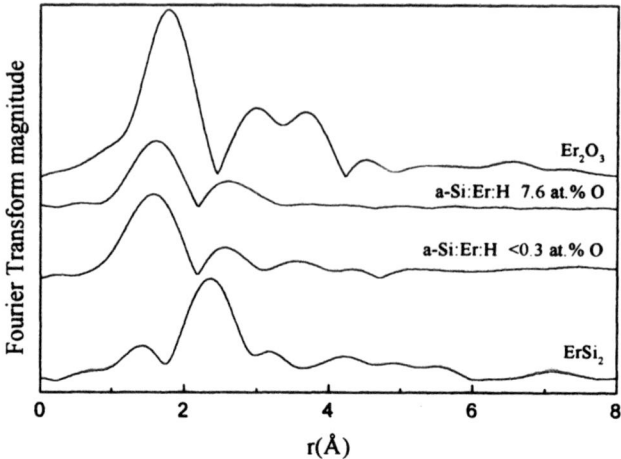

Figure 3. Fourier transform magnitudes of the EXAFS functions $\chi(k)$ multiplied by k^2. The similarity between a-Si:Er:O:H and Er_2O_3 is evident.

A discussion of the details of the second shell is beyond the scope of this paper. It is worth to mention only that for all the samples r_{Si} was in the range 3.10 to 3.19 Å, and the coordination number around 6 to 10.

DISCUSSION

The present results show clearly that in as-deposited luminescent a-Si:Er:H the immediate chemical neighborhood of erbium is very similar to that in Er_2O_3. This had already been proposed from photoluminescence [13] and Mössbauer spectroscopy [8] measurements. Moreover, some interesting features are also present:
a) The coordination number of the first oxygen shell is always much smaller than 6, the erbium coordination in Er_2O_3. The remainder oxygen is bonded to silicon forming a

Table I. Parameters of the first shell for samples with increasing oxygen contents. The error for the coordination number is estimated in 10 – 20%, and for the interatomic separation is below 0.02 Å

Sample	Coordination Number	Average Er-O separation (Å)	σ^2
A	2.2	2.12	0.0034
B	2.5	2.14	0.0034
C	2.7	2.12	0.0073
D	2.7	2.09	0.0034
E	2.7	2.07	0.0022
F	3.2	2.14	0.0033
G	3.6	2.14	0.0097
Er_2O_3	6	2.26	

substoichiometric a-Si:O:H alloy.

b)   The Er-O average interatomic distance is significantly smaller than in Er_2O_3, and decreases as the average shell coordination increases from 2 to 2.7, then starts to increase again. A very small decrease in this distance for Er in crystalline silicon had already been noticed [6].

c)   The Er-Si average second neighbor separation is significantly (up to 0.5 Å) larger than the first neighbor distance in $ErSi_2$. This indicates that indeed silicon atoms form the second shell for erbium atoms already surrounded by oxygen rather than the occurrence of two different erbium sites.

d)   The Debye-Waller factors obtained in the simulations are very small, indicating that erbium sits in a site with a high degree of local order, comparable to that of Er_2O_3. This is not surprising if we consider the fact that a-Si:H itself presents a relatively high degree of short range order.

e)   We have prepared several samples in our conventional RF-sputtering deposition system with special care to have no oxygen contamination. However, even these samples present the same general erbium EXAFS pattern: an oxygen first shell with low coordination number and a silicon second shell. It seems that erbium is so reactive with oxygen that any contamination results in an oxygen cage in a-Si:H. In such samples the infra-red local modes associated with Si-O vibrations can not be detected, which indicates that virtually all the oxygen present in the sample is bonded to erbium.

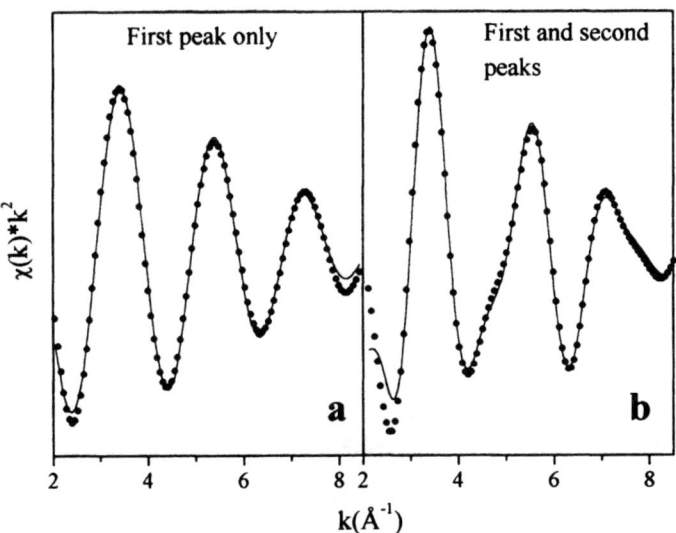

Figure 4. Simulation of the filtered backtransforms for the sample containing 7.6 at. % oxygen. In (a) only the first peak corresponding to oxygen was used. In (b) the two peaks were used, keeping fixed the parameters obtained from (a). The dots represent the backtransforms and the lines are the simulations.

CONCLUSIONS

We have studied the chemical environment of erbium in samples of a-Si:H and a-Si:O:H which present Er^{3+} luminescence at room temperature as-deposited. In all samples the general features are the same: the first shell consists of oxygen atoms and the second shell is made of silicon. The Er-O distance is smaller than in Er_2O_3 and is minimum for a coordination of 2.7. We were unable to prepare samples without the oxygen first neighbor shell surrounding the erbium atoms. This has an important implication for the interpretation of luminescence data. It has been shown that the temperature quenching in a-Si:Er:H is much less important than in crystalline silicon [3], even in samples without intentional oxygen contamination. It turns out that any residual oxygen (normally present in a-Si:H prepared in conventional deposition systems) tends to be bound to erbium and thus enhance the photoluminescence efficiency. It is well known that erbium is a strong oxygen getter in crystalline silicon [14]. Samples with lower Er-O separation may have the probability of non-radiative de-excitation to the silicon network reduced.

ACKNOWLEDGMENTS

The authors acknowledge I. Chambouleyron for critical reading of the manuscript. This work was partially supported by PRONEX "Fotônica em Telecomunicações", CNPq, FINEP, FAPESP contract 97/05097-8 and PADCT. The XAFS beamline was partially supported by FAPESP contract 95/6439-4

REFERENCES

1. Rare Earth Doped Semiconductors, edited by G. S. Pomrenke, P. B. Klein, and D. W. Langer, (Mater. Res. Soc. Proc. 301, Pittsburgh, PA 1993); Rare Earth Doped Semiconductors II, edited by A. Polman, S. Coffa and R. Schwartz, (Mater. Res. Soc. Proc. 422, Pittsburgh, PA 1996).
2. S. Coffa, G. Franzò, F. Priolo, A. Polman and R. Serna, Phys. Rev. B 49, 16313 (1994).
3. J. H. Shin, R. Serna, G. N. van den Hoven, A. Polman, W. G. J. H. M. van Sark and A. M. Vredenberg, Appl. Phys. Lett. 68, 997 (1996).
4. M. S. Bressler, O. B. Gusev, V. Kh. Kudoyarova, A. N. Kuznetsov, P. E. Pak, E. I. Terukov, I. E. Yassievitch, B. P. Zakharchenya, W. Fuhs and A. Sturm, Appl. Phys. Lett. 67, 3599 (1995).
5. A. R. Zanatta, L. A. O. Nunes and L. R. Tessler, Appl. Phys. Lett. 70, 511 (1997).
6. D. L. Adler, D. C. Jacobson, D. J. Eaglesham, M. A. Marcus, J. L. Benton, J. M. Poate and P. H. Citrin, Appl. Phys. Lett. 61, 2128 (1992).
7. A. Terrasi, G. Franzò, S. Coffa, F. Priolo, F. D'Acapito and S. Mobilio, Appl. Phys. Lett. 70, 1712 (1997).
8. V. F. Masterov, F. S. Nasredinov, P. P. Seregin, V. Kh. Kudoyarova, A. N. Kuznetsov and E. I. Terukov, Appl. Phys. Lett. 72, 728 (1998).
9. L. R. Tessler and A. C. Iñiguez, in Amorphous Silicon Technology –1998, (Mater. Res. Soc. Proc., Pittsburgh, PA 1998).
10. H. Tolentino, J.C. Cezar, D.Z. Cruz, V. Compagnon-Cailhol, E. Tamura, M.C. Martins Alves J. Synchrotron Rad. 5, (1998) in press.
11. International workshop on standards and criteria in x-ray absorption spectroscopy, Physics B 158, 701 (1989).
12. T. Ressler, J. Physique IV 7, C2-269 (1997).
13. L. R. Tessler and A. R. Zanatta, J. Non-Cryst. Sol., (1998) in press.
14. A. Polman, J. Appl. Phys. 82, 1 (1997).

AN EXAFS STUDY OF COMPOSITIONAL HOMOGENEITY IN SOL-GEL PROCESSED POTASSIUM TANTALUM NIOBATES

A. P. WILKINSON, J. XU, AND S. PATTANAIK
School of Chemistry and Biochemistry, Georgia Institute of Technology, Atlanta, GA 30332-0400

ABSTRACT

A series of $K(Ta_{1-x}Nb_x)O_3$ solid-solution samples were prepared by direct reaction of oxides and carbonates, a homogeneous sol-gel process, and an inhomogeneous sol-gel process. The inhomogeneous sol-gel samples were prepared by the prehydrolysis of separate $KTaO_3$ and $KNbO_3$ precursor solutions followed by mixing of the resulting sols, drying and calcination. An examination of these samples by EXAFS at the Ta L_{III} and Nb K-edges, i) illustrated the difficulty of obtaining homogeneous solid solutions *via* direct reaction of oxides/carbonates, and ii) showed that the homogeneity of a solid-solution could be controlled by varying the hydrolysis procedure used during its sol-gel synthesis.

INTRODUCTION

The compositional homogeneity of a material can have a profound effect on its properties. In the case of ferroelectrics, short length scale compositional fluctuations have been implicated in the relaxor behavior of materials such as $Pb(Mg_{1/3}Nb_{2/3})O_3$ (PMN),[1-6] and variations of composition within bulk samples will influence the temperature dependence of the dielectric constant for solid solutions such as $K(Ta_{1-x}Nb_x)O_3$ (KTN) and $Pb(Zr_{1-x}Ti_x)O_3$ (PZT). This arises because the ferroelectric to paraelectric phase transition temperature is composition dependent.[7]

Typically, solution-processing routes, such as alkoxide based sol-gel methods, are employed with the goal of obtaining materials that display good compositional homogeneity. While there have been reports that sol-gel methods do not always deliver compositionally homogeneous materials,[8-10] we are not aware of any prior efforts to employ sol-gel chemistry as a tool for introducing controlled amounts of compositional inhomogeneity into a material. In this paper, we present results from our initial experiments on the manipulation of compositional homogeneity in the perovskite-solid-solution KTN by variation of the hydrolysis procedure used during the sol-gel synthesis of these materials.

EXPERIMENTAL

Sample preparation

The $K(Ta_{1-x}Nb_x)O_3$ samples were synthesized using three distinct approaches: i) direct reaction of oxides and carbonates, ii) calcining gels that had been prepared by the hydrolysis of homogeneous solutions containing potassium, niobium and tantalum precursors (subsequently referred to as homogeneous sol-gel samples), and iii) calcining gels that had been prepared by mixing prehydrolyzed $KNbO_3$ and $KTaO_3$ precursor solutions (subsequently referred to as inhomogeneous sol-gel samples). The $Nb(OEt)_5$ and $Ta(OEt)_5$ used during these experiments were prepared according to literature procedures,[11,12] the potassium acetate was obtained from Fischer scientific, and both the 2-methoxyethanol and acetylacetone were obtained from Aldrich.

Homogeneous sol-gel samples: Potassium acetate was dissolved in 2-methoxyethanol. The solution was briefly refluxed and then concentrated by distillation. Stoichiometric amounts of $Nb(OEt)_5$ and $Ta(OEt)_5$ dissolved in 2-methoxyethanol were added to the potassium acetate solution. The mixture was briefly refluxed and the solution reduced in volume by distillation. Acetylacetone was then added to slow down the hydrolysis of the alkoxides. A 4:1 water : metal alkoxide ratio was used for the hydrolysis. For the 50:50 solid solution sample, a gel formed after ~30 hours. The gel was dried in a vacuum oven overnight at 110 °C. Portions of resulting xerogel were then heat treated in air. The same method was used for both the $K(Ta_{0.67}Nb_{0.33})O_3$, and $K(Nb_{0.50}Ta_{0.50})O_3$ samples.

Inhomogeneous KTN 50:50 sol-gel samples: Separate KN and KT precursor solutions were prepared using the method outlined for the KTN precursor solution. The concentrations of the KT and KN precursors were determined gravimetrically. Prior to the hydrolysis of these solutions, 0.5 moles of acteylacetone per mole of metal alkoxide were added to each of the precursors. A 4:1 water : metal alkoxide ratio was used for the hydrolysis. A solution of the required water in 2-methoxyethanol was added to both the KN and KT precursors dropwise. The KN appeared to react more rapidly than the KT, as indicated by the viscosity of the two sols. The hydrolyzed KT precursor was then added to an equal amount of the hydrolyzed KN precursor while the later was a high viscosity sol. The mixture was stirred for several minutes and then dried in a vacuum oven. Portions of the resulting xerogel were heated in air.

Direct reaction of oxides and carbonates: Two samples of $KTa_{0.5}Nb_{0.5}O_3$, subsequently referred to as A and B, were prepared *via* this route. Stoichiometric amounts of K_2CO_3, Nb_2O_5 and Ta_2O_5 were mixed and ground together. For sample A, the resulting powder was loose fired at 1000 °C for 20 hours. For sample B a pellet was pressed and heated to 1000 °C for 10 hours, the material was then reground formed into a pellet and heated to 1100 °C for 10 hours. The material was reground, pressed into a pellet and fired for a further 10 hours at 1000 °C. EDX was used to check the compositional homogeneity of the two samples. Reference samples of $KTaO_3$ and $KNbO_3$ were also prepared by the direct reaction of carbonates and oxides.

EXAFS measurements

X-ray absorption spectra of the samples were recorded at both the Nb-K and Ta L_{III} edges in transmission mode at beamline X11A, National Synchrotron Light Source, Brookhaven National Laboratory, with the storage ring operating at 2.58 GeV and 150-350 mA. Finely powdered samples were diluted with boron nitride to give $\mu x \approx 1.5$, and packed into aluminum sample holders with Kapton windows. A Si(111) double crystal monochromator and a 1 x 10 mm x-ray beam were employed for all of the measurements. Three ion chambers were used as detectors so that a reference spectrum from a metal foil could be recorded at the same time each sample was examined. All the samples were cooled to ~ -50 °C using liquid nitrogen before any data was taken (temperature measured in the middle of the sample on its surface). The monochromator was detuned by 20% for the Nb K-edge and 40% for the Ta L_{III} edge spectra respectively. For all but one of the samples (Sample A), data was taken to a k_{max} of 20 and 18 Å^{-1} for the Nb and Ta edges respectively. All data were processed using EXAFSPAK.[13] FEFF6 was used to calculate appropriate phase and amplitude functions for the EXAFS data analysis.[14]

RESULTS

A comparison of both the FT magnitudes and $k^3\chi(k)$ for the Ta L_{III}-edge data from ceramic samples A, B and the reference material $KTaO_3$ (KT) is shown in Figure 1. Nb K-edge data for samples A and B are compared with the reference material $KNbO_3$ (KN) in Figure 2. For the Ta L_{III} edge data there is a striking similarity between both the FT magnitudes and $k^3\chi(k)$ for the KT reference and sample A, and the Nb K-edge data for sample A strongly resemble those for the KN reference material. The data for ceramic sample B differ markedly from those observed for both the KT and KN reference materials. Fitting these data to estimate the number of tantalums and niobiums in the first shell of transition metal ions around the absorbing atom is complicated by the strong correlation between thermal parameters and occupancies even when the total number of tantalum and niobium ions is constrained to be six. However, with any reasonable assumed values for the thermal parameters it is clear that sample A is highly inhomogeneous with approximately 5 Ta and 1 Nb ions around each absorbing tantalum, and 5 Nb and 1 Ta ions around each absorbing Nb atom. A high level of inhomogeniety in this sample was also indicated by our EDAX measurements. The phases and amplitudes used in fitting all of the KTN EXAFS data were calculated using FEFF6 and a perovskite structural model containing 3 Nb and 3 Ta ions in the first shell of transition metals around the absorber. A comparison of the data for ceramic sample B with that measured for the "homogeneous" sol-gel KTN 50:50 sample and a preliminary fitting of the data for sample B indicates that it is close to a random solid solution.

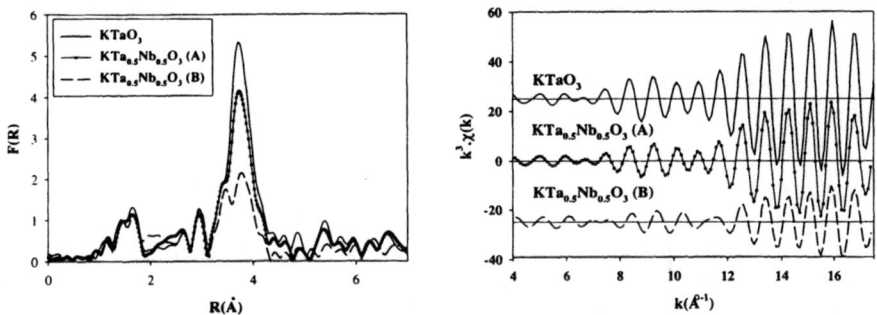

Figure 1. A comparison of the Ta L_{III} edge FT magnitudes and $k^3\chi(k)$ for $KTa_{0.5}Nb_{0.5}O_3$ samples A and B, and $KTaO_3$

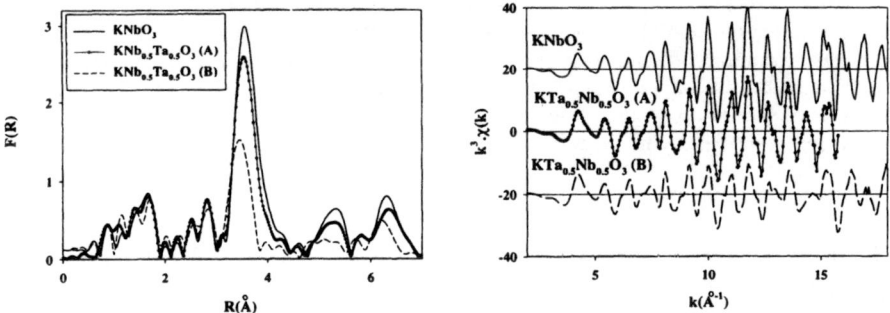

Figure 2. A comparison of the Nb K-edge FT magnitudes and $k^3\chi(k)$ for $KTa_{0.5}Nb_{0.5}O_3$ samples A and B, and $KNbO_3$. k_{max} ~15.4 Å^{-1} was used for all of the Fourier transforms.

A comparison of the FT magnitudes and $k^3\chi(k)$ for the KTN 50:50 and KTN 67:33 "homogeneous" sol-gel samples with those observed for the KTN 50:50 "inhomogeneous" sol-gel sample is presented in Figure 3. All of the data in this figure are for materials heated to 1000°C. The data for the KTN 50:50 "inhomogeneous" sample is clearly different from that for the KTN 50:50 "homogeneous" sample and most closely resembles that for the "homogeneous" KTN 67:33 sample. This suggests that each tantalum in the "inhomogeneous" sol-gel sample had on average approximately 4 Ta and 2 Nb ions in its immediate coordination environment. A preliminary fitting of these data, while broadly consistent with this, indicates that the compositional inhomogeneity in the "inhomogeneous" sol-gel sample is not quite that pronounced.

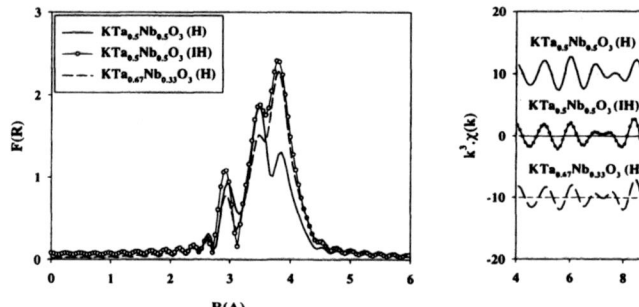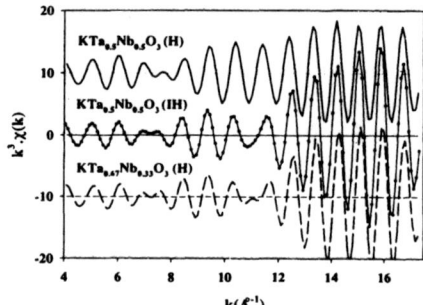

Figure 3. A comparison of the Ta L_{III} edge Fourier filtered FT magnitudes and $k^3\chi(k)$ for the KTN 50:50 and KTN 67:33 "homogeneous" sol-gel samples with the those observed for the KTN 50:50 "inhomogeneous" sample.

CONCLUSIONS

The data presented in Figures 1 and 2, along with our preliminary fitting results clearly indicate the difficulty of obtaining homogeneous solid-solution products by direct reaction of oxides and carbonates. Sample A is massively inhomogeneous and is not far from a mixture of KN and KT. Even after three grinding and calcinations steps the product (sample B) is probably not as homogeneous as that obtained by the "homogeneous" sol-gel procedure.

The data presented in Figure 3 show that there is a memory of the hydrolysis procedure used in the preparation of the KTN 50:50 sol-gel samples, and that variations in processing chemistry can be used to control the solid solutions homogeneity. The observed level of inhomogeneity in the sample prepared using the "inhomogeneous" hydrolysis procedure was not as great as we at first expected, perhaps indicating that there is an M-O-M link breaking and remaking process going on during the gel drying process.

ACKNOWLEDGEMENTS

We are grateful for financial support under NSF award DMR-9623890. The EXAFS data collection was performed using beam line X11A at the NSLS.

REFERENCES

1   V. A. Isupov, Sov. Phys. - Technical Physics **1**, 1846-1849 (1956).

2   B. N. Rolov, Sov. Phys. - Solid State **6**, 1676-1678 (1965).

3   G. A. Smolensky, J. Phys. Soc. Jpn. **S28**, 26-37 (1970).

4   N. Setter and L. E. Cross, J. Appl. Phys. **51**, 4356-4360 (1980).

5   V. V. Kirillov and V. A. Isupov, Ferroelectrics **5**, 3-9 (1973).

6   L. E. Cross, Ferroelectrics **76**, 241-267 (1987).

7   F. Jona and G. Shirane, *Ferroelectric Crystals* (Dover, 1993).

8   S. S. Sengupta, L. Ma, D. L. Adler, and D. A. Payne, J. Mater. Res. **10**, 1345-1348 (1995).

9   C. D. E. Lakeman, Z. Xu, and D. A. Payne, J. Mater. Res. **10**, 2042-2051 (1995).

10   B. A. Tuttle, T. J. Headley, B. C. Bunker, R. W. Schwartz, T. J. Zender, C. L. Hernandez, D. C. Goodnow, R. J. Tissot, and J. Michael, J. Mater. Res. **7**, 1876-1882 (1992).

11   D. C. Bradley, W. Wardlaw, and A. Whitley, J. Chem. Soc., 726-728 (1955).

12   D. C. Bradley, B. N. Chakravarti, and W. Wardlaw, J. Chem. Soc., 2381-2384 (1956).

13   G. N. George and I. J. Pickering, *EXAFSPAK - a Suite of Computer Programs for Analysis of X-ray Absorption Spectra*, SSRL, Stanford University, CA, USA, (1993).

14   S. I. Zabinsky, J. J. Rehr, A. Ankubinov, R. C. Albers, and M. J. Eller, Phys. Rev. B **52**, 2995 (1995).

X-RAY ABSORPTION SPECTROSCOPY STUDIES OF ELECTROCHEMICALLY DEPOSITED THIN OXIDE FILMS

M. Balasubramanian[*], C.A. Melendres[*], A.N. Mansour[+] and S. Mini[#]

[*] Argonne National Laboratory, Argonne, IL 60439
[+] Naval Surface Warfare Center, West Bethesda, MD 20817
[#] Northern Illinois University, DeKalb, IL 60115

ABSTRACT

We have utilized "in situ" X-ray Absorption Fine Structure Spectroscopy to investigate the structure and composition of thin oxide films of nickel and iron that have been prepared by electrodeposition on a graphite substrate from aqueous solutions. The films are generally disordered. Structural information has been obtained from the analysis of the data. We also present initial findings on the local structure of heavy metal ions, e.g. Sr and Ce, incorporated into the electrodeposited nickel oxide films. Our results are of importance in a number of technological applications, among them, batteries, fuel cells, electrochromic and ferroelectric materials, corrosion protection, as well as environmental speciation and remediation.

INTRODUCTION

Radioactive contamination of piping systems and storage tanks generally emanates from radionuclides that have been incorporated into corrosion products and scale deposits that coat the surface of the metal alloys which constitute the materials of construction of the pipes and tanks. Knowledge of both the local structure of the corrosion scales by itself and that of the heavy metal ions that are incorporated into the corrosion scales is essential in developing cost-effective clean-up processes via chemical and electrochemical methods[1]. Of particular interest are the corrosion scales on base metals such as Fe, Ni and Cr which are the main constituents of the steels used in nuclear waste storage tanks and piping systems. Corrosion processes are basically electrochemical in nature, and consequently electrochemical deposition is an excellent method to simulate the formation of scale deposits. For example, it is widely accepted that the corrosion film on nickel consists mainly of $Ni(OH)_2$ in the passive region and $NiOOH$ in the transpassive region[2]. The corrosion films can be well simulated by electrochemical deposition from an aqueous solution containing Ni^{2+} ions. Cathodic deposition results in $Ni(OH)_2$, while anodic deposition leads to the production of $NiOOH$[3]. Incorporation of heavy metal ions into the corrosion scales can occur by any one of the following processes: co-precipitation, lattice substitution into vacancies or interstitial sites, surface adsorption or absorption in pores that ultimately results in the co-deposition of the impurity atom with the oxide film on a surface. In this paper we report the use of X-ray Absorption Fine Structure (XAFS) to obtain structural and electronic information on both the host oxide films of Ni and Fe, as well as the incorporated heavy metal ions, Sr and Ce. The use of high energy x-rays enables the measurements to be carried out "in situ"; i.e., the films can be studied in the presence of solution and under potential control in an appropriately designed spectro-electrochemical cell. As XAFS does not rely on long range order to provide structural information, highly disordered materials (which is often the case for corrosion films) can be studied. The X-ray Absorption Near Edge Structure (XANES) contains valuable information on the oxidation state of the probed atom.

Mat. Res. Soc. Symp. Proc. Vol. 524 © 1998 Materials Research Society

EXPERIMENTAL DETAILS

The host metal oxide (hydroxide) films were prepared either by cathodic or anodic deposition from aqueous solutions of the metal ion of interest onto a graphite electrode. Thus, $Ni(OH)_2$ films were prepared by cathodic electrolysis of $Ni(NO_3)_2$ solution at constant current similar to that followed by Capehart et al.[4], NiOOH films were formed by anodic deposition at positive potentials close to oxygen evolution from Ni(II) sulfate plus acetate solution, as was done by Chen and Noufi[3]; iron oxide films were prepared by constant potential electrolysis from dilute solutions of Fe (II) in borate buffer solution[6]. Incorporation of Sr and Ce in nickel oxide films was carried out by co-deposition from a solution containing both Ni^{2+} as well as the impurity ions. A saturated calomel electrode (SCE) was used as reference for the electrochemical experiments. The electrochemical equipment consisted of a Princeton Applied Research (PAR) model 173 potentiostat/galvanostat, a PAR model 179 universal programmer and a Hewlett Packard model 7045 XY recorder. A specially fabricated spectro-electrochemical cell was used for simultaneous electrochemical and "in situ" XAFS measurements.

XAFS measurements were performed at beamlines X-11A, X-10C and X-23A2 of the National Synchrotron Light Source (NSLS) using a Si (111), Si (220) or Si (311) double crystal monochromator, respectively. Care was taken to minimize harmonic contamination. The monochromator was calibrated using appropriate pure metal foils. Measurements were performed either in standard transmission mode or in the fluorescence mode depending on the absorption edge step-height obtained for the various films. Care was taken to minimize thickness effect in transmission measurements and fluorescence distortion in fluorescence measurements. A suitable Z-1 filter was often used in conjunction with a Stern-Heald Soller slit assembly to minimize the elastic and Compton scattered background components in fluorescence measurements. Alternatively, in some cases a 13-element Ge solid state detector was used. Both "in situ" (in the presence of solution at open circuit or with an applied potential) and " ex situ" (in air) measurements were carried out on the electrodeposited films.

RESULTS AND DISCUSSION

(1) Nickel Oxide Thin Films

$Ni(OH)_2$ was deposited from a 0.1 M $Ni(NO_3)_2$ solution at a constant current density of 8 mA/cm^2 for a period of 5 to 15 minutes. The greenish, adherent deposit was found to be α-$Ni(OH)_2$ by XAFS analysis. Ni atoms were coordinated to about 6 oxygen first near neighbors at 2.05 Å and to 6 Ni second near neighbors at 3.10 Å. The electrochemically prepared α-$Ni(OH)_2$ is more disordered than β-$Ni(OH)_2$ which we prepared chemically. Fig. (1) compares the XAFS $\chi(k)$ function and the corresponding Fourier Transforms (FT) for the two materials. These findings are in good agreement with the results of others[4].

NiOOH films were prepared by anodic deposition at a constant potential of 1.10 V (vs. SCE) from a solution of 0.05 M $NiSO_4$ and 0.5 M Na_2SO_4. The $\chi(k)$ functions and the corresponding FT's are illustrated in Fig. (2). The first shell of the as-deposited film can be fit using two Ni-O bond distances at 1.88 Å and 2.05 Å. In conjunction with findings on reference compounds the former distance can be associated with a Ni^{4+} species and the latter distance with a Ni^{2+} species[5]. The film was further charged in 1M KOH solution to convert more of the Ni^{2+} species to $Ni^{4+.}$ In order to understand the electrochemistry of this system the charged film was discharged at a rate of 1 mA/cm^2 in 1M KOH. The discharged film contained only Ni^{2+} species and Ni atoms are coordinated to a first shell of 6 oxygen atoms at a distance of 2.05 Å. The conversion of Ni^{4+} to Ni^{2+} is evident in the FT, which shows that the FT peaks shift to larger r-values on discharge, as expected for a Ni^{2+} species. This discharged film can be oxidized to

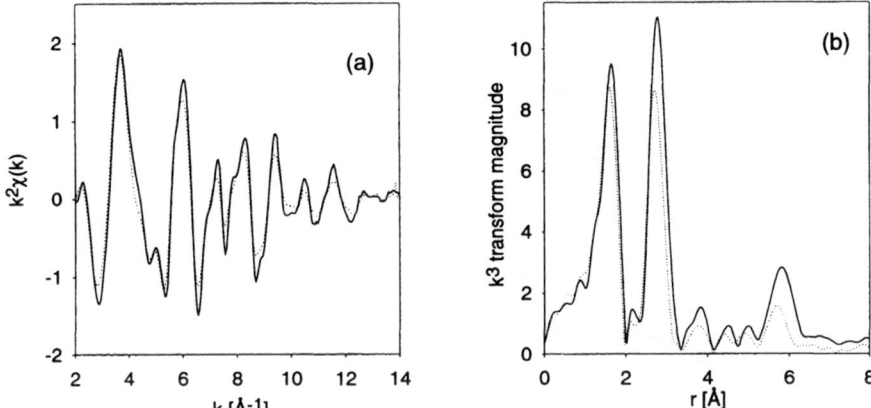

Figure 1: (a) Comparison of the XAFS $\chi(k)$ function for electrodeposited α-Ni(OH)$_2$ (dotted) and chemically prepared β-Ni(OH)$_2$ (solid), (b) Corresponding Fourier transforms

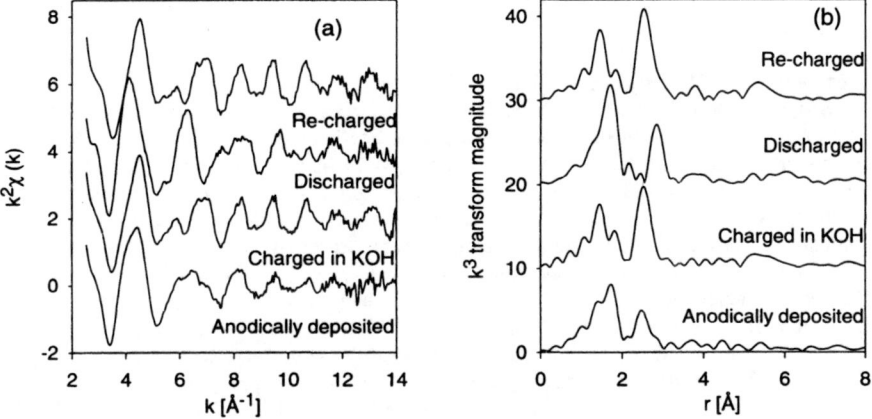

Figure 2: (a) Comparison of XAFS $\chi(k)$ spectra of nickel oxide thin film formed by anodic deposition and taken through oxidation-reduction cycles, (b) Corresponding Fourier Transform

regain some of the Ni^{4+} by recharging in KOH. Detailed data analysis and other findings such as the self-discharge behavior of anodic films are beyond the scope of this paper and will be published elsewhere[7].

(2) Iron Oxide Thin Films

The composition and structure of the corrosion films on iron remain a subject of much debate. We report preliminary XAFS measurements performed on anodically deposited iron oxide films. Iron oxide thin films were prepared following the procedure of Markovac and Cohen[6] using dilute solutions of Fe^{2+} in a borate buffer. Anodic deposition was carried out from a deaerated solution of 0.005 M $FeSO_4$ + 0.3 M H_3BO_3 + 0.075 M $Na_2B_4O_7$ borate buffer solution by holding the potential at 0.6 V (vs. SCE) for 8 hours and 10 minutes. The XANES spectrum of the film formed is shown in Fig. (3a). This film was subsequently reduced at −1.05 V (vs. SCE) and the spectrum shown in Fig. (3b) was obtained. The spectrum of bcc Fe foil is shown in Fig. (3c) for comparison. It is evident that a part of the film has been converted to bcc Fe. This can be seen more clearly in the Fig. (4), which shows the XAFS $\chi(k)$ spectra of the anodic film, the reduced film and bcc Fe foil. The $\chi(k)$ function of the anodic film diminishes very rapidly in intensity with increasing k. This indicates that the anodic film is very disordered. Preliminary fits to the data suggest that the anodic film is some form of FeOOH. More detailed experimentation and data analysis is required to elucidate the exact structure of the anodic film. The XAFS spectrum of the reduced film is distinctly different from the as-deposited anodic film. The reduced film shows spectral features that are similar to bcc Fe in the high-k region. However, the low-k region is distinctly different from bcc Fe. This difference in the low-k region indicates that some form of oxide is also present in the reduced film. Oxygen is a good backscatter of low-energy electrons but does not backscatterer significantly at high-k. Consequently, the low-k region of the XAFS spectra is different from bcc Fe. Detailed electrochemistry, XAFS data collection and data analysis of the iron oxide films will be reported subsequently.

(3) Incorporation of Strontium in Nickel Oxide Thin Films

Sr was co-deposited with both NiOOH and $Ni(OH)_2$ from aqueous solutions containing Ni^{2+} ions and 0.005 to 0.1 M strontium acetate. Anodically co-deposited samples were prepared at an applied potential of 1.1 V (vs. SCE) for about 8 hours while the cathodic films were prepared at a constant cathodic current density of 8 mA/cm^2 for about 15 minutes. Fig. (5) shows the Sr XAFS of an anodic film compared with a pure strontium oxide (hydroxide) which was prepared by passage of a constant cathodic current of 8 mA/cm^2 through a solution of 0.1 M strontium acetate. Both measurements were performed "ex situ". The XAFS in both cases look similar and the corresponding FT's show only one distinct peak. This shell can be fit to about 8 to 10 oxygen first near neighbors at 2.62 (2) Å. This distance is similar to the Sr-O bond distance of 2.62 Å found for Sr^{2+} in aqueous solutions[8]. Similarly, cathodically co-deposited samples of Sr and $Ni(OH)_2$ also show only a single Sr-O first neighbor shell at 2.61 (2) Å . In all cases no intercalation or lattice substitution of Sr^{2+} as manifested by a Sr-Ni bond distance could be detected in the XAFS. Incorporation of Sr^{+2} into $Ni(OH)_2$ and NiOOH films via electrodeposition appears to occur by a co-precipitation process that results in a separate phase. The absence of Sr-Ni bonds suggests that intercalation or lattice substitution into the nickel oxide lattice is negligible. Furthermore, the absence of higher shell peaks in Sr XAFS shows that the co-precipitated phase is very disordered. More details of the data analysis can be found in reference (7).

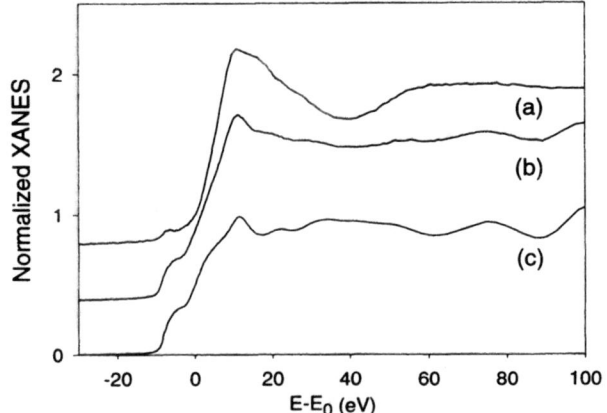

Figure 3: Fe K-edge XANES of
(a) Anodically deposited film at +0.6 V (SCE)
(b) Sample (a) cathodically reduced at -1.05 V (SCE)
(c) Fe foil

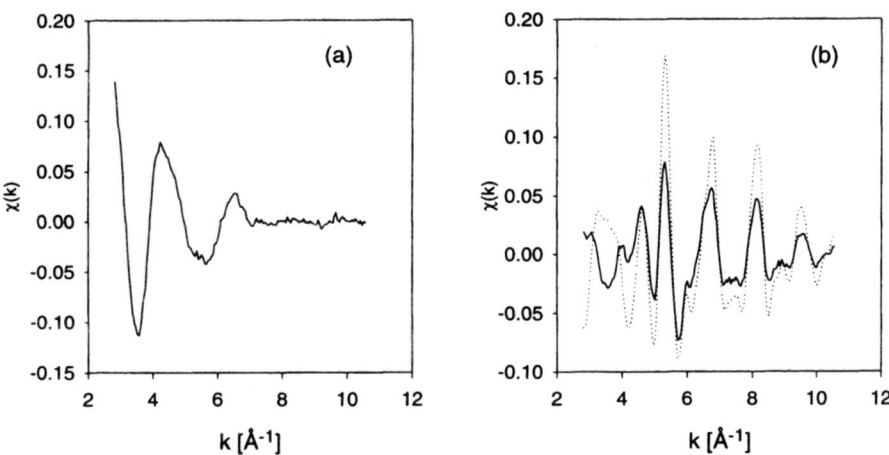

Figure 4: Fe K-edge $\chi(k)$ spectra of
(a) Anodically deposited film at +0.6 V vs. SCE
(b) Fe foil, taken at ~ 80 K (dotted) and cathodic reduction
of anodic film at -1.05 V vs. SCE, taken at 298 K (solid)

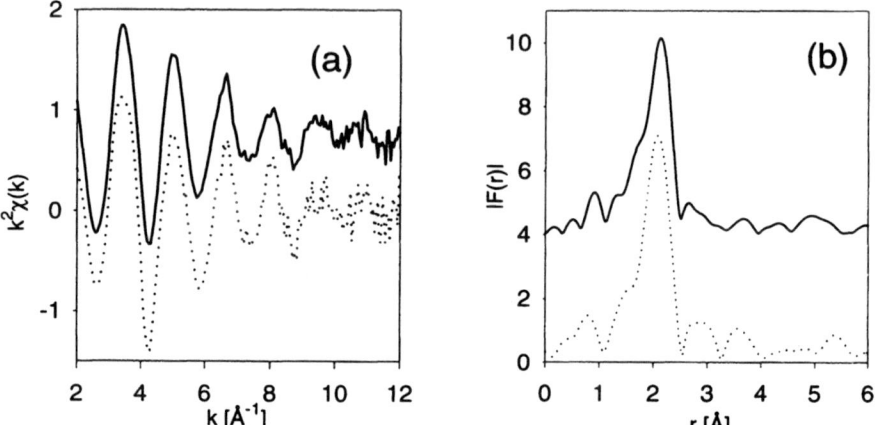

Figure 5: (a) Sr K-edge XAFS $\chi(k)$ function in cathodically deposited pure Sr-oxide film (dotted) and Sr anodically co-deposited with NiOOH (solid) (b) Corresponding Fourier transforms

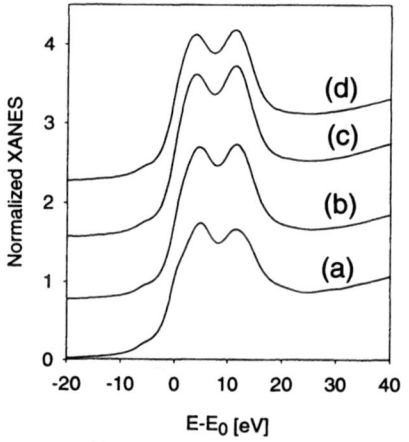

Figure 6: Ce L_3- XANES of
(a) CeO_2
(b) $Ce(OH)_4$
(c) Ce in NiOOH
(d) Pure anodic Ce oxide film

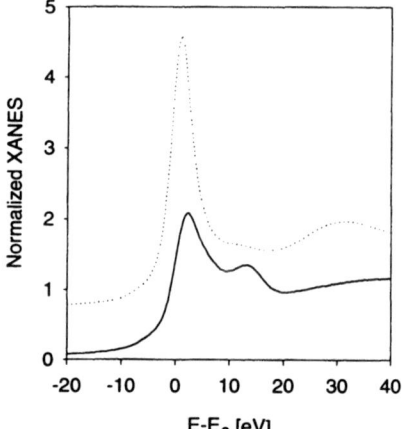

Figure 7: Ce L_3- XANES of
0.08 M Ce(III) nitrate solution (dotted), and Ce in $Ni(OH)_2$ (solid)

344

(4) Incorporation of Cerium in Nickel Oxide Thin Films

The general shape of the near-edge spectra of Ce at the L_3-edge is a good fingerprint of the valence state of Ce in its oxides/hydroxides. Ce(III) compounds exhibit an intense white line at the threshold energy while Ce(IV) compounds show a double peak structure. We have studied the incorporation of Ce in both NiOOH and Ni(OH)$_2$ by co-depositing Ce with the nickel oxides from solutions containing Ce^{3+} and Ni^{2+} ions. Fig. (6c) shows the "in situ" XANES spectrum of Ce incorporated into NiOOH by anodic co-deposition at 1.1 V (vs. SCE) from a solution of 0.05 M NiSO$_4$ + 0.5 M sodium acetate + 0.04 M cerium (III) acetate for 9 hours. The double peak indicates the presence of Ce^{4+} in the film by comparison with the CeO$_2$ and Ce(OH)$_4$ standards (Fig. 6a and b, respectively). We could also deposit a pure cerium oxide film by constant potential electrolysis at 1.1 V (vs. SCE) in 0.08 M Ce(III) acetate solution. The XANES of this film shown in Fig. (6d) is quite similar to Ce in NiOOH film. By using laser Raman measurements, x-ray diffraction and XAFS analysis we find that the structure of the anodic films is similar to Ce(OH)$_4$. We also find that Ce(OH)$_4$ is not really a hydroxide but rather a disordered form of hydrous cerium oxide. This material has the same medium range structure (~ 4 Å) as crystalline CeO$_2$ but has much higher structural disorder. More details can be found in references (7 and 9).

Fig. (7) shows the XANES of Ce in Ni(OH)$_2$ and that of 0.08 M Ce(III) acetate solution. The co-deposited film was prepared from solution containing 0.01 M Ce(III) acetate solution and 0.1 M Ni(NO$_3$)$_2$. The XANES of Ce in the co-deposited film shows a doublet at threshold. The first peak is larger and broader than in Ce(IV) compounds (Fig. 6). This indicates that the Ce in the film is a superposition of Ce^{3+} and Ce^{4+} components. This is a surprising result because one expects Ce to be present as Ce^{3+} on cathodic deposition. This can be explained, however, if the Ce^{3+} species cathodically co-deposited with Ni(OH)$_2$ is easily oxidized by dissolved oxygen. The outer parts of the film, in contact with solution, may have lost contact with the electrode and would then be oxidized to Ce^{4+}. Further details of this investigation will be published elsewhere[7, 9].

ACKNOWLEDGEMENTS

This research was supported by the U. S. Department of Energy (DOE), Office of Environmental Management and Office of Energy Research under contract No. W-31-109-ENG-38. We thank Dr. J. Woicik and the staff at beamline X-10C of NSLS for help with some of the experiments.

REFERENCES

(1) R. P. Allen and H. W. Arrowsmith, Materials Performance **18**, 216 (1979).

(2) C. A. Melendres and M. Pankuch, J. Electroanal. Chem. **333**, 103 (1992).

(3) Y. D. Chen and R. N. Noufi, J. Electrochem. Soc. **131**, 1447 (1984).

(4) T. W. Capehart, D. A. Corrigan, R. S. Lonell, K. I. Pandya and R. W. Hoffman, Appl. Phys. Lett. **58**, 865 (1991).

(5) A. N. Mansour and C. A. Melendres, Physica B **208&209**, 583 (1995).

(6) V. Markovac and M. Cohen, J. Electrochem. Soc. **114**, 678 (1967).

(7) M. Balasubramanian, C. A. Melendres and A. N. Mansour, to be published in J. Electrochem Soc.

(8) D. M. Pfund, J. G. Darab, J. L. Fulton and Y. Ma, J. Phys. Chem. **98**, 13102 (1994).

(9) M. Balasubramanian, C. A. Melendres and A. N. Mansour, submitted to Thin Solid Films.

Synchrotron X-ray Absorption Studies of Atomic-Level Alloying in Immiscible Mixtures

J.-H. He, P.J. Schilling, and E. Ma
Center for Advanced Microstructures and Devices (CAMD) and Department of Mechanical
Engineering, Louisiana State University, Baton Rouge, LA 70803

ABSTRACT

An X-ray absorption beamline has been developed recently at the electron storage ring of
the LSU Center for Advanced Microstructures and Devices. Using Extended X-ray Absorption
Fine Structure (EXAFS) and X-ray Absorption Near Edge Structure (XANES), we have studied
the local atomic environments in immiscible mixtures processed by high-energy ball milling, a
mechanical alloying technique involving heavy deformation. By examining the local coordination
and bond distances, it is concluded that atomic-level alloying can indeed be induced between Cu
and Fe through milling at room temperature, forming substitutional fcc and bcc solid solutions. In
addition to single-phase regions, a two-phase region consisting of fcc/bcc solutions has been found
after milling at both room temperature and liquid nitrogen temperature. In contrast to the Cu-Fe
system, solid solution formation is not detectable in milled Ag-Fe and Cu-Ta mixtures. This work
demonstrates the power of synchrotron EXAFS/XANES experiments in monitoring
nonequilibrium alloying on the atomic level. At the same time, the results provide direct
experimental evidence of the capability as well as limitations of high-energy ball milling to form
alloys in positive-heat-of-mixing systems.

INTRODUCTION

As seen in equilibrium phase diagrams, a large number of binary systems exhibit very
limited mutual solid solubility (e.g., Cu-Fe) or even negligible liquid solubility (e.g., Ag-Fe).
Such systems are characterized by a relatively large positive heat of mixing. These immiscible
(either in solid or in both solid and liquid state) elements are difficult to alloy in solid state, as
diffusional processes naturally lead to phase separation. Among solid-state processing techniques,
it is well known that mechanical alloying, a high-energy ball milling technique involving repeated
deformation, fracturing, and cold welding of powder particles, is capable of producing intimate,
ultrafine mixtures of immiscible elements. Recently, there has even been evidence indicating that
solid solutioning, i.e., alloying on atomic level, is achievable using this nonequilibrium technique
in positive-heat-of-mixing systems such as Cu-Fe [1-4] and Co-Cu [5].

To characterize the phases present in transformation products, the most widely used
techniques are transmission electron microscopy (TEM) and conventional x-ray diffraction (XRD).
However, these techniques sometimes encounter difficulties in the analysis of milled powders.
TEM and high-resolution TEM observations, in addition to requiring tedious sample preparation,
only reveal atomic arrangements for submicron-sized local regions in individual particles [6]. In
XRD patterns, Bragg peaks of the constituent elements tend to overlap because these peaks are
significantly broadened due to the small domain sizes and high level of strains in these heavily
deformed powders [2-5]. Also, the Bragg peaks of one of the elements can fade away due to
absorption effects (if the two elements are very different in atomic number)[7] or the formation of
small coherent regions [5] rather than resulting from true alloying. It is hence often not a trivial
task to prove unequivocally atomic-level alloying in such milled mixtures [2-5, 8-11].

X-ray absorption techniques such as EXAFS should be very helpful in overcoming these
problems. The technique is atom specific and looks at the absorption edge of each element and
hence avoids the problem of peak overlapping. It is also short-range-order sensitive and directly
probes the local environment of the absorbing atom. In addition, EXAFS signal comes from a
relatively large amount of powder particles, and as such provides bulk-averaged information and
hence an overall picture in a statistical sense. EXAFS is also powerful because it is available from
a large number of elements. Other techniques capable of providing atomic-level information, such

347

as Mössbauer spectroscopy and magnetization measurements, are restricted to magnetic elements [8-11].

In the following, we present some of the x-ray absorption results of three ball-milled systems, Cu-Fe, Ag-Fe, and Cu-Ta. These data have been acquired using the recently-completed beamline at the electron storage ring of the LSU Center for Advanced Microstructures and Devices (CAMD).

EXPERIMENTS

Commercial pure Cu (99% purity, -325 mesh), Ag (99.9% purity, 4-7 micron), Fe (99.9% purity, 6-9 micron) and Ta (99% purity, -325 mesh) powders were used as the starting material. The ball milling experiments were performed using a SPEX 8000 shaker mill/mixer either at room temperature or at liquid nitrogen temperature (external cooling). Elemental powders were mixed at the desired composition and sealed with milling balls in a vial under a purified argon atmosphere. The typical ball to powder weight ratio used was 4:1 and the typical milling duration was 20 hours. EXAFS measurements at Cu and Fe K edges and Ag L_{III} edge, in both transmission and total electron yield modes, were performed at LSU-CAMD synchrotron operating with an electron beam energy of 1.3 GeV and an injection current of 180 mA. Standard software packages (CDXAS and UWXAFS) were used for data reduction, which followed a standard procedure of background removal, extraction of the EXAFS signal, $\chi\ (k)$, Fourier transform of $\chi\ (k)$, and inverse transform to isolate the EXAFS contribution from a selected region in real space [12,13].

RESULTS AND DISCUSSIONS

Cu-Fe

Selected absorption data at Fe-K and Cu-K edges are presented in Figure 1 for the elements and representative mixtures after milling. Fourier transforms of the EXAFS signals of Cu-Fe

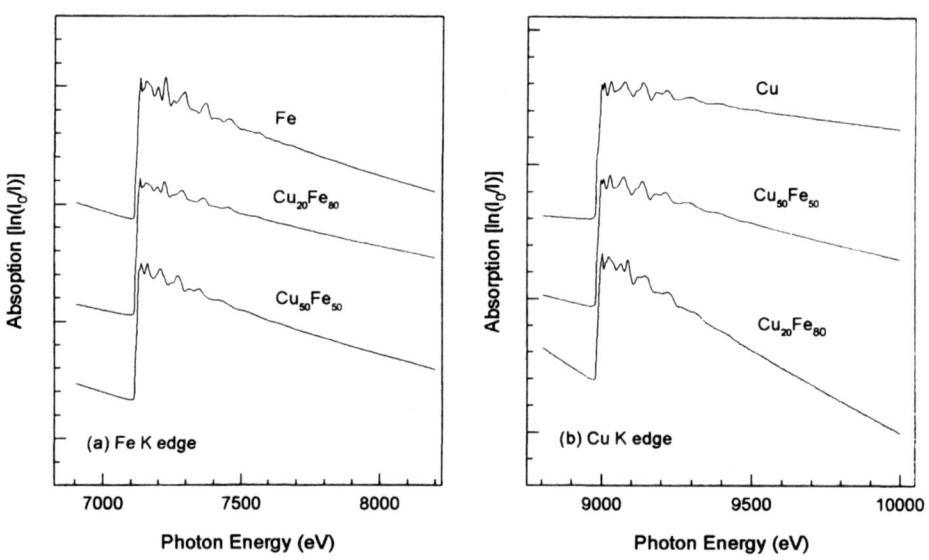

Fig. 1   X-ray absorption spectra of some representative powders in the Cu-Fe system.

powders before and after milling are shown in Figure 2. In the left column, pure Fe shows a clear bcc fingerprint. When Fe is milled with 20%Cu, Fe signature remains unchanged. However, when Cu is increased to 50%, the Fourier transform of Fe-K edge no longer match that of pure bcc Fe but instead bears an fcc signature similar to that of pure Cu, indicating that Fe is taking on the fcc coordination in an fcc solid solution. Similarly, in the right column, the Cu-K edge transforms exhibit fcc features for both pure Cu and Cu with 50% Fe. However, Cu in the 80%Fe mixture closely resembles the "fingerprint" of bcc Fe, suggesting bcc coordination. In other words, Cu has dissolved into Fe forming substitutional bcc $Cu_{20}Fe_{80}$ solid solution. Further information regarding local environment is provided by quantitative analysis. Nearest-neighbor structural parameters (coordination number of like and unlike atoms and bond distances) obtained from the best fits (Figure 3) for ball-milled $Cu_{50}Fe_{50}$ are shown in Table I. It is seen that both Cu and Fe are now surrounded by both like and unlike neighbors. Such quantitative simulation confirms atomic-level alloying and rules out the possibility that Fe is in fcc coordination (or Cu in bcc coordination) not because of alloying but mainly because of formation of fcc Fe (or bcc Cu) domains coherent with Cu (or Fe) [4,14].

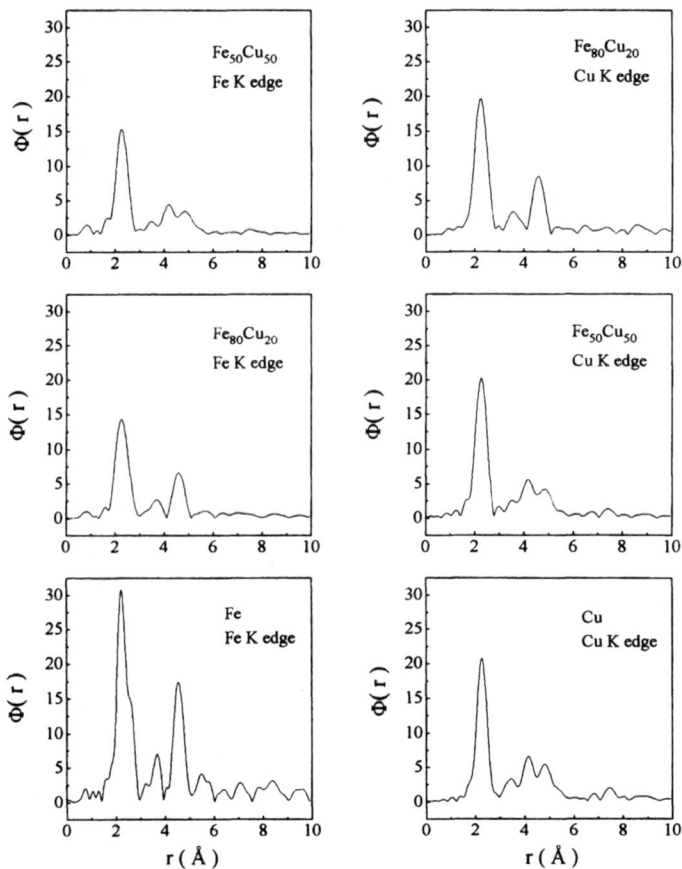

Fig. 2   Fourier transforms of Cu and Fe K edge EXAFS of as-received and ball-milled Cu-Fe powders.

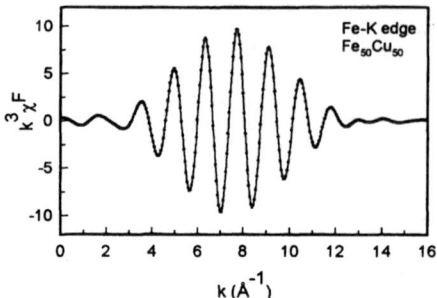
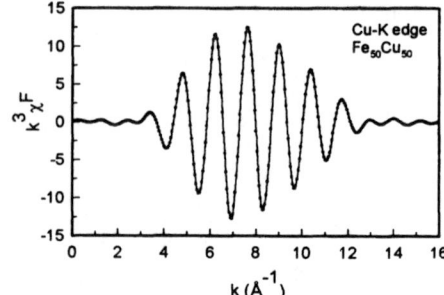

Fig. 3   Inverse Fourier transforms of the first peaks of $Cu_{50}Fe_{50}$. Dotted curves are
results of fitting.

Table I   Structral parameters obtained from *ab initio* simulations of EXAFS
for ball-milled $Fe_{50}Cu_{50}$.

Central Atom	N_{unlike}	N_{like}	r (Å)	σ_1^2 ($Å^2$)	σ_2^2 ($Å^2$)
Cu	5.9±0.7	6.1±0.7	2.571±0.002	0.0087±0.0001	0.0082±0.0001
Fe	5.6±1.4	6.0±1.4	2.583±0.005	0.0097±0.0002	0.0092±0.0003

Here, σ_1^2 and σ_2^2 are the mean square deviation about the average bond distance r of
unlike pair and like pair, respectively.

As discussed above, both fcc and bcc solutions have been obtained in the Cu-Fe system. It is
of interest to study how the crystal structure depends on the overall mixture composition. Shown
in Fig. 4 are XANES spectra at Cu-K and Fe-K edges for $Cu_{100-x}Fe_x$ milled at liquid nitrogen
temperature. It can be seen that bcc solution forms on the Fe rich side whereas fcc solution forms
on the Cu rich side. The transition between the two structures is not abrupt, rather it occurs
through a two-phase (fcc/bcc) region spanning a composition range of approximately x=60-70.

Ag-Fe and Cu-Ta

EXAFS data obtained for ball-milled Ag-Fe show very different alloying behavior. The
Fourier transforms are displayed in Figure 5 for $Ag_{50}Fe_{50}$ as an example. Both the Ag and Fe
edge transforms are compared with those of unmilled elemental fcc Ag and bcc Fe. Apparently Fe
and Ag give almost identical "signatures" before and after milling. Simulation of the inverse
transform of the first structural peak in the Fe transform shows that the best fit is obtained when all
nearest neighbors are Fe atoms with a bond distance very close to that of elemental Fe (2.482Å).
Adding unlike (Ag-Fe) pairs degrades the quality of the fit and leads to unrealistic bond distances
[15]. Cu-Ta behaves in a similar way [16], as suggested by the fitting results shown in Table II.
In both systems, alloying is insignificant, if it happens at all.

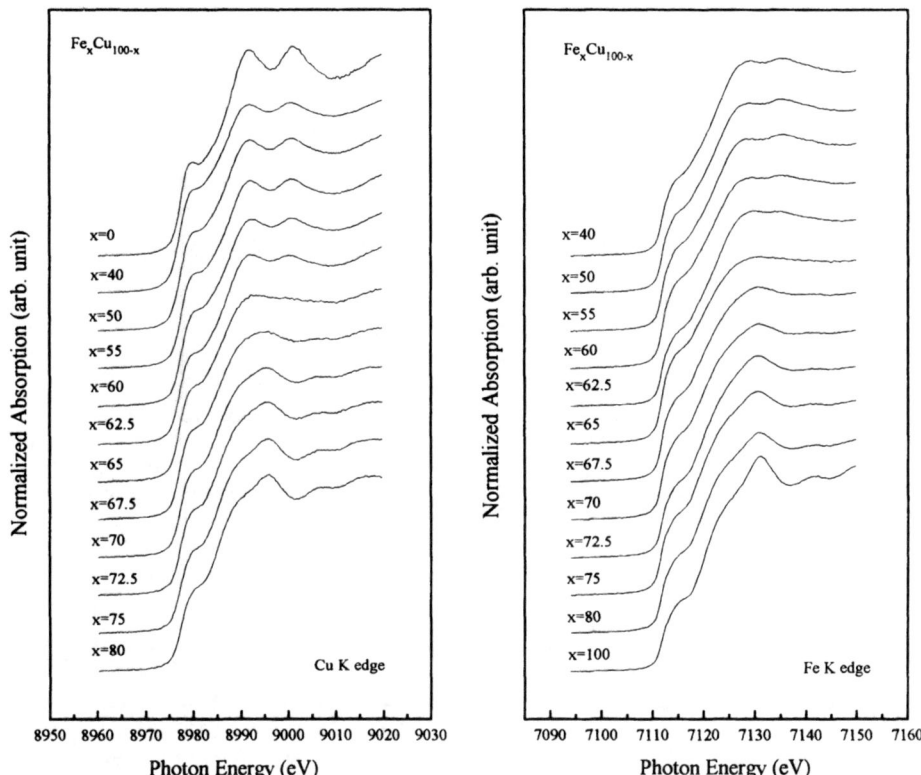

Fig. 4   XANES spectra at Cu-K and Fe-K edge for $Cu_{100-x}Fe_x$ milled at the liquid nitrogen temperature.

SUMMARY

EXAFS/XANES analysis suggests that true alloying between Cu and Fe forming fcc and bcc solid solutions can indeed be achieved at low temperatures using mechanical milling. A two-phase region persists in milled $Cu_{100-x}Fe_x$ around x=60-70. Mechanically induced alloying appears to be insignificant in systems with large positive heat of mixing and/or pronounced liquid immiscibility. EXAFS and XANES are powerful techniques for monitoring atomic-level alloying in immiscible mixtures.

This work was supported by the National Science Foundation, Grant No. DMR-9613865.

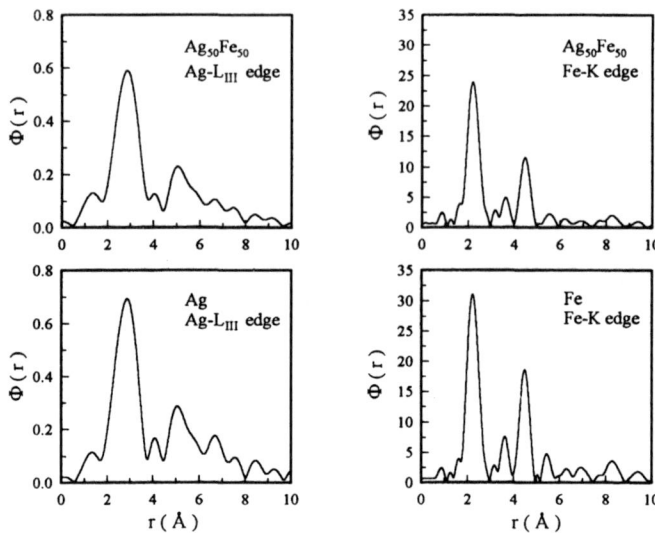

Fig. 5   Fourier transforms of Ag-L$_{III}$ and Fe-K edge EXAFS of Ag, Fe, and Ag$_{50}$Fe$_{50}$ powders.

Table II   Structral parameters obtained from *ab initio* simulations of EXAFS for ball-milled Cu$_{30}$Ta$_{70}$ and Ag$_{50}$Fe$_{50}$

System	Central atom	N$_{like}$	r (Å)	σ^2 (Å^2)
Cu$_{30}$Ta$_{70}$	Cu	10.3±1.2	2.548±0.008	0.0122±0.0010
Ag$_{50}$Fe$_{50}$	Fe	7.5±0.5	2.482±0.007	0.0052±0.0004

Here, σ^2 is the mean square deviation about the average bond distance.

REFERENCES

1. A.R. Yavari, P.J. Desre and T. Benameur, Phys. Rev. Lett. **68**, 2235 (1992).
2. J. Eckert, J.C. Holzer, C.E. Krill III and W.L. Johnson, J. Appl. Phys. **73**, 2794 (1993).
3. E. Ma, M. Atzmon and F.E. Pinkerton, J. Appl. Phys. **74**, 955 (1993).
4. O. Drbohlav and A.R. Yavari, Acta. Metall. Mater. **43**,1799 (1995).
5. C. Gente, M. Oehring and R. Bormann, Phys. Rev. **B48**, 13244 (1993).
6. J.Y. Huang, Y.K. Wu, A.Q. He and H.Q. Ye, Nanostructured Materials, **4**, 293 (1994); also **4**, 1 (1994).
7. H.P. Klug and L.E. Alexander, X-ray Diffraction Procedures, 2nd ed. (John Wiley, NY, 1974), Chap. 7, p. 531.
8. J. Kuyama, H. Inui, S. Imaoka, S. Nasu, K. Ishihara and P.H. Shingu, J. J. Appl. Phys. **30**, L854 (1991).
9. J. Jiang, U. Gonser, C. Gente and R. Bormann, Appl. Phys. Lett. **63**, 1056 (1993).
10. P. Macri, P. Rose, R. Frattini, S. Enzo, G. Principi, W. Hu, and N. Cowlam, J. Appl. Phys. 76, 4061 (1994).
11. U. Herr, J. Jing, U. Gonser and H. Gleiter, Sol. Stat. Comm. **76**, 197 (1990).
12. J. Wong, W.E. Nixon and J.W. Mitchell, J. Appl. Phys. **71**, 150 (1992).
13. E.A. Stern, M. Newville, B. Ravel, Y. Yacoby and D. Haskel, Physica B, in press.
14. P.J. Schilling, J. He, J. Cheng and E. Ma, Appl. Phys. Lett. **68**, 767 (1996).
15. E. Ma, J.H. He and P.J. Schilling, Phys. Rev. B **55**, 5542 (1997).
16. J. Xu, J.-H. He and E. Ma, Met. Mater. Trans. **28A**, 1569 (1997).

CHARACTERIZATION OF SILVER BINDING IN CRYPTOMELANE BY X-RAY ABSORPTION SPECTROSCOPY

R. Ravikumar*,1, D. W. Fuerstenau*, and G. A. Waychunas**,2
* Department of Materials Science and Mineral Engineering, University of California at Berkeley,
 Berkeley, CA 94720
** Center for Materials Research, Stanford University, Stanford, CA 94305
1 Siemens Microelectronics Inc., 1580 Route 52, B630/Z33A, Hopewell Junction, NY 12533
2 Lawrence Berkeley National Laboratory, Earth Sciences Division, 1 Cyclotron Road, Mail Stop 90-1116, Berkeley, CA 94720

Abstract

Using silver K-edge extended X-ray absorption fine structure (EXAFS) spectroscopy, two different samples of silver-containing manganese oxide were analyzed in the fluorescence mode. For the first sample, silver ions from solution were sorbed onto one synthetic manganese oxide phase, namely cryptomelane ($K_xMn_8O_{16}$, where $1<x<2$). The second sample was a silver-manganese oxide from Colorado. From the EXAFS analysis, silver was found to occupy two different sites in the synthetic sample. The natural samples from Colorado also exhibited a very similar coordination distances as the synthetic samples. In the low temperature spectrum of the synthetic sample at 10 K, the Ag-O peak was found to be missing and the amplitude of the Ag-Ag peak was approximately three times larger than the corresponding room temperature sample.

Introduction

Silver-manganese oxides are minerals that occur in the class of secondary ore bodies where silver is transported from its source and is subsequently trapped in a manganese oxide matrix. Silver-manganese oxide ores are present in parts of the western United States, Mexico, South America, and Sumatra. The amount of silver in these ore bodies varies between a few parts per million (ppm) to thousands of ppm. However, these ores are not compatible with conventional metallurgical treatment for the extraction of silver. Currently, there is no economically viable process to extract the silver from these manganese oxide ore bodies. Cryptomelane ($K_xMn_8O_{16}$, where $1<x<2$), a phase of manganese oxide with the hollandite structure, is quite widely present in parts of Colorado, with silver-bearing cryptomelane ore bodies containing up to a maximum of 1 wt% of silver. The structure of cryptomelane is made up of $[MnO_6]^{8-}$ octahedra in an edge and corner sharing geometry with regular holes or tunnels that are the sites for the potassium ions. The purpose of this research is to investigate the kind of sites silver occupies in the manganese oxide structure by X-ray absorption spectroscopy.

EXAFS spectra can be used to determine interatomic distances, the identity of the near neighbor atoms, thermal and static disorders associated with structural vibrations and irregularities [1, 2]. At a particular X-ray absorption edge, like silver, only the environments around the absorber atoms are sampled, yielding element-specific local structure.

Experimental

The X-ray absorption work for the silver K-edge (25514 eV) was conducted at beam-line 7-3 at the Stanford Synchrotron Radiation Laboratory (SSRL) during its regular running conditions (≈ 3 GeV and average current of 50 mA). The energy was scanned at 10 eV increment between 25285 and 25495 eV, and then over the edge (25495-25545 eV) at steps of 0.35 eV. For the EXAFS region (k between 1.62 and 16.2), the increment was 0.05 k. The energy selection was made using a Si (220) double-crystal monochromator. The spectra were collected in the fluorescence mode using a 13-element germanium detector and the XAS-Collect software program[1] at SSRL was used to obtain the data. During data collection, the monochromator was detuned by 50% to minimize the contributions from any higher-order harmonics. Multiple scans were collected for the various samples that were used.

There were three kinds of samples for the X-ray absorption experiments. The first kind were model compounds that included both silver metal and silver oxide. The second kind was the synthetic silver-loaded cryptomelane that was prepared at a variety of solution chemistry conditions. The samples were prepared as in the previously mentioned sorption experiments [7], after which the solid-liquid separation was performed in a Sorvall RC-5B centrifuge at 10000 rpm. This wet solid sample was used for the experiments. The third kind of sample was a natural silver-manganese oxide ore that was used in an earlier extraction investigation by Pesic et al. [3] and the details about this sample source and characteristics are listed there.

A Teflon sample holder with slots of dimension 30 mm X 5 mm was used to mount the sample in front of the X-ray beam. The solid samples (silver oxide and the silver-loaded synthetic sample) were applied as a thin uniform layer on a filter paper which was secured to the sample holder using MYLAR tape. As the natural sample had a smaller silver concentration (a few ounces/ton of ore), a thicker sample containing the silver-manganese oxide was used. All sample spectra were collected in the fluorescence mode with the sample holder oriented at a 45° angle to the incident X-ray beam. The data were collected at room temperature except for one silver-cryptomelane sample, where the data collection was also done at 10 K.

Results and Discussions

The X-ray edge spectrum for the standard samples silver metal foil and silver oxide showed that for silver metal the absorbance starts to increase at 25480 eV, and by 25500 eV the peak in the absorbance was reached. However, for silver oxide, the rise in the absorbance starts around 25500 eV. The edge spectra from the synthetic silver-loaded cryptomelane and the natural ore samples were compared to the standard samples. From the position of the start of the absorption peak, it is possible to say that silver is present as a monovalent ion in the manganese oxide matrix. Also, the relatively small rise in the absorbance for the natural ore indicates the dilute concentration of silver in these samples.

From the crystal structure of cryptomelane [4], the phase and amplitude functions for silver occupying the tunnel sites in a manganese oxide framework were derived using the Feff 6.20 program[2]. The shells corresponding to the different atom pairs are listed in Table I. The EXAFS data analysis was performed using the EXAFSPAK software package[3].

The complete X-ray absorption spectra collected at 10 K for the silver-loaded cryptomelane sample is shown in Figure 1. The edge energy, E_o, was chosen to be the first inflection point of the absorption edge. For the extracted EXAFS function a k^3 weighting is used. The radial distribution function (RDF) was obtained by performing the fourier transform of the weighted EXAFS function. Performing a back transform over a selected R range to k-space yields a filtered EXAFS signal. For all the EXAFS analysis, various filtered spectra were used. Shells corresponding to the various atoms in the structure were added in stages by increasing the range of the filter, and the optimal solution using the EXAFSPAK software[3] was obtained, where there was a good fit for both the EXAFS function and the RDF. The fitted spectra are calculated

---

[1] developed at SSRL.
[2] developed by J. Mustre de Leon, J. J. Lehr, R. C. Albers, and S. Zaminsky.
[3] developed at SSRL.

Table I. The model extracted values from Feff 6.2 of the amplitude ratios and the effective atomic distances in Angstroms, for the different atom pairs in the closest shell of cryptomelane that would contribute to the Ag-EXAFS function.

Atom pair	amplitude ratio	R-effective, Angstroms
Ag-Ag	100.0	2.87
Ag-O	100.0	2.87
Ag-O	99.1	2.88
Ag-O	36.2	3.17
Ag-O	27.3	3.45
Ag-Mn	35.3	3.57
Ag-Mn	33.3	3.63
Ag-Mn	49.6	3.99
Ag-Mn	43.8	4.14

using the phase and amplitude functions for Ag-Ag, Ag-Mn, and Ag-O atom pairs as generated by Feff 6.2. Figures 2a and 2b show the optimal solution for the EXAFS function and the RDF obtained for the silver-loaded cryptomelane sample at 10 K. The same procedure was performed to obtain the optimal solution with the synthetic and the natural ore samples. These spectra were obtained at room temperature. As all spectra were similar for the synthetic samples prepared at various surface chemistry conditions, these data sets were combined and processed again as one silver-loaded cryptomelane sample. The results of the optimal solution showing the filtered and the fitted spectra for the silver-loaded cryptomelane sample is shown in Figure 3a, and the corresponding RDF in Figure 3b. The results containing the number and type of atoms, bond distances, Debye-Waller factor and other parameters for the silver-loaded cryptomelane sample at 10K, at room temperature, and the natural ore sample are summarized in Tables II-IV.

Table II. The optimal solution obtained for the silver-loaded cryptomelane sample at 10 K. The values in the parenthesis correspond to the standard deviation and also signifies that those terms were floating variables for obtaining the solution.

Type of atom	bond distance Angstroms	number of atoms	σ^2 (Debye-Waller) Å^2	ΔE_0 eV
Silver	2.885 (0.003)	1.07 (0.09)	-0.0001 (0.0004)	-2.816
Manganese	3.745 (0.015)	17.42 (7.2)	0.0335 (0.006)	-2.816
Manganese	4.292 (0.011)	0.97 (0.50)	0.004 (0.003)	-2.816

Table III. The optimal solution obtained for the silver-loaded cryptomelane sample at room temperature. The values in the parenthesis correspond to the standard deviation and also signifies that those terms were floating variables for obtaining the solution.

Type of atom	bond distance Angstroms	number of atoms	σ^2 (Debye-Waller) Å^2	ΔE_0 eV
Silver	2.831 (0.005)	2.36 (0.13)	0.014	-4.335
Oxygen	2.231 (0.007)	2.23 (0.16)	0.014	-4.335
Manganese	3.671 (0.006)	5.57 (0.24)	0.014	-4.335

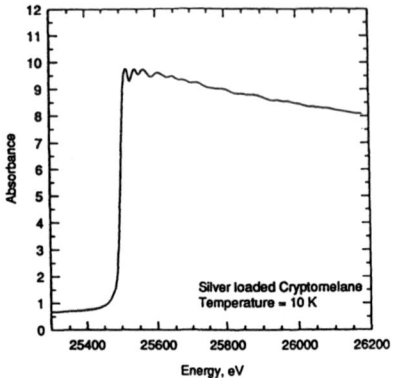

Figure 1. The complete X-ray absorption spectra for silver-containing cryptomelane collected at 10 K.

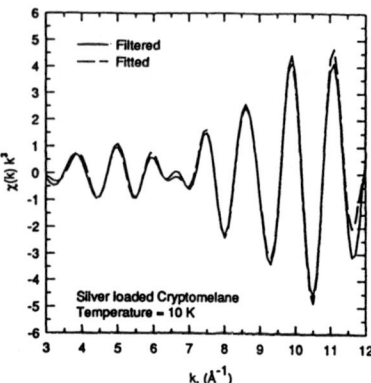

Figure 2a. The filtered and fitted EXAFS function for the silver-containing crypto-melane sample at 10 K. The fitted curve corresponds to the optimal solution containing the phase and amplitude functions of the back scattering atoms from the structure.

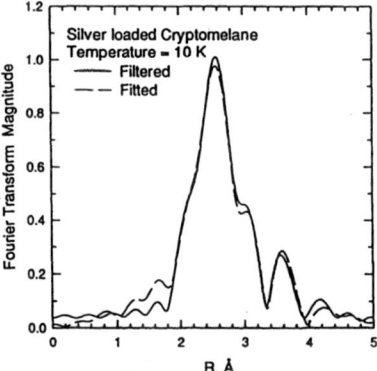

Figure 2b. The filtered and fitted radial distribution function for the silver-containing cryptomelane sample at 10 K. The fitted curve corresponds to the optimal solution containing the phase and amplitude functions of the back scattering atoms from the structure.

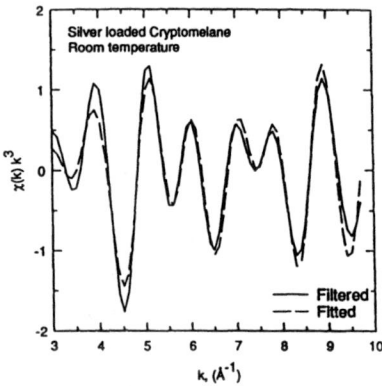

Figure 3a. The filtered and fitted EXAFS function for the silver-containing crypto-melane sample at room temperature. The fitted curve corresponds to the optimal solution containing the phase and amplitude functions of the back scattering atoms from the structure.

Table IV. The optimal solution obtained for the silver-containing natural ore sample at room temperature. The values in the parenthesis correspond to the standard deviation and also signifies that those terms were floating variables for obtaining the solution.

Type of atom	bond distance Angstroms	number of atoms	σ^2 (Debye-Waller) Å^2	ΔE_0 eV
Silver	2.786 (0.005)	2.10 (0.09)	0.008	-2.253
Oxygen	2.23 (0.008)	2.04 (0.19)	0.008	-2.253
Manganese	3.56	4.04 (0.32)	0.014	-2.253

For the silver-loaded cryptomelane sample investigated at 10 K, since the thermal vibration is low compared to the room temperature samples, the contributions are mainly from the high atomic number atoms of the matrix, namely silver and manganese. The silver-silver distance corresponds to the tunnel positions, and the manganese atoms are from the octahedral framework of the structure. For the room temperature sample, apart from the contributions of silver and manganese, there is a silver-oxygen peak corresponding to a distance of 2.23 Å. From the cryptomelane structure [4] and assuming a perfectly centered silver ion in the tunnels, there are eight oxygen atoms around 2.88 Å and four oxygen atoms around 3.45 Å. The obtained bond distance is much shorter than these regular positions.

In order to understand the silver-oxygen coordination distance, data from the literature containing silver-oxygen species in various coordinations and their corresponding distances are plotted in Figure 4 [5]. From Figure 4, a 2.23 Å silver-oxygen distance corresponds to a mixture

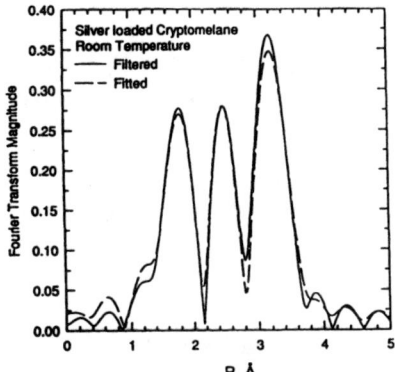

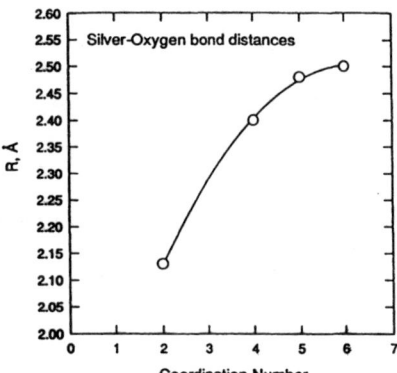

Figure 3b. The filtered and fitted radial distribution function for the silver-containing cryptomelane sample at room temperature. The fitted curve corresponds to the optimal solution containing the phase and amplitude functions of the back scattering atoms from the structure.

Figure 4. Silver-oxygen bond distances as a function of the coordination number.

of two and four-fold coordinations. Dalba et al. [6] in their study of glasses doped with silver oxide had similar two and four-fold coordinated silver with oxygen that had a bond distance between 2.23 Å and 2.3 Å. For the silver-cryptomelane system, the shorter distance may correspond to two different bonding situations. The first scenario corresponds to silver ions being present at both the surface of the manganese oxide and at the center of the tunnels, with the surface silver ions coordinated to the oxygen atoms. The second is the situation corresponding to the silver atoms occupying a off-center position in the tunnels (ac plane that is perpendicular to the tunnel b-axis) but still retaining the silver-silver distance along the tunnel axis direction. This kind of position is likely and similar findings has been reported by Post [4] for the smaller sized sodium and strontium ions occupying similar off-center positions in the tunnels. However, it is not possible to differentiate between the two different scenarios.

The shorter silver-oxygen bond does not show up prominently in the sample analyzed at 10 K. The possible explanation for this may be that the silver-silver peak is strong, that the silver-oxygen peak may be buried under it. Also, the small shoulder observed in the RDF may suggest that this indeed may be the situation.

Conclusions

The synthetic cryptomelane samples were loaded with silver ions from the solution and were a good model compound to understand the nature of complex silver-manganese oxide ores. From the X-ray absorption experiments, the silver is present as a monovalent ion in the structure and may occupy two different sites in the oxide. The first corresponds to silver in an off-center position in the tunnels of cryptomelane. The second corresponds to silver occupying the tunnel sites (at the center) and also present on the surface of the manganese oxide as well. For both the off-center as well as the surface positions, silver is in a mixture of two and four-fold coordination with the oxygen atoms. Because silver predominantly occupies the tunnel sites of the manganese oxide, it is difficult to remove by simple extraction methods.

Acknowledgment

The authors would like to acknowledge SSRL for providing the beam time and Professor B. Pesic of University of Idaho, Moscow, for providing the natural silver-manganese oxide ore sample.

References

1. B. K. Teo, EXAFS : Basic principles and data analysis, (Springer-Verlag, New York, 1986), pp. 1-182.
2. G. E. Brown Jr. and G. A. Parks, Rev. of Geophys. **27**, 519 (1989).
3. B. Pesic and J. E. Wey, Trans. AIME **280**, 1846 (1986).
4. J. E. Post, PhD thesis, Arizona State University, 1981.
5. T. Yamaguchi, G. Johansson, B. Holmberg, M. Maeda, and H. Ohtaki, Acta Chem. Scandinavica A **38**, 437 (1984).
6. G. Dalba, P. Fornasini, F. Rocca, E. Bernieri, E. Burattini, and S. Mobilio, J. Non-Crystalline Solids **91**, 153 (1987).
7. R. Ravikumar and D. W. Fuerstenau, in Aqueous Chemistry and Geochemistry of Oxides, Oxyhydroxides, and Related Materials, edited by J. A. Voigt, T. E. Wood, B. C. Bunker, W. H. Casey, and L. J. Crossey (Mater. Res. Soc. Proc. **432**, Pittsburgh, PA, 1997) pp. 243-248.

ELECTRONIC AND SPATIAL STRUCTURE STUDIES OF CADMIUM AND ZINC DIALKYLDITHIOCARBAMATE MOLECULES IN NONAQUEOUS SOLUTIONS USED IN THE PROCESSES SPRAY PYROLYSIS

S.B. ERENBURG, N.V. BAUSK, S.M. ZEMSKOVA, S.V. TKACHEV, L.N. MAZALOV
Institute of Inorganic Chemistry SB RAS, Lavrentiev av. 3, Novosibirsk 630090, RUSSIA.
simon@che.nsk.su

ABSTRACT

CdK EXAFS and ^{113}Cd NMR spectra, ZnK EXAFS and XANES spectra were measured for solutions of cadmium and zinc dialkyldithiocarbamates in organic solvents with different donor numbers: tributylphosphine, methylene chloride, benzene, dibutylsulfide, pyridine, dimethylsulfoxide and for some model compounds. The parameters of the local surroundings of the Cd and Zn atoms for complex forms in solutions were determined using EXAFS . Spatial structure models of the complex forms in a metal chelate - nonaqueous solvent system are suggested. It is established that coordination of nitrogen atoms by cadmium atoms in pyridine solutions is realized for both $Cd[(C_2H_5)_2NCS_2]_2$ and $Cd[(n-C_4H_9)_2NCS_2]_2$. For the tributylphosphine solution of $Cd[(n-C_4H_9)_2NCS_2]_2$ and $Zn[(n-C_4H_9)_2NCS_2]_2$, additional coordination of the phosphorus atoms of the solvent molecules by the metal atoms are established.

INTRODUCTION

Solvation is a complex phenomenon that involves energy and structural changes in a solute - solvent system. In monographs [1, 2] solvation is treated as a donor-acceptor reaction accompanied by a thermal effect, which may be expressed via the donor number of the solvent. The result of this process is coordination of solvent molecules to metal ions and withdrawal of the ligand anion into the solution. However, these ideas were developed on the basis of the data obtained for solutions of metal salts in nonaqueous media and are not applicable to chelate compounds in which the anion is a chelate ligand. Conventional models of physicochemical properties of such metal complexes in solutions use interactions of two types: 1) chemical interaction of the metal atom with ligands of the first coordination sphere and 2) nonspecific interaction of the inner sphere of the metal complex with medium (solvent) as dielectric continuum. For electroneutral complexes, the interaction inside the complex (in the inner sphere) is much greater than that with the medium (solvent). Another model used for these systems is based on the idea of outer-spheric coordination of solvent molecules by complex molecules [3].

There are cases when organic solvent molecules are directly coordinated by the metal atoms of a molecular compound, for example, alkylbenzenes - antimony trichloride in solutions, where characteristic fast exchange of free and bonded solvent molecules has been detected [4, 5]. On the other hand, examples of solid metal complexes with organic solvent molecules have been found : $[Hg((i-C_3H_7)OCS_2)(P(c-C_6H_{11})_3)_2(CH_2Cl_2)](ClO_4)$ and $Ni(C_2H_5OCS_2)B*A$ (B = 1,10-phenanthroline, 2,2' -Bipy, en; A = $(C_2H_5)_2O$), but a nature of the coordination of solvent molecules is not discussed [6, 7].

It seems important to investigate formation of metal chelate complexes having hydrophobic substituents of different lengths in series of nonaqueous solvents with varying donating abilities.

Examples of such chelates are metal dithiocarbamates, whose specific feature is the ability of some part of strained four-membered metallocycles to be rearranged into bridged chain cycles, which leads to additional coordination sites in the chelate molecule.

Organic solutions of zinc and cadmium dithiocarbamates are used in spray pyrolysis to prepare films of zinc and cadmium sulfides. Thus thin films of CdS of satisfactory quality are formed when a pyridine solution of cadmium diethyldithiocarbamate $Cd[(C_2H_5)_2NCS_2]_2$ (below CdL_2) is sprayed at T = 300-400°C [8]. It was found experimentally that not all organic solvents are suitable for these purposes, and empirical rules for solvent selection were formulated.

In the NMR studies of the state of metal chelates in nonaqueous media, it was shown that the nuclear chemical shift of the central atom changes considerably depending on the nature of the solvent used (δ varies by more that 100 ppm) [9, 10]; this is attributed to the energy and structural changes in the first coordination sphere of the central atom.

EXAFS spectroscopy has been successfully used in studies of aqueous solutions of metal ions [11]. In [12] it is reported that the rate constant of ligand exchange for metal complex hydrates in aqueous solutions correlates with the Debye - Waller factors obtained from the EXAFS data. In [13] EXAFS measurements provided information about the structure of the inner and outer hydrate shells for some metal ions in aqueous solutions. The works investigating the state of different types of complexes in nonaqueous solutions generally use IR and optical spectroscopy. Few works use the NMR of heavy nuclei and still fewer studies employ EXAFS, XANES, and X-ray fluorescence spectroscopy [14-16].

EXPERIMENTAL

The chelates CdL_2 and $Cd[(n-C_4H_9)_2NCS_2]_2$ (below CdL^1) were prepared according to the standard procedure by the exchange reaction of $Cd(CH_3COO)_2$-$2.5H_2O$ and NaL or NaL^1 [17]. Synthesis of mixed-ligand complexes CdL_2B (B = 1,10-phenanthroline, 2,2'-Bipy) is described in [18]. $Zn[(n-C_4H_9)_2NCS_2]_2$ (below ZnL^1) were prepared according to the standard procedure, too. Organic solvents of reagent grade [chloroform, benzene, acetonitrile, acetone ethylacetate, dimethylformamide (DMF), dimethylsulfoxide (DMSO), pyridine (Py), dibutylsulfide (DBS) and tributylphosphine (TBP)] were used without further purification.

NMR spectra were recorded on a CXP-300 pulse spectrometer with an operating frequency of 66.57 MHz for ^{113}Cd. The chemical shifts δ(positive downfield) were determined relative to the external standard - aqueous 1.5 M solution of $Cd(ClO_4)_2$.

The CdK EXAFS, ZnK EXAFS and XANES spectra were recorded using synchrotron radiation of the VEPP-3 ring (G. Budker Institute of Nuclear Physics, Siberian Branch, Russian Academy of Sciences). Ring settings during the measurements: energy 2 GeV, current 100 mA. The CdK and ZnK absorption spectra were measured in the transmission mode using Ar/He and Xe ionization chambers as monitoring and final detectors, respectively. A channel-cut Si(111) single crystal was used as a monochromator. During the measurements, the station was monitored using microcomputers via CAMAC modules.

The data obtained were processed with the EXCURV-92 package. The phase and amplitude characteristics were calculated in the $X\alpha$ approximation using the package procedures. For analyzing the local surroundings of Cd and Zn atoms, Fourier-filtered data were fitted with k^2 weighing in the range of photoelectron wave vectors from 3 to 10 Å^{-1}. In the fitting procedure, the error of determination of interatomic distances was not higher than 0.02 Å in all cases. The factor of amplitude damping S_0^2 caused by many-electron effect was determined from the results of the fitting procedure for solid model complexes with known structure. Solid complexes CdL_2,

$CdL_2(2,2'$-Bipy$)$, $CdL_2(1,10$-phenanthroline$)$ and ZnL_2^1, $ZnL_2^1(1,10$-phenanthroline$)$ were used as model compounds. Various models of local surroundings of Cd and Zn atoms were checked to examine the possibility of further coordination of solvent molecules by the metal.

RESULTS AND DISCUSSION

Cadmium dialkyldithiocarbamate molecules in solutions

To obtain primary information about the structure of solvate complexes formed by dissolving metal chelates in nonaqueous solvents with different donor numbers, we recorded and analyzed the ^{113}Cd NMR spectra of cadmium dialkyldithiocarbamates in solvents with low (methylene chloride, benzene), medium (ethylacetate), and high (Py, TBP) donor numbers. NMR spectra of cadmium complexes in the following solvent mixtures methylene chloride-benzene, methylene chloride-ethylacetate, methylene chloride-pyridine, methylene chloride-TBP, chloroform-TBP were studied.

The ^{113}Cd NMR spectra of some cadmium diethyldithiocarbamate complexes with nitrogen-containing heterocycles in chloroform were investigated in [19]. The characteristic shift of cadmium in these complexes containing two sterically strained four-member cycles SCSCd is ~330-340 ppm [19]. Addition of nitrogen-containing donors leads to increased screening of the central nucleus and to an upfield shift of the cadmium signal by approximately 100 ppm. Here we study the ^{113}Cd NMR spectra of solutions of some cadmium dibutyldithiocarbamate complexes with nitrogen-containing heterocycles (1,10-phenanthroline, 2,2'-Bipy) in methylene chloride. In agreement with the results of [19], these systems show an upfield shift of the Cd signal by 100 ppm relative to the NMR signal for Cd due to increased screening of the central nucleus caused by the addition of nitrogen-containing donors to CN = 6.

The chemical shift $\delta^{113}Cd$ for dialkyldithiocarbamates in different solvents changes up to 160 ppm. In solvents with pronounced donating ability (DMF, Py, DMSO), the ^{113}Cd NMR signal is shifted upfield by ~160 ppm, which may be due to coordination of solvent molecules to the cadmium chelate.

When pyridine is added to the CH_2Cl_2 solution of the CdL_2, the interaction of dissolved CdL_2 with pyridine manifests itself quite distinctly. The rate of exchange increases with concentration of pyridine in solution, and methylene chloride, evidently, also participates in solvate formation. When tributylphosphine is added to the solution, the results are interpreted more unambiguously. As in the case of pyridine, the rate of exchange increases with the concentration of free TBP in solution. The presence of Cd - P spin - spin coupling allows one to establish formation of inner-spheric complexes containing one and two TBP molecules, the former complex being more labile. The ^{31}P - ^{113}Cd spin-spin coupling constant is 1629 Hz for the complex with two TBP molecules and 1523 Hz for the complex with one TBP molecule; this corresponds to the value given in [20].

CdK EXAFS spectra were measured for the complexes in the solid state and for their organic solutions. We checked whether the asymmetric surroundings formed by the sulfur atoms around the Cd atoms, characteristic for the solid state, are preserved in a pyridine solution of CdL_2^1. As a result of the fitting procedure, it was established that the four sulfur atoms in the solution of the complex lie at equal distances from the cadmium atoms. The results of the fitting procedure are considerably improved when one or two nitrogen atoms, which may be coordinated to cadmium atoms in solution, are introduced in sequence into the model: the fitting index F characterizing the consistency of the model with experimental spectra [16] decreased by factors of 1.7 and 2.0, respectively. According to the results of the fitting procedure, the Cd-S and Cd-N interatomic distances are 2.61 and 2.36 Å, respectively, for the model with one nitrogen atom and 2.63 and 2.40 Å, respectively, for the model with two nitrogen atoms. Analogous results were obtained for pyridine solutions of CdL_2. These results are in agreement with the distances calculated for CdL_2(2,2'-Bipy) and CdL_2(1,10-phenanthroline) by XRD analysis in [21]. The data obtained indicate that coordination of nitrogen atoms by cadmium atoms in a pyridine solution is realized for both CdL_2 and CdL_2^1 and that replacement of L by L^1 does not produce a significant effect on the Cd-N distances (**Fig. 1**).

It was interesting to verify that benzene solutions contain $[CdL_2]_2$ dimers, serving as a basis for packing the complex in the solid state, as indicated by the NMR data. For this purpose, during the fitting of EXAFS data, we analyzed a model of surroundings of cadmium atoms including five sulfur atoms: two pairs of sulfur atoms lying at close distances and one sulfur atom of the bridging bidentate diethyldithiocarbamate, which is coordinated to the neighboring Cd atom in the dimer, lying at a distance of 2.77 Å. This model gave a fitting index that is 1.8 times smaller than that for the monomer model including four sulfur atoms at equal distances. The fitting procedure for this model showed that two sulfur atoms are at a distance of 2.60 Å, two at a distance of 2.55 Å, and one at a distance of 2.75 Å. These values are close to analogous distances in CdL_2: 2.60, 2.63 Å; 2x2.52 Å; 2.77 Å (XRD data [22]). Variation of the number of sulfur atoms in the surroundings of the cadmium atoms gives 2.0, 2.13, and 1.18 for Cd - S distances of 2.60, 2.55, and 2.75 Å, respectively.

The resulting interatomic distances and coordination numbers support the existence of dimers in benzene solutions. If the Cd-Cd atomic pair were identified at a distance of 3.30 Å, as in the solid complex, this would be an unambiguous support of the existence of dimers in solution. However, since the dimer is relatively free to make vibration with all degrees of freedom in solution, and due to high dynamic components of the Debye-Waller factors, this fact may not be unambiguously established from EXAFS data with the existing level of noise. For this solution, a model including the CdL_2 monomer in which the benzene molecule is coordinated by the cadmium atom was also tested.

This model improved the fitting index only by a factor of 1.3. Analysis of EXAFS data with the assumption that the solution contains simultaneously monomers and dimers coordinating solvent molecules is not possible because of many parameters in the fitting procedure.

The amplitude of the main maximum on the radial distribution function obtained by processing EXAFS data for TBP solution of CdL^1 increases by 25% compared to other solutions, indicating that TBP molecules are coordinated by metal atoms. However, the introduction of one or two phosphorus atoms in addition to the four sulfur atoms decreases the fitting index by only 20%. The fitting gave Cd-S and Cd-P distance of 2.56 and 2.73 Å, respectively. The discrepancy

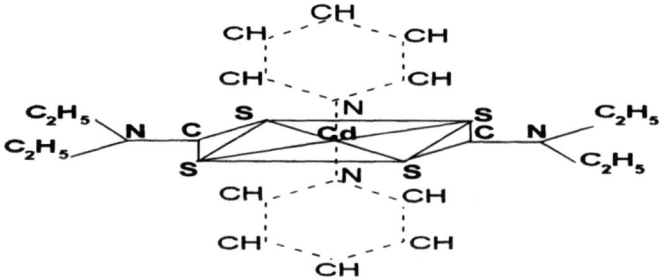

Fig. 1. The scheme of coordination by cadmium of pyridine molecules in solution.

362

between the model and the experimental data may be caused by the fact that the available version of the EXCURV-92 package uses the harmonic approximation to describe the root-mean displacement of atoms. In solution, coordination of phosphorus atoms by Cd atoms occurs at a distance of 2.73 Å that is slightly longer than the characteristic distance of chemical bonding in a solid complex which is equal to 2.57 Å according to the data presented in [20]. The form of the potential well in which the phosphorus atoms lie in this case can differ from harmonic.

Zinc dialkyldithiocarbamate molecules in solutions

ZnK EXAFS and XANES spectra for solutions of complexes ZnL'_2, $ZnL'_2(1,10$-phenanthroline) in the TBP, benzene, Py which are characterized by different donor number were obtained. The models of spatial structure of complex forms for the zinc chelates-nonaqueous solvents have been suggested. It was determined that $[ZnL'_2]_2$ molecules are dimeric in benzene solution, but its

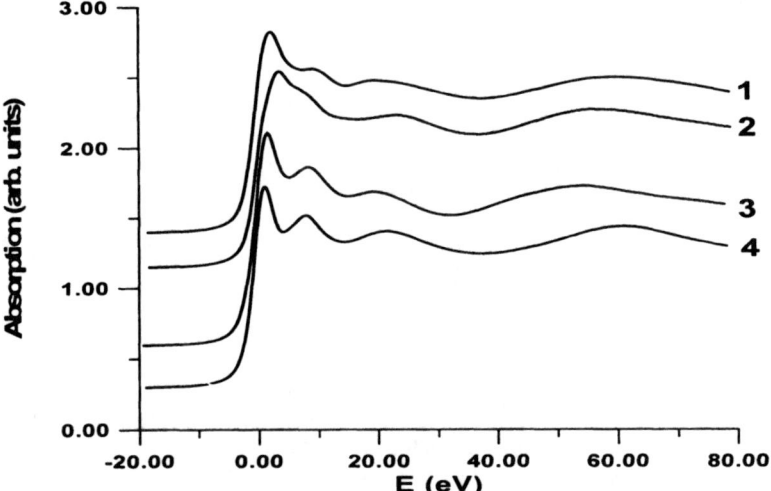

Fig. 2. ZnK XANES spectra for solid $Zn[(n-C_4H_9)_2NCS_2]_2$ - (1), for $Zn[(n-C_4H_9)_2NCS_2]_2$ in solvents: tributilphosphine - (2), pyridine - (3), benzene - (4).

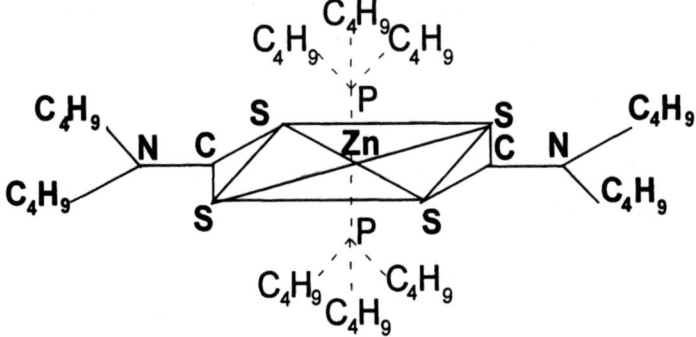

Fig. 3. The scheme of coordination by zinc of tributilphosphine molecules in solution.

spatial structure is differ substantially in comparison with the solid state one. Zinc coordinates five sulfur atoms at a distance of 2,36 Å, Zn-Zn distance is 3,80 Å, the whole molecule geometry is fixed harder than in solid state [23]. The amplitude of the main maximum on the Zn radial distribution function for the TBP and Py solutions differs from that of the benzene solution. These changes give an indication of possible coordination of Zn atoms by TBP and pyridine. As illustrated in **Fig. 2**, the XANES spectra shape of ZnL_2^1 in TBP solution differs from the other ones indicating a change of the local symmetry around the Zn atoms in this solution and confirming our supposition about coordination of TBP molecules by Zn (**Fig. 3**) simile to $ZnL_2P(alkyl)_3$ [24]. At the same time, coordination of solvent molecules by $ZnL_2^1(1,10$-phenanthroline) in benzene and pyridine is not observed.

ACKNOWLEDGMENT

This work was supported by RFBR grant No. 96-03-33166.

REFERENCES

1. V. Gutman, Coordination Chemistry in Non-Aqueous Solutions, Springer, New York (1968).
2. K. Burger, Solvation, Ionic, and Complex Formation Reactions in Non-Aqueous Solvents: Experimental Methods for Their Investigation, Elsevier, Amsterdam (1983).
3. V.M. Nekipelov and K.I. Zamaraev, Coord. Chem. Rev., **61**, 185 (1985).
4. H. Schmidbaur, R. Novak, B. Huber, and G. Muller, Organomet., **6**, 2666 (1987).
5. V.G.Torgov, G.V.Veryovkin, V.V.Denisov, S.V.Tkachev, Zh. Neorg. Khim., **30**, 1012 (1985).
6. B.F. Abrahams, D. Dakternieks, B.F. Hoskins, and G.Winter, Aust. J. Chem., **41**, 757 (1988).
7. G. N. Natu, S. B. Kulkarni, and P. S. Dhar, Thermochim. Acta, **54**, 297 (1982).
8. L. F. Zharovskii, L. V. Zav'yalova, M. Ya. Rakhlin, and S. V. Svechnikov, Poluprovodn. Tekh. Mikroelektron., No. 29, 64 (1979).
9. M.F. Summers, Coord. Chem. Rev., **86**, 43 (1988).
10. B.F. Abrahams, D. Dakternieks, and G. Winter, Inorg. Chim. Acta, **162**, 211 (1989).
11. H. Ohtaki and T. Radnai, Chem. Rev., **93**, 1157 (1993).
12. T. Miyanaga and T. Fujikawa, L Phys. Soc. Jpn., **63**, 1036 (1994).
13. A. Munoz-Paez, S. Diaz, P.J. Perez, et al., Physica B, **208**, 395 (1995).
14. S.B.Erenburg, L.N.Mazalov, N.V.Bausk, M.K.Drozdova, Russian J. Struct. Chem., **35**, 517 (1994).
15. L.N. Mazalov and S.B. Erenburg, Russian J. Struct. Chem., **35**, 548 (1994).
16. S.B. Erenburg, L.N. Mazalov, N.V. Bausk, M.K. Drozdova, V.G. Torgov, Russian J. Struct. Chem. **36**, 941 (1995).
17. D. Coucouvanis, Prog. Inorg. Chem., **11**, 234 (1970).
18. S.V. Larionov, V.N. Kirichenko, S.M. Zemskova, I.M. Oglezneva, Koord. Khim., **16**, 79 (1990).
19. S.M. Zemskova, I.M. Oglezneva, M.A. Fedotov, et al., Izv. Sib. Otd. Akad. Nauk SSSR, Ser. Khim., No. 5, 89-93 (1990).
20. D. Zeng, M.J. Hampden-Smith, T.M. Alam, A.L. Rheingold, Polyhedron, **13**, 2715 (1994).
21. L.A. Glinskaya, R.F. Klevtsova, S.M. Zemskova, Russian J. Struct. Chem., **33**, 91 (1992).
22. E.A. Shugam and V.M. Agre, Kristallografiya, **13**, 253 (1968).
23. M. Motevally, P. O'Brien, J.R. Walsh and I.M. Watson, Polyhedron, **15**, 2801 (1996).
24. D. Zeng, M.J. Hampden-Smith and T.M. Alam, Polyhedron, **13**, 2715 (1994).

SURFACES OF SEMI-FLUORINATED BLOCK COPOLYMERS STUDIED USING NEXAFS

J. GENZER [1], E. SIVANIAH [1], E. J. KRAMER [1], J. WANG [2], H. KÖRNER [2], M.-L. XIANG [2], S. YANG [2], C. K. OBER [2], K. CHAR [2], M. K. CHAUDHURY [3], B. M. DEKOVEN [4], R. A. BUBECK [4], D. A. FISCHER [5],[6], S. SAMBASIVAN [6]

[1] Materials Department, University of California at Santa Barbara, Santa Barbara, CA 93106
[2] Department of Materials Science and Engineering, Cornell University, Ithaca, NY 14853
[3] Department of Chemical Engineering, Lehigh University, Bethlehem, PA 18015
[4] Analytical Sciences Laboratory, The Dow Chemical Company, Midland, MI 48674
[5] National Institute of Standards and Technology, Gaithersburg, MD 20899
[6] National Synchrotron Light Source, Brookhaven National Laboratory, Upton, NY 11973

ABSTRACT

The molecular orientation within a surface liquid crystalline layer made up of semi-fluorinated side-groups $[-CO-(CH_2)_{x-1}-(CF_2)_yF]$ (SF groups) attached to the isoprene block of a styrene-isoprene diblock copolymer was determined by analyzing the partial electron yield C-edge NEXAFS signal. The results show that in contrast to the bulk, where the SF groups lie parallel to the diblock copolymer lamellae and thus parallel to the surface, the surface SF groups make an average angle with the surface normal of between 29 and 46° depending on x and y.

INTRODUCTION

To develop polymers with low surface energies, it is common to introduce a fluorinated group onto a polymer backbone in order to create a fluorinated surface coating. Commercially available fluorinated ester side chain acrylic and methacrylic polymers are typical low surface energy coating materials [1]. Small amounts of fluorinated polymer can also be mixed with, or chemically bound to the surface of another polymer to dramatically alter its surface properties [2]. One critical problem, surface reconstruction in contact with polar liquids, still has not been solved and this limits the practical application of these materials.

A uniformly organized trifluoromethyl ($-CF_3$) array would create a surface with the lowest possible surface tension. Self-assembled monolayers (SAM), produced by self-organization of randomly oriented molecules from a solution or vapor onto a surface, can result in such arrays [3] but are impractical for large-scale applications, especially for coatings on elastomers. An effective approach which avoids SAM techniques for production of a uniform $-CF_3$ surface is to harness the self-assembly behavior of a class of fluorinated materials, the liquid crystalline semi-fluorinated alkanes. In order to prevent surface reconstruction processes from occurring, it would be ideal also for a side chain to micro-phase-separate (*i.e.* have a long enough fluorocarbon unit to be immiscible with the polymer backbone and hydrocarbon side chain). Otherwise, the surface molecular composition would not be a uniform $-CF_3$ structure but would be a mixture of $-CF_3$, $-CF_2-$, $-CH_2-$, and $-CH_3$ groups. Bearing this idea in mind, self-organization of a liquid crystalline mesogen on the nanometer length scale could play an important role in the design of stable, low surface energy materials.

Styrene-isoprene diblock copolymers were synthesized, in which semi-fluorinated $[-CO-(CH_2)_{x-1}-(CF_2)_yF]$ side-groups were attached to the isoprene block [4]. We refer to such copolymers as semi-fluorinated diblock copolymers (SFDs). Our previous experiments indicated that the morphology of SFDs is governed by the ratio of the different blocks (LC or amorphous -

lamellar, cylinder) and the structure within the LC block is controlled by the lengths of the side-groups. Namely, depending on the x/y ratio either smectic-A, smectic-B (S_B) or isotropic phases exist. The relationship between the order of the mesogenic semi-fluorinated (SF) side-groups and the microphase formation has been investigated by small angle X-ray scattering, transmission electron microscopy, and time resolved X-ray diffraction [4]. It was found that the lamellar plane is oriented parallel to the substrate surface, while the SF side-groups prefer to pack so that they lie parallel to the lamellar plane. A detailed analysis of the X-ray data demonstrated that the SF side-groups arrange in a liquid crystalline hexagonal S_B phase (similar to precursors). Contact angle measurements revealed a very low surface energy consistent with close packing of $-CF_3$ groups at the surface [4]. These results imply that the SF side groups do not have the same orientation at the surface as they do in the bulk. These surfaces proved highly resistant to reconstruction in water, a resistance that might possibly originate in the stabilizing influence of a S_B LC surface ordering of the SF side-groups.

It is thus of great interest to probe the near-surface orientation of the LC side-groups in SFDs using a truly surface sensitive technique. Near-edge X-ray absorption fine-structure (NEXAFS) has proven to be a powerful tool for studying molecular orientation of a variety of materials [5]. The ability to probe structures at the surface and in the bulk of the material at the same time, along with its high sensitivity to the character and orientation of chemical bonds make NEXAFS superior to other surface methods for this purpose.

EXPERIMENT

The same block copolymer, which had styrene and isoprene (60% 1,2 and 40% 3,4 units) blocks with degrees of polymerization of 395 and 107, respectively, was used as the basic backbone to which the SF side-groups were attached. SF side-groups with various values of x and y (FB-y-x) were attached to the pendent double bonds of the isoprene using synthesis methods reported previously [4]. Due to the bulkiness of the fluorinated side chains, the volume fractions of the styrene and SF isoprene were approximately equal which led to the formation of a microphase separated lamellar morphology of the block copolymer. Thin (ca. 50 nm) films of SFDs were prepared by spin-coating solutions of the block copolymers in α,α,α-trifluorotoluene onto silicon wafers. The samples were then annealed in vacuum at 150°C for 4 hours to perfect the lamellar morphology, which in thicker films forms parallel to the substrate.

The NEXAFS experiments were carried out on the NIST/DOW materials characterization end-station at the National Synchrotron Light Source at Brookhaven National Laboratory. The U7A beamline is equipped with toroidal mirror spherical grating monochromator. The incident photon energy resolution and intensity were 0.2 eV and 5×10^{10} photon/s, respectively, for an incident photon energy of 300 eV and a typical storage ring current of 500 mA. The materials characterization end-station is equipped with a heating/cooling stage positioned on a goniometer, which controls the orientation of the sample with respect to the polarization vector of the X-rays. A differentially pumped ultra-high vacuum compatible proportional counter is used for collecting the fluorescence yield (FY) signal. In addition, the partial-electron-yield (PEY) signal is collected using a channeltron electron multiplier with an adjustable entrance grid bias (EGB). A crude depth profiling within the top 5 nm is possible by increasing the negative EGB on the channeltron detector, at the highest bias thus selecting only the Auger electrons which have suffered negligible energy loss. The monochromator energy resolution and photon energy were calibrated by comparing the transmission spectrum from gas-phase carbon monoxide with electron energy-loss reference data. To eliminate the effect of incident beam intensity fluctuations and monochromator absorption features, the FY and PEY signals were normalized by the incident beam intensity obtained from the photo yield of a clean gold grid.

RESULTS

NEXAFS involves the resonant x-ray excitation of a K or L shell electron to an unoccupied low-lying antibonding molecular orbital of σ symmetry, σ* [5]. The initial state K or L shell excitation gives NEXAFS its element specificity, while the final-state unoccupied molecular orbitals provide NEXAFS with its bonding or chemical selectivity. A measurement of the intensity of NEXAFS spectral features enables the identification of chemical bonds and a determination of their relative population density within the sample. Because of the fixed geometry and the fact that the σ → σ* excitations are governed by dipole rules, the resonances are polarized, that is, their intensity varies as a function of the direction of the electric vector **E** of the incident X-ray relative to the symmetry of the molecule. Because sharp core level excitations for C and F occur in the soft X-ray spectral region, NEXAFS is an ideal technique for probing molecular orientations of SFDs.

To resolve the molecular orientation of SFD surfaces, NEXAFS experiments were first carried out on a sample consisting of a well-ordered self-assembled semi-fluorinated monolayer (SAM-F) whose structure $[-O_{1.5}Si-(CH_2)_2-(CF_2)_8F]$ closely resembles that of the SF-LC side-groups in SFDs. These experiments provided important benchmarks for interpreting NEXAFS results on the SFDs.

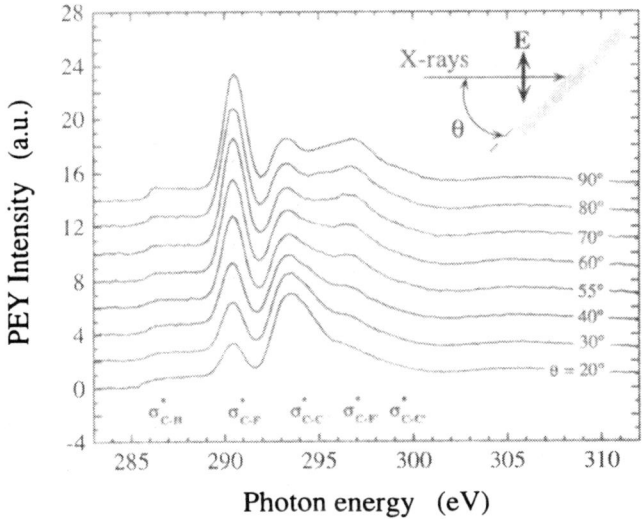

Fig. 1 PEY NEXAFS spectra of SAM-F sample at EGB = -150 V and various sample orientations with respect to the X-ray beam, θ. The inset shows schematically the sample geometry.

Figure 1 shows the PEY NEXAFS spectra at the C 1s edge from SAM-F sample measured at EGB of -150 V and 8 different angles θ between the sample normal and the polarization vector of the X-ray beam. The dotted lines denote the positions of the σ → σ* transitions for the C-H, C-F, and C-C bonds. The fact that the intensities originating from these transitions change with varying angle θ (as θ increases the intensity corresponding to σ* of the C-F bond increases while that of the C-C bond decreases) indicates that the sample is well oriented. By recording the PEY

NEXAFS spectra at 4 different EGBs at each θ we were able to "depth-profile" the sub-surface structure of the SAM-F sample. Following the method proposed by Outka and coworkers [6] the PEY NEXAFS spectra were fitted to a series of Gaussians and a step corresponding to the excitation edge of carbon. Similar experiments were carried out on the SFDs with various combinations of x and y. In particular, two series of SFDs were studied in which either x was kept constant and y was varied or *vice versa*.

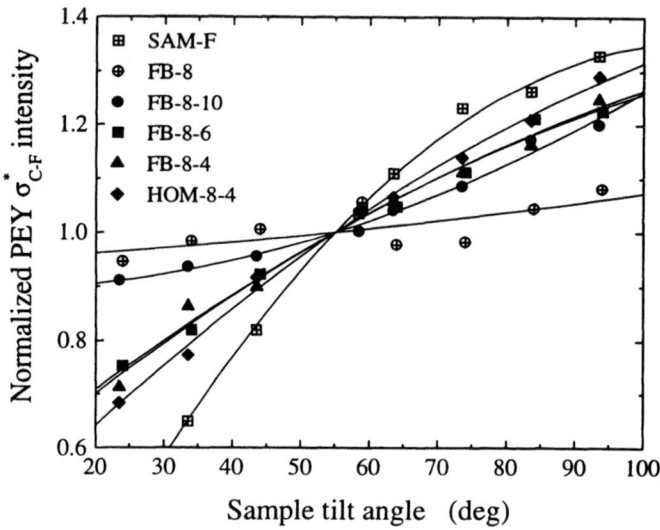

Fig. 2 Normalized PEY NEXAFS intensities *vs.* sample tilt angle from σ* C-F bond in C 1s for SF-LC groups in SFDs with the FB-8-x structure. HOM-8-4 and FB-8 refer to a sample with no PS block and a sample with no -CH$_2$- spacer in the SF group, respectively. The lines are guides to the eye.

Figure 2 shows the normalized PEY σ* C-F bond NEXAFS intensities *vs.* θ from in C 1s for SF-LC groups in SFDs with the FB-8-x structure. The actual PEY NEXAFS intensities have been normalized with respect to the intensity at the "magic angle", θ = 55°. HOM-8-4 and FB-8 refer to a sample with no PS block and a sample with no -CH$_2$- spacer in the SF group, respectively. Clearly, the fact that the intensity of the σ* C-F signal changes when the sample tilt angle is varied indicates that the SF-LC groups at the surface of the SFDs are oriented. Bearing in mind that the steeper the slope of the intensity *vs.* sample tilt angle, the smaller is the deviation of the molecule main axis from the substrate normal, the results in Fig. 2 illustrate that in samples where x is held constant, the average tilt angle τ of the SF-LC groups on the surface relative to the sample normal increases with increasing x. The results in Fig. 2 reveal that τ for the FB-8-x copolymers is larger than τ for the SAM-F sample but smaller than τ of the non-LC FB-8 polymer. The fact that the results for FB-8-x samples are close to those for HOM-8-4 shows that the constraints of the lamellar block copolymer structure have a minor effect on the SF-side group orientation. Systematic measurements of surface orientation have also been carried out on the series of SFDs, in which x was kept constant and y was varied. Figure 3 depicts the normalized PEY NEXAFS intensities *vs.* θ from σ* C-F bond in C 1s for SF-LC groups in SFDs with the FB-y-10 structure. As previously, the PEY NEXAFS intensities have been normalized with respect to the intensity at $\theta = 55°$. The results in Fig. 3 show that at constant x, τ decreases modestly with increasing y. Similar to the case of FB-8-x structures the average tilts in FB-y-10 samples are larger than in SAM-F and smaller than in FB-8. The results presented in Figs. 2 and 3 show that an increase in the length of the SF group leads to smaller average angle τ. Similarly, decreasing the length of the hydrocarbon spacer leads to smaller τ.

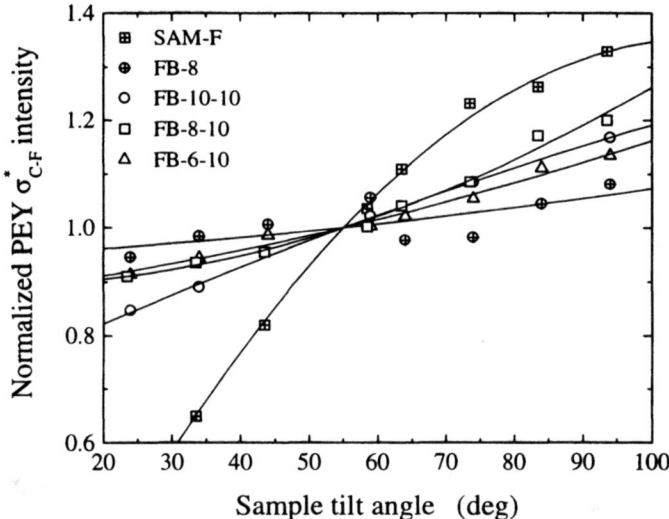

Fig. 3 Normalized PEY NEXAFS intensities *vs.* sample tilt angle from σ^* C-F bond in C 1s for SF-LC groups in SFDs with the FB-y-10 structure. HOM-8-4 and FB-8 refer to a sample with no PS block and a sample with no -CH$_2$- spacer in the SF group, respectively. The lines are guides to the eye.

In order to quantify the results shown in Figs. 2 and 3, the SF chains were modeled using the "building block model" (BBM) [5,6], in which we assumed that the –CH$_2$- spacer assumes a planar zigzag structure 2/1 (*e.g.*, polyethylene) and the –CF$_2$- part has a helical structure 15/7 (*e.g.*, poly(tetrafluoroethylene)). In our implementation, the original BBM was modified such that the NEXAFS intensity originating from a σ^* orbital was corrected for the energy loss of the Auger electrons originating from the sub-surface regions of the sample [7]. The total intensity from the sample was then given as a weighted sum of the particular intensities from all corresponding σ^*s. By analyzing several data sets collected from the same sample and different EGBs we were able to determine the average tilt angles of the hydrocarbon and fluorocarbon parts of the SF chain, τ_{CH2} and τ_{CF2}, respectively. The results of the analysis are shown in Table I. The values in Table I reinforce the information from Figs. 2 and 3. Namely at constant y τ_y increases with increasing x (*cf.* Fig. 2) and decreases with increasing y at constant x (*cf.* Fig. 3). Even though the values of τ_{CH2} follow generally the same trends as τ_{CF2}, there are some deviations (FB-8-10 in the FB-y-10 series). Overall, the results from the PEY NEXAFS analysis reveal that the average tilt of the SF-LC chain is dictated by both the LC ordering, originating from the SF group, and the flexibility of the hydrocarbon spacer.

Tab. I Average tilt angles of the hydrocarbon (τ_{CH2}) and fluorocarbon (τ_{CF2}) chains in the SFDs

sample	X	y	τ_{CH2} (deg)	τ_{CF2} (deg)
SAM-F	2	8	39 ± 2	4 ± 2
FB-8	-	8	-	51 ± 5
FB-8-10	10	8	63 ± 5	43 ± 3
FB-8-6	6	8	39 ± 4	35 ± 2
FB-8-4	4	8	33 ± 9	33 ± 3
HOM-8-4	4	8	10 ± 7	29 ± 3
FB-10-10	10	10	52 ± 6	38 ± 3
FB-6-10	10	6	45 ± 5	46 ± 1

CONCLUSIONS

NEXAFS is used to study the molecular ordering on surfaces of styrene-isoprene diblock copolymer films in which the isoprene block is modified by attaching semi-fluorinated [-CO-$(CH_2)_{x-1}$-$(CF_2)_yF$] side groups that exhibit liquid crystalline order. The PEY NEXAFS signal reveals that in contrast to the bulk, where the SF-LC groups lie parallel with the styrene/isoprene lamellae, the SF-LC groups on the surfaces of SFDs are on average only slightly tilted from the sample normal. In samples where y is held constant, this average tilt, τ, increases with increasing x. Similarly, in samples with constant x, increasing y leads to smaller τ.

ACKNOWLEDGMENTS

This research was supported by the Office of Naval Research, Grant No. N00014-92-J-1246 and the Division of Materials Research, NSF Polymers Program, Grant No. DMR92-23099. Partial support from Division of Materials Research, NSF Polymer Program, Grant No. DMR93-214573 is also appreciated.

REFERENCES

1. A. G. Pittman, in Fluoropolymers, Vol. 25, edited by L. A. Wall (Wiley: New York, 1972) p.419; D. L. Schmidt, *et al.*, Nature **368**, 39 (1994).

2. S. S. Hwang, C. K. Ober, S. Perutz, D. Iyengar, L. A. Schneggenburger and E. J. Kramer, Polymer **36**, 1321 (1995); D. R. Iyengar, S. M. Perutz, C.-A. Dai, C. K. Ober and E. J. Kramer, Macromolecules **29**, 1229 (1996).

3. A. Ulman, An Introduction to Ultrathin Organic Films from Langmuir-Blodgett to Self Assembly (Academic Press: New York, 1991); J. D. Swalen, *et al.*, Langmuir **3**, 932 (1987); C. D. Bain and G. M. Whitesides, Angew. Chem. **101**, 522 (1989); G. M. Whitesides and P. E. Labinis, Langmuir **6**, 87 (1990).

4. J. Wang, G. Mao, C. K. Ober and E. J. Kramer, Macromolecules **30**, 1906 (1997).

5. J. Stöhr, NEXAFS Spectroscopy (Springer-Verlag, Berlin, 1992).

6. D. Outka, J. Stöhr, J. Rabe and J. D. Swalen, J. Chem. Phys. **88**, 4076 (1994).

7. J. Genzer, E. J. Kramer, J. Wang, H. Körner, C. K. Ober, B. M. DeKoven, R. A. Bubeck and D. A. Fischer, to be published.

STRUCTURAL CHANGES IN IRON(II) POLYMERIC COMPLEXES UPON THERMALLY AND OPTICALLY INDUCED SPIN TRANSITION DETERMINED BY EXAFS SPECTROSCOPY

S.B. ERENBURG, N.V. BAUSK, L.G. LAVRENOVA, Yu.G. SHVEDENKOV, L.N. MAZALOV
Institute of Inorganic Chemistry SB RAS, Lavrentiev av. 3, Novosibirsk 630090, RUSSIA.
simon@che.nsk.su

ABSTRACT

Changes in the electronic and spatial structure of polymeric Fe(II) complexes with 1,2,4-triazoles and various anions upon spin transition was studied using EXAFS and XANES spectroscopy. Spin transition and structural changes were induced by variations of the anion, dilution with Zn, under heating or the action of light. In all complexes, the spin transition is accompanied by drastic changes in the local environment of Fe atoms. The increase in spin transition temperature for the complexes with variable anions ClO_4^-, I^-, Br^-, BF_4^-, NO_3^- was found to correlate with changes in the Fe-N distances and changes in bond covalence determined from the chemical shifts in Mössbauer spectra. High spin metastable long life states were detected and studied in the polymeric complex $Fe(atrz)_3(ClO_4)_2$. It was established that the changes in structure of polymeric complexes upon the transition to a metastable high spin state under the action of light differ from those in the thermally induced spin transition. Such differences are determined by mutual influence of Fe atoms in high spin and low spin states in polymeric chains.

INTRODUCTION

In recent years, materials with molecular structure and low-dimensional systems having, for example, a chain structure with bridging chemical bonds have attracted the attention of researchers due to their possible applications in electronic technologies. Such materials may be useful as two-dimensional matrices for recording and storing binary information which can be retrieved and erased. Compounds which change their spin states under the influence of temperature and pressure or under irradiation with light are especially promising for these purposes.

Thermally induced spin transitions were discovered more than sixty years ago by Cambi et al [1] and studied in detail by König and co-workers [2] and Gütlich and co-workers [3,4]. An abrupt change in the population of electronic states and, correspondingly, of the spin state of the transition metal ion with the increase in temperature was revealed for the coordination compounds of 3d elements which have the molecular symmetry near O_h and the electronic configurations from d^4 to d^7. In such compounds, the ground state may be low-spin (LS) or high-spin (HS), depending on the temperature or pressure.

At a low temperature transition metal ions spin state can be changed reversibly by irradiating of light of a certain wavelength. This phenomenon was discovered by Decurtins and Gütlich and co-workers in 1984 and termed light-induced excited spin state trapping (LIESST) [5]. The excited state can be maintained for a long time (>100 hours) at temperatures below 40 K (**Fig.1**). Light-induced HS states were observed in a great number of Fe (II) spin crossover complexes [4], but direct structural information about metastable HS states was not obtained.

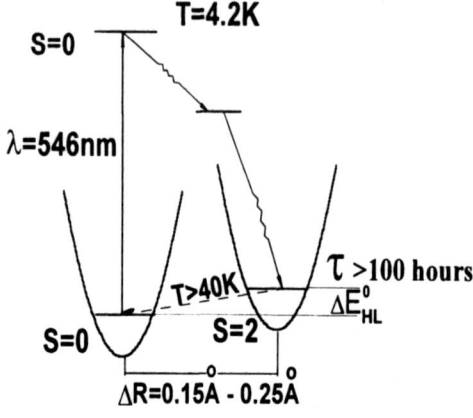

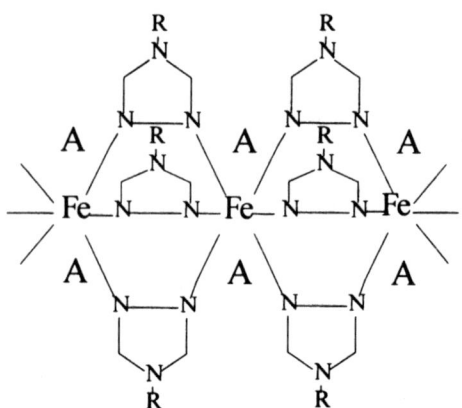

Fig. 1. The scheme of LIESST mechanism for our experiments.

Fig. 2. Assumed structure of the complexes Fe(trz)$_3$A$_2$ (R = H$_2$), Fe(atrz)$_3$A$_2$ (R = NH$_2$)

Spin transitions are exhibited by systems in which the crystal field has a magnitude comparable to spin pairing energy. If the difference in energy between the HS and LS states is of the order of kT, the spin transition can be observed. For octahedral compounds of Fe(II), the LS state of the symmetry $^1A_{1g}$ is provided by the filled configuration $(t_{2g})_6$ and the HS state of the symmetry $^5T_{2g}$, by the configuration $(t_{2g})_4(e_g)_2$ with four unpaired spins. The e_g-orbitals are more antibonding than the t_{2g} orbitals. Therefore the metal-ligand bond lengths are larger in the HS state than in the LS state. This increase in bond length ranges from 0.1 Å to 0.25 Å for different compounds of Fe(II) [2].

EXPERIMENT

At the Institute of Inorganic Chemistry of the Russian Academy of Sciences methods were developed for the synthesis of polynuclear compounds of Fe(II) of the composition FeL$_3$A$_2$ (**Fig.2**) and phases diluted with other transition metals which have the composition Fe$_x$Me$_{1-x}$(atrz)$_3$(NO$_3$)$_2$ (Ligand (L) is 1,2,4-triazole (trz) or 4-amino-1,2,4-triazole (atrz), A is ClO$_4^-$, Br$^-$, BF$_4^-$, NO$_3^-$, Me is Ni, Zn, Mn) [6-8].

The magnetic properties of the samples were measured by Faraday method in the temperature range 78-400 K. In most cases, such compounds exhibit an abrupt reversible spin transition (S=0, S=2) with a wide hysteresis loop (**Table I**) accompanied by a change in optical properties.

Complexes with Br$^-$ and diluted phases with NO$_3^-$ can find application as materials for optical memory where information can be recorded, for example, with the help of local heating by several degrees (above 300 K) with a thin laser beam and stored at room temperature. Such a two-dimensional picture can be read and then erased by cooling. Since no one has, so far, succeeded in obtaining well crystallized polymeric compounds of this type, EXAFS (Extended X-ray Absorption Fine Structure) spectroscopy gives almost the only possibility for characterizing their structural parameters.

TABLE I. The temperatures of spin transitions LS↔HS (S=0↔S=2) upon heating (T↑) and cooling (T↓) for $Fe(L)_3A_2$.

Anion	T↑(K)	T↓(K)	Ligand
ClO_4^-	175	175	atrz
Br^-	302	284	atrz
BF_4^-	335	307	atrz
NO_3^-	342	310	atrz
ClO_4^-	266	263	trz
BF_4^-	397	370	trz
NO_3^-	355	345	trz

FeK-, ZnK- and BrK- edge EXAFS and XANES spectra were measured at temperatures above and below the spin-transition. Measurements of EXAFS and XANES spectra were performed using synchrotron radiation of the VEPP-3 storage ring at the Bunker Institute of Nuclear Physics in Novosibirsk. The typical operating conditions of the storage ring were as follows: electron beam energy E = 2×10^9 eV, current I = 50-100 mA. The higher order reflections were filtered by a flat Au-coated mirror. The spectra were recorded in the transmission mode using a channel-cut Si(111) monochromator and ionization chambers filled with an Ar/He mixture.

A procedure and experimental devices were devised for measurement of EXAFS and XANES spectra at liquid helium temperature simultaneous with optical treatment of the samples. Measurements of EXAFS and XANES spectra of the polymeric spin-transition compounds of the composition $Fe(atrz)_3(NO_3)_2$, $Fe(atrz)_3SiF_6$, $Fe(atrz)_3(ClO_4)_2$ and the mononuclear spin-transition compound $Fe(Phen)_2(NCS)_2$ (Phen = 1,10-phenantroline) at 4.2 K were performed without an optical treatment of the samples and after their irradiation with a Hg green line 546 nm with various exposition within 90 minutes. The spectra were obtained using fluorescence detection mode as well as transmission mode. The x-ray fluorescence intensity were measured by Xe-filled electroluminescent detector in a current mode and monitored using ionization chambers filled with Ar/He mixtures.

The obtained experimental data were preliminarily treated by the standard method, which involves extrapolating the near-edge absorption to the post-threshold region, constructing the smooth part of the absorption coefficient in the post-threshold region, and calculating the normalised oscillation part of the absorption coefficient $\chi(k)$. Further data processing was performed using the program package EXCURV92 [9]. The k^3- weighting was carried out for photoelectron wave vectors in the interval 3 Å^{-1} - 13 Å^{-1}. The phase and amplitude characteristics were calculated in an X_α-DW approximation. The absolute values of distances from Fe and Zn to the surrounding atoms were determined by fitting the Fourier-filtered functions $k^3\chi(k)$. Because filtration was performed within 3.5 Å in the R- space, multiple scattering was not taken into account. The energy-independent amplitude factor S_0^2 intended to take care of the reduction in amplitude due to multiple excitations etc. was set equal to 0.8.

RESULTS AND INFERENCES

Thermally induced spin transitions

The fitting model involved the nitrogen and carbon atoms of the triazole ring: N(1) atoms, which are coordinated to the absorbing metal atom, N(2) atoms, which are coordinated to the neighboring metal atoms, and C atoms, which are nearest to the absorbing metal atom. Only the interatomic distances Fe-N, Fe-C or Zn-N, Zn-C and angles between the triazole ring and Fe-N and Zn-N bonds were varied while the geometry of the triazole rings was treated as fixed. The procedure produced good convergence in all cases, except for the FeK data for Zn- containing samples in the LS-state (**Fig. 3**). The failure of the model with fixed triazole ring geometry in this latter case may be explained by deformational distortions of the triazole ring. The fitted angles between the triazole ring planes and the Fe-N and Zn-N bonds directions did not exceed 3°.

The BrK EXAFS spectra which were measured at room temperature could not be treated because of very low oscillations amplitude. The latter points to significant structural and dynamic disordering of the bromide anions. The Fourier-transform magnitude at 78 K demonstrates that two layers of environmental atoms for Br appear within 2-3 Å.

In all studied polynuclear compounds there is a considerable increase in the Fe-N distances upon the transition from the LS to the HS state, as observed earlier for mononuclear complexes, while the Zn-N distances remain unchanged. The Fe-N and Fe-C interatomic distances in the

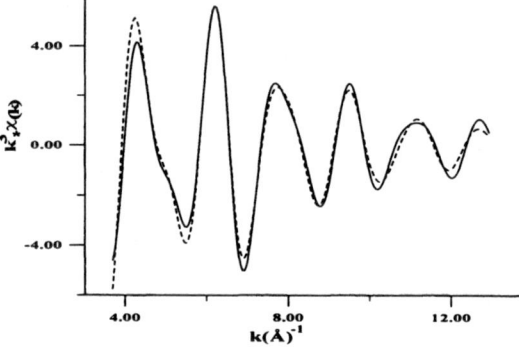

Fig. 3. Experimental function $k^3\chi(k)$ for Fe(atrz)$_3$(ClO$_4$)$_2$ (solid line) and fitted by three-sphere model theoretical function (dashed line).

TABLE II. The values of the distances between Fe and Zn atoms and the atoms of the triazole ring - R (±0.01 Å.). N$_1$ is the nearest nitrogen atom, N$_2$ is the atom coordinated to the next metal atom in the chain, C is the nearest carbon atom, T is the temperature of EXAFS measurements.

Compound	Spin state	T(K)	R(Å) Fe-N$_1$	R(Å) Fe-N$_2$	R(Å) Fe-C
Fe(atrz)$_3$(ClO$_4$)$_2$	LS	78	1.98	2.99	2.96
	HS	300	2.16	3.16	3.14
Fe(atrz)$_3$Br$_2$*H$_2$O	LS	78	1.97	2.98	3.01
	HS	370	2.17	3.13	3.13
Fe(atrz)$_3$(BF$_4$)$_2$	LS	78	1.95	2.97	2.94
	HS	370	2.19	3.19	3.15
Fe(atrz)$_3$(NO$_3$)$_2$	LS	78	1.95	2.97	2.94
	HS	370	2.18	3.18	3.15
Fe$_{0.8}$Zn$_{0.2}$(atrz)$_3$(NO$_3$)$_2$	LS	300	1.97		
	HS	370	2.18	3.16	3.13
Fe$_{0.33}$Zn$_{0.66}$(atrz)$_3$(NO$_3$)$_2$	LS	78	2.04		
	HS	300	2.18	3.18	3.18
	Spin state	T(K)	R(Å) Zn-N$_1$	R(Å) Zn-N$_2$	R(Å) Zn-C
Fe$_{0.8}$Zn$_{0.2}$(atrz)$_3$(NO$_3$)$_2$	LS	300	2.13	3.12	3.09
	HS	370	2.14	3.14	3.11
Fe$_{0.33}$Zn$_{0.66}$(atrz)$_3$(NO$_3$)$_2$	LS	78	2.15	3.15	3.12
	HS	300	2.14	3.14	3.11

region up to 3.5 Å were found to be changed by amounts from 0.12 Å to 0.24 Å (**Table II**). The effect of Zn atoms in the chain structure as the Zn content increases results in deformations in the LS state and an increase in the Fe-N distances as compared with the initial complex.

The decrease of spin transition temperature for the complexes with variable anions NO_3^-, BF_4^-, Br^-, I^-, ClO_4^- and for Zn-diluted complexes was found to correlate with increase of the Fe-N distances and changes in bond covalency determined from the chemical shifts in Mössbauer spectra [10] and the analysis of relative intensities of FeL_α and FeL_β X-ray fluorescence spectra[11,12], as well as with shifts of diffusion reflection spectra frequencies. The decrease of Tc appears to be associated with the increase in the Fe-N distances which can be caused by a decreasing "chemical pressure" due to the anion-cation interaction.

Optically induced spin transitions

Since the rate constant of the low-temperature relaxation (tunneling) $K_{HL}(T{\rightarrow}0)$ exponentially increases with ΔE^0_{HL} (**Fig.** 1)and correlates with T_c [4], we have chosen for the experiment polymeric complexes: $Fe(atrz)_3(NO_3)_2$, $Fe(atrz)_3SiF_6$ and $Fe(atrz)_3(ClO_4)_2$, with substantially different temperatures of the equilibrium spin transition ($T\uparrow,T\downarrow$): of (342 K,310 K); (255 K,241 K) and (175 K,175 K), respectively, and the mononuclear compound $Fe(Phen)_2(NCS)_2$ ($T\uparrow$, $T\downarrow$ = 180 K).

The FeK EXAFS- and XANES spectra of the Fe(II) compounds were measured at 300K and at 4.2K without an optical treatment and after irradiation by Hg line 546nm with various exposition within 90 minutes using fluorescence detection mode as well as transmission mode.

The existence of the $Fe(Phen)_2(NCS)_2$ high-spin (HS) metastable long life light-induced spin state was confirmed. It was determined that structural parameters of this compound at LIEESST state coincide within the experimental errors with those for HS state at 300K.

It was found that the optical irradiation of the $Fe(atrz)_3(ClO_4)_2$ compound at 4.2K leads to significant changes of its structural parameters which nevertheless remain quite different in comparison with those at equilibrium HS state. Fourier transform magnitudes, obtained by treating the FeK- edge EXAFS data for the polymeric $Fe(atrz)_3(ClO_4)_2$ complex are presented in **Fig. 4**. A detailed analysis of the data obtained using EXCURV92 computation and equilibrium structural parameters of the HS and low-spin (LS) states shown that the metastable state can not be treated as a mixture of equilibrium states. Fitting the corresponding Fourier-filtered functions $k^3\chi(k)$ reveals that the optical irradiation of this compound causes LS-HS transition of a significant part of the Fe atoms which is accompanied by the Fe-N distance increase up to 2.16Å. At the same time it was found that Fe atoms remaining at the LS state are characterized by strong local surrounding geometry distortions and the average Fe-N distance decrease to 1.93Å. Such types of distortions determined by mutual influence of HS and LS Fe atoms in polymeric chains can produce a resistance to completing the LS-HS transition.

An influence of the Fe atoms in the metastable HS state on the Fe atoms remaining in the LS state seems to be analogous to those for the Zn atoms on the Fe atoms in Zn-diluted compounds. As has been shown above, in the diluted compounds Zn-N distances do not change upon the spin transition to the LS state and are close to the Fe-N distances in the HS state.

Thus the process of a non-equilibrium light-induced transition to the HS state appears to be fundamentally different for molecular complexes with a weak bonding between molecules and the polymeric compounds with strong chemical bonds along the polymeric chains. The long-life metastable HS state of the $Fe(ATP)_3(NO_3)_2$ and $Fe(ATP)_3SiF_6$ has not been found in any detectable amount after optical irradiation at 4.2K may be because of their short time of life which are exponentially decreasing while the spin transition temperature increasing. An

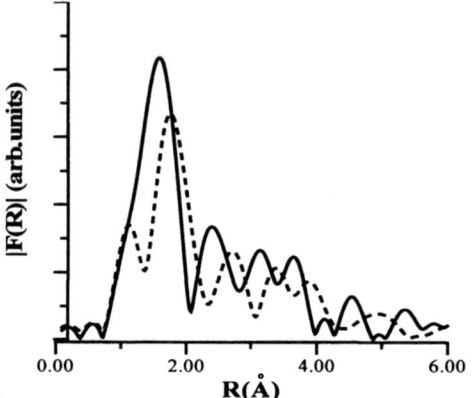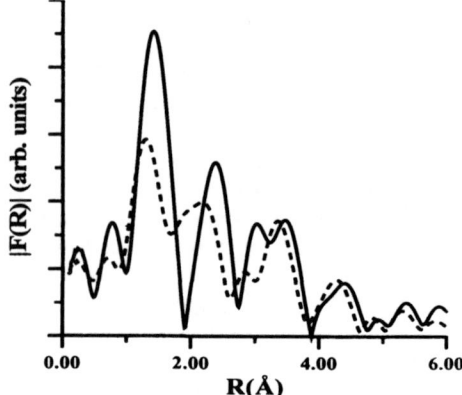

FIGURE 4. Fourier transform magnitude $|FT(R)|$), obtained by treating the FeK EXAFS data for Fe(atrz)$_3$(ClO$_4$)$_2$: **left** - in equilibrium at 78 K (LS state, solid line) and at 300 K (HS state, dashed line). **right** - at 4 K before optical treatment (solid line) and after irradiation (dashed line).

amplitude changes in the FeK EXAFS data detected after optical irradiation can be explained by the chain geometry distortions caused by the HS states appearing and their following fast relaxation.

The structural changes upon the light-induced spin transition was detected by the direct method for the first time showing that EXAFS spectroscopy is very powerful tool to study the LIEEST peculiarities in various compounds.

ACKNOWLEDGMENT

This work was supported by the Russian FBR (Project No.96-0332948).

REFERENCES

1. L. Cambi, L. Szegö, Ber. Dtsch. Chem. Ges. 64 (1931) 259; ibid. 66 (1933) 656.
2. E. König, Struct. Bonding (Berlin) 76 (1991) 51.
3. P. Gütlich, A. Hauser, Coord.Chem.Rev. 97 (1990)1.
4. P. Gütlich, A. Hauser, H. Spiering, Angew. Chem. Int. Ed. Engl. 33 (1994) 2024.
5. S. Decurtins, P. Gütlich, C.P. Köhler, H. Spiering, A. Hauser, Chem. Phys. Lett. 105 (1984) .
6. L.G. Lavrenova, V.N. Ikorskii, V.A.Varnek et al, Soviet J. Coord. Chem. 12 (1986) 119.
7. L.G. Lavrenova, V.N. Ikorskii, V.A.Varnek et al, Soviet J. Coord. Chem. 16 (1990) 349.
8. L.G. Lavrenova, N.G. Yudina, V.N. Ikorskii, Polyhedron 14 (1995) 1333.
9. N. Binsted, J.W. Campbell, S.J. Gurman and P.C. Stephenson, SERC Daresbury Lab. Rep. (1991).
10.S.B. Erenburg, N.V.Bausk, L.G.Lavrenova, V.A. Varnek, L.N. Mazalov, Solid State Ionics 101-103 (1997) 571-577.
11. N.V. Bausk, S.B. Erenburg, L.G. Lavrenova et al, J.Structural Chemistry 36 (1995) 925.
12. N.V. Bausk, S.B. Erenburg, L.N. Mazalov et al, J.Structural Chemistry 35 (1994) 509.

AUTHOR INDEX

Ackermann, G.D., 221, 227
Ade, Harald, 25
Agui, A., 169
Allen, J.W., 179
Alves, M.C. Martin, 327
Anders, Simone, 25
Armand, Pascale, 261
Armentrout, Guy, 241
Atkins, R.M., 251
Ayre, C., 227

Baggot, E., 121
Balasubramanian, M., 339
Barbee, Jr., Troy W., 145, 279, 285
Baruchel, J., 59
Batson, P.J., 215
Bausk, N.V., 359, 371
Beck, U., 95
Bedrossian, P.J., 185
Belikova, G.S., 161
Bello, A.F., 279, 285
Bilello, J.C., 81, 109, 115
Bliss, D.F., 65
Boyd, P., 185
Brennan, S., 245
Brissonneau, L., 121
Brown, J.T., 3
Brummond, William A., 241
Bubeck, R.A., 365
Butler, G.C., 43

Cai, Z., 233
Carter, Jr., C.H., 71
Casalis, L., 203
Celestre, R., 297
Cerrina, F., 215
Chabert, L., 59
Chang, C.H., 55, 95
Char, K., 365
Chaudhury, M.K., 365
Chaumont, J., 309
Chiaramonti, A., 81
Chung, H., 65
Chung, Jin-Seok, 49, 153, 233
Citrin, P.H., 251
Clark, A., 309
Clerc, C., 309
Coon, P., 227
Coulthard, I., 273, 291
Cox, David E., 261
Cummins, T.R., 185, 197

DeKoven, B.M., 365
Denlinger, J.D., 179
Deters, E.M., 139
Doh, S.J., 321
Dominguez, D.D., 127
Douillard, L., 315

Dowben, P.A., 191
Dudley, M., 65, 71
Duraud, J.P., 315

Ebbinghaus, Bart B., 241
Erenburg, S.B., 359, 371
Espeso, J.I., 59

Farmer, Joseph C., 241
Fischer, D.A., 365
Fischer-Colbrie, A., 245
Flink, C., 297
Foran, G.J., 303, 309
Fratzl, P., 89
Fuerstenau, D.W., 353
Fujimoto, H., 227

Gammon, W.J., 197
Garcia, J.M., 89
Gautier-Soyer, M., 101, 315
Genzer, J., 365
Ghosh, S., 245
Glodis, P.F., 251
Glover, C.J., 303, 309
Goldman, A.I., 139
Goodman, K.W., 197
Gota, S., 101, 315
Gould, Thomas H., 241
Gozzo, F., 227
Gregoratti, L., 203
Guiot, E., 101
Gunion, R., 185
Gunther, S., 203
Guo, J., 169
Guo, Y., 65
Guvenilir, A., 37, 43

Haase, J.D., 37
Haeffner, D.R., 139
Hagans, P.L., 127
Han, Yujie, 77
Hansen, J.L., 309
He, J-H., 347
Henriot, M., 101
Hershberger, J., 81, 109
Hieslmair, H., 297
Holman, Hoi-Ying N., 17
Hsu, L-S., 179
Hu, Y., 273
Huang, W., 71
Huang, X.R., 65, 71
Huang, Z., 65
Hurle, D.T.J., 65
Hussain, Z., 221, 227

Ice, Gene E., 49, 153, 233
Imaizumi, M., 297
Iñiguez, Ana Carola, 327

Isa, S., 153
Istratov, A., 297

Jacobson, D.C., 251
Je, J.H., 321
Jiang, Jianhua, 77
Jiao, Jinghua, 77
Johnson, A.L., 221, 227

Kamijo, N., 31
Karouta, F., 309
Kegel, I., 89
Kihara, H., 31
Kinney, P.D., 227
Kiskinova, M., 203
Klepeis, J.E., 279, 285
Körner, H., 365
Kotthaus, J.P., 89
Kramer, E.J., 365
Kramer, M.J., 139
Kurtis, K.E., 3
Kustas, F., 109
Kvardakov, V.V., 59
Kycia, S., 139

Laderman, S., 245
Lai, B., 233
Lamble, G., 297
Lang, J.C., 139
Larson, Jr., D.J., 65
Larson, P., 185
Lavelle, B., 121
Lavrenova, L.G., 371
Le Fèvre, P., 315
Lee, J.M., 321
Lee, T.W., 309
Lefebvre, S., 101
Lin, Lanying, 77
Liu, Xunlang, 77
Lorke, A., 89
Lorusso, G.F., 215

Ma, E., 347
Ma, Yanjun, 169
MacDowell, A.A., 55, 297
Magnan, H., 315
Mansour, A.N., 339
Marcus, M.A., 251
Margulies, L., 139
Marsi, M., 203
Martin, Michael C., 11, 17
Martinuzzi, S., 297
Matsouli, I., 59
Mazalov, L.N., 161, 359, 371
McCallum, R.W., 139
McDowell, D.L., 43
McHugo, Scott A., 153, 241, 297
McKinney, Wayne R., 11, 17
Melendres, C.A., 339
Metzger, T.H., 89, 95
Meyer-Ilse, W., 3
Miller, M., 185

Mini, S., 339
Mishra, S.R., 185, 197
Moller, H.J., 297
Monteiro, P.J.M., 3
Moon, Y., 309
Murai, K., 31

Naftel, S.J., 273, 291
Nagel, D.J., 127
Naghedolfeizi, M., 233
Negri, R., 185
Niu, L., 251
Noh, D.Y., 321
Nordgren, J., 169
Northrup, P.A., 251
Nylandsted-Larsen, A., 309

Ober, C.K., 365
Odom, R.W., 221
Ohtani, K., 31
Okotrub, A.V., 161

Padmore, Howard A., 25, 55, 221, 227, 297
Patel, J.R., 55, 95
Pattanaik, S., 333
Peisl, J., 89
Perichaud, I., 297
Perry, Dale L., 17, 241
Peterson, E., 185
Petroff, P.M., 89
Petuskey, W.T., 133
Piamonteze, Cínthia, 327
Pianetta, P., 245
Prasad, V., 65
Price, David L., 261
Principe, E.L., 221

Qadri, S.B., 127

Raghothamachar, B., 65
Ravikumar, R., 353
Rek, Z.U., 81, 109, 115
Renner, T., 227
Ridgway, M.C., 303, 309
Rinio, M., 297

Saboungi, Marie-Louise, 261
Sambasivan, S., 365
Sathe, C., 169
Scheinfein, Michael, 25
Schilling, P.J., 347
Sham, T.K., 273, 291
Sheridan, B., 227
Shimazaki, A., 245
Shvedenkov, Yu.G., 371
Singh, S., 215
Sivaniah, E., 365
Skelton, E.F., 127
Solak, H., 215
Sopori, B., 297
Sparks, C.J., 153
Stammler, Thomas, 25

Steele, W., 227
Steiner, K.A., 133
Stock, S.R., 37, 43
Stöhr, Joachim, 25
Suzuki, Y., 31

Takaura, N., 245
Tamura, S., 31
Tan, H.H., 309
Tessler, Leandro R., 327
Thompson, Albert C., 55, 153, 241, 297
Tian, Yulian, 77
Tkachev, S.V., 359
Tobin, J.G., 185, 197
Tolentino, H., 327
Triplett, B., 227
Turskaya, T.N., 161

Underwood, J.H., 215
Uritsky, Y.S., 227

Vahlas, C., 121
Van Buuren, T., 279, 285
van der Laan, G., 197
van Konynenburg, Richard, 241
Vescovo, E., 191
Vetter, W.M., 71

Waddill, G.D., 185, 197
Waldfried, C., 191
Waldhauer, A., 245

Wang, J., 365
Wang, S., 71
Wang, Zhouguang, 77
Waychunas, G.A., 353
Weber, E.R., 297
Werner, M., 297
Whitacre, J.F., 115
Wiemer, C., 245
Wilkinson, A.P., 333
Williams, J.J., 139
Witt, J.R., 37

Xiang, M-L., 365
Xu, J., 333

Yalisove, S.M., 109, 115
Yamaguchi, M., 297
Yang, Hongning, 169
Yang, Nancy, 241
Yang, S., 365
Yasumoto, M., 31
Ynzunza, R., 227
Yoon, E., 309
Yoshida, K., 31
Yu, K.M., 303
Yun, W.B., 233

Zaitz, M.A., 245
Zemskova, S.M., 359
Zhao, Z.B., 81
Zinke-Allmang, M., 273

SUBJECT INDEX

absorption edge (see XAFS, EXAFS)
  Ag K, 353
  Ag L$_{III}$, 347
  As K, 309
  Br K, 371
  C K, 25, 161
  Cd K, 359
  Ce L$_{III}$, 339
  Co K, 273
  Cu K, L$_2$ or L$_3$, 279, 285, 347, 315
  Er L$_{III}$, 327
  Fe K, 297, 321, 339, 347, 371
  Ge K, 309
  Mo K, 43
  Nb K, 333
  Ni K, 339
  Pd K, 43
  Pt L$_{3,2}$ or M$_{3,2}$, 291
  Si K or L$_{2,3}$, 273, 291
  Sr K, 339
  Ta L$_{III}$, 333
  Zn K, 359, 371
  Zr K, 43
Ag, 353
  Ag-Fe, 347
Al, 37, 55, 127, 153, 215
  AlGaN, 215
  Al$_2$O$_3$, 101, 203, 315
  AuAl$_2$, 179
alkali silica reaction, 3
alloying, 347
amorphous phases, 109, 169, 261, 303,
    309, 327
analysis speed, throughput, 221
Au, 297
  AuAl$_2$, 179

B$_4$C, 109
bacteria, 17

C, 25, 145, 153, 161, 169
  SiC/C, 233
  WC/C, 145
Ca, 241
  CaOH, 3
Ce, 241
chemical
  analysis, 227
  mapping, 203
  state detection sensitivity, 221
  vapor deposition (CVD), 121, 309
Co, 347
  CoSi$_2$, 273
concrete, 3
corrosion, 3
Cr, 17, 297

cryptomelane (see K$_x$Mn$_8$O$_{16}$), 353
crystal growth, 65, 77
Cu, 43, 185, 245, 279, 297, 347
  clusters, 315
  Cu/Al, 31
  Cu/Cr, 279, 285
  Cu/Mo, 279
  Cu/Re, 279
  Cu/Ta, 279
  Cu-Ta, 347
  Cu/W, 279
Czochralski, 65

D, 127
deformation, 43, 81
dialkyldithiocarbamate, 359
diffraction, 37, 43, 55, 59, 65, 71, 77, 81, 89,
    101, 115, 121, 127, 133, 139, 153, 161
  anomalous, 261
  energy dispersive, 37, 43, 121, 127, 133
  grazing indidence (see grazing incidence)
  high
    pressure, 133
    temperature, 139
  topography (see x-ray topography)
dislocations, superscrew, 71
disorder, 261, 303

edge facets, 65
electrochemistry, 127
electromigration, 55, 127, 215
electron spectroscopy (see photo- and ESCA)
epitaxial lift-off, 309
Er, 327
ESCA microscopy, 203
EXAFS, 145, 273, 303, 309, 315, 321, 327, 333,
    347, 353, 359, 371
  NEXAFS, 365
  transmission, 309

fatigue, 37
Fe, 245, 321
  Fe-Ag, 347
  oxide, 117, 101, 153, 297, 321, 339
  FeBO$_3$, 59
films (see thin films)
fluorescence, 121, 245, 309
  EXAFS, 145, 321
  standing wave, 145
focusing, 31, 59
Fresnel zone plate (FZP), 31
FTIR, 11

GaAs, 303, 309
  InAs/GaAs, 89
  In$_{0.47}$Ga$_{0.53}$As, 309

Gd, 191, 197, 241
  GeO_2, 261
  Ge/Si, 203
glass, 145, 261
grain
    orientation, 37, 49, 55
    subdivision, 43
grazing incidence, 321
    x-ray
        diffraction, 89, 101
        scattering, 95, 109, 115
growth kinetics, 315

hard disk coatings, 25
HfO_2, 241
high-energy ball milling, 347

immiscible elements, 347
In
    InAs/GaAs, 89
    InGaAs, 309
    InP, 65, 77, 309
infrared, 11
    microspectroscopy, 17
interfaces, 145, 203
    buried, 145
    ion implanted, 95
    structure, 145
ion implantation, 95, 303, 309

$KCa_2Nb_3O_{10}$, 133
    $K_xMn_8O_{16}$, 353
    $K(Ta_{1-x}Nb_x)O_3$, 333

Laue pattern, 37, 43, 49, 55, 127
layered perovskite, 133
liquid crystal, 365
low(-)
    dielectric constant, 169
    energy electron diffraction (LEED), 101
luminescence, 327

magnetic linear dichroism, 185, 197
magnetoacoustic effect, 59
mechanical alloying, 347
micro(-)
    ESCA, 203
    infra-red, 11, 17
    XPS, 221
microbeams, 11, 31, 37, 43, 49, 55, 127, 153
microflorescence, 153, 241, 297
microscope, 3, 241
microspectroscopy, 11, 17
mineral surfaces, 17
Mn
    Mn-O, 353
    Mn-O-Ag, 353
Mo, 81
    MoO_3, 203
molecular-beam-epitaxy (MBE), 101, 309
multilayer(s), 31, 145, 233, 279, 285
    Fresnel zone plate, 31

neutron diffraction, 261
NEXAFS, 25, 365
NH_4AP, 161
Ni, 121, 245, 297
    $Ni(OH)_2$, 339
    Ni/Si, 203
    Ni/SiO/Si, 203

optical fibers, 251
order, 261
oxidation, 101, 321

passivated aluminum interconnect, 127
Pd, 127
photoelectron
    microscopy, 25, 203
    spectroscopy, 185
photoemission
    angle-resolved, 179
    resonant, 197
    (XPS), 169, 191, 197, 215, 221, 227
polarity, 65
polychromatic x-rays, 37, 43, 49, 55, 65, 71, 81
    topography, 65, 71, 77, 81
polycrystal, 37, 43, 49, 55, 121, 297
polymer, 25, 365, 371
Pt
    $PtGa_2$, 179
    Pt/Si, 291
pyrochlore, 241

quantum dots, 89

Reflection high-energy electron diffraction (RHEED), 101
rocking curve, 77
Rb
    Rb_2O, 261
    $(Rb_2O)_x(GeO_2)_{1-x}$, 261

scattering
    anomalous x-ray, 261
    diffuse, 95
Si, 49, 95, 245, 327
    amorphous, 327
        SiC, 109
    $CoSi_2$, 273
    Ge/Si, 203
        Si_xGe_{1-x}, 309
    ion implanted, 95
        SiO_2, 251
        SiOH, 3
    polycrystalline, 297
    porous, 291
    Pt/Si, 291
    Ta/Si, 81
    Ti_5Si_3, 139
    wafer cleanliness, 245
Sn, 203
solar cell, 297
sol-gel, 333

spectrometer, crystal, 153
spectromicroscopy, 25, 215, 227
spin(-)
    polarized
        band structure, 191
        photoemission, 191
    spectroscopy, 185
    transition, 371
spray pyrolysis, 359
sputtering, 115, 279, 285, 327
standing waves, acoustic, 59
strain, 43, 81, 89, 101, 109, 191
stress relaxation, 81
sulfate attack,, 3
super-Coster-Kronig (sCK) process, 197
surface, 17, 65, 89, 203, 315, 365
    contamination, 245
    mineral, 17
    orientation, 365

Ta, 115, 145
    Ta/Si, 81
    Ta-Cu, 347
texture, 37, 121
thin films, 25, 81, 101, 109, 115, 121, 185,
    197, 215, 273, 327, 339, 365
three-phase boundary (TPB), 65
Ti, 241
    TiO_2, 241
    TiO_2/MoO_3, 203
    Ti_5Si_3, 139
torsion, 43

trace elements, 233, 245, 251, 297
transmission
    electron microscopy (TEM), 115, 279, 285
    Laue patterns, 37, 43
tribology, 25
twins, 65, 77

W, 81,145
    WC/C, 145
    WO_3, 153
wafer cleanliness, 245
wear, 25
white lines, 279, 285

x-ray
    absorption, 169, 251, 339
        near edge structure (see XANES)
        spectroscopy (see EXAFS, XAFS), 273,
            279, 285, 297
    anomalous scattering, 65
    diffraction (see defraction)
    fluorescence, 153, 233, 241, 245, 297
        microtomography, 233
    microbeam (see microbeam)
    photoelectron microscopy, 25
    photoemission spectroscopy, 169, 221, 227
    topography, 59, 65, 71, 77, 81, 115
XAFS, 339 (see also EXAFS)
XANES, 273, 291, 339, 347, 359, 371
XPS (see x-ray photoemission)

Zn, 245

383

CPSIA information can be obtained at www.ICGtesting.com
Printed in the USA
LVOW06s1352260514

386972LV00010BA/121/P